DECENTRALIZED CONTROL OF COMPLEX SYSTEMS

Dragoslav D. Šiljak
Santa Clara University
Santa Clara, California

Dover Publications, Inc.
Mineola, New York

Bibliographical Note

This Dover edition, first published in 2012, is an unabridged republication of the work originally published in 1991 by Academic Press, Inc., Boston. The author has provided a new Errata List for this edition.

Library of Congress Cataloging-in-Publication Data

Šiljak, Dragoslav D.
Decentralized control of complex systems / Dragoslav D. Šiljak. — Dover ed.
p. cm.
Originally published: Boston : Academic Press, 1991. With new errata list.
Summary: "The interconnection of computer subsystems presents a unique variety of control problems. This heavily cited reference book provides a graph-theoretic framework for structural modeling of complex systems and explores stabilization, optimization, output feedback, and static and dynamic controllers. Additional topics include overlapping decompositions and reliability design. An extensive appendix provides graph algorithms. 1991 edition: — Provided by publisher.
Includes bibliographical references and index.
ISBN-13: 978-0-486-48614-7 (pbk.)
ISBN-10: 0-486-48614-1 (pbk.)
1. System analysis. 2. Control theory. I. Title.

QA402.S48 2012
003—dc23

2011024858

www.doverpublications.com

To Ana and Matija
for love and fun

Simple perceptions, or impressions and ideas, are such as admit of no distinction nor separation. The complex are the contrary to these, and may be distinguished into parts.

David Hume
"A Treatise of Human Nature"

Contents

Preface

Complexity is a central problem in modern system theory and practice. Because of our intensive and limitless desire to build and to control ever larger and more sophisticated systems, the orthodox concept of a high performance system driven by a central computer has become obsolete. New emerging notions are subsystems, interconnections, distributed computing, neural networks, parallel processing, and automated factories, to mention a few. It is becoming apparent that a "well-organized complexity" is the way of the future.

The notion of complexity is used broadly, in widely different contexts, and often with subjective connotations. A great deal of our research in modeling and control of complex systems indicates that the fundamental characteristics of complexity are dimensionality, uncertainty, and information structure constraints. For us, these three features will be of overriding concern and will serve as major motivating factors for development of decentralized control theory for complex systems.

We regard a dynamic system as an interconnection of subsystems, which may be known as physical entities, or are a purely mathematical artifice to be identified by a suitable partitioning algorithm. In either case, we can take advantage of special structural features of a decomposed system and come up with a substantial reduction of *dimensionality* in control problems. Presumably, each subsystem can be considered independently and the individual solutions can be combined in some way to get a solution for the overall problem. While this recipe has been around for quite some time, it is only recently that the decomposition approach is emerging as a powerful design concept. We shall provide for the first time a comprehensive presentation of this approach to control of complex systems, exposing, at the same time, critical conceptual insights and significant numerical simplifications available in this context.

We customarily assume that neither the internal nor external nature of complex systems can be described precisely in deterministic or stochastic terms. Essential *uncertainties* of both the structured (parametric) and unstructured variety are present in the subsystems and their interconnections as well as information channels and controller configurations. For this reason, we emphasize, and perhaps overemphasize, the superior *robustness* of decentralized control laws with respect to plant and controller uncertainties and, in particular, unpredictable structural perturbations whereby the subsystems are disconnected and again connected in various ways during operation. We also include the multiple controller configurations to cope with controller failures and to achieve a satisfactory *reliability* of decentralized control schemes. Our concern for robustness and reliability is justified by the fact that more often than not the normal operating mode of a complex system is a failure mode!

Our ability to study and master large complex systems is greatly enhanced by the advent of modern computing machinery. This reliance on computing power is likely to increase with our persistent efforts to build high-performance control systems at lower costs. Therefore, it is absolutely essential that we develop control theory that can accommodate the new and decisive trend in computer technology toward *distributed parallel computing* with highly reliable low-cost multiprocessor architectures. In complex systems, where databases are developed around the plants with distributed sources of data, a need for fast control action in response to local inputs and perturbations dictates use of distributed (that is, decentralized) information and control structures. We require a theory for synthesizing control laws under *decentralized information structure constraints*, which is what this book is all about.

The plan of the book is as follows. Chapter 1 provides a graph-theoretic framework for structural modeling of complex systems. This is a natural environment for decentralized control problems, because they are primarily structural. Furthermore, screening candidate structures for controllability, observability, and fixed modes in this context is computationally attractive, especially when large systems are involved: the outcomes are generic and only binary operations are required. In Chapter 2, we present the results concerning robust stabilization *via* decentralized state feedback. Optimization is the subject of Chapter 3, where we emphasize robustness in terms of gain and phase margins and gain reduction tolerance in each control channel. Output feedback is considered in the context of interconnected observers (Chapter 4) and directly using controllers of a static or dynamic variety (Chapter 5). Manipulative power of graphs is exploited in Chapters 6 and 7 to discover hidden structures conducive to decentralized control. Both lower block triangular and weak coupling forms are considered, which offer a considerable savings in computational effort involved in building controllers and observers for complex systems. Overlapping decompositions and the underlying Inclusion Principle are presented in Chapter 8, which enlarge to a great extent the scope and applicability of decentralized control. It is in reliability design (Chapter 9) where decentralized control with overlapping information constraints is the basic tool for making the multiple controller concept work. In this last chapter, a unified synthesis of robust decentralized control is achieved with respect to both the plant and controller failures.

At the end of each chapter we include a Notes and References section where we provide comments and cite relevant literature to describe the evolution of the ideas, as well as to broaden the range of the presented results. References for each individual chapter are listed in the Bibliography section following the Notes and References section. Omitted references are either well known, or are regretfully excluded to keep the list at a resonable length.

Finally, we provide graph algorithms in the Appendix. Depth-first search, input and output reachability, structural controllability and observability, and lower block triangular and nested epsilon decompositions are all included and explained. Utility of our theoretical results depends a great deal on how efficient these computer algorithms are, and we shall provide a discussion of their complexity.

My first thanks go to Masao Ikeda and Erol Sezer for a fruitful, gratifying and, above all, exciting collaboration, which provided a wealth of results for this book. As if it were not enough, they read the final manuscript and offered numerous comments and suggestions.

I am grateful to Gan Ladde, Rade Krtolica, Yuzo Ohta, Yoshi Hayakawa, Paolo Fergola, Lello Tenneriello, Eugene Kaszkurewicz, Srdjan Stanković, as well as my former students Dave White, Magda Metwally, Val Pichai, Migdat Hodžić, Don Gavel, and Doug LaMont for exploring with me the new avenues as well as the backroads of complex systems.

My special thanks are due to Fran Karmitz for typing the manuscript in her peerless fashion after deciphering my innumerable and often confusing revisions.

Finally, I wish to gratefully acknowledge the support of the National Science Foundation and the U.S. Department of Energy. Their continual and generous assistance made our ambitious broad-range research in complex systems come to fruition.

Saratoga, California
December 1990

Dragoslav D. Šiljak

Chapter 1 | Structured Systems

In the mathematical modeling of physical systems, one is invariably confronted with a dilemma: to use a more accurate model which is harder to manage, or to work with a simpler model which is easier to manipulate but with less confidence. A hierarchy of models with increasing complexity and fidelity is more often than not the best approach. When the number of variables is large, it is prudent, if not imperative, to start the analysis with simple structural models that offer relatively easy ways to identify unsuitable system configurations, causes for lack of desired properties, and straightforward remedies.

"Structure" is a widely used and often misused term in system theory and practice. For us, however, a *structure* is a graph, which is a precise mathematical object. In this way, a structural analysis of dynamic systems is carried out within the rich and rigorous framework of graph theory. Our objective in this chapter is to set up the graph-theoretic approach and present basic results concerning the existence of suitable structures for decentralized control. The end product is a collection of efficient and reliable algorithms for testing candidate schemes for decentralized control and estimation.

Before proceeding to technical developments, let us discuss the intuitive ideas underlying the approach. With a linear system we associate a directed graph (digraph) in an obvious way: to each variable we assign a vertex of a digraph, and an edge is present whenever the corresponding coefficient in the system matrix, which relates a pair of variables, is different from zero. Then, the central problem is to determine if there is a path to every state vertex from at least one input vertex, that is, if the system is *input reachable*. The dual concept is *output reachability*, which is the property that each state reaches at least one output. These concepts were formulated as the basic structural requirements for controllability and observability. They apply to linear and nonlinear systems alike, they are parameter independent, and there are very efficient algorithms to test them. A crucial advantage offered by the reachability concepts is their manipulative power, which can be used in decompositions of large dynamic systems (see Chapters 6 and 7).

When the property of generic rank is added to input reachability, one arrives at *structural controllability.* This notion is a better graph-theoretic approximation of controllability than reachability, but it is achieved at the price of liability to parameter dependencies of the generic rank condition (see Examples 1.33 and 1.34). A way to simplify this problem and, at the same time, decrease the computational effort involved in checking the rank condition, is to manipulate first the system into hierarchically ordered input reachable subsystems using graphs (see Chapter 6). Then, testing for structural controllability of the overall system is reduced to verifying the same property of low-order subsystems.

Finally, we consider *structurally fixed modes*, which were defined to study the important existence problem of control laws under arbitrary structure constraints including output and decentralized control schemes. Again, as expected, the presence of structurally fixed modes, which cannot be moved by any choice of system parameters, is determined by a reliable algorithm involving only binary computations. This is of special importance to us, because unstable structurally fixed modes make stabilization impossible no matter how we choose the feedback parameters in a decentralized control law.

1.1. Graphs and Dynamic Systems

Let us consider a linear dynamic system described by the equations

$$\mathbf{S}\colon\ \dot{x} = Ax + Bu, \qquad y = Cx, \tag{1.1}$$

where $x(t) \in \mathbf{R}^n$ is the state, $u(t) \in \mathbf{R}^m$ is the input, and $y(t) \in \mathbf{R}^\ell$ is the output of $\mathbf{S}$ at time $t \in \mathbf{R}$, and $A = (a_{ij})$, $B = (b_{ij})$, and $C = (c_{ij})$ are constant matrices of dimensions $n \times n$, $n \times m$, and $\ell \times n$, respectively.

A precise way to represent the structure of $\mathbf{S}$ is to use the interconnection (adjacency) matrix specified by the following:

1.1. DEFINITION. The interconnection matrix of $\mathbf{S}$ is a binary $(n +$

$m + \ell) \times (n + m + \ell)$ matrix $E = (e_{ij})$ defined as

$$E = \begin{bmatrix} \bar{A} & \bar{B} & 0 \\ 0 & 0 & 0 \\ \bar{C} & 0 & 0 \end{bmatrix}, \tag{1.2}$$

where the matrices $\bar{A} = (\bar{a}_{ij})$, $\bar{B} = (\bar{b}_{ij})$, and $\bar{C} = (\bar{c}_{ij})$ have the elements

$$\bar{a}_{ij} = \begin{cases} 1, & a_{ij} \neq 0, \\ 0, & a_{ij} = 0, \end{cases} \qquad \bar{b}_{ij} = \begin{cases} 1, & b_{ij} \neq 0, \\ 0, & b_{ij} = 0, \end{cases} \qquad \bar{c}_{ij} = \begin{cases} 1, & c_{ij} \neq 0, \\ 0, & c_{ij} = 0. \end{cases} \tag{1.3}$$

The submatrices $\bar{A}$, $\bar{B}$, $\bar{C}$ of E are Boolean representations of the original system matrices A, B, C. This conversion of matrix elements to binary values makes the interconnection matrix E a useful modeling tool for the study of qualitative properties of the system **S**, which are independent of specific numerical values taken by system parameters (Šiljak, 1977a). In this way, we can also handle parameter uncertainties caused by modeling errors or operating failures in the system.

While the interconnection matrix E is useful in computations because only binary operations are involved, for qualitative interpretations of structural properties of **S** the equivalent concept of a *directed graph (digraph)* is often preferred. We recall (Deo, 1974) that a digraph is the ordered pair $\mathbf{D} = (\mathtt{V}, \mathtt{E})$, where $\mathtt{V}$ is a nonempty finite set of *vertices* (points, nodes), and $\mathtt{E}$ is a relation in $\mathtt{V}$, that is, $\mathtt{E}$ is a set of ordered pairs (v_j, v_i), which are the directed *edges* (lines, arcs) connecting the vertices of **D**.

Using Definition 1.1 of the matrix $E = (e_{ij})$, we formulate the following:

1.2. DEFINITION. The digraph $\mathbf{D} = (\mathtt{V}, \mathtt{E})$ of a system **S** has the vertex set $\mathtt{V} = \mathtt{U} \cup \mathtt{X} \cup \mathtt{Y}$, where $\mathtt{U} = \{u_1, u_2, \dots, u_m\}$, $\mathtt{X} = \{x_1, x_2, \dots, x_n\}$, and $\mathtt{Y} = \{y_1, y_2, \dots, y_\ell\}$ are nonempty sets of input, state, and output vertices of **D**, respectively, and $\mathtt{E}$ is the edge set such that $(v_j, v_i) \in \mathtt{E}$ if and only if $e_{ij} = 1$.

We notice that digraph **D** of **S** contains only the edges (u_j, x_i), (x_j, x_i), and (x_j, y_i), which reflects our basic assumption about the system **S**: there are no connections among inputs, outputs, between inputs and outputs, *etc.* (Šiljak, 1977a, b).

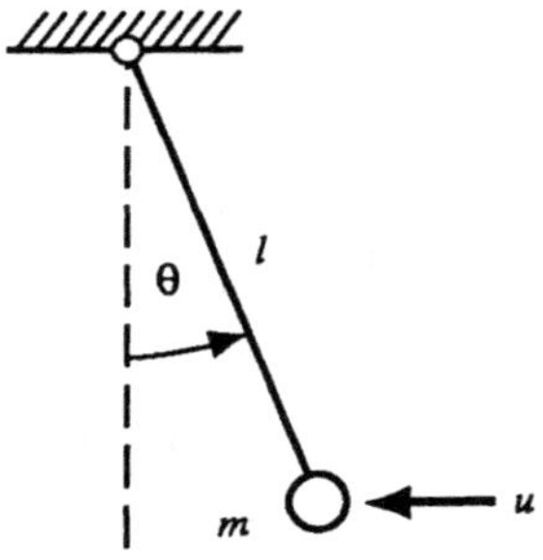

Fig. 1.1. Pendulum.

1.3. EXAMPLE. To illustrate how graphs are associated with dynamic systems, let us consider the motion of a frictionless pendulum shown in Figure 1.1, which is described by the equation

$$m\ell\ddot{\theta} + mg\theta = u, \tag{1.4}$$

where $\theta(t)$ is the angle from the vertical to the rod at time t, $\ddot{\theta}(t) \equiv d^2\theta(t)/dt^2$ is the acceleration of the bob at time t, ℓ is the length of the rod (rigid and without mass), and m is the mass of the bob subject to a force (input) $u(t)$ at time t. Choosing the state vector $x = (x_1, x_2)^T$, where $x_1(t) = \theta(t)$ and $x_2(t) = d\theta(t)/dt$, we get the system $\mathbf{S}$ representing the pendulum as

$$\mathbf{S}\colon\ \begin{aligned} \dot{x} &= \begin{bmatrix} 0 & 1 \\ \alpha & 0 \end{bmatrix} x + \begin{bmatrix} 0 \\ \beta \end{bmatrix} u, \\ y &= [1 \quad 0]\, x, \end{aligned} \tag{1.5}$$

where $\alpha = -g/\ell$, $\beta = 1/m\ell$, and

$$A = \begin{bmatrix} 0 & 1 \\ \alpha & 0 \end{bmatrix}, \qquad B = \begin{bmatrix} 0 \\ \beta \end{bmatrix}, \qquad C = [1 \quad 0], \tag{1.6}$$

are the system matrices. From (1.5), it is clear that we consider $\theta(t)$ as a measurable state of $\mathbf{S}$, that is, as the output $y(t)$ of $\mathbf{S}$. The interconnection matrix E is given as

$$E = \left[\begin{array}{cc|c|c} 0 & 1 & 0 & 0 \\ 1 & 0 & 1 & 0 \\ \hline 0 & 0 & 0 & 0 \\ \hline 1 & 0 & 0 & 0 \end{array}\right], \tag{1.7}$$

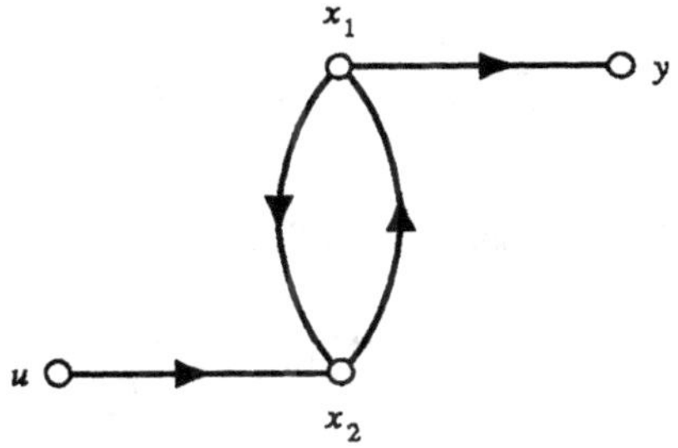

Fig. 1.2. System digraph.

where dashed lines delineate the matrices $\bar{A}$, $\bar{B}$, and $\bar{C}$. The corresponding digraph $\mathbf{D}$ is shown in Figure 1.2.

It is important to note that if the system parameters m and ℓ are changed to m_1 and ℓ_1, the interconnection matrix E and, thus, the digraph $\mathbf{D}$, remain the same. That is, E and $\mathbf{D}$ represent the structure of $\mathbf{S}_1$ as well, where

$$\mathbf{S}_1\colon\ \dot{x}_1 = \begin{bmatrix} 0 & 1 \\ \alpha_1 & 0 \end{bmatrix} x_1 + \begin{bmatrix} 0 \\ \beta_1 \end{bmatrix} u, \qquad y = [1 \quad 0]\, x_1, \tag{1.8}$$

and $\alpha_1 = -g/\ell_1$, $\beta_1 = 1/m_1\ell_1$. We say that the systems $\mathbf{S}$ and $\mathbf{S}_1$ are *structurally equivalent*.

In addition to the variation of system parameters, structural equivalence allows for a relabeling of the inputs, states, and outputs by permutation matrices. More precisely, let us consider two systems

$$\mathbf{S}_1\colon\ \dot{x}_1 = A_1 x_1 + B_1 u, \qquad y = C_1 x_1 \tag{1.9}$$

and

$$\mathbf{S}_2\colon\ \dot{x}_2 = A_2 x_2 + B_2 u, \qquad y = C_2 x_2, \tag{1.10}$$

where the state vectors x_1 and x_2 have the same dimension n. We form the Boolean triples $(\bar{A}_1, \bar{B}_1, \bar{C}_1)$ and $(\bar{A}_2, \bar{B}_2, \bar{C}_2)$ and formulate:

1.4. DEFINITION. Two systems $\mathbf{S}_1$ and $\mathbf{S}_2$ are said to be structurally equivalent if there exist (nonsingular) pemutation matrices P_A, P_B, and P_C

such that

$$\bar{A}_2 = P_A \bar{A}_1 P_A^T, \qquad \bar{B}_2 = P_A \bar{B}_1 P_B, \qquad \bar{C}_2 = P_C \bar{C}_1 P_A^T. \tag{1.11}$$

1.5. EXAMPLE. On the basis of Definition 1.4, the system

$$\begin{aligned} \mathbf{S}_2\colon\ \dot{x}_2 &= \begin{bmatrix} 0 & \alpha_2 \\ 1 & 0 \end{bmatrix} x_2 + \begin{bmatrix} \beta_2 \\ 0 \end{bmatrix} u, \\ y &= [0 \quad 1]\, x_2, \end{aligned} \tag{1.12}$$

where $\alpha_2 = -g/\ell_2$, and $\beta_2 = 1/m_2\ell_2$, is structurally equivalent to the system $\mathbf{S}_1$ of (1.8) independent of the nonzero numerical values α_1, α_2, β_1, β_2 taken by the system parameters α and β. The permutation matrices

$$P_A = \begin{bmatrix} 0 & 1 \\ 1 & 0 \end{bmatrix}, \qquad P_B = P_C = \begin{bmatrix} 1 & 0 \\ 0 & 1 \end{bmatrix} \tag{1.13}$$

satisfy the relations (1.11) for the two systems $\mathbf{S}_1$ and $\mathbf{S}_2$.

The invariance of structural equivalence to parameter variations is the underlying feature of graph-theoretic analysis, which is appealing in modeling of physical systems, especially those having large dimensions where uncertainty about system parameter values is always present. The properties of a system $\mathbf{S}$ established by its interconnection matrix E or digraph $\mathbf{D}$ are at the same time valid for systems structurally equivalent to $\mathbf{S}$. This fact not only provides a way to cope with uncertainty in the modeling process, but also serves as a cornerstone in robust designs of control and estimation schemes for large scale systems. This aspect of structural modeling will be exploited throughout this book.

Another appealing aspect of structural modeling of complex systems is the binary nature of the interconnection matrices and digraphs. Tests of intrinsic characteristics of such systems, when formulated in structural terms, are well-posed numerical problems that can be executed with speed and reliability by standard computing machinery. This feature of structural modeling is of our immediate interest.

1.2. Input and Output Reachability

The existence and actual design of control laws for a dynamic system depend crucially upon the well-known fundamental properties of complete controllability and observability (Kalman, 1963). Complete controllability means that any initial state of a given dynamical system can be transferred to the zero state in a finite length of time by a suitable input, while complete observability implies that any state of the system can be determined in a finite length of time from subsequent inputs and outputs. In most situations, these properties are all one needs to construct controllers for linear constant plants.

For a system **S** of (1.1), the standard tests for complete controllability and observability are

$$\operatorname{rank}\left[B \;\; AB \;\ldots\; A^{n-1}B\right] = n \tag{1.14}$$

and

$$\operatorname{rank}\left[C \;\; CA \;\ldots\; CA^{n-1}\right] = n, \tag{1.15}$$

where n is the dimension of the state space of **S** (that is, the order of **S**). In systems of large dimensions, computing the rank conditions (1.14) and (1.15) is an ill-posed numerical problem. Furthermore, when the tests fail, there is no indication how the rank deficiency can be removed even in relatively simple situations. For these reasons, the notions of *input* and *output reachability*, which represent the crucial ingredients in controllability and observability, have been defined by Šiljak (1977a, b). The reachability properties are numerically attractive, because they can be established by binary computations alone.

To define the concept of input and output reachability, we need several well-known notions from the theory of digraphs (Harary, 1969). We consider a digraph $\mathbf{D} = (\mathtt{V}, \mathtt{E})$ of **S**, and recall that if a collection of distinct vertices $v_1, v_2, \ldots, v_k$ together with the edges $(v_1, v_2), (v_2, v_3), \ldots, (v_{k-1}, v_k)$ are placed in sequence, then the ordered set $\{(v_1, v_2), (v_2, v_3), \ldots, (v_{k-1}, v_k)\}$ is a (directed) *path* from v_1 to v_k. We say that v_i is *reachable* from v_j if there is a path from v_j to v_i. A *reachable set* $\mathtt{V}_i(v_j)$ *of a vertex* v_j is a set $\mathtt{V}_i$ of vertices v_i reachable from the vertex $v_j \in \mathtt{V}$. Carrying this a step further, we define a *reachable set* $\mathtt{V}_i(\mathtt{V}_j)$ *of a set* $\mathtt{V}_j$ as a set of vertices v_i reachable from at least one vertex $v_j \in \mathtt{V}_j$. Therefore, $\mathtt{V}_i(\mathtt{V}_j)$ is the union of the sets $\mathtt{V}_i(v_j)$ for all $v_j \in \mathtt{V}_j$. An *antecedent set* $\mathtt{V}_j(\mathtt{V}_i)$ *of a set* $\mathtt{V}_i$ is a set of vertices v_j from which at least one vertex $v_i \in \mathtt{V}_i$ is reachable.

1.6. DEFINITION. A system **S** with a digraph $\mathbf{D} = (\mathtt{U} \cup \mathtt{X} \cup \mathtt{Y},\ \mathtt{E})$ is input reachable if $\mathtt{X}$ is a reachable set of $\mathtt{U}$.

The "directional dual" of input reachability is the following:

1.7. DEFINITION. A system **S** with a digraph $\mathbf{D} = (\mathtt{U} \cup \mathtt{X} \cup \mathtt{Y}, \mathtt{E})$ is output reachable if $\mathtt{X}$ is an antecedent set of $\mathtt{Y}$.

Definition 1.6 means that if no component u_j of the input vector u can reach (influence) a state x_i either directly or *via* other states of **S**, then there is no way to choose an input function $u(t)$ to drive the state x_i to zero, which implies that the system is uncontrollable independently of numerical values of the nonzero system parameters. In a dual manner, Definition 1.7 means that if a state x_j does not reach (influence) any component y_i of the output vector y either directly or indirectly by connecting the other states of **S**, then it is impossible to determine the state x_j at any given time from the future values of the output y and the knowledge of the input u. In this case, a structural deficiency (not the values of system parameters) prevents observability of **S**.

In order to establish input and output reachability of a given system **S**, we start with the digraph **D** and recall (Harary, 1969) that the *reachability matrix* $R = (r_{ij})$ of $\mathbf{D} = (\mathtt{V}, \mathtt{E})$ is defined as follows:

$$r_{ij} = \begin{cases} 1, & \text{if } v_i \text{ is reachable from } v_j, \\ 0, & \text{otherwise.} \end{cases} \tag{1.16}$$

In other words, $r_{ij} = 1$ if and only if there exists a path of nonzero length from v_j to v_i; the length of a path being the number of individual edges between v_j and v_i.

To determine the reachability matrix for a given digraph, we need the following result which was obtained by Festinger *et al.* (1950):

1.8. THEOREM. Let $E = (e_{ij})$ be the $s \times s$ interconnection matrix corresponding to a digraph $\mathbf{D} = (\mathtt{V}, \mathtt{E})$, and let $P = (p_{ij})$ be an $s \times s$ matrix such that $P = E^d$ where $d \in \{1, 2, \dots, s\}$. Then, p_{ij} is the total number of distinct sequences $(v_j, .), \dots, (., v_i)$ of length d in **D**.

Proof. We can prove this theorem by induction (Berztiss, 1975). We show that the theorem is true for $d = 1$, and then show that it is true for $d + 1$ whenever it is true for d. For $d = 1$, the theorem is actually the definition of the matrix E. For $d + 1$, $e_{ik}p_{kj} = p_{kj}$ if (v_k, v_i) is an edge of **D**, and $e_{ik}p_{kj} = 0$ if it is not. The total number of sequences of length $d+1$ having the form $(v_j, .), \dots, (., v_k), (v_k, v_i)$ is equal to $\sum_{k=1}^{s} e_{ik}p_{kj}$, which is the ijth element of E^{d+1}. Q.E.D.

The reachability matrix can be computed by the following corollary to Theorem 1.8:

1.9. COROLLARY. If $R = (r_{ij})$ is the $s \times s$ reachability matrix of $\mathbf{D}$ and

$$Q = E + E^2 + \cdots + E^s, \tag{1.17}$$

then $r_{ij} = 1$ if and only if $q_{ij} \neq 0$.

This corollary does not offer an efficient procedure to compute the reachability matrix. Instead of algebraic manipulations involved in (1.17), one can use the Boolean operations $\wedge$ and $\vee (1 \wedge 1 = 1,\ 1 \wedge 0 = 0 \wedge 1 = 0 \wedge 0 = 0;\ 0 \vee 0 = 0,\ 0 \vee 1 = 1 \vee 0 = 1 \vee 1 = 1)$ to get the reachability matrix as

$$R = E \vee E^2 \vee \cdots \vee E^s, \tag{1.18}$$

where $E^k = E^{k-1} \wedge E$. For two $s \times s$ binary matrices $A = (a_{ij})$ and $B = (b_{ij})$, the Boolean operations $C = A \wedge B$ and $D = A \vee B$ are defined by $C = (c_{ij})$ and $D = (d_{ij})$ as

$$c_{ij} = \bigvee_{k=1}^{s} (a_{ik} \wedge b_{kj}), \qquad d_{ij} = a_{ij} \vee b_{ij}. \tag{1.19}$$

A more efficient Boolean-type algorithm for computing the matrix R was proposed by Warshall (1962):

1.10. ALGORITHM.

BEGIN $E^o = E$
  FOR $i \leftarrow 1$ TO s DO
  FOR $j \leftarrow 1$ TO s DO
    IF $e^o_{kj} = 1$ THEN
    FOR $k \leftarrow 1$ TO s DO
      $e^o_{kj} = e^o_{kj} \vee e^o_{ki}$
END

The IF statement means that after the iteration with $i = \ell \leq s$, $e^o_{kj} = 1$ if and only if there exists a path from j to k that contains only the points from the set $\{j, k, 1, 2, \ldots, \ell\}$. This algorithm has many useful generalizations (Deo, 1974). A still faster algorithm for computing the reachability matrix is provided by the *depth-first search* on a graph formulated by Hopcroft and Tarjan (1971) and subsequently developed by Tarjan (1972).

The reachability algorithms are reviewed in Notes and References (Section 1.7), and an efficient graph-theoretic algorithm, which is based upon the depth-first search, is described in the Appendix.

Let us now turn our attention to input and output reachability. We note that E^d for the adjacency matrix E of the system $\mathbf{S}$, which has the form (1.2), can be calculated as

$$E^d = \begin{bmatrix} \bar{A}^d & \bar{A}^{d-1}B & 0 \\ 0 & 0 & 0 \\ \bar{C}\bar{A}^{d-1} & \bar{C}\bar{A}^{d-1}\bar{B} & 0 \end{bmatrix}. \tag{1.20}$$

Using this expression in (1.18), one gets the reachability matrix as

$$R = \begin{bmatrix} F & G & 0 \\ 0 & 0 & 0 \\ H & \theta & 0 \end{bmatrix}, \tag{1.21}$$

where the binary matrices F, G, H, and θ have dimensions obvious from (1.20). Of course, the matrix R could have been obtained by the Algorithm 1.10, but the structure of the matrix would not have been explicit in terms of the system matrices.

From (1.21), the following result is automatic:

1.11. THEOREM. A system $\mathbf{S}$ is input reachable if and only if the binary matrix G has no zero rows, and it is output reachable if and only if the binary matrix H has no zero columns.

1.12. REMARK. From R of (1.21), one gets for free the additional information about reachability properties involving only inputs and outputs. In (Šiljak, 1978), a system $\mathbf{S}$ is considered input–output reachable if Y is a reachable set of U, and U, in turn, is an antecedent set of Y. The system has this property if and only if the binary matrix θ of (1.21) has neither zero rows nor zero columns.

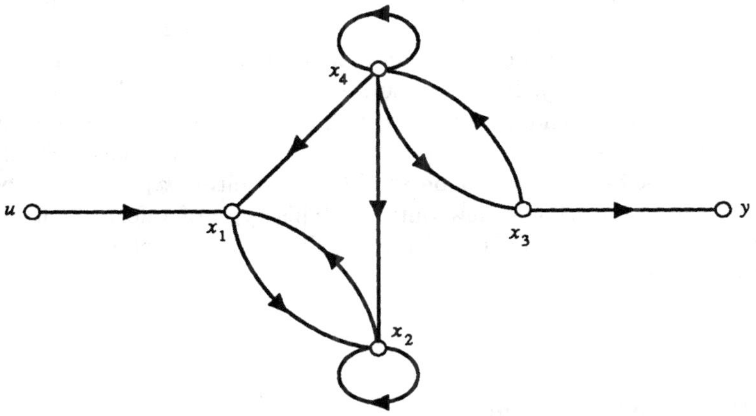

Fig. 1.3. Input and output unreachable digraph.

1.13. EXAMPLE. To illustrate the application of Theorem 1.11, The reachability matrix can be computed by the following corollary to let us consider a system digraph shown in Figure 1.3. The interconnection matrix E is

$$
\begin{array}{c} \\ E = \end{array}
\begin{array}{c}
\begin{array}{cccccc} x_1 & x_2 & x_3 & x_4 & u & y \end{array} \\
\left[\begin{array}{cccc:c:c}
0 & 1 & 0 & 1 & 1 & 0 \\
1 & 1 & 0 & 1 & 0 & 0 \\
0 & 0 & 0 & 1 & 0 & 0 \\
0 & 0 & 1 & 1 & 0 & 0 \\
\hdashline
0 & 0 & 0 & 0 & 0 & 0 \\
\hdashline
0 & 0 & 1 & 0 & 0 & 0
\end{array}\right]
\end{array}
\begin{array}{c} \\ x_1 \\ x_2 \\ x_3 \\ x_4 \\ u \\ y \end{array}, \qquad (1.22)
$$

and the corresponding reachability matrix R is computed as

$$
\begin{array}{c} \\ R = \end{array}
\begin{array}{c}
\begin{array}{cccccc} x_1 & x_2 & x_3 & x_4 & u & y \end{array} \\
\left[\begin{array}{cccc:c:c}
1 & 1 & 1 & 1 & 1 & 0 \\
1 & 1 & 1 & 1 & 1 & 0 \\
0 & 0 & 1 & 1 & 0 & 0 \\
0 & 0 & 1 & 1 & 0 & 0 \\
\hdashline
0 & 0 & 0 & 0 & 0 & 0 \\
\hdashline
0 & 0 & 1 & 1 & 0 & 0
\end{array}\right]
\end{array}
\begin{array}{c} \\ x_1 \\ x_2 \\ x_3 \\ x_4 \\ u \\ y \end{array}. \qquad (1.23)
$$

By comparing (1.23) with (1.21), we see that there are two zero rows in G of (1.23) and the system is not input reachable. From Figure 1.3, it is obvious that the input u does not reach the states x_3 and x_4. We also see that H has two zero columns and, therefore, the states x_1 and x_2 do not reach the output y. It is easy to see what should be done to recover input and output reachability of **S**. A connection should be made from the existing input (or a new added input) to either x_3 or x_4 (or both). Similarly, the output reachability would be established if either x_1 or x_2 (or both) were measured to become new outputs. This type of information makes the graph-theoretic methods useful in the design of control and estimation schemes, especially for complex systems with a large number of variables.

1.3. Partitions and Condensations

Problems involving input and output reachability can be greatly simplified by partitioning of the given directed graph into subgraphs with assigned reachability properties. By condensing each subgraph to a vertex of a new digraph, one can greatly increase the conceptual insight while, at the same time, reduce the computational effort required in the graph-theoretic analysis of system structures. The process of partitioning and condensing large graphs has been effective in diverse fields, especially in communication networks, social sciences, and transportation systems. The use of the graph-theoretic decompositions in control theory is of a recent interest (see Chapters 6 and 7).

Let us recall several useful notions about a directed graph $\mathbf{D} = (\mathtt{V}, \mathtt{E})$ that are not necessarily related to a dynamic system. When a vertex v_i is reachable from a vertex v_j, we use the notation $v_j \mathbf{R} v_i$. We note that $\mathbf{R}$ is a relation, and it is transitive because $v_j \mathbf{R} v_k$ and $v_k \mathbf{R} v_i$ imply $v_j \mathbf{R} v_i$. We say that a pair of vertices $(v_j,\ v_i)$ is *strongly connected* if $v_j \mathbf{R} v_i$ and $v_i \mathbf{R} v_j$, which we denote by $v_j \tilde{\mathbf{R}} v_i$. It is obvious that $\tilde{\mathbf{R}}$ is an *equivalence relation* on $\mathtt{V}$. Furthermore, we can *partition* $\mathtt{V}$ into equivalence classes under $\tilde{\mathbf{R}}$, unless $\mathbf{D}$ is strongly connected, that is, every vertex in $\mathbf{D}$ is reachable from every other vertex, and $\mathtt{V}$ is the only equivalence class under $\tilde{\mathbf{R}}$.

If $\mathbf{D} = (\mathtt{V}, \mathtt{E})$ is a digraph and $\mathtt{V}_k$ is a subset of $\mathtt{V}$, then the digraph $\mathbf{D}_k = [\mathtt{V}_k, (\mathtt{V}_k \times \mathtt{V}_k) \cap \mathtt{E}]$ is a *subgraph* of $\mathbf{D}$. That is, a subgraph $\mathbf{D}_k = (\mathtt{V}_k,\ \mathtt{E}_k)$ has the vertex set $\mathtt{V}_k$ as a subset of $\mathtt{V}$ and the edge set $\mathtt{E}_k$ which contains all the edges of $\mathtt{E}$ joining the vertices of $\mathtt{V}_k$.

We also recall that a *cycle* is a path of length greater than one which begins and ends on the same vertex. A special edge $(v_i,\ v_i)$ is not considered as a cycle and is termed a *self-loop*. A digraph $\mathbf{D} = (\mathtt{V}, \mathtt{E})$ is *acyclic* if it has no cycles.

1.14. DEFINITION. A subgraph $\mathbf{D}_k = [\mathtt{V}_k, (\mathtt{V}_k \times \mathtt{V}_k) \cap \mathtt{E}]$ is a strong component of $\mathbf{D}$ if it is strongly connected and there are no two points $v_i \in \mathtt{V}_k$ and $v_j \notin \mathtt{V}_k$ which lie on the same cycle in $\mathbf{D}$.

Equivalence classes and strong components of a digraph are isomorphic concepts. They induce a partition in a given digraph and serve as a basis for subsequent reductions of the digraph by the process of condensation (*i.e.*, clustering, aggregation), which preserves the reachability property of the original digraph.

Let us consider a digraph $\mathbf{D} = (\mathtt{V}, \mathtt{E})$ and let $\{\mathtt{V}_1, \mathtt{V}_2, \dots, \mathtt{V}_N\}$ be the equivalence classes of $\mathtt{V}$ under $\tilde{\mathbf{R}}$. Obviously, the sets $\mathtt{V}_k$ are pairwise disjoint, that is, $\mathtt{V}_i \cap \mathtt{V}_j = \emptyset$ for all $i, j = 1, 2, \dots, N$, $i \neq j$. To each equivalence class $\mathtt{V}_k$ corresponds a strong component $\mathbf{D}_k$ of $\mathbf{D}$. Then, we have:

1.15. DEFINITION. Given a digraph $\mathbf{D} = (\mathtt{V}, \mathtt{E})$, define

$$\begin{aligned} \mathtt{V}^* &= \{\mathtt{V}_k\colon\ \mathtt{V}_k \text{ equivalence class of } \mathbf{D}\}, \\ \mathtt{E}^* &= \{(\mathtt{V}_j, \mathtt{V}_i)\colon\ v_j \in \mathtt{V}_j,\ v_i \in \mathtt{V}_i,\ (v_j, v_i) \in \mathtt{E},\ \mathtt{V}_j \neq \mathtt{V}_i\}. \end{aligned} \tag{1.24}$$

The digraph $\mathbf{D}^* = (\mathtt{V}^*, \mathtt{E}^*)$ is the condensation of $\mathbf{D}$.

In other words, the condensation $\mathbf{D}^*$ of $\mathbf{D}$ is the digraph whose vertices v_k^* are the equivalence classes $\mathtt{V}_k$ and whose edges $(v_j^*, v_i^*) \in \mathtt{E}^*$ are determined by the following rule: there is an edge from the vertex $v_j^* \in \mathtt{V}^*$ to the vertex $v_i^* \in \mathtt{V}^*$ in the new digraph $\mathbf{D}^*$ if and only if there is in $\mathbf{D}$ at least one edge from a vertex of $\mathtt{V}_j$ to a vertex of $\mathtt{V}_i$. In Figure 1.4, a graph $\mathbf{D}$ is shown together with its strong components $\mathbf{D}_1$ and $\mathbf{D}_2$, as well as the condensation $\mathbf{D}^*$.

The next step toward a formulation of the desired partition is the following (Harary *et al.* 1965):

1.16. THEOREM. Given any digraph $\mathbf{D}$, its condensation $\mathbf{D}^*$ is acyclic.

Proof. Suppose not. Then, there is a cycle $(v_{k_1}^*, v_{k_2}^*)$, $(v_{k_2}^*, v_{k_3}^*)$, ..., $(v_{k_\ell}^*, v_{k_1}^*)$ in $\mathbf{D}^*$. By the construction rule of $\mathbf{D}^*$, there exist two vertices $v'_{k_i}, v''_{k_i} \in \mathtt{V}_{k_i}$, $i = 1, 2, \dots, N$, such that (v'_{k_1}, v''_{k_1}), (v'_{k_2}, v''_{k_2}), ..., $(v'_{k_\ell}, v''_{k_1}) \in \mathtt{E}$. Since in each strong component $\mathbf{D}_{k_i}$ corresponding to the equivalence class $\mathtt{V}_{k_i}$, there is a path from v'_{k_i} to v''_{k_i}, we conclude that the original digraph has a cycle containing the vertices $\{v'_{k_1}, v''_{k_1}, v'_{k_2}, v''_{k_2},$

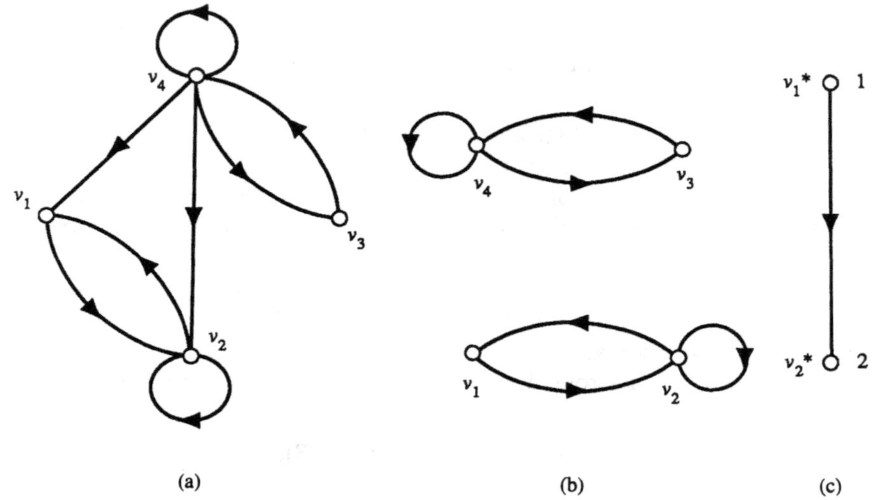

Fig. 1.4. (a) Digraph; (b) strong components; (c) condensation.

$\ldots, v'_{k_\ell}, v''_{k_\ell}\}$. But this is a contradiction because the vertices $\{v'_{k_i}, v''_{k_i}\}$ and $\{v'_{k_j}, v''_{k_j}\}$ belong to two distinct equivalence classes $\mathtt{V}_{k_i}$ and $\mathtt{V}_{k_j}$. Q.E.D.

Now, every acyclic digraph is a *partial order* because the corresponding relation on its vertices is irreflexive, asymmetric, and transitive. This means that we can assign an integer n_i to each node v_i^* of the condensation $\mathbf{D}^* = (\mathtt{V}^*, \mathtt{E}^*)$, which are called *levels* of $\mathbf{D}^*$. Furthermore, $\mathbf{D}^*$ can always be made to have an ascending level assignment, as illustrated in Figure 1.4c, where $n_1 = 1$ and $n_2 = 2$. In order to establish this fact rigorously, we recall that the *outdegree* of a vertex v_i, which is denoted by $\mathrm{od}(v_i)$, is the number of edges from v_i. The *indegree* of v_i, written $\mathrm{id}(v_i)$, is the number of edges to v_i. We prove the following (Harary *et al.* 1965):

1.17. THEOREM. An acyclic digraph $\mathbf{D} = (\mathtt{V}, \mathtt{E})$ has at least one vertex $v_i \in \mathtt{V}$ such that $\mathrm{od}(v_i) = 0$, and at least one vertex $v_j \in \mathtt{V}$ such that $\mathrm{id}(v_i) = 0$.

Proof. Assume that $\mathbf{D}$ has no vertices of outdegree zero. Then, we can start at any vertex v_{k_1} and form an arbitrarily long sequence of edges $(v_{k_1}, v_{k_2}), (v_{k_2}, v_{k_3}), \ldots \in \mathtt{E}$. Since the number of vertices of $\mathbf{D}$ is finite, we must repeat a vertex v_{k_i}, which means that $\mathbf{D}$ is not acyclic. This is a contradiction. Q.E.D.

Finally, we establish (Harary *et al.* 1965):

1.18. THEOREM. For any digraph $\mathbf{D} = (\mathtt{V}, \mathtt{E})$, the following statements are equivalent:

(*i*) $\mathbf{D}$ is acyclic.
(*ii*) There exists an order of the vertices of $\mathbf{D}$ so that the corresponding adjacency matrix $E = (e_{ij})$ is lower triangular.
(*iii*) There exists a level assignment for $\mathbf{D}$ such that if $(v_j, v_i) \in \mathtt{V}$ then $n_j < n_i$.

Proof. (*i*) $\Rightarrow$ (*ii*). By Theorem 1.17, $\mathbf{D}$ has a vertex of indegree zero. One such vertex is labeled v_1, and we form a new digraph $\mathbf{D} - v_1$ by truncating the vertex v_1 and its edges from $\mathbf{D}$. The digraph $\mathbf{D} - v_1$ has no cycles because $\mathbf{D}$ is acyclic and, therefore, has a vertex v_2 with indegree zero. By repeating the truncation process to form $\mathbf{D} - \{v_1, v_2\}$, and locating v_3 such that $\text{id}(v_3) = 0$, we eventually exhaust all the vertices of $\mathbf{D}$. By construction, there is no edge $(v_j, v_i) \in \mathtt{E}$ if $j \geq i$, which implies that $e_{ij} = 0$ for $j \geq i$, and the interconnection matrix E is lower triangular.

(*ii*) $\Rightarrow$ (*iii*). Suppose E of $\mathbf{D}$ is lower triangular. We assign to each vertex v_i, which correspond to the ith row and column of E, the level i. Since $e_{ij} = 0$ for $j \geq i$, there is no edge from any lower level to any of the upper levels.

(*iii*) $\Rightarrow$ (*i*). Suppose $\mathbf{D}$ has a cycle containing v_j and v_i with level assignment j and i such that $j < i$. Then, there is a path from v_j to v_i and a path from v_i to v_j, which is a contradiction, because $\mathbf{D}$ has an ascending level assignment and there are no edges from lower levels to upper levels of $\mathbf{D}$. This means that there are no paths from v_i to v_j, which is a contradiction to the assumption that $\mathbf{D}$ has a cycle. Q.E.D.

Since by Theorem 1.16 the condensation $\mathbf{D}^*$ of a digraph $\mathbf{D}$ is acyclic, we can use the implications of Theorem 1.18 with regard to reachability properties of digraphs appearing in dynamic systems. Let us consider the digraph $\mathbf{D} = (\mathtt{U} \cup \mathtt{X} \cup \mathtt{Y}, \mathtt{E})$, and define the *state truncation* $\mathbf{D}_x = (\mathtt{X}, \mathtt{A})$, which is obtained from $\mathbf{D}$ by removing all the input vertices $\mathtt{U}$ and the output vertices $\mathtt{Y}$ as well as the edges in $\mathtt{E}$ connected to the vertices of $\mathtt{U}$ and $\mathtt{Y}$. The adjacency matrix of $\mathbf{D}_x$ is the Boolean $n \times n$ matrix $\bar{A} = (\bar{a}_{ij})$ defined as a submatrix of E in (1.2). We form the condensation $\mathbf{D}_x^* = (\mathtt{X}^*, \mathtt{A}^*)$ and the interconnection $N \times N$ matrix $\bar{A}^* = (\bar{a}_{ij}^*)$, where N is the number of strong components of $\mathbf{D}_x$. Since $\mathbf{D}_x^*$ is acyclic, its vertices can be ordered so that the interconnection matrix $\bar{A}^*$ is lower triangular:

$$\bar{A}^* = \begin{bmatrix} 0 & & & & & \\ \bar{a}_{21}^* & 0 & & & \bigcirc & \\ \bar{a}_{31}^* & \bar{a}_{32}^* & 0 & & & \\ \cdots & \cdots & \cdots & \cdots & & \\ \bar{a}_{N1}^* & \bar{a}_{N2}^* & \bar{a}_{N3}^* & \cdots & \bar{a}_{N,N-1}^* & 0 \end{bmatrix}. \tag{1.25}$$

Partition of the input and output sets $\mathtt{U}$ and $\mathtt{Y}$ of $\mathbf{D}$ is not crucial at present and can be done in any convenient way. We assume that $\mathtt{U}$ and $\mathtt{Y}$ are partitioned in M and L pairwise disjoint sets $\mathtt{U} = \{\mathtt{U}_1, \mathtt{U}_2, \ldots, \mathtt{U}_M\}$ and $\mathtt{Y} = \{\mathtt{Y}_1, \mathtt{Y}_2, \ldots, \mathtt{Y}_L\}$. We already have the partition $\mathtt{X} = \{\mathtt{X}_1, \mathtt{X}_2, \ldots, \mathtt{X}_N\}$, where $\mathtt{X}_i$ corresponds to the ith strong component. Now, we can form a condensation $\mathbf{D}^* = (\mathtt{U}^* \cup \mathtt{X}^* \cup \mathtt{Y}^*, \mathtt{E}^*)$ whose vertices x_i^*, u_i^*, and y_i^* are the subsets $\mathtt{X}_i$, $\mathtt{U}_i$, and $\mathtt{Y}_i$. Of course, $\mathtt{U}_i$ and $\mathtt{Y}_i$ are not necessarily equivalence classes in $\mathbf{D}$, but $\mathtt{X}_i$'s are. This fact will be used soon.

With $\mathbf{D}^*$ we associate the interconnection matrix

$$E^* = \begin{bmatrix} \bar{A}^* & \bar{B}^* & 0 \\ 0 & 0 & 0 \\ \bar{C}^* & 0 & 0 \end{bmatrix}, \tag{1.26}$$

where the Boolean matrix $\bar{B}^* = (\bar{b}_{ij}^*)$ has $\bar{b}_{ij}^* = 1$ if and only if there is an edge in $\mathbf{D}$ from a vertex of $\mathtt{U}_j$ to a vertex in the equivalence class $\mathtt{X}_i$. Likewise, $\bar{C}^* = (\bar{c}_{ij}^*)$ and $\bar{c}_{ij}^* = 1$ if and only if there is an edge from $\mathtt{X}_j$ to $\mathtt{Y}_i$. The matrix $\bar{A}^*$ is lower triangular as shown in (1.25).

Finally, we arrive at a simple but important result:

1.19. THEOREM. A digraph $\mathbf{D} = (\mathtt{U} \cup \mathtt{X} \cup \mathtt{Y}, \mathtt{E})$ is input (output) reachable if and only if its condensation $\mathbf{D}^* = (\mathtt{U}^* \cup \mathtt{X}^* \cup \mathtt{Y}^*, \mathtt{E}^*)$ is input (output) reachable.

Proof. We prove only the input reachability part of the theorem, because the output reachability part is its directional dual. If $\mathbf{D}^*$ is input reachable then $\mathtt{X}^*$ is the reachable set of $\mathtt{U}^*$, and there is a path to each point of $\mathtt{X}^*$ from a point of $\mathtt{U}^*$. Since $\mathbf{D}_x^*$ is the condensation of $\mathbf{D}_x$, it follows from reachability of components that there is a path to each point of $\mathtt{X}$ from a point of $\mathtt{U}$. That is, $\mathtt{X}$ is the reachable set of $\mathtt{U}$ and $\mathbf{D}$ is input reachable. Conversely, if $\mathbf{D}^*$ is not input reachable, then $\mathtt{X}^*$ is not the reachable set of $\mathtt{U}^*$, and there are points of $\mathtt{X}^*$ that cannot be reached by any point of $\mathtt{U}^*$.

Obviously, by the definition of condensation, those points of $\mathtt{X}$ that correspond to the unreachable points of $\mathtt{X}^*$ cannot be reached by any point of $\mathtt{U}$ and $\mathbf{D}$ is not input reachable. Q.E.D.

With an abuse of Definitions 1.6 and 1.7, we refer to input and output reachability of the digraphs $\mathbf{D}$ and $\mathbf{D}^*$ instead of the system $\mathbf{S}$. This removes ambiguity arising from the fact that both the digraph $\mathbf{D}$ and its condensation $\mathbf{D}^*$ belong to the same system $\mathbf{S}$. This change, however, should cause no confusion since input (output) reachability is defined unambiguously in a given digraph.

1.20. EXAMPLE. To illustrate Theorem 1.19 by a simple example, let us consider the digraph $\mathbf{D}$ of Figure 1.5a. Since $\mathbf{D}_x$ corresponding to $\mathbf{D}$ is given in Figure 1.4a, the strong components of $\mathbf{D}_x$ are $\mathbf{D}_1$ and $\mathbf{D}_2$ shown in Figure 1.4b. The condensation $\mathbf{D}^*$ of Figure 1.5b indicates that the digraph $\mathbf{D}$ is input reachable, but it is not output reachable: The two inputs u_1^* and u_2^* reach the two super-states x_1^* and x_2^* of $\mathbf{D}^*$, while the super-state x_2^* does not reach the output y^*.

We should note here that if we had lumped the two input vertices u_1 and u_2 into a single super-vertex u_1^*, then the condensation $\mathbf{D}_1^*$ would have been that of Figure 1.6. Information on input reachability would have been unchanged, but the fact that each super-state has its own independent input would have not been obvious as in $\mathbf{D}^*$ of Figure 1.5b. How to decompose best the inputs once the strong components are identified will be explained in Chapter 6, when we present the hierarchical decomposition scheme for complex systems.

Results, which are formulated on digraphs, can always be rephrased in terms of interconnection matrices, and Theorem 1.19 is no exception. We have:

1.21. COROLLARY. A system $\mathbf{S}$ with an aggregate interconnection matrix E^* is input reachable if and only if the composite matrix

$$E_{\mathrm{I}}^* = \begin{bmatrix} \bar{A}^* & \bar{B}^* \end{bmatrix}, \tag{1.27}$$

has no zero rows, and it is output reachable if and only if the composite matrix

$$E_{\mathrm{O}}^* = \begin{bmatrix} \bar{A}^* \\ \bar{C}^* \end{bmatrix}, \tag{1.28}$$

has no zero columns.

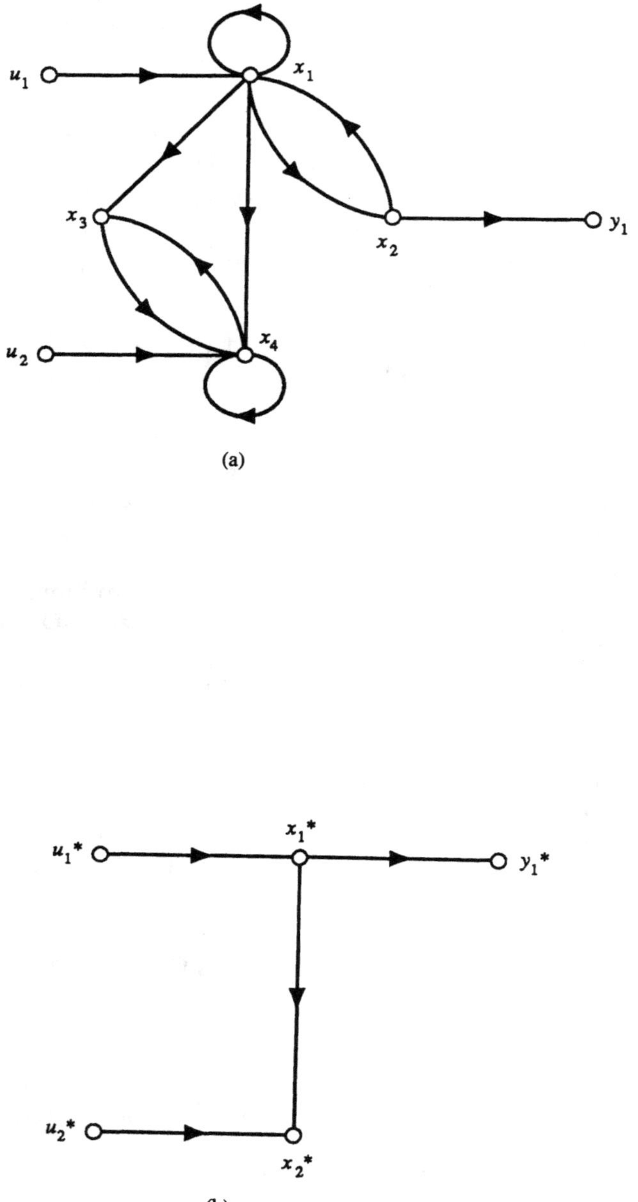

Fig. 1.5. (a) Digraph; (b) condensation.

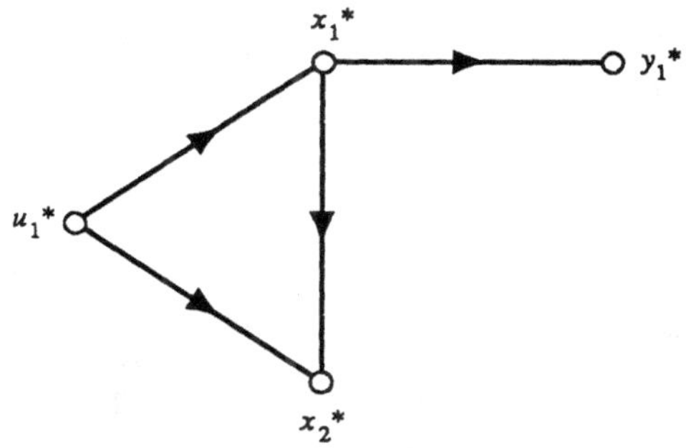

Fig. 1.6. Condensation.

Of course, it is assumed that the matrix $\bar{A}^*$ in (1.27) and (1.28) has already a lower triangular form as in (1.25). Then, Corollary 1.21 is obvious from Theorem 1.19. For the example of Figure 1.5, this means that the matrices E_{I}^* and E_{O}^* are obtained as

$$E_{\mathrm{I}}^* = \left[\begin{array}{cc|c} 0 & 0 & 1 \\ 1 & 0 & 1 \end{array}\right] \tag{1.29}$$

and

$$E_{\mathrm{O}}^* = \left[\begin{array}{cc} 0 & 0 \\ 1 & 0 \\ \hline 1 & 0 \end{array}\right]. \tag{1.30}$$

As expected, by applying Corollary 1.21 we conclude from (1.29) that the system $\mathbf{S}$ is input reachable, but it is not output reachable because of a zero column in (1.30).

Algorithms for determining strong components of a graph have a long history. Among the pioneers in this area is Harary (1959) who used Boolean multiplication of matrices to arrive at the reachability matrix that reveals the strong components of the graph. A way to speed up the calculation of the reachability matrix is to use the popular Algorithm 1.10 of Warshall (1962). Improvements of this algorithm were proposed by Purdom (1970) and Munro (1971), who solved for strong components as a subproblem of the more general problem of transitive closure, which consists in determining all paths through a directed graph. These improvements were only slight having $O(n^2)$ operations at best, where n is the number of the nodes in the digraph. The most efficient algorithm is the "depth-first search" of Tarjan (1972), which detects the strong components in $O(\max\{n, e\})$ time for a graph of n nodes and e edges. Gustavson (1976) and Duff and Reid (1978a, b) have implemented Tarjan's algorithm with efficiency superior to existing codes. A version of Tarjan's algorithm is described in the Appendix.

In partitioning a system digraph into strong components, our interest focused on simplifying the interpretation and computation of input and output reachability. It is more or less obvious that this type of partition induces a decomposition of the associated system into interconnected subsystems and, therefore, have potential applications in "piece-by-piece" stabilization and control of large systems (Vidyasagar, 1980). The use is limited, however, because partitions of this kind ignore the inputs and outputs of the system, and the lack of a suitable distribution of inputs and outputs over the subsystems may prevent a decentralized control design. More sophisticated decomposition schemes, with special attention to inputs and outputs, are described in Chapters 6 and 7.

1.4. Structural Controllability and Observability

For control problems to be properly formulated so that a useful solution can be expected, the plant should be controllable and observable. This fundamental result of Kalman (1963) has been further strengthened by the fact that controllability and observability are robust properties in linear constant plants (Lee and Markus, 1961). That is, the set of all controllable pairs (A, B) is open and dense in the space of all such pairs. This fact was exploited by Lin (1974) in formulating his concept of *structural controllability*, which states that all uncontrollable systems structurally equivalent to a structurally controllable system are atypical. Not only is this concept consistent with physical reality in the sense that values of system parameters are never known precisely, but it is also helpful in testing the properties of controllability using only binary computations. It is a well-known fact that testing controllability is a difficult numerical problem (Nour-Eldin, 1987), and structural controllability offers a computable alternative especially for systems of high dimensions. The same facts hold for observability for it is the dual of controllability.

In order to present the concept of structural controllability, let us first introduce the notion of a *structured matrix* $\tilde{M}$.

1.22. DEFINITION. An $n \times m$ matrix $\tilde{M} = (\tilde{m}_{ij})$ is said to be a structured matrix if its elements $\tilde{m}_{ij}$ are either fixed zeros or independent free parameters.

For example, a 2 × 3 structured matrix is

$$\tilde{M} = \begin{bmatrix} * & * & 0 \\ * & 0 & * \end{bmatrix}. \tag{1.31}$$

To relate a numerical $n \times m$ matrix $M = (m_{ij})$ to a structured matrix $\tilde{M} = (\tilde{m}_{ij})$ we define $\mathbf{n} = \{1, 2, \dots, n\}$, $\mathbf{m} = \{1, 2, \dots, m\}$ and state the following:

1.23. DEFINITION. A numerical matrix M is said to be admissible with respect to a structured matrix $\tilde{M}$, that is, $M \in \tilde{M}$, if and only if $\tilde{m}_{ij} = 0$ implies $m_{ij} = 0$ for all $i \in \mathbf{n}$ and $j \in \mathbf{m}$.

The matrix

$$M = \begin{bmatrix} 1 & 2 & 0 \\ 0 & 0 & -1 \end{bmatrix} \tag{1.32}$$

is admissible with respect to $\tilde{M}$ of (1.31), but neither of the matrices

$$\begin{bmatrix} 0 & \beta & 0 \\ \alpha & 0 & 1 \end{bmatrix}, \quad \begin{bmatrix} \alpha & \beta & 0 \\ \alpha & 0 & 0 \end{bmatrix} \tag{1.33}$$

is admissible, where α and β are independent free parameters.

To state the definition of structural controllability and observability, let us associate with a system

$$\begin{aligned} \mathbf{S}\colon\ \dot{x} &= Ax + Bu, \\ y &= Cx, \end{aligned} \tag{1.1}$$

a structured system $\tilde{\mathbf{S}} = (\tilde{A}, \tilde{B}, \tilde{C})$ so that $(A, B, C) \in (\tilde{A}, \tilde{B}, \tilde{C})$. Structural controllability of the system $\tilde{\mathbf{S}}$ is defined *via* the pair $(\tilde{A}, \tilde{B})$ as in (Lin, 1974):

1.24. DEFINITION. A pair of matrices $(\tilde{A}, \tilde{B})$ is said to be structurally controllable if there exists a controllable pair (A, B) such that $(A, B) \in (\tilde{A}, \tilde{B})$.

Similarly,

1.25. DEFINITION. A pair of matrices $(\tilde{A}, \tilde{C})$ is said to be structurally observable if there exists an observable pair (A, C) such that $(A, C) \in (\tilde{A}, \tilde{B})$.

In order to state the necessary and sufficient conditions for structural controllability and observability of $\tilde{\mathbf{S}}$, we need the notion of *generic rank* (or, *term rank*) of a structured matrix $\tilde{M}$, which we denote by $\tilde{\rho}(\tilde{M})$. Simply, the generic rank of $\tilde{M}$ is the maximal rank that $\tilde{M}$ can achieve by choosing numerical values for indeterminate elements of $\tilde{M}$. Therefore, a matrix $\tilde{M}$ has full generic rank if and only if there exists a matrix M of full rank such that $M \in \tilde{M}$.

To make the idea of the generic rank precise, we need certain notions from algebraic geometry (Wonham, 1979). Let us assume that a matrix $\tilde{M}$ has elements in $\mathbf{R}$, there are ν indeterminate entries, and the rest of the entries are fixed zeros. Then, with $\tilde{M}$ we can associate a parameter space $\mathbf{R}^\nu$ such that every data point $p \in \mathbf{R}^\nu$ defines a matrix $M \in \tilde{M}(p)$, which is obtained by replacing the arbitrary entries $\tilde{m}_{ij}$ of $\tilde{M}$ by the corresponding elements of $p = (p_1, p_2, \ldots, p_\nu)^T$. In a physical problem, it is important to know that if a matrix M has a certain rank at a nominal parameter vector p°, it has the same rank at a vector p close to p°, which corresponds to small deviations of the parameters from their nominal values. Most often, it turns out that the rank holds true for all $p \in \mathbf{R}^\nu$ except at the points p that lie on an algebraic surface in $\mathbf{R}^\nu$ and which are, therefore, atypical. An arbitrarily small perturbation of such points restores the rank of M.

Let us denote by $\varphi_k(p_1, p_2, \ldots, p_\nu)$, $k \in \boldsymbol{K}$, $\boldsymbol{K} = \{1, 2, \ldots, K\}$, a set of K polynomials generated by all the rth order minors of M, and consider a variety $\mathcal{V} \subset \mathbf{R}$ which is the locus of common zeros of the polynomials φ_k:

$$\mathcal{V} = \{p \in \mathbf{R}^\nu\colon\ \varphi_k(p_1, p_2, \ldots, p_v) = 0,\ k \in \boldsymbol{K}\}. \tag{1.34}$$

The variety $\mathcal{V}$ is proper if $\mathcal{V} \neq \mathbf{R}^\nu$ and nontrivial if $\mathcal{V} \neq 0$. We say that the rank r of the matrix M holds generically relative to $\mathcal{V}$ when the parameter values, which make the rank of M smaller than r, all lie on a proper variety $\mathcal{V}$ in $\mathbf{R}^\nu$. In other words, the variety $\mathcal{V}$ is either the whole space $\mathbf{R}^\nu$ in which case $\tilde{\rho}(\tilde{M}) < r$; or the complement $\mathcal{V}^c$ of $\mathcal{V}$ is open and dense in $\mathbf{R}^\nu$ and, therefore, generic. What this means is that, if the rank condition fails at $p^\circ \in \mathbf{R}^\nu$, then the condition can be restored by an arbitrarily small perturbation of the parameter vector p.

Conditions for structural controllability of $\tilde{\mathbf{S}}$ are stated in terms of the matrix pair $(\tilde{A}, \tilde{B})$. To formulate these conditions, we need the following two definitions introduced by Lin (1974):

1.26. DEFINITION. A composite matrix $[\tilde{A}\ \tilde{B}]$ has Form I if there exists an $n \times n$ permutation matrix P such that

$$P[\tilde{A}\ \tilde{B}]\begin{bmatrix} P^T & 0 \\ 0 & I \end{bmatrix} = \begin{bmatrix} \tilde{A}_{11} & 0 & 0 \\ \tilde{A}_{21} & \tilde{A}_{22} & \tilde{B} \end{bmatrix}, \tag{1.35}$$

where $\tilde{A}_{11}$ is $q \times q$, $\tilde{A}_{22}$ is $(n-q) \times (n-q)$, $\tilde{B}_2$ is $(n-q) \times m$, and $0 < q < n$.

In view of Definition 1.4, we consider a composite matrix $[\tilde{A}\ \tilde{B}]$ of a system $\tilde{\mathbf{S}}$ to be in Form I if $\tilde{\mathbf{S}}$ is structurally equivalent to a system having a composite matrix which appears on the right side of Equation (1.35).

1.27. DEFINITION. A composite matrix $[\tilde{A}\ \tilde{B}]$ has Form II if

$$\tilde{\rho}([\tilde{A}\ \tilde{B}]) < n. \tag{1.36}$$

The following result concerning structural controllability was initiated by Lin (1974) and established later by Shields and Pearson (1976) and Glover and Silverman (1976):

1.28. THEOREM. A pair $(\tilde{A}, \tilde{B})$ is structurally controllable if and only if $[\tilde{A}\ \tilde{B}]$ has neither Form I nor Form II.

We do not prove this theorem here. It appears as an easy Corollary 1.57 to Theorem 1.52 which is concerned with a more general problem of structurally fixed modes.

An automatic dual to Theorem 1.28 is the following:

1.29. THEOREM. A pair $(\tilde{A}, \tilde{C})$ is structurally observable if and only if $[\tilde{A}^T\ \tilde{C}^T]$ has neither Form I nor Form II.

We observe that Forms I and II are purely structural forms concerning the distribution of fixed zeros and indeterminate elements of the matrices $\tilde{A}$ and $\tilde{B}$, which suggests that they can be tested using graph-theoretic concepts. Before we present graph-theoretic conditions for structural controllability and observability, let us look at several examples.

1.30. EXAMPLE (Kailath, 1980). Two inverted penduli of lengths ℓ_1 and ℓ_2 having the bobs of mass m are mounted on a cart of mass M as shown in Figure 1.7. For small angle deviations $\theta_1(t)$ and $\theta_2(t)$ the equations of motion are

$$\begin{aligned} \mathbf{S}\colon\ M\dot{v} + mg\theta_1 + mg\theta_2 &= u, \\ m(\dot{v} + \ell_1\ddot{\theta}_1) - mg\theta_1 &= 0, \\ m(\dot{v} + \ell_2\ddot{\theta}_2) - mg\theta_2 &= 0, \end{aligned} \tag{1.37}$$

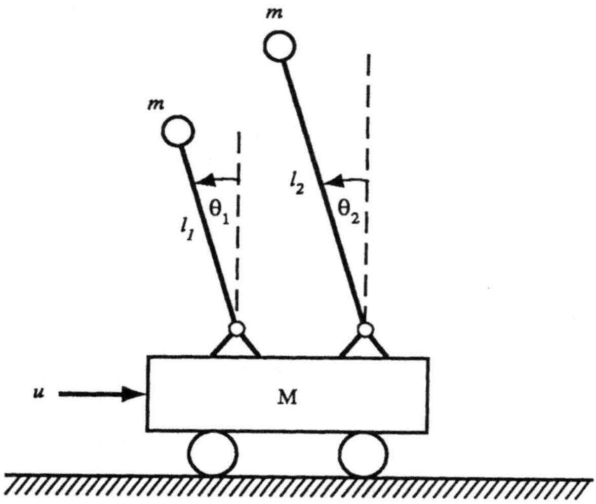

Fig. 1.7. Inverted penduli on a cart.

where $v(t)$ is the velocity of the cart and $u(t)$ is the external force acting on the cart. Is the system **S** controllable by $u(t)$? That is, can the penduli be kept in the upright position ($\theta_1 \approx 0,\ \theta_2 \approx 0$) by moving the cart?

Choosing the state vector $x \in \mathbf{R}^4$ as $x = (\theta_1,\, \dot\theta_1,\, \theta_2,\, \dot\theta_2)^T$ and eliminating $\dot v$, we get from (1.37) the state equations as

$$\mathbf{S}\colon\ \dot x = Ax + bu, \tag{1.38}$$

where

$$A = \begin{bmatrix} 0 & 1 & 0 & 0 \\ a_{21} & 0 & a_{23} & 0 \\ 0 & 0 & 0 & 1 \\ a_{41} & 0 & a_{43} & 0 \end{bmatrix}, \qquad b = \begin{bmatrix} 0 \\ b_2 \\ 0 \\ b_4 \end{bmatrix} \tag{1.39}$$

and

$$\begin{aligned} a_{21} &= \frac{(M+m)g}{M\ell_1}, & a_{23} &= \frac{mg}{M\ell_1} \\ a_{41} &= \frac{mg}{M\ell_2}, & a_{43} &= \frac{(M+m)g}{M\ell_2} \\ b_2 &= -\frac{1}{M\ell_1}, & b_4 &= \frac{1}{M\ell_2}. \end{aligned} \tag{1.40}$$

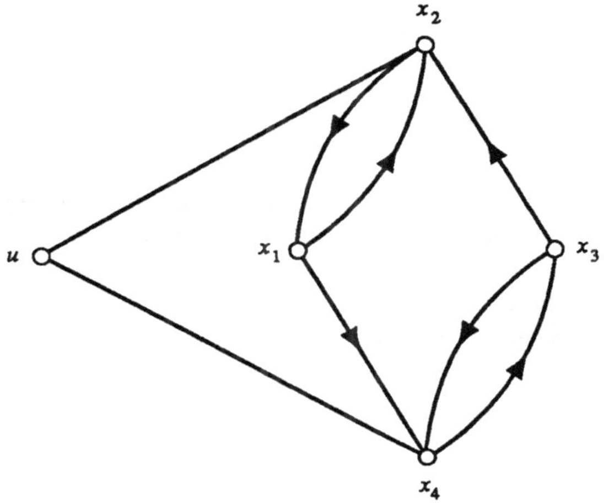

Fig. 1.8. System digraph.

We compute

$$\det([b \ \ Ab]) = -(\alpha - \beta)^2, \tag{1.41}$$

where

$$\alpha = \frac{(M+m)g}{M^2\ell_1} + \frac{mg}{M^2\ell_2}, \qquad \beta = \frac{(M+m)g}{M^2\ell_2} + \frac{mg}{M^2\ell_1}. \tag{1.42}$$

Therefore, $\det([b \ \ Ab]) = 0$ if and only if $\alpha = \beta$, that is,

$$\ell_1 = \ell_2. \tag{1.43}$$

In other words, the system **S** is uncontrollable when the lengths of the two penduli are equal, which is physically difficult, if not impossible, to achieve. Controllability of **S** is generic. The property fails on a proper variety $\mathcal{V}$ in the space of physical parameters, which is defined by condition (1.43).

To confirm the above finding in terms of structural controllability, we first present the system digraph **D** on Figure 1.8. It is clear that the system **S** is input reachable, that is, the composite matrix $[\tilde{A} \ \ \tilde{b}]$ is not in Form I. If we write the structured composite matrix

$$[\tilde{A} \ \ \tilde{b}] = \begin{bmatrix} 0 & \otimes & 0 & 0 & 0 \\ * & 0 & \otimes & 0 & * \\ 0 & 0 & 0 & \otimes & 0 \\ * & 0 & * & 0 & \otimes \end{bmatrix}, \tag{1.44}$$

we observe that the elements $\otimes$ guarantee

$$\tilde{\rho}([\tilde{A}\ \tilde{b}]) = 4, \tag{1.45}$$

and $[\tilde{A}\ \tilde{b}]$ is not in Form II. The system $\tilde{\mathbf{S}}$ is structurally controllable as predicted.

It is important to note, however, that in testing structural controllability, we ignored the fact that elements of the matrices A and b of (1.39) are not independent. In certain cases, ignoring the parameter dependencies may cause errors since a system may be structurally controllable when, in fact, it is not controllable (see Example 1.34 and Remark 1.35).

Let us now go back to Theorem 1.29 and discuss the meaning of Forms I and II.

1.31. REMARK. It is easy to see that the system $\tilde{\mathbf{S}}$ has Form I if and only if it is not input reachable. If there is a set $\mathtt{X}_k$ of state nodes in $\mathbf{D}$ which cannot be reached from the input nodes $\mathtt{U}$, we can transform the matrix $[\tilde{A}\ \tilde{B}]$ into the form (1.35) by permuting its rows and columns until the first k rows and columns of the transformed matrix correspond to the set $\mathtt{X}_k$. Conversely, if $[\tilde{A}\ \tilde{B}]$ has the form (1.35), then the state nodes corresponding to the submatrix $\tilde{A}_{11}$ cannot be reached from any of the input nodes of $\mathbf{D}$. Therefore, to test the system $\tilde{\mathbf{S}}$ for Form I is equivalent to testing $\mathbf{D}$ for input reachability, which is an easy computational problem (see Appendix). In case of rank deficiency (1.36), the system $\tilde{\mathbf{S}}$ has Form II, which is equivalent to saying that the corresponding digraph $\mathbf{D}$ has a *dilation* (Lin, 1974): A digraph $\mathbf{D} = (\mathtt{U} \cup \mathtt{X},\ \mathtt{E})$ is said to have a dilation if there exists a subset $\mathtt{X}_k \subseteq \mathtt{X}$ such that the number of distinct vertices of $\mathbf{D}$ from which a vertex in $\mathtt{X}_k$ is reachable is less than the number of vertices of $\mathtt{X}_k$. In matrix terms, this means that the corresponding structured $n \times (n+m)$ matrix $[\tilde{A}\ \tilde{B}]$ contains a zero submatrix of order $r \times (n + m + 1 - r)$ for some r such that $1 \leq r \leq n$. A numerical procedure to determine generic rank deficiency, that is, the existence of dilation, is given in the Appendix.

Uncontrollability due to lack of input reachability needs no explanation. It is a simple disconnection of inputs from states. Although input unreachability is a simple cause of uncontrollability, in physical systems, it is the basic one. The other is the generic rank deficiency which we illustrate next.

1.32. EXAMPLE. A very simple example of an uncontrollable system due to rank deficiency is

$$A = \begin{bmatrix} 0 & 0 \\ 0 & 0 \end{bmatrix}, \qquad b = \begin{bmatrix} b_1 \\ b_2 \end{bmatrix}. \tag{1.46}$$

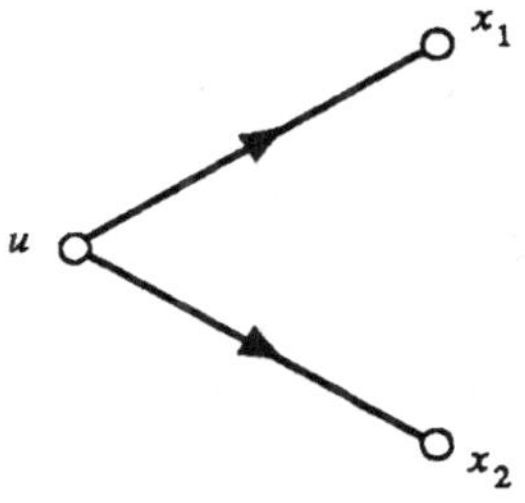

Fig. 1.9. Dilation.

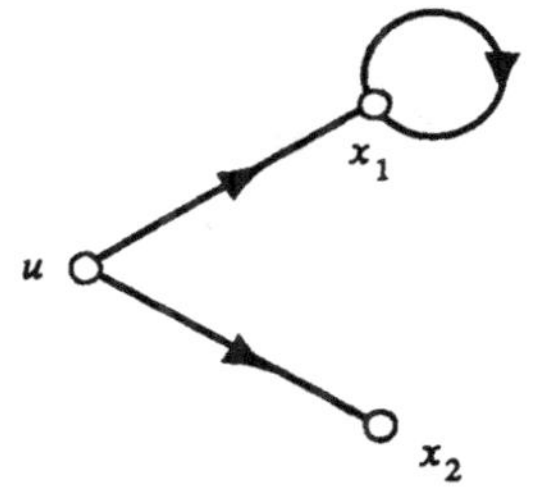

Fig. 1.10. No dilation.

The corresponding matrix

$$[\tilde{A}\ \tilde{b}] = \begin{bmatrix} 0 & 0 & * \\ 0 & 0 & * \end{bmatrix}, \tag{1.47}$$

has Form II. The digraph **D** is shown in Figure 1.9, which proves that we have input reachability, but $\tilde{\rho}([\tilde{A}\ \tilde{b}]) < 2$. The digraph **D** has a dilation. It is of interest to note that if we add a self-loop as in Figure 1.10, the modified digraph $\hat{\mathbf{D}}$ has no dilation.

1.33. EXAMPLE. If L_1 and L_2 are pure inductors in the network of Figure 1.11, then the state equation is

$$\mathbf{S}\colon\ \dot{x} = \begin{bmatrix} 0 & 0 & 0 \\ 0 & 0 & 0 \\ 0 & 0 & -1/RC \end{bmatrix} x + \begin{bmatrix} 1/L_1 \\ 1/L_2 \\ 1/RC \end{bmatrix} u, \tag{1.48}$$

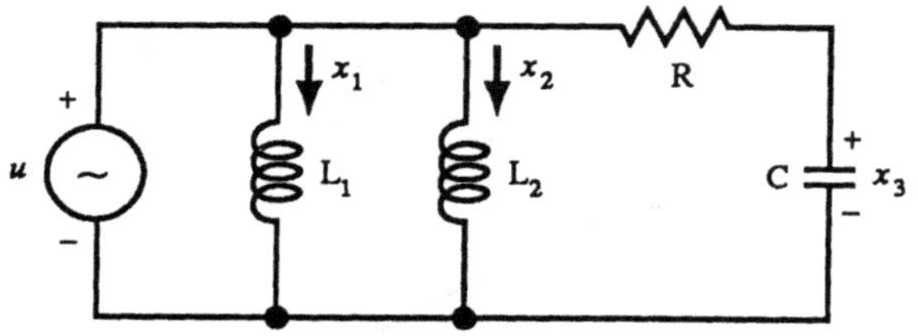

Fig. 1.11. Electrical network.

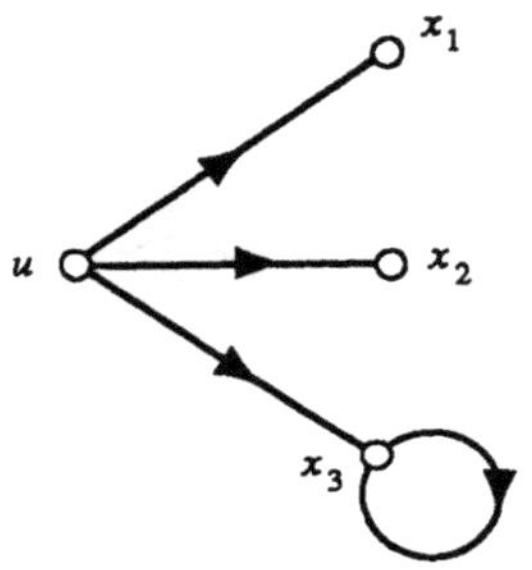

Fig. 1.12. Digraph with dilation.

and the composite matrix $[\tilde{A}\ \tilde{b}]$ is obtained as

$$[\tilde{A}\ \tilde{b}] = \left[\begin{array}{ccc|c} 0 & 0 & 0 & * \\ 0 & 0 & 0 & * \\ \hline 0 & 0 & * & * \end{array}\right], \tag{1.49}$$

resulting in $\tilde{\rho}([\tilde{A}\ \tilde{b}]) < 3$. The digraph **D** of Figure 1.12 has a dilation, because the matrix in (1.49) has a zero submatrix of order 2×3.

On the other hand, if the inductors, L_1 and L_2 have resistance r_1 and r_2, then the state equation is

$$\mathbf{S}\colon\ \dot{x} = \begin{bmatrix} -r_1/L_1 & & \bigcirc \\ & -r_2/L_2 & \\ \bigcirc & & -1/RC \end{bmatrix} x + \begin{bmatrix} 1/L_1 \\ 1/L_2 \\ 1/RC \end{bmatrix} u, \tag{1.50}$$

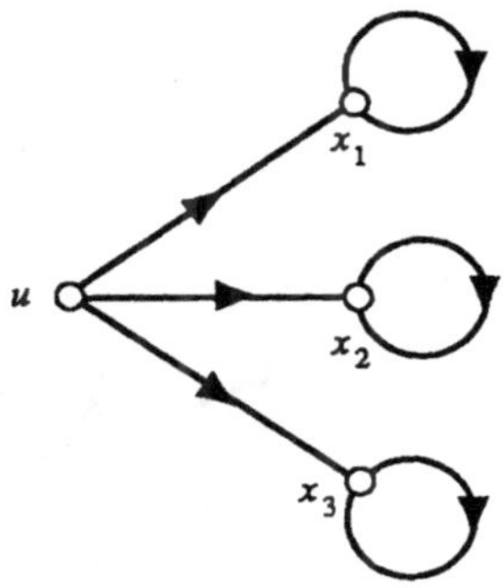

Fig. 1.13. Digraph without dilation.

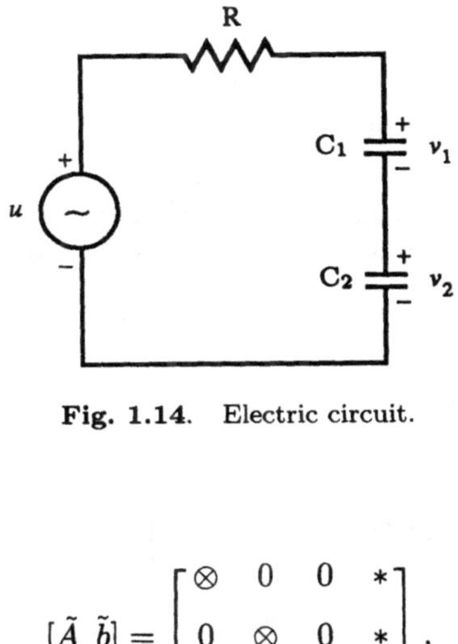

Fig. 1.14. Electric circuit.

and

$$[\tilde{A}\ \tilde{b}] = \begin{bmatrix} \otimes & 0 & 0 & * \\ 0 & \otimes & 0 & * \\ 0 & 0 & \otimes & * \end{bmatrix}, \tag{1.51}$$

indicating $\tilde{\rho}([\tilde{A}\ \tilde{b}]) = 3$. The corresponding digraph **D** of Figure 1.13 has no dilation. Since **D** is also input reachable, the system **S** of (1.50) is structurally controllable.

1.34. EXAMPLE. A notorious cause of uncontrollability is an erroneous choice of state variables. For a circuit in Figure 1.14, we select

$$x_1 = v_1, \qquad x_2 = v_2 - \frac{C_1}{C_2} v_1, \tag{1.52}$$

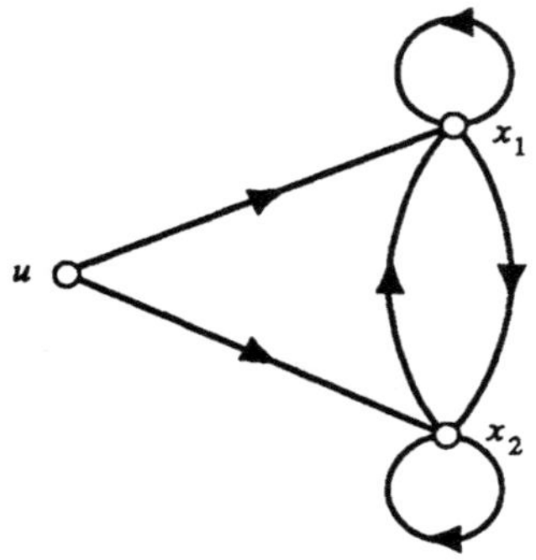

Fig. 1.15. Digraph $\check{\mathbf{D}}$.

and get the system

$$\mathbf{S}:\ \dot{x} = \begin{bmatrix} -(1/RC_1) - (1/RC_2) & 0 \\ 0 & 0 \end{bmatrix} x + \begin{bmatrix} 1/RC_1 \\ 0 \end{bmatrix} u, \tag{1.53}$$

which is both input unreachable and rank deficient as expected, because the choice (1.52) is not appropriate (the circuit has only one state). If we choose the states

$$\check{x}_1 = v_1, \qquad \check{x}_2 = v_2, \tag{1.54}$$

which is also wrong, we get

$$\check{\mathbf{S}}:\ \dot{\check{x}} = \begin{bmatrix} -1/RC_1 & -1/RC_1 \\ -1/RC_2 & -1/RC_2 \end{bmatrix} \check{x} + \begin{bmatrix} 1/RC_1 \\ 1/RC_2 \end{bmatrix} u. \tag{1.55}$$

Now, the digraph $\check{\mathbf{D}}$ for the system $\check{\mathbf{S}}$ in Figure 1.15 is both input reachable and has no dilation and, therefore, implies that $\check{\mathbf{S}}$ is structurally controllable when, in fact, $\check{\mathbf{S}}$ is an uncontrollable system. The entries in $\check{\mathbf{S}}$ are interdependent, which is the fact that the digraph $\check{\mathbf{D}}$ ignores and, thus, it fails to identify the uncontrollability of the system $\check{\mathbf{S}}$. The central issue seems to be a suitable choice of the state variables so that the system matrices contain the fewest generically dependent parameters. Obviously, the representation $\mathbf{S}$ contains no such parameters, while the representation $\check{\mathbf{S}}$ does. In fact, $\check{\mathbf{S}}$ has only two independent parameters. The parametrization problem has been considered in (Corfmat and Morse, 1976; Anderson and Hong, 1982).

1.35. REMARK. From Example 1.34 we learn that parameter dependencies may invalidate the conclusions obtained by structural analysis, which is a drawback of the concept of structural controllability and observability. We should note, however, that our primary goal is to use the concept of input and output reachability to formulate decomposition schemes for large dynamic systems (see Chapters 6 and 7). For this purpose, the manipulative capabilities of graphs are superior to any other approach. Equally important is the fact that controllability and observability (not necessarily structural!) of a decomposed system can be checked *via* low-order subsystems.

It remains to be discussed which algorithms can be used to test generic rank of the composite structured matrix $[\tilde{A} \;\; \tilde{B}]$. There has been a long history of the problem of finding the maximum number of nonzero elements of a matrix, no two of which lie in the same row or column. The permutation that puts these nonzero elements on the diagonal has been given various names such as selecting the maximum transversal, finding the complete, perfect, or maximum matching in bipartite graphs, determining a distinct representative, seeking an output set, etc. A survey of the relevant results and algorithms is available in a paper by Duff (1977) who provides an algorithm (Duff, 1981a) and a code in FORTRAN (Duff, 1981b). See also the discussion of the computational complexity of various algorithms presented by Duff (1981a). A version of Duff's algorithm is given in the Appendix.

1.5. Plant and Feedback Structures

An essential assumption in our studies is that a plant is composed of interconnected subsystems. This assumption is made for two distinct reasons. First, physical plants are made up of parts and by identifying individual parts, or groups of parts, as subsystems, we can take advantage of the structural features of the system in a control design. Second, we may ignore the physical boundaries of components and decompose a large system into "mathematical" subsystems which have no obvious physical identity. This type of decomposition can bring about a considerable savings in computer memory and time when systems of high dimensions are involved. Both kinds of partitions of dynamic systems, which are termed *physical* and *mathematical* decompositions (Šiljak, 1978), will be of central interest throughout this book. At this point, we only want to identify basic structures that are common in modeling of complex systems.

Let us consider again the linear constant system

$$\mathbf{S}\colon\ \dot{x} = Ax + Bu,\qquad y = Cx, \tag{1.1}$$

which we assume to be partitioned into N interconnected subsystems

$$\begin{aligned}\mathbf{S}\colon\ \dot{x}_i &= A_i x_i + B_i u_i + \sum_{j=1}^{N} (A_{ij} x_j + B_{ij} u_j),\\ y_i &= C_i x_i + \sum_{j=1}^{N} C_{ij} x_j, \qquad i \in \boldsymbol{N},\end{aligned} \tag{1.56}$$

with decoupled subsystems identified as

$$\begin{aligned}\mathbf{S}_i\colon\ \dot{x}_i &= A_i x_i + B_i u_i,\\ y_i &= C_i x_i, \qquad i \in \boldsymbol{N},\end{aligned} \tag{1.57}$$

where $x_i(t) \in \mathbf{R}^{n_i}$, $u_i(t) \in \mathbf{R}^{m_i}$, $y_i(t) \in \mathbf{R}^{\ell_i}$ are the state, input, and output of the subsystem $\mathbf{S}_i$ at a fixed time $t \in \mathbf{R}$, and $\boldsymbol{N} = \{1, 2, \ldots, N\}$. At present we are interested in *disjoint* decompositions, that is,

$$\begin{aligned}\mathbf{R}^n &= \mathbf{R}^{n_1} \times \mathbf{R}^{n_2} \times \ldots \times \mathbf{R}^{n_N},\\ \mathbf{R}^m &= \mathbf{R}^{m_1} \times \mathbf{R}^{m_2} \times \ldots \times \mathbf{R}^{m_N},\\ \mathbf{R}^\ell &= \mathbf{R}^{\ell_1} \times \mathbf{R}^{\ell_2} \times \ldots \times \mathbf{R}^{\ell_N}\end{aligned} \tag{1.58}$$

or,

$$n = \sum_{i=1}^{N} n_i, \qquad m = \sum_{i=1}^{N} m_i, \qquad \ell = \sum_{i=1}^{N} \ell_i \tag{1.59}$$

and

$$\begin{aligned}x &= \left(x_1^T, x_2^T, \ldots, x_N^T\right)^T,\\ u &= \left(u_1^T, u_2^T, \ldots, u_N^T\right)^T,\\ y &= \left(y_1^T, y_2^T, \ldots, y_N^T\right)^T\end{aligned} \tag{1.60}$$

are the state, input, and output vectors of $\mathbf{S}$.

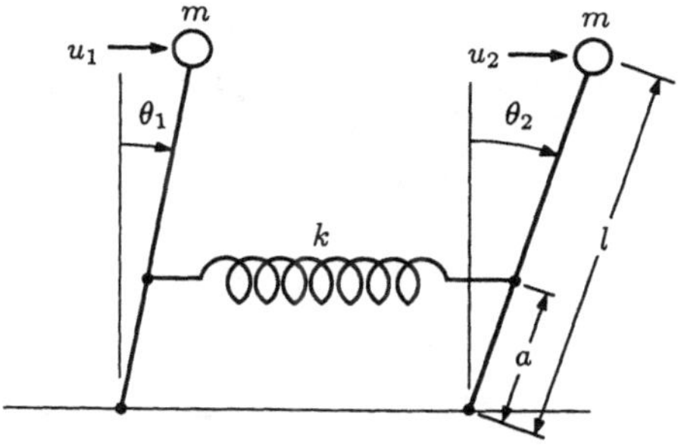

Fig. 1.16. Inverted penduli.

1.36. EXAMPLE. To illustrate a *physical* decomposition of a dynamic system, let us consider two identical penduli that are coupled by a spring and subject to two distinct inputs as shown in Figure 1.16. The equations of motion are

$$\begin{aligned} m\ell^2\ddot{\theta}_1 &= mg\ell\theta_1 - ka^2(\theta_1 - \theta_2) + u_1, \qquad & y_1 &= \theta_1, \\ m\ell^2\ddot{\theta}_2 &= mg\ell\theta_2 - ka^2(\theta_2 - \theta_1) + u_2, \qquad & y_2 &= \theta_2. \end{aligned} \tag{1.61}$$

Choosing the state vector as $x = (\theta_1,\ \dot{\theta}_1,\ \theta_2,\ \dot{\theta}_2)$ and denoting $u = (u_1,\ u_2)^T$, $y = (y_1,\ y_2)^T$, (1.61) can be rewritten as

$$\mathbf{S}\colon\ \dot{x} = \left[\begin{array}{cc|cc} 0 & 1 & 0 & 0 \\ \frac{g}{\ell} - \frac{ka^2}{m\ell^2} & 0 & \frac{ka^2}{m\ell^2} & 0 \\ \hline 0 & 0 & 0 & 1 \\ \frac{ka^2}{m\ell^2} & 0 & \frac{g}{\ell} - \frac{ka^2}{m\ell^2} & 0 \end{array}\right] x + \left[\begin{array}{c|c} 0 & 0 \\ \frac{1}{m\ell^2} & 0 \\ \hline 0 & 0 \\ 0 & \frac{1}{m\ell^2} \end{array}\right] u,$$

$$y = \left[\begin{array}{cc|cc} 1 & 0 & 0 & 0 \\ \hline 0 & 0 & 1 & 0 \end{array}\right] x. \tag{1.62}$$

The digraph **D** of the system **S** is shown in Figure 1.17.

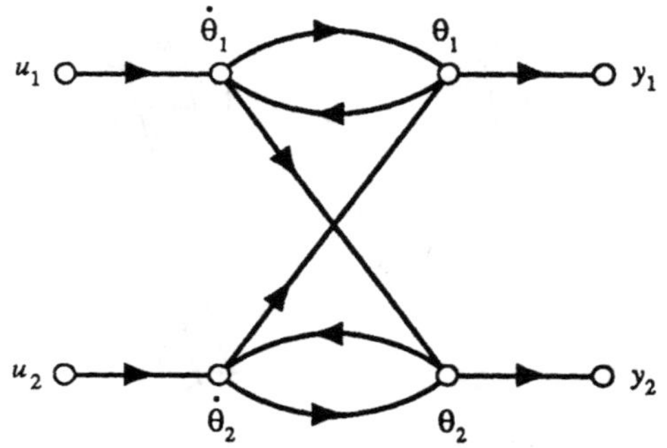

Fig. 1.17. System digraph.

A physical decomposition of **S** along the dashed lines produces a representation (1.56) as

$$\mathbf{S}\colon\ \dot{x}_1 = \begin{bmatrix} 0 & 1 \\ \alpha & 0 \end{bmatrix} x_1 + \begin{bmatrix} 0 \\ \beta \end{bmatrix} u_1 + e \begin{bmatrix} 0 & 0 \\ -\gamma & 0 \end{bmatrix} x_1 + e \begin{bmatrix} 0 & 0 \\ \gamma & 0 \end{bmatrix} x_2,$$

$$\dot{x}_2 = \begin{bmatrix} 0 & 1 \\ \alpha & 0 \end{bmatrix} x_2 + \begin{bmatrix} 0 \\ \beta \end{bmatrix} u_1 + e \begin{bmatrix} 0 & 0 \\ \gamma & 0 \end{bmatrix} x_1 + e \begin{bmatrix} 0 & 0 \\ -\gamma & 0 \end{bmatrix} x_2, \quad (1.63)$$

$$y_1 = [1 \quad 0] x_1,$$

$$y_2 = [1 \quad 0] x_2,$$

which is an interconnection of two subsystems

$$\mathbf{S}_1\colon\ \dot{x}_1 = \begin{bmatrix} 0 & 1 \\ \alpha & 0 \end{bmatrix} x_1 + \begin{bmatrix} 0 \\ \beta \end{bmatrix} u_1,$$

$$\mathbf{S}_2\colon\ \dot{x}_2 = \begin{bmatrix} 0 & 1 \\ \alpha & 0 \end{bmatrix} x_2 + \begin{bmatrix} 0 \\ \beta \end{bmatrix} u_2, \quad (1.64)$$

$$y_1 = [1 \quad 0]\, x_1,$$

$$y_2 = [1 \quad 0]\, x_2,$$

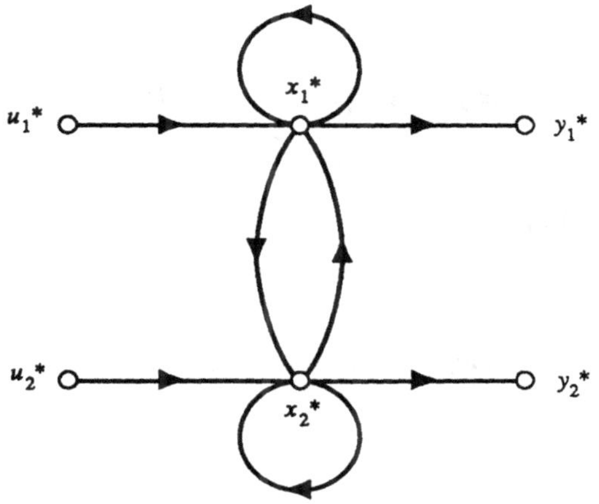

Fig. 1.18. Condensation.

each representing a single decoupled pendulum. The subsystem state vectors are obviously $x_1 = (\theta_1, \dot{\theta}_1)^T$ and $x_2 = (\theta_2, \dot{\theta}_2)^T$, and the system parameters are $\alpha = g/\ell$, $\beta = 1/m\ell^2$, and $\gamma = a^2k/m\ell^2$. The interconnected system **S** of (1.63) can be represented by a condensation digraph $\mathbf{D}^*$ in an obvious way as shown in Figure 1.18. It is important to note that in forming $\mathbf{D}^*$ we used the individual penduli as subsystems rather than strong components as specified by Definition 1.15. Since the entire state truncation $\mathbf{D}_x$ of the system digraph **D** of Figure 1.17 is a strong component, no decomposition of **S** in the sense of Section 1.3 is possible. The physical and mathematical decompositions are two distinct concepts. More on physical decompositions will be said in Chapter 8, while mathematical decompositions is the subject of Chapter 6 and 7 of this book.

An important feature in this example is that the matrices B_{ij} and C_{ij} of (1.56) are all zero. According to the terminology introduced by Šiljak (1978), the fact that $B_{12} = 0$, $B_{21} = 0$ (each input u_1 and u_2 applies to the respective subsystem $\mathbf{S}_1$ and $\mathbf{S}_2$ only) makes the system **S** of (1.63) an *input decentralized system.* Since $C_{12} = 0$ and $C_{21} = 0$, **S** is also an *output decentralized system.*

On various occasions it is convenient to use a compact notation for the composite system **S** of (1.56):

$$\mathbf{S}\colon \quad \begin{aligned} \dot{x} &= A_D x + B_D u + A_C x + B_C u, \\ y &= C_D x + C_C x, \end{aligned} \tag{1.65}$$

where

$$A_D = \text{diag}\{A_1, A_2, \ldots, A_N\},$$
$$B_D = \text{diag}\{B_1, B_2, \ldots, B_N\}, \tag{1.66}$$
$$C_D = \text{diag}\{C_1, C_2, \ldots, C_N\},$$

and the coupling block matrices are

$$A_C = (A_{ij}), \qquad B_C = (B_{ij}), \qquad C_C = (C_{ij}). \tag{1.67}$$

The collection of N decoupled subsystems is described by

$$\begin{aligned} \mathbf{S}_D\colon\ \dot{x} &= A_D x + B_D u, \\ y &= C_D x, \end{aligned} \tag{1.68}$$

which is obtained from (1.65) by setting the coupling matrices to zero.

An important special class of input and output decentralized systems $\mathbf{S} = (A_D, B_D, C_D, A_C)$ is represented by

$$\begin{aligned} \mathbf{S}\colon\ \dot{x} &= A_D x + B_D u + A_C x, \\ y &= C_D x, \end{aligned} \tag{1.69}$$

where $B_C = 0$ and $C_C = 0$. The inverted penduli of Example 1.36 belong to this class of models.

With the notion of a decentralized plant firmly in place, we are ready to formulate the decentralized constraints on feedback control laws. We step back first and recall the well-known and useful fact that the behavior of a linear plant $\mathbf{S}$ of (1.1) can be altered efficiently by feedback without changing the plant itself. The principle of feedback is to choose the inputs of the plant as functions of the outputs or states and achieve stability, optimality, robustness, etc., of the closed-loop system (*e.g.*, Kailath, 1980).

A linear constant control law for the original plant $\mathbf{S}$ of (1.1) can be chosen as the *static output feedback*

$$u = -Ky, \tag{1.70}$$

where the constant output gain matrix $K = (k_{ij})$ has dimension $m \times \ell$ and feedback coefficients k_{ij}. By applying the control law (1.70) to $\mathbf{S}$, we obtain the closed-loop system

$$\hat{\mathbf{S}}\colon\ \dot{x} = (A - BKC)x. \tag{1.71}$$

The structure of $\hat{\mathbf{S}}$ can be effectively studied by introducing the feedback interconnection matrix $\bar{K} = (\bar{k}_{ij})$ with elements defined as

$$\bar{k}_{ij} = \begin{cases} 1, & k_{ij} \neq 0, \\ 0, & k_{ij} = 0. \end{cases} \tag{1.72}$$

The interconnection matrix E of (1.2) with $\bar{K}$ included is the interconnection matrix $\hat{E}$ of the closed-loop system $\hat{\mathbf{S}}$:

$$\hat{E} = \begin{bmatrix} \bar{A} & \bar{B} & 0 \\ 0 & 0 & \bar{K} \\ \bar{C} & 0 & 0 \end{bmatrix}. \tag{1.73}$$

When a *dynamic output feedback* is used, we define for each input u_i of $\mathbf{S}$ in (1.1) an index set

$$\boldsymbol{J}_i = \{j : \bar{k}_{ij} = 1\}, \qquad i \in \boldsymbol{M}, \tag{1.74}$$

where $\boldsymbol{M} = \{1, 2, \ldots, m\}$, and apply the time-invariant dynamic controllers of the form

$$\begin{aligned} \mathbf{C}_i\colon\ \dot{z}_i &= F_i z_i + \sum_{j \in \boldsymbol{J}_i} g_{ij} y_j, \\ u_i &= -h_i^T z_i - \sum_{j \in \boldsymbol{J}_i} k_{ij} y_j, \qquad i \in \boldsymbol{M}, \end{aligned} \tag{1.75}$$

where $z_i(t) \in \mathbf{R}^{r_i}$ is the state of the controller $\mathbf{C}_i$, and F_i, g_{ij}, h_i, and k_{ij} are constant matrices, vectors, and scalars.

A special case of the output feedback constraints is the *decentralized feedback structure*, where the sets $\mathtt{U} = \{u_1, u_2, \ldots, u_m\}$ and $\mathtt{Y} = \{y_1, y_2, \ldots, y_\ell\}$ are partitioned into N disjoint subsets

$$\mathtt{U} = \bigcup_{r=1}^{N} \mathtt{U}_r, \qquad \mathtt{Y} = \bigcup_{r=1}^{N} \mathtt{Y}_r, \tag{1.76}$$

and where the feedback structure constraint is such that $\bar{k}_{ij} = 1$ if and only if $u_i \in \mathtt{U}_r$ and $y_j \in \mathtt{Y}_r$ for some $r \in \boldsymbol{N}$.

When no decomposition is used, we describe **S** as

$$\mathbf{S}^N\colon\ \dot{x} = Ax + \sum_{i=1}^{N} B_i u_i, \qquad y_i = C_i x, \qquad i \in \boldsymbol{N}, \tag{1.77}$$

where $x(t) \in \mathbf{R}^n$ is the state of $\mathbf{S}^N$; $u_i(t) \in \mathbf{R}^{m_i}$ and $y_i(t) \in \mathbf{R}^{\ell_i}$, $i \in \boldsymbol{N}$, are the inputs and outputs of $\mathbf{S}^N$; and A, B_i, C_i are constant matrices of appropriate dimensions. By defining the block matrices

$$B^N = [B_1\ B_2\ \dots\ B_N], \qquad C^N = [C_1\ C_2\ \dots\ C_N], \tag{1.78}$$

and using the decentralized control law

$$u_i = -K_i y_i, \qquad i \in \boldsymbol{N}, \tag{1.79}$$

or equivalently,

$$u = -K_D y, \qquad K_D = \text{diag}\{K_1, K_2, \dots, K_N\}, \tag{1.80}$$

we obtain the closed-loop system

$$\hat{\mathbf{S}}^N\colon\ \dot{x} = (A - B^N K_D C^N)x\ . \tag{1.81}$$

When dynamic output feedback is used with decentralized constraints, then controllers of the following type are used:

$$\mathbf{C}_i\colon\ \dot{z}_i = F_i z_i + G_i y_i, \qquad u_i = -H_i z_i - K_i y_i, \qquad i \in \boldsymbol{N}, \tag{1.82}$$

which can be written in a compact form as a single decentralized controller $\mathbf{C}_D = (F_D, G_D, H_D, K_D)$ defined as

$$\mathbf{C}_D\colon\ \dot{z} = F_D z + G_D y, \qquad u = -H_D z - K_D y, \tag{1.83}$$

where

$$z = (z_1^T, z_2^T, \dots, z_N^T)^T, \qquad y = (y_1^T, y_2^T, \dots, y_N^T), \qquad u = (u_1^T, u_2^T, \dots, u_N^T), \tag{1.84}$$

and

$$\begin{aligned} F &= \text{diag}\{F_1,\, F_2,\, \ldots,\, F_N\}, \qquad & G &= \text{diag}\{G_1,\, G_2,\, \ldots,\, G_N\}, \\ H &= \text{diag}\{H_1,\, H_2,\, \ldots,\, H_N\}, \qquad & K &= \text{diag}\{K_1,\, K_2,\, \ldots,\, K_N\}. \end{aligned} \tag{1.85}$$

Then, the closed-loop system is defined as a composite system

$$\mathbf{S}^N \,\&\, \mathbf{C}_D\text{:} \quad \begin{bmatrix} \dot{x} \\ \dot{z} \end{bmatrix} = \begin{bmatrix} A - B^N K_D C^N & -B^N H_D \\ G_D C^N & F_D \end{bmatrix} \begin{bmatrix} x \\ z \end{bmatrix}. \tag{1.86}$$

We recognize the fact that in case of the open-loop system $\mathbf{S}^N$ of (1.77), which is characterized by the triple $(A,\, B^N,\, C^N)$, the system is not decomposed into subsystems and the system matrix A is considered as a whole. It is only feedback (static or dynamic) that is decentralized. In modeling of complex systems this is not enough, and we require total decentralization of both the plant and the control. In this way, we are able to capture the effects of the essential uncertainty residing in the interconnection structure on the behavior of the overall system. Furthermore, decompositions and decentralizations, when combined together, can bring about a considerable increase in speed of computation (on- and off-line), as well as reliability of operation in control of complex systems (see Chapter 9). For these reasons, we devote our attention almost entirely to situations when the decentralized control law (1.80) is applied to the fully decentralized system $\mathbf{S}$ of (1.69), which is characterized by the quadruple $(A_D,\, B_D,\, C_D,\, A_C)$. The closed-loop system is described as

$$\hat{\mathbf{S}}\text{:} \quad \dot{x} = (A_D - B_D K_D C_D)x + A_C x. \tag{1.87}$$

In case local dynamic controllers $\mathbf{C}_i$ are used, we combine $\mathbf{S}$ and $\mathbf{C}_D$ to get the composite closed-loop system as

$$\mathbf{S} \,\&\, \mathbf{C}_D\text{:} \quad \begin{bmatrix} \dot{x} \\ \dot{z} \end{bmatrix} = \begin{bmatrix} A_D - B_D K_D C_D + A_C & -B_D H_D \\ G_D C_D & F_D \end{bmatrix} \begin{bmatrix} x \\ z \end{bmatrix}. \tag{1.88}$$

Obviously, a decomposed system with input and output decentralization $\mathbf{S}_D = (A_D,\, B_D,\, C_D,\, A_C)$ is a more restrictive structure than the "quasi-decentralized system" $\mathbf{S}^N = (A,\, B^N,\, C^N)$ without decompositions, as it is always possible to describe the former class as a special case of the later class of systems, but not the other way around. The fully decentralized structure with interconnection uncertainties makes the control design more difficult involving a variety of conditions (most of them sufficient) for robust stabilizability of interconnected systems. There is, however, an essential necessary condition for stabilizability of both classes of systems, given in terms of *structurally fixed modes*, which is derived in the following section.

1.6. Structurally Fixed Modes

Since the early work of McFadden (1967) and Aoki (1972), the stabilization problem under decentralized information structure constraints has attracted considerable attention. The concept of decentralized fixed modes has been introduced by Wang and Davison (1973) to obtain necessary and sufficient conditions for stabilizability of decentralized systems, where only the feedback is decentralized but the plant is treated as a whole. Later, Corfmat and Morse (1976) and Anderson and Clements (1981) achieved a more refined characterization of the fixed modes and a deeper insight into the problem of decentralized control of the same type of linear systems. A number of characterizations of decentralized fixed modes have been later proposed by Anderson (1982), Vidyasagar and Viswanadham (1982), and Davison and Özgüner (1983), which have led to existence conditions involving difficult numerical problems of rank or eigenvalue computations.

As in the case of an uncontrollable or unobservable mode, a decentralized fixed mode can originate from two distinct sources: it is either a consequence of a perfect matching of system parameters (in which case, a slight change of parameters can eliminate the mode), or it is due to a special structure of the system (in which case, the mode remains fixed no matter how much the parameters are changed as long as the original structure is preserved). From a physical point of view, only the latter type of fixed modes are important, not only because it is very unlikely to have an exact matching of parameters, but also because it is not possible to know whether such a matching occurs or not. This type of *structurally fixed modes* have been defined and characterized by Sezer and Šiljak (1981a, b) for arbitrary feedback patterns including output and decentralized control.

An obvious advantage offered by the concept of structurally fixed modes is numerical. Since existence of such modes is a qualitative structural property of the system, it can be characterized in terms of directed graphs leading to computationally attractive tests involving binary computations only (Pichai *et al.* 1983, 1984; Linnemann, 1983).

Consider again the system $\mathbf{S} = (A, B, C)$ described by

$$\mathbf{S}\colon\ \dot{x} = Ax + Bu,\qquad y = Cx, \tag{1.1}$$

and define a set of constant output feedback matrices

$$\mathcal{K} = \{K \in \mathbf{R}^{m\times\ell}\colon\ k_{ij} = 0 \text{ if } \bar{k}_{ij} = 0\}, \tag{1.89}$$

which can be used in the control law

$$u = -Ky. \tag{1.70}$$

By $\Lambda(\bullet)$ we denote the set of eigenvalues of a given matrix.

1.37. DEFINITION. The set

$$\mathcal{L}_{\bar{K}} = \bigcap_{K\in\mathcal{K}} \Lambda(A - BKC), \tag{1.90}$$

is the set of fixed modes of the system $\mathbf{S} = (A, B, C)$ with respect to a feedback structure $\bar{K}$.

1.38. REMARK. This definition is a generalization of the notion of decentralized fixed modes given by Wang and Davision (1973) to the case of arbitrary feedback structure constraints. The decentralized fixed modes of $\mathbf{S}^N$ is the set

$$\mathcal{L}_{\bar{K}_D} = \bigcap_{K_D\in\mathcal{K}_D} \Lambda(A - B^N K_D C^N), \tag{1.91}$$

where K_D is a block diagonal matrix of (1.85) and $\mathcal{K}_D$ is a class of all such matrices.

Stabilizability of $\mathbf{S}$ under arbitrary structure constraints specified by the controller $\mathbf{C}$, which is the union of the individual controllers

$$\mathbf{C}_i\colon\ \dot{z}_i = F_i z_i + \sum_{j\,\in\,\boldsymbol{J}_i} g_{ij} y_j,\qquad u_i = -h_i^T z_i - \sum_{j\,\in\,\boldsymbol{J}_i} k_{ij} y_j,\qquad i \in \boldsymbol{M}, \tag{1.75}$$

is determined by the following:

1.39. THEOREM. The system $\mathbf{S} = (A, B, C)$ can be stabilized by the controller $\mathbf{C}$ if and only if the set of fixed modes $\mathcal{L}_{\bar{K}}$ contains no elements with nonnegative real parts.

Proof. Define an $N = m$ channel system $\check{\mathbf{S}}^N = (\check{A}, \check{B}^N, \check{C}^N)$, where

$$\check{A} = A, \quad \check{B}^N = B, \quad \check{C}_N = (\check{C}_1^T, \check{C}_2^T, \ldots, \check{C}_m^T)^T, \tag{1.92}$$

and $\check{C}_i$ consists of those rows of C with indices in $\boldsymbol{J}_i$ of (1.74). Then, for any $K \in \mathcal{K}$,

$$\begin{aligned} A - BKC &= A - \sum_{i=1}^{m} \sum_{j=1}^{\ell} b_i k_{ij} c_j^T \\ &= A - \sum_{i=1}^{m} b_i \left(\sum_{j \in \boldsymbol{J}_i} k_{ij} c_j \right) \\ &= \check{A} - \sum_{i=1}^{m} \check{b}_i \, \check{k}_{ij} \, \check{C}_i, \end{aligned} \tag{1.93}$$

where $\check{k}_{ij}^T = (k_{ij_1}, k_{ij_2}, \ldots)$, $j_1, j_2, \ldots, \in \boldsymbol{J}_i$, $i \in \boldsymbol{M}$, and b_i and c_i^T denote the ith column of B and jth row of C, respectively. From (1.93), it is clear that $\mathcal{L}_{\bar{K}} = \check{\mathcal{L}}_{\bar{K}_D}$, where $\check{\mathcal{L}}_{\bar{K}_D}$ is the set of decentralized fixed modes of $\check{\mathbf{S}}^N$, and the proof follows from the main theorem of Wang and Davison (1973) on decentral stabilizability of the system $\check{\mathbf{S}}^N$. Q.E.D.

To complete our consideration of stabilizability of $\mathbf{S}$, we give an algebraic characterization of the fixed modes $\mathcal{L}_{\bar{K}}$ of (1.90). For this purpose, we define the sets $\boldsymbol{M} = \{1, 2, \ldots, m\}$, $\boldsymbol{L} = \{1, 2, \ldots, \ell\}$, and for an arbitrary subset $\boldsymbol{I} = \{i_1, i_2, \ldots, i_p\}$ of $\boldsymbol{M}$, we form a set

$$\boldsymbol{J} = \bigcup_{i \in \boldsymbol{M} - \boldsymbol{I}} \boldsymbol{J}_i, \tag{1.94}$$

where $\boldsymbol{J}_i = \{j_1, j_2, \dots, j_q\}$. For example, if $m = 5$ and $\ell = 4$,

$$\bar{K} = \begin{bmatrix} 1 & 1 & 0 & 0 \\ 0 & 1 & 0 & 1 \\ 0 & 1 & 1 & 0 \\ 1 & 0 & 0 & 1 \\ 0 & 0 & 1 & 1 \end{bmatrix}, \tag{1.95}$$

and $\boldsymbol{I} = \{1, 2, 3\}$, then $\boldsymbol{J} = \boldsymbol{J}_4 \cup \boldsymbol{J}_5 = \{1, 4\} \cup \{3, 4\} = \{1, 3, 4\}$. Now, introducing the matrices

$$B_I = (b_{i_1}, b_{i_2}, \dots, b_{i_p}), \quad C_J = (c_{j_1}, c_{j_2}, \dots, c_{j_q})^T, \tag{1.96}$$

and recalling that $\rho(\bullet)$ denotes the rank of a given matrix, we prove the following:

1.40. THEOREM. A complex number λ is a fixed mode of the system $\mathbf{S} = (A, B, C)$ with respect to a feedback structure $\bar{K}$ if and only if there exists a subset $\boldsymbol{I} \subset \boldsymbol{M}$ such that

$$\rho\left(\begin{bmatrix} A - \lambda I & B_I \\ C_J & \end{bmatrix}\right) < n. \tag{1.97}$$

Proof. Consider the system $\check{\mathbf{S}}^N$ defined in the proof of Theorem 1.39. By Theorem 4.1 of Anderson and Clements (1981), $\lambda \in \check{\mathcal{L}}_{\bar{K}_D}$ if and only if there exists an $\boldsymbol{I} \subset \boldsymbol{M}$ such that

$$\rho\left(\begin{bmatrix} \check{A} - \lambda I & \check{B}_I \\ \check{C}_{M-I} & 0 \end{bmatrix}\right) < n, \tag{1.98}$$

where $\check{C}_{M-I} = (\check{C}^T_{i'_1}, \check{C}^T_{i'_2}, \dots, \check{C}^T_{i'_{m-p}})^T$ with $\{i'_1, i'_2, \dots, i'_{m-p}\} = \boldsymbol{M} - \boldsymbol{I}$. Noting that any c_i^T is a row of $\check{C}_{M-I}$ if and only if it is a row of C_j, we establish the equivalence of (1.97) and (1.98). The proof follows from the fact that $\mathcal{L}_{\bar{K}} = \check{\mathcal{L}}_{\bar{K}_D}$ as shown in the proof of Theorem 1.39. Q.E.D.

Although Theorem 1.40 provides a complete characterization of fixed modes, in practice it is impossible to identify all such modes because of inexact knowledge of the system parameters. For this reason, structurally fixed modes are significant because they occur as a result of physical disconnections between the system and some controllers (or parts of a single controller), and remain fixed unless the control structure is modified.

Given a structured system $\tilde{\mathbf{S}} = (\tilde{A}, \tilde{B}, \tilde{C})$, we call a system $\mathbf{S} = (A, B, C)$ admissible if $(A, B, C) \in (\tilde{A}, \tilde{B}, \tilde{C})$. Then we are interested in the following:

1.41. DEFINITION. A structured system $\tilde{\mathbf{S}} = (\tilde{A}, \tilde{B}, \tilde{C})$ is said to have structurally fixed modes with respect to a feedback structure $\bar{K}$ if every admissible system $\mathbf{S} = (A, B, C)$ has fixed modes with respect to $\bar{K}$.

In the spirit of structural controllability and observability (Definitions 1.24 and 1.25) the above definition can be restated as: A structured system $\tilde{\mathbf{S}} = (\tilde{A}, \tilde{B}, \tilde{C})$ has no structurally fixed modes with respect to a feedback structure $\bar{K}$ if there exists an admissible system $\mathbf{S} = (A, B, C)$ which has no fixed modes with respect to $\bar{K}$. In terms of structural equivalence (Definition 1.4) this means that a system $\tilde{\mathbf{S}} = (\tilde{A}, \tilde{B}, \tilde{C})$ has structurally fixed modes with respect to $\bar{K}$ if every system structurally equivalent to $\tilde{\mathbf{S}}$ has fixed modes with respect to $\bar{K}$. In this way, the notion of structurally fixed modes is concerned with an equivalence class of systems rather than a particular member of the class and, therefore, keeps the atypical fixed modes, which arise from perfect matching of parameters, out of considerations.

1.42. EXAMPLE. To illustrate the notion of structurally fixed modes, let us consider a system $\mathbf{S} = (A, B, C)$ defined as

$$A = \begin{bmatrix} 1 & 2 & 0 \\ 3 & 4 & 3 \\ 0 & 2 & 1 \end{bmatrix}, \quad B = \begin{bmatrix} 0 & 0 \\ 0 & 1 \\ 1 & 0 \end{bmatrix}, \quad C = \begin{bmatrix} 0 & 1 & 0 \\ 1 & 0 & 0 \\ 0 & 1 & 1 \end{bmatrix}, \tag{1.99}$$

which is controllable and observable. Let the feedback structure constraint be specified as

$$\bar{K} = \begin{bmatrix} 1 & 0 & 0 \\ 0 & 1 & 1 \end{bmatrix}, \quad K = \begin{bmatrix} k_{11} & 0 & 0 \\ 0 & k_{22} & k_{23} \end{bmatrix}. \tag{1.100}$$

Then, the matrix

$$A - BKC = \begin{bmatrix} 1 & 2 & 0 \\ 3 - k_{22} & 4 & 3 - k_{23} \\ 0 & 2 - k_{11} & 1 \end{bmatrix}, \tag{1.101}$$

has an eigenvalue at $\lambda = 1$ independent of the feedback gains k_{ij}, which is by Definition 1.37 a *fixed mode*. If, however, the nonzero element $a_{33} = 1$

of A in (1.99) is slightly perturbed by a small amount ϵ to obtain the new matrix

$$A_\epsilon = \begin{bmatrix} 1 & 2 & 0 \\ 3 & 4 & 3 \\ 0 & 2 & 1+\epsilon \end{bmatrix}, \tag{1.102}$$

then it is easy to show that the resulting system $\mathbf{S}_\epsilon = (A_\epsilon, B, C)$ has no fixed modes with respect to the feedback constraints (1.100). Thus, the fixed mode of $\mathbf{S}$ is a result of a perfect matching of some of the nonzero elements of the matrix A. If, on the other hand, the matrix A of (1.99) were

$$A = \begin{bmatrix} 0 & 2 & 0 \\ 3 & 4 & 3 \\ 0 & 2 & 0 \end{bmatrix}, \tag{1.103}$$

then $\mathbf{S}$ would have a fixed mode at $\lambda = 0$ no matter how the nonzero elements of the matrices A, B and C were perturbed, although it would always remain jointly controllable and observable. In this case, the fixed mode $\lambda = 0$ is caused entirely by the structure of the system, and is independent of the values taken by the nonzero parameters of A, B and C; it is a *structurally fixed mode.*

We now prove a theorem of Sezer and Šiljak (1981a, b) that provides an algebraic characterization of structurally fixed modes. To avoid trivialities, we assume that the system $\tilde{\mathbf{S}} = (\tilde{A}, \tilde{B}, \tilde{C})$ is structurally controllable and observable.

1.43. THEOREM. A system $\tilde{\mathbf{S}} = (\tilde{A}, \tilde{B}, \tilde{C})$ has structurally fixed modes with respect to a feedback structure $\bar{K}$ if and only if either of the following conditions holds:

(*i*) There exists a subset $\boldsymbol{I} \subset \boldsymbol{M}$ and a permutation matrix P such that

$$P^T \tilde{A} P = \begin{bmatrix} \tilde{A}_{11} & 0 & 0 \\ \tilde{A}_{21} & \tilde{A}_{22} & 0 \\ \tilde{A}_{31} & \tilde{A}_{32} & \tilde{A}_{33} \end{bmatrix}, \qquad P^T \tilde{B}_{I} = \begin{bmatrix} 0 \\ 0 \\ \tilde{B}_3^I \end{bmatrix}, \tag{1.104}$$

$$\tilde{C}_J P = [\tilde{C}_1^J \quad 0 \quad 0].$$

(*ii*) There exists a subset $\boldsymbol{I} \subset \boldsymbol{M}$ such that

$$\tilde{\rho}\left(\begin{bmatrix} \tilde{A} & \tilde{B}_I \\ \tilde{C}_J & 0 \end{bmatrix}\right) < n. \tag{1.105}$$

Before we present a proof of this theorem, we need some preliminary considerations. First, by $\mathbf{R}^{\nu_A}$, $\mathbf{R}^{\nu_B}$, and $\mathbf{R}^{\nu_C}$, we denote the parameter spaces of structured matrices $\tilde{A}$, $\tilde{B}$, and $\tilde{C}$, and ν_A, ν_B, ν_C are the number of unspecified elements of $\tilde{A}$, $\tilde{B}$, and $\tilde{C}$, respectively. Also, we denote a typical point $p \in \mathbf{R}^{\nu}$ by $p = (p_A, p_B, p_C)$, where $\mathbf{R}^{\nu} = \mathbf{R}^{\nu_A} \times \mathbf{R}^{\nu_B} \times \mathbf{R}^{\nu_C}$. Next, we present a definition and a simple lemma of Davison (1977), which we give without proof:

1.44. DEFINITION. A matrix $\tilde{A}$ is said to be degenerate if, for all $p_A \in \mathbf{R}^{\nu_A}$, $\tilde{A}$ has all of its eigenvalues at the origin.

1.45. LEMMA. If a matrix $\tilde{A}$ is nondegenerate, then the set $\{p_A \in \mathbf{R}^{\nu_A}:\ \tilde{A}$ has repeated nonzero eigenvalues$\}$ is contained in a proper variety $\mathcal{V}_A \subset \mathbf{R}^{\nu_A}$.

We now prove a result established by Hosoe and Matsumoto (1979) using a somewhat different approach:

1.46. LEMMA. Let $\tilde{A}$ be nondegenerate, $\mathcal{V}_A$ be as defined in Lemma 1.45, and let $\mathcal{V}_{AB} \subset \mathbf{R}^{\nu_A} \times \mathbf{R}^{\nu_B}$ be a proper variety which contains $\mathcal{V}_A \times \mathbf{R}^{\nu_B}$. Suppose that, for all $(p_A, p_B) \notin \mathcal{V}_{AB}$, there exists a $\lambda \in \mathbf{C}$ such that $\lambda \neq 0$ and

$$\tilde{\rho}\left([\tilde{A} - \lambda I \;\; \tilde{B}]\right) < n. \tag{1.106}$$

Then, $[\tilde{A} \;\; \tilde{B}]$ is of Form I (Definition 1.26).

Proof. Let $\mathcal{V}_{AB}$ be defined by the polynomial $\varphi(p_A, p_B)$. We first claim that $\tilde{B}$ has some fixed zero rows. Otherwise, for any $p_A = p_A^*$ such that $\varphi(p_A^*, p_B) \not\equiv 0$, since $\rho(A^* - \lambda I) \geq n - 1$ for all $\lambda \neq 0$, we can choose a $p_B = p_B^*$ such that $\varphi(p_A^*, p_B^*) \neq 0$ and $\rho([A^* - \lambda I \;\; B^*]) \geq n$ for all $\lambda \neq 0$, contradicting the assumption. Therefore, there exists a permutation matrix P_1 such that

$$P_1^T \tilde{A} P_1 = \begin{bmatrix} \tilde{A}_{11} & \tilde{A}_{12} \\ \tilde{A}_{21} & \tilde{A}_{22} \end{bmatrix}, \qquad P_1^T \tilde{B} = \begin{bmatrix} 0 \\ \tilde{B}_2 \end{bmatrix}, \tag{1.107}$$

where $\tilde{A}_{ii}$ is $n_i \times n_i$, $i = 1, 2$, with $n_1 + n_2 = n$, and $\tilde{B}_2$ has no fixed zero rows. If $\tilde{A}_{12} = 0$, the proof follows. Suppose $\tilde{A}_{12} \neq 0$. If $\tilde{A}_{22}$ is degenerate, fix $p_{A_{21}}, p_{A_{22}}$ and p_B such that $\varphi(p_{A_{11}}, p_{A_{12}}, p^*_{A_{21}}, p^*_{A_{22}}, p^*_B) \not\equiv 0$ and thus define a proper variety $\mathcal{V}_{A_{11}A_{12}}$ in the parameter space of $(\tilde{A}_{11},\ \tilde{A}_{12})$. Since

$A_{22}^* - \lambda I$ is nonsingular for all $\lambda \neq 0$, for all $(p_{A_{11}}, p_{A_{12}}) \notin \mathcal{V}_{A_{11}A_{12}}$,

$$\tilde{\rho}\left([\tilde{A}_{11} - \lambda I \quad \tilde{A}_{12}]\right) < n_1, \qquad \text{for some } \lambda \neq 0. \tag{1.108}$$

If $\tilde{A}_{22}$ is nondegenerate, since $\tilde{B}_2$ contains no zero rows, $p_{A_{21}}$, $p_{A_{22}}$ and p_B can be fixed such that $\rho([A_{22}^* - \lambda I \quad B_2^*]) \geq n_2$ for all $\lambda \neq 0$, so that again (1.108) must hold for all $(p_{A_{11}}, p_{A_{12}}) \notin \mathcal{V}_{A_{11}A_{12}}$. Obviously, (1.108) also implies that $\tilde{A}_{11}$ is nondegenerate, and that $\mathcal{V}_{A_{11}A_{12}}$ contains $\mathcal{V}_{A_{11}} \times \mathbf{R}^{\nu_{A_{12}}}$, where $\mathcal{V}_{A_{11}}$ is defined as in Lemma 1.45.

We now continue the proof by induction on n. For $n = 2$, or $n > 2$, $n_1 = 1$, the proof is obvious. For $n > n_1 \geq 2$, the induction hypothesis and (1.108) imply that $[\tilde{A}_{11} \quad \tilde{A}_{12}]$ is of Form I, that is, there exists a permutation matrix P_2 which puts $[\tilde{A}_{11} \quad \tilde{A}_{12}]$ into Form I. Then, the permutation matrix $P = P_1 \operatorname{diag}\{P_2, I_{n_2}\}$ permutes $[\tilde{A} \quad \tilde{B}]$ into Form I. Q.E.D.

1.47. LEMMA. If there exists a $p \in \mathbf{R}^\nu$ such that the rank of the matrix

$$\tilde{M}(\lambda) = \begin{bmatrix} \tilde{A} - \lambda I & \tilde{B} \\ \tilde{C} & 0 \end{bmatrix}, \tag{1.109}$$

is no less than n for all $\lambda \in \mathbf{C}$, then the set $\mathcal{P} = \{p \in \mathbf{R}^\nu\colon\ \tilde{\rho}[\tilde{M}(\lambda)] < n$ for some $\lambda\}$ is a proper variety in $\mathbf{R}^\nu$.

Proof. Let $\varphi_0(\lambda; p_A) = \det(\tilde{A} - \lambda I)$, and let $\varphi_1(\lambda; p)$ denote the sum of the squares of all other $n \times n$ principal minors of $\tilde{M}(\lambda)$. Then, the set $\mathcal{P}$ is exactly the set of $p \in \mathbf{R}^\nu$ for which φ_0 and φ_1 have common zeros or, equivalently, for which the resultant φ_2 of φ_0 and φ_1 is zero. Since $\varphi_2 \not\equiv 0$ by assumption, the proof follows. Q.E.D.

1.48. LEMMA. Suppose $[\tilde{A} \quad \tilde{B}]$ and $[\tilde{A}^T \quad \tilde{C}^T]$ are not of Form I. Let $\mathcal{V} \subset \mathbf{R}^\nu$ be a proper variety, and let $\mathcal{L} = \{\lambda_1, \lambda_2, \ldots, \lambda_k\}$, where $\lambda_i \neq 0$ for all i. Then there exists a $p \notin \mathcal{V}$ such that

$$\tilde{\rho}[\tilde{M}(\lambda)] \geq \begin{cases} n, & \lambda \neq 0, \\ n+1, & \lambda \in \mathcal{L}, \end{cases} \tag{1.110}$$

where $\tilde{M}(\lambda)$ is defined in (1.109).

Proof. We claim that there exists a $p \notin \mathcal{V}$ such that $\tilde{\rho}[\tilde{M}(\lambda)] \geq n$ for $\lambda \neq 0$. If $\tilde{A}$ is degenerate, this is obvious. If $\tilde{A}$ is nondegenerate, let $\mathcal{V}_A$ be as defined in Lemma 1.45, and let the proper variety $\mathcal{V} \cup (\mathcal{V}_A \times$

$\mathbf{R}^{\nu_B} \times \mathbf{R}^{\nu_C}$) be defined by the polynomial $\varphi(p_A, p_B, p_C)$. Fix $p_C = p_C^*$ such that $\varphi(p_A, p_B, p_C^*) \not\equiv 0$, and thus define a proper variety $\mathcal{V}_{AB} \subset \mathbf{R}^{\nu_A} \times \mathbf{R}^{\nu_B}$ which contains $\mathcal{V}_A \times \mathbf{R}^{\nu_B}$. If the claim were not true, then for all $(p_A, p_B) \notin \mathcal{V}_{AB}$ there would exist a $\lambda \neq 0$ such that $\tilde{\rho}([\tilde{A} - \lambda I \;\; \tilde{B}]) < n$; and Lemma 1.46 would imply that $[\tilde{A} \;\; \tilde{B}]$ has Form I, contradicting the assumption. Therefore, the claim must be true. Using the same argument as in the proof of Lemma 1.47, we can show that

$$\{p \notin \mathcal{V}\colon\; \tilde{\rho}[\tilde{M}(\lambda)] < n \text{ for some } \lambda \neq 0\} \subset \mathcal{V}_1, \tag{1.111}$$

where $\mathcal{V}_1$ is a proper variety in $\mathbf{R}^{\nu}$.

On the other hand, it is easy to show that the set $\{p \in \mathbf{R}^{\nu_A}\colon\; \det(\tilde{A} - \lambda I) = 0 \text{ for some } \lambda_i \in \mathcal{L}\}$ is contained in a proper variety $\mathcal{V}'_A$. Also, $\tilde{C}(\tilde{A} - \lambda I)^{-1}\tilde{B} \not\equiv 0$, for otherwise we would have $\tilde{C}\tilde{A}^i\tilde{B} = 0$ for $i = 1, 2, \ldots,$ which can easily be shown to imply that both $[\tilde{A} \;\; \tilde{B}]$ and $[\tilde{A}^T \;\; \tilde{C}^T]$ have Form I. Thus, letting $\mathcal{V}'_2 = \mathcal{V}'_A \times \mathbf{R}^{\nu_B} \times \mathbf{R}^{\nu_C}$, we can show that $\{p \notin \mathcal{V}_2\colon\; \tilde{C}(\tilde{A} - \lambda I)^{-1}\tilde{B} = 0 \text{ for some } \lambda_i \in \mathcal{L}\}$ is contained in a proper variety $\mathcal{V}_3 \subset \mathbf{R}^{\nu}$. Now, it is obvious that, for all $p \notin \mathcal{V}_2 \cup \mathcal{V}_3$,

$$\tilde{\rho}[\tilde{M}(\lambda)] = \tilde{\rho}\left(\begin{bmatrix} \tilde{A} - \lambda I & \tilde{B} \\ 0 & \tilde{C}(\tilde{A} - \lambda I)^{-1}\tilde{B} \end{bmatrix}\right) \geq n + 1 \quad \text{for all } \lambda \in \mathcal{L}. \tag{1.112}$$

Combining this result with (1.111), we conclude that for all $p \notin \mathcal{V}_1 \cup \mathcal{V}_2 \cup \mathcal{V}_3$ condition (1.110) holds. Q.E.D.

We need one more lemma before we give the proof of Theorem 1.43.

1.49. LEMMA. Suppose

$$\tilde{\rho}\left(\begin{bmatrix} \tilde{A} & \tilde{B} \\ \tilde{C} & 0 \end{bmatrix}\right) \geq n. \tag{1.113}$$

Then, either the set $\{p \in \mathbf{R}^{\nu} :\; \tilde{\rho}[\tilde{M}(\lambda)] < n \text{ for some } \lambda\}$ is contained in a proper variety in $\mathbf{R}^{\nu}$, where $\tilde{M}(\lambda)$ is given by (1.109); or there exists a permutation matrix P such that the matrices $P^T\tilde{A}P$, $P^T\tilde{B}$, and $\tilde{C}P$ have the form in (1.104).

Proof. Let $\mathcal{V}$ be a proper variety on which $\tilde{\rho}[\tilde{M}(0)] < n$. If $\tilde{A}$ is degenerate, then the set $\{p \in \mathbf{R}^{\nu}\colon\; \tilde{\rho}[\tilde{M}(\lambda)] < n \text{ for some } \lambda \in \mathbf{C}\}$ is contained in $\mathcal{V}_1$ and the proof follows. If $\tilde{A}$ is nondegenerate, let $\mathcal{V}_A$ be as in Lemma 1.45 and let $\mathcal{V} = \mathcal{V}_1 \cup (\mathcal{V}_A \times \mathbf{R}^{\nu_B} \times \mathbf{R}^{\nu_C})$. If there exists a $p \in \mathbf{R}^{\nu} - \mathcal{V}$ such that

$\tilde{\rho}[\tilde{M}(\lambda)] \geq n$, then the proof follows from Lemma 1.47. Otherwise, for all $p \in \mathbf{R}^\nu - \mathcal{V}$ we must have a $\lambda \neq 0$ such that $\tilde{\rho}[\tilde{M}(\lambda)] < n$. Using the same argument as in the first part of the proof of Lemma 1.48, we can show that both $[\tilde{A} \ \tilde{B}]$ and $[\tilde{A}^T \ \tilde{C}^T]$ are of Form I, which in turn implies that there exists a permutation P such that

$$P^T\tilde{A}P = \begin{bmatrix} \tilde{A}_{11} & 0 & 0 & 0 \\ \tilde{A}_{21} & \tilde{A}_{22} & 0 & 0 \\ \tilde{A}_{31} & \tilde{A}_{32} & \tilde{A}_{33} & 0 \\ \tilde{A}_{41} & \tilde{A}_{42} & \tilde{A}_{43} & \tilde{A}_{44} \end{bmatrix}, \quad P^T\tilde{B} = \begin{bmatrix} 0 \\ 0 \\ \tilde{B}_3 \\ \tilde{B}_4 \end{bmatrix}, \tag{1.114}$$

$$\tilde{C}P = [\, \tilde{C}_1 \quad 0 \quad \tilde{C}_3 \quad 0 \,].$$

In particular, P can be chosen such that the pairs

$$\left(\begin{bmatrix} \tilde{A}_{33} & 0 \\ \tilde{A}_{43} & A_{44} \end{bmatrix}, \begin{bmatrix} \tilde{B}_3 \\ \tilde{B}_4 \end{bmatrix} \right), \quad \left(\begin{bmatrix} \tilde{A}_{11} & 0 \\ \tilde{A}_{31} & A_{33} \end{bmatrix}^T, \quad [\tilde{C}_1 \ \ \tilde{C}_3]^T \right), \tag{1.115}$$

are not of Form I. In (1.114), let $n_i \times n_i$ matrices $\tilde{A}_{ii}$ consist of those rows and columns of $\tilde{A}$ with indices in $\boldsymbol{N}_i \subset \boldsymbol{N} = \{1, 2, \ldots, n\}$, $i = 1, 2, 3, 4$. Note that some $\boldsymbol{N}_i$ might be empty.

We now claim that $\boldsymbol{N}_3 = \emptyset$. If not, fix all the nonzero elements of $\tilde{A}$, $\tilde{B}$, and $\tilde{C}$ except those of $\tilde{A}_{33}$, $\tilde{B}_3$, and $\tilde{C}_3$ such that $\varphi(p^*_{A_{11}}, \cdots, p_{A_{33}}, \cdots, p^*_{A_{44}};\ p_{B_3}, p^*_{B_4};\ p^*_{C_1}, p^*_{C_3}) \not\equiv 0$; and thus define a proper variety $\mathcal{V}_{A_{33}B_3C_3}$ in the parameter space of $(\tilde{A}_{33}, \tilde{B}_3, \tilde{C}_3)$, where φ is the polynomial that defines $\mathcal{V}$. Let $\mathcal{L}$ be the set of nonzero eigenvalues of A^*_{11}, A^*_{22}, and A^*_{44}. Since $[\tilde{A}_{33} \ \tilde{B}_3]$ and $[\tilde{A}^T_{33} \ \tilde{C}^T_3]$ are not of Form I, Lemma 1.48 implies that there exists $(p^*_{A_{33}}, p^*_{B_4}, p^*_{C_3}) \notin \mathcal{V}_{A_{33}B_3C_3}$ such that

$$\rho\left(\begin{bmatrix} A^*_{33} - \lambda I & B^*_3 \\ C^*_3 & 0 \end{bmatrix} \right) \geq \begin{cases} n_3, & \lambda \neq 0, \\ n_3 + 1, & \lambda \in \mathcal{L}. \end{cases} \tag{1.116}$$

Now, for any $\lambda \neq 0$, if $\lambda \notin \mathcal{L}$, then A^*_{11}, A^*_{22} and A^*_{44} are nonsingular; so that (1.116) implies $\rho[M^*(\lambda)] \geq n$. If $\lambda \in \mathcal{L}$, then, since $\rho(A^*_{11} - \lambda I) + \rho(A^*_{22} - \lambda I) + \rho(A^*_{44} - \lambda I) = n_1 + n_2 + n_4 - 1$, again (1.116) implies $\rho[M^*(\lambda)] \geq n$. This, however, contradicts the assumption that, for all $p \notin \mathcal{V}$, $\tilde{\rho}[\tilde{M}(\lambda)] < n$ for some $\lambda \neq 0$. Thus, we must have $\boldsymbol{N}_3 = \emptyset$.

Finally, to complete the proof, we claim that $\boldsymbol{N}_2 \neq \emptyset$. To prove the claim, assume the contrary, that is,

$$P^T \tilde{A} P = \begin{bmatrix} \tilde{A}_{11} & 0 \\ \tilde{A}_{41} & \tilde{A}_{44} \end{bmatrix}, \qquad P^T \tilde{B} = \begin{bmatrix} 0 \\ \tilde{B}_4 \end{bmatrix},$$
$$\tilde{C} P = [\, \tilde{C}_1 \quad 0 \,], \tag{1.117}$$

and recall that: (i) $\tilde{A}$ is nondegenerate, (ii) for all $p \notin \mathcal{V}$, there exists a $\lambda \neq 0$ such that $\tilde{\rho}[\tilde{M}(\lambda)] < n$, and (iii) $[\tilde{A}_{44} \;\; \tilde{B}_4]$ and $[\tilde{A}_{11}^T \;\; \tilde{C}_1^T]$ are not in Form I. If $\tilde{A}_{44}$ is degenerate, then (i) implies that $\tilde{A}_{11}$ is nondegenerate; and (ii), implies that for all $(p_{A_{11}}, p_{C_1})$, except possibly those values that lie on a proper variety in the parameter space of $(\tilde{A}_{11}, \; \tilde{C}_1)$, there exists a $\lambda \neq 0$ such that $\tilde{\rho}([\tilde{A}_{11}^T - \lambda I \;\; \tilde{C}_1^T]) < n_1$. But then, by Lemma 1.46, $[\tilde{A}_{11}^T \;\; \tilde{C}_1^T]$ must be of Form I, contradicting (iii). If $\tilde{A}_{44}$ is nondegenerate, then, since $[\tilde{A}_{44} - \lambda I \;\; \tilde{B}_4]$ is not of Form I, Lemma 1.46 implies that $\tilde{\rho}([\tilde{A}_{44}, \; -\lambda I \;\; \tilde{B}_4]) \geq n_2$ for almost all $(p_{A_{44}}, p_{B_4})$, which, together with (ii) leads to the same contradiction. Hence, $\boldsymbol{N}_2 \neq \emptyset$. Q.E.D.

Proof of Theorem 1.43. If $\tilde{\mathbf{S}} = (\tilde{A}, \tilde{B}, \tilde{C})$ has structurally fixed modes, it follows from Definition 1.41 and Theorem 1.40 that, for all $p \in \mathbf{R}^{\nu}$, there exists a λ and a subset $\boldsymbol{I} \subset \boldsymbol{M}$ (both depending on p) such that $\tilde{\rho}[\tilde{M}_I(\lambda)] < n$, where $\tilde{M}_I(\lambda)$ is the matrix defined by (1.109) with $\tilde{B}$ and $\tilde{C}$ replaced by $\tilde{B}_I$ and $\tilde{C}_J$, respectively. This, however, is equivalent to saying that there exists a particular $\boldsymbol{I} \subset \boldsymbol{M}$ for which, for all data points (p_A, p_{B_I}, p_{B_J}) in the parameter space of $(\tilde{A}, \tilde{B}_I, \tilde{C}_J)$ except possibly those that lie on a proper variety, there exists a λ such that $\tilde{\rho}[\tilde{M}_I(\lambda)] < n$, and Lemma 1.49 completes the proof of the *necessity* part of the theorem.

The *sufficiency* part of the theorem, on the other hand, follows immediately from Theorem 1.40 on noting that in case (i), $\Lambda(\tilde{A}_{22})$, and in case (ii), $\lambda = 0$, are fixed modes. Q.E.D.

Because of our interest in decentralized control, we shall interpret the results of this section in the context of an N-channel system $\mathbf{S}^N = (A, B^N, C^N)$. To discuss the special case of *structurally fixed modes under decentralized constraints* (Remark 1.38), we start with the following:

1.50. EXAMPLE. Let us consider a two-channel system $\mathbf{S}^N$ defined as

$$A = \begin{bmatrix} 0 & 1 & 0 \\ 1 & 1 & 0 \\ 0 & 0 & 1 \end{bmatrix}, \quad B_1 = C_2^T = \begin{bmatrix} 1 \\ 0 \\ 0 \end{bmatrix}, \quad B_2 = C_1^T = \begin{bmatrix} 0 \\ 0 \\ 1 \end{bmatrix}, \tag{1.118}$$

which is controllable and observable. For $K_D = \text{diag}\{k_1,\ k_2\}$, the matrix

$$A - B^N K_D C^N = \begin{bmatrix} 0 & 1 & -k_1 \\ 1 & 1 & 0 \\ -k_2 & 0 & 1 \end{bmatrix} \tag{1.119}$$

has an eigenvalue at $\lambda = 1$ independent of k_1 and k_2, which is by definition a fixed mode of $\mathbf{S}^N$. However, if we perturb the coefficient $a_{33} = 1$ of A to get

$$A_\epsilon = \begin{bmatrix} 0 & 1 & 0 \\ 1 & 1 & 0 \\ 0 & 0 & 1+\epsilon \end{bmatrix}, \tag{1.120}$$

where ϵ is an arbitrarily small (but fixed) number, it is easy to see that the resulting system $\mathbf{S}_\epsilon^N = (A_\epsilon,\ B^N,\ C^N)$ would have no fixed modes. The fixed mode of $\mathbf{S}^N$ is caused by a perfect matching of the nonzero elements of A. If, on the other hand, the matrix in (1.118) were

$$A_1 = \begin{bmatrix} 0 & 1 & 0 \\ 1 & 0 & 0 \\ 0 & 0 & 0 \end{bmatrix}, \tag{1.121}$$

then the system $\mathbf{S}^N$ would have a fixed mode at $\lambda = 0$ no matter how the nonzero elements of the triple $(A_1,\ B^N,\ C^N)$ were perturbed, although it would always stay centrally controllable and observable. This is the case of structurally fixed modes under decentralized constraints.

To state the corollary to Theorem 1.43 regarding the *decentralized version* of structurally fixed modes, let us assume that $\mathbf{S}^N = (A,\ B^N,\ C^N)$ defined by

$$\mathbf{S}^N\colon\ \dot{x} = Ax + \sum_{i=1}^{N} B_i u_i,$$
$$y_i = C_i x_i, \qquad i \in \boldsymbol{N}, \tag{1.77}$$

and

$$B^N = [B_1\ B_2\ \dots\ B_N], \quad C^N = [C_1\ C_2\ \dots\ C_N], \tag{1.78}$$

is jointly structurally controllable and observable; that is, the pairs $(\tilde{A},\ \tilde{B}^N)$ and $(\tilde{A},\ \tilde{C}^N)$ are structurally controllable and observable, respectively. We also introduce a subset of indices $\boldsymbol{K} = \{i_1,\ i_2,\ \dots,\ i_k\} \subset \boldsymbol{N}$ to define the block matrices

$$B^K = [B_{i_1}\ B_{i_2}\ \dots\ B_{i_k}], \quad C^K = \left[C_{i_1}^T\ C_{i_2}^T\ \dots\ C_{i_k}^T\right], \tag{1.122}$$

and state the following (Sezer and Šiljak, 1981a, b):

1.51. COROLLARY. A system $\tilde{\mathbf{S}}^N = (\tilde{A},\ \tilde{B}^N,\ \tilde{C}^N)$ has structurally fixed modes with respect to the decentralized control

$$u = -K_D y, \quad K_D = \operatorname{diag}\{K_1,\ K_2,\ \dots,\ K_N\}, \tag{1.80}$$

if and only if either of the following conditions hold:

(*i*) There exists a subset $\boldsymbol{K} \subset \boldsymbol{N}$ and a permutation matrix P such that

$$P^T \tilde{A} P = \begin{bmatrix} \tilde{A}_{11} & 0 & 0 \\ \tilde{A}_{12} & \tilde{A}_{22} & 0 \\ \tilde{A}_{13} & \tilde{A}_{32} & \tilde{A}_{33} \end{bmatrix}, \quad P^T \tilde{B}^K = \begin{bmatrix} \tilde{B}_1^K \\ \tilde{B}_2^K \\ \tilde{B}_3^K \end{bmatrix}, \quad P^T \tilde{B}^{N\text{-}K} = \begin{bmatrix} 0 \\ 0 \\ \tilde{B}_3^{N\text{-}K} \end{bmatrix},$$

$$\tilde{C}^K P = [\tilde{C}_1^K\ \ 0\ \ 0], \quad \tilde{C}^{N\text{-}K} P = \left[\tilde{C}_1^{N\text{-}K} \tilde{C}_2^{N\text{-}K}\ \tilde{C}_3^{N\text{-}K}\right]. \tag{1.123}$$

(*ii*) There exists a subset $\boldsymbol{K} \subset \boldsymbol{N}$ such that

$$\tilde{\rho}\left(\begin{bmatrix} \tilde{A} & \tilde{B}^K \\ \tilde{C}^{N\text{-}K} & 0 \end{bmatrix}\right) < n. \tag{1.124}$$

We observe that both conditions of Theorem 1.43 and its Corollary 1.51 are purely structural conditions concerning the distribution of indeterminate and fixed zero elements of the system matrices, which suggests that they can be tested using graph theoretical concepts. This we consider next.

We need a few notions from graph theory. A digraph $\mathbf{D}_S = (\mathtt{V}_S,\ \mathtt{E}_S)$ where $\mathtt{V}_S \subseteq \mathtt{V}$ and $\mathtt{E}_S \subseteq \mathtt{E}$ is called a subgraph of $\mathbf{D} = (\mathtt{V}, \mathtt{E})$. If $\mathtt{V}_S = \mathtt{V}$, $\mathbf{D}_S$ is said to span $\mathbf{D}$. The union of digraphs $\mathbf{D}_1 = (\mathtt{V}_1,\ \mathtt{E}_1)$ and $\mathbf{D}_2 = (\mathtt{V}_2,\ \mathtt{E}_2)$ is the digraph $\mathbf{D}_1 \cup \mathbf{D}_2 = (\mathtt{V}_1 \cup \mathtt{V}_2,\ \mathtt{E}_1 \cup \mathtt{E}_2)$. $\mathbf{D}_1$ and $\mathbf{D}_2$ are said to be disjoint if $\mathtt{V}_1 \cap \mathtt{V}_2 = \emptyset$. When a constant output feedback

$$u_i = -\sum_{j \in \boldsymbol{J}_i} k_{ij} y_j, \tag{1.70$'$}$$

satisfying the constraints specified by a binary matrix $\bar{K}$, is applied to the system $\mathbf{S} = (A, B, C)$, we accordingly add feedback edges $\mathtt{E}_K$ to $\mathbf{D}$ to obtain the digraph $\hat{\mathbf{D}} = (\mathtt{V}, \mathtt{E} \cup \mathtt{E}_K)$ of the closed-loop system, where

$$\mathtt{E}_K = \{(y_j, u_i)\colon\ i \in \boldsymbol{M}\,, \quad j \in \boldsymbol{J}_i\}. \tag{1.125}$$

The corresponding closed-loop interconnection matrix $\hat{E}$ is given as

$$\hat{E} = \begin{bmatrix} \bar{A} & \bar{B} & 0 \\ 0 & 0 & \bar{K} \\ \bar{C} & 0 & 0 \end{bmatrix}. \tag{1.73}$$

The following result is a graph-theoretic characterization of structurally fixed modes.

1.52. THEOREM. A system $\tilde{\mathbf{S}} = (\tilde{A}, \tilde{B}, \tilde{C})$ has no structurally fixed modes with respect to a feedback structure $\bar{K}$ if and only if both of the following two conditions hold:

- (i) Each state vertex $x_k \in \mathtt{X}$ is contained in a strong component of $\hat{\mathbf{D}}$ which includes an edge from $\mathtt{E}_K$.
- (ii) There exists a set of mutually disjoint cycles $\mathbf{C}_r = (\mathtt{V}_r, \mathtt{E}_r)$, $r = 1, 2, \ldots, R$, in $\hat{\mathbf{D}}$ such that

$$\mathtt{X} \subset \bigcup_{r=1}^{R} \mathtt{V}_r. \tag{1.126}$$

We first state a couple of simple lemmas without proofs, which were established by Shields and Pearson (1976).

1.53. LEMMA. Let $\tilde{M}$ be a $p \times q$ matrix. Then $\tilde{\rho}(\tilde{M}) \geq r$, $r \leq \min\{p, q\}$, if and only if $\tilde{M}$ contains at least r nonzero elements located in distinct rows and columns.

1.54. LEMMA. Let $\tilde{M}$ be a $p \times q$ matrix. Then, $\tilde{\rho}(\tilde{M}) < r$, $1 \leq r \leq \min\{p, q\}$, if and only if $\tilde{M}$ contains a zero submatrix of order no less than $s \times (p + q - r - s + 1)$ for some s such that $p - r < s \leq p$.

We also need the following result which is straightforward consequence of Lemma 1.53:

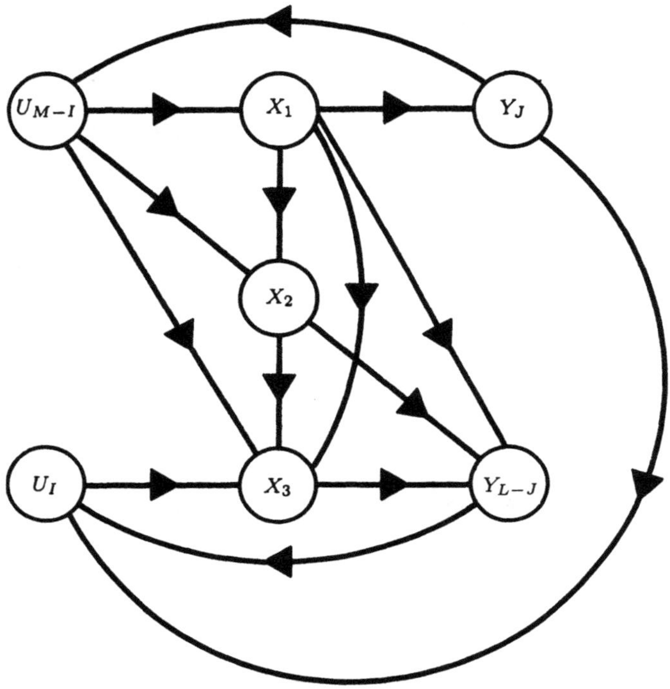

Fig. 1.19. Digraph $\check{\mathbf{D}}$.

1.55. LEMMA. Let $\mathbf{D}$ be a digraph with n vertices, and let $\bar{M}$ be its adjacency matrix. Then, $\mathbf{D}$ is spanned by a disjoint union of cycles if and only if $\tilde{\rho}(\tilde{M}) = n$.

Now, we modify the digraph $\hat{\mathbf{D}}$ by adding self-loops at every input and output vertex to obtain a new digraph $\check{\mathbf{D}} = (\mathrm{V}, \mathrm{E} \cup \mathrm{E}_K \cup \check{\mathrm{E}})$, where $\check{\mathrm{E}} = \{(v, v)\colon\ v \in \mathrm{U} \cup \mathrm{Y}\}$, and proceed to the proof of the theorem.

Proof of Theorem 1.52. We show that conditions (i) and (ii) of Theorem 1.52 are negations of conditions (i) and (ii) of Theorem 1.43.

If (i) of Theorem 1.43 holds, then $\hat{\mathbf{D}}$ has a structure shown in Figure 1.19, where X_1, X_2, X_3 are identified by the diagonal blocks $\tilde{A}_{11}$, $\tilde{A}_{22}$, $\tilde{A}_{33}$ in (1.104). From Figure 1.19, we see that $\hat{\mathbf{D}}$ has at least one strong component with all vertices in X_2, which contains no feedback edges. To prove the converse, let $\hat{\mathbf{D}}_k = (\mathrm{V}^k, \mathrm{E}^k \cup \mathrm{E}_K^k)$, where $\mathrm{V}^k = \mathrm{U}^k \cup \mathrm{X}^k \cup \mathrm{Y}^k$, $k = 1, 2, \ldots, K$, denote the strong components of $\hat{\mathbf{D}}$ ordered so that no vertex in V^k reaches a vertex in $\mathbf{V}^j$ if $j < i$, $i = 2, 3, \ldots, n$ (Theorem 1.18). Let $\hat{\mathbf{D}}_k$ be a strong component that includes no feedback edges, that is, one with $\mathbf{E}_K^k = \emptyset$. Then, obviously, $\mathbf{U}^k = \emptyset$ and $\mathbf{Y}^k = \emptyset$. Defining $\mathbf{X}_1 = \mathbf{X}^1 \cup \mathbf{X}^2 \cup \ldots \cup \mathbf{X}^{k-1}$, $\mathbf{X}_2 = \mathbf{X}^k, \mathbf{X}_3 = \mathbf{X}^{k+1} \cup \mathbf{X}^{k+2} \cup \ldots \cup \mathbf{X}^K$, and $\mathbf{U}_I = \mathbf{U}^{k+1} \cup \mathbf{U}^{k+2} \cup \ldots \cup \mathbf{U}^K$, it is easy to see that $\hat{\mathbf{D}}$ has the structure shown in Figure 1.19. This completes the proof of the first part of the theorem.

To prove part *(ii)* of the theorem, we first notice that condition *(ii)* of Theorem 1.43 is equivalent to the modified graph $\check{\mathbf{D}}$ being spanned by a disjoint union of cycles. Due to Lemma 1.55, the latter condition is, in turn, equivalent to the structured version of the adjacency matrix of $\check{\mathbf{D}}$,

$$\tilde{E} = \begin{bmatrix} \tilde{A} & \tilde{B} & 0 \\ 0 & \tilde{I}_m & \tilde{K} \\ \tilde{C} & 0 & \tilde{I}_\ell \end{bmatrix}, \qquad (1.127)$$

being of full generic rank. Thus, it suffices to prove that part *(ii)* of Theorem 1.43 holds for some $\boldsymbol{I} \subset \boldsymbol{M}$ if and only if $\tilde{\rho}(\tilde{E}) < n+m+\ell$. This means that if

$$\tilde{\rho}\left(\begin{bmatrix} \tilde{A} & \tilde{B}_I \\ \tilde{C}_J & 0 \end{bmatrix}\right) < n \qquad (1.105)$$

holds for some $\boldsymbol{I} = \{i_1, i_2, \ldots, i_p\}$ and $\boldsymbol{J} = \{j_1, j_2, \ldots, j_q\}$, then Lemma 1.54 implies that the matrix in (1.105) contains a zero submatrix of total size no less than $n+p+q+1$, where by total size of a matrix we mean the sum of the number of rows and the number of columns. Partitioning $\tilde{E}$ in (1.127) as

$$\tilde{E} = \begin{bmatrix} \tilde{A} & \tilde{B}_I & \tilde{B}_{M\text{-}I} & 0 & 0 \\ 0 & \tilde{I}_p & 0 & \tilde{K}_J & \tilde{K}_{I,\ L\text{-}J} \\ 0 & 0 & \tilde{I}_{m-p} & \tilde{K}_{M\text{-}I,\ J} & \tilde{K}_{M\text{-}I,\ L\text{-}J} \\ \tilde{C}_J & 0 & 0 & \tilde{I}_q & 0 \\ \tilde{C}_{L\text{-}J} & 0 & 0 & 0 & \tilde{I}_{\ell-q} \end{bmatrix}, \qquad (1.128)$$

and noting that $\hat{K}_{M\text{-}I,\ L\text{-}J} = 0$, we observe that $\tilde{E}$ contains a zero submatrix of total size no less than $n+p+q+1+(m-p)+(\ell-q) = n+m+\ell+1$; and Lemma 1.54 implies that $\tilde{\rho}(\tilde{E}) < n+m+\ell$.

Conversely, if $\tilde{\rho}(\tilde{E}) < n+m+\ell$, then $\tilde{E}$ contains a zero submatrix of total size $n+m+\ell+1$. Let this submatrix consist of r_x rows and c_x columns corresponding to the state vertices, r_u rows and c_u columns corresponding to the input vertices with indices in sets $\boldsymbol{R}_u \subset \boldsymbol{M}$ and $\boldsymbol{C}_u \subset \boldsymbol{M}$, and r_y rows and c_y columns corresponding to the output vertices with indices in sets $\boldsymbol{R}_y \subset \boldsymbol{L}$ and $\boldsymbol{C}_y \subset \boldsymbol{L}$. Thus,

$$r_x + r_u + r_y + c_x + c_u + c_y \geq n + m + \ell + 1. \tag{1.129}$$

Note that by structure of $\tilde{E}$,

$$\boldsymbol{R}_u \cap \boldsymbol{C}_u = \emptyset, \quad \boldsymbol{R}_y \cap \boldsymbol{C}_y = \emptyset. \tag{1.130}$$

Let $\boldsymbol{I} = \boldsymbol{C}_u$. Since $\tilde{K}_{R_u C_y} = 0$, we have $\boldsymbol{J} \subset \boldsymbol{L} - \boldsymbol{C}_y$. Therefore,

$$\tilde{\rho}\left(\begin{bmatrix} \tilde{A} & \tilde{B}_I \\ \tilde{C}_J & 0 \end{bmatrix}\right) \leq \tilde{\rho}\left(\begin{bmatrix} \tilde{A} & \tilde{B}_I \\ \tilde{C}_{L\text{-}C_y} & 0 \end{bmatrix}\right). \tag{1.131}$$

On the other hand, by (1.130), $\boldsymbol{L} - \boldsymbol{C}_y \supset \boldsymbol{R}_y$, so that the $(n+\ell-c_y) \times (n+c_u)$ matrix on the right side of (1.131) contains a zero submatrix of total size at least $r_x + r_y + c_x + c_u$. Using (1.129) and the fact that $m \geq r_u + c_u$, as implied by (1.130), we have

$$\begin{aligned} r_x + r_y + c_x + c_u &\geq n + m + \ell + 1 - r_u - c_y \\ &\geq n + (\ell - c_y) + c_u + 1, \end{aligned} \tag{1.132}$$

so that the matrix in (1.131) has generic rank less than n. Q.E.D.

1.56. EXAMPLE. Theorem 1.52 not only provides a very convenient graphical test for structurally fixed modes, but may also be useful in deciding what additional feedback links should be introduced in case the existing ones are not sufficient for all the modes. As an example, consider a structured system $\tilde{\mathbf{S}} = (\tilde{A}, \tilde{B}, \tilde{C})$ specified by

$$\tilde{A} = \begin{bmatrix} 0 & * & 0 \\ * & * & * \\ 0 & * & 0 \end{bmatrix}, \quad \tilde{B} = \begin{bmatrix} 0 & 0 \\ 0 & * \\ * & 0 \end{bmatrix}, \quad \tilde{C} = \begin{bmatrix} 0 & * & 0 \\ * & 0 & 0 \\ 0 & 0 & * \end{bmatrix}. \tag{1.133}$$

Let the feedback structure constraints be specified as

$$\bar{K} = \begin{bmatrix} 1 & 0 & 0 \\ 0 & 1 & 1 \end{bmatrix}, \tag{1.134}$$

that is,

$$\tilde{K} = \begin{bmatrix} * & 0 & 0 \\ 0 & * & * \end{bmatrix}. \tag{1.135}$$

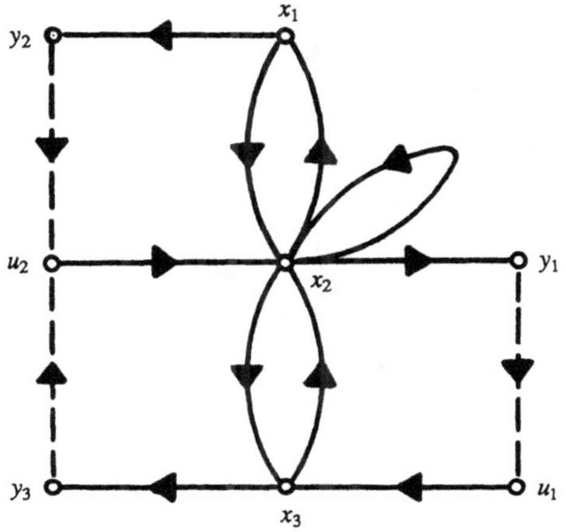

Fig. 1.20. Digraph $\hat{\mathbf{D}}$.

The digraph $\hat{\mathbf{D}}$ associated with $\hat{\mathbf{S}}$ is shown in Figure 1.20, where the feedback edges of $\mathbf{E}_K$ are identified by dashed lines. From this figure we find that not all state vertices of $\hat{\mathbf{D}}$ can be covered by disjoint cycles, so that by condition (*ii*) of Theorem 1.52, $\hat{\mathbf{S}}$ has a structurally fixed mode. On the other hand, a quick inspection of Figure 1.20 shows that if the feedback edge (y_2, u_1) were added to $\hat{\mathbf{D}}$ to get a modified structure $\hat{\mathbf{D}}'$ as in Figure 1.21, both conditions of Theorem 1.52 would be satisfied without even using the existing edges of $\mathbf{E}_K$, indicating that feedback structure constraints such as

$$\bar{K}'' = \begin{bmatrix} 0 & 1 & 0 \\ 0 & 0 & 0 \end{bmatrix}, \tag{1.136}$$

resulting in a digraph $\hat{\mathbf{D}}''$ of Figure 1.22, would suffice to control $\tilde{\mathbf{S}}$ in almost all cases.

Finally, we should comment on the implications of our Theorem 1.52 concerning Theorems 1.28 and 1.29 on structural controllability and observability. It is easy to see that under the full output feedback and $C = I_n$, condition (*i*) of Theorem 1.52 is equivalent to system $\mathbf{S}$ being input reachable, that is, the pair $[\tilde{A}\ \ \tilde{B}]$ is not of Form I. Furthermore, from the proof of part (*ii*) of the theorem, it follows that under the same feedback constraints the existence of the appropriate cycle family is equivalent to the rank condition $\tilde{\rho}([\tilde{A}\ \ \tilde{B}]) = n$, that is, $[\tilde{A}\ \ \tilde{B}]$ not being of Form II. Therefore, we have the

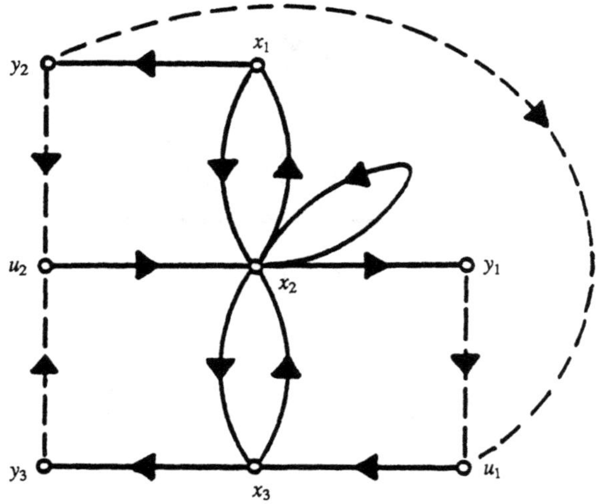

Fig. 1.21. Digraph $\hat{\mathbf{D}}'$.

following easy corollary (Linnemann, 1983) to Theorem 1.52, which provides the known characterization of structural controllability (Lin, 1974; Shields and Pearson, 1976; Glover and Silverman, 1976; Hosoe and Matsumoto, 1979):

1.57. COROLLARY. A system $\tilde{\mathbf{S}} = (\tilde{A}, \tilde{B}, \tilde{C})$ has no structurally fixed modes with respect to the full state feedback, that is, the pair $(\tilde{A}, \tilde{B})$ is structurally controllable if and only if both of the following conditions hold:

(*i*) $\tilde{\mathbf{S}}$ is input reachable.

(*ii*) $\tilde{\rho}([\tilde{A} \;\; \tilde{B}]) = n$.

Conditions (*i*) and (*ii*) are equivalent to saying that the composite matrix $[\tilde{A} \;\; \tilde{B}]$ has neither Form I nor Form II, respectively, which is the condition of Theorem 1.28. The dual of Corollary 1.57 is Theorem 1.29.

1.7. Notes and References

The basic ingredients in control design of simple and complex systems alike are controllability and observability (Kalman, 1963). Yet, even in medium size linear systems, numerical tests of these properties are either unreliable or require an excessive computational effort. For these reasons,

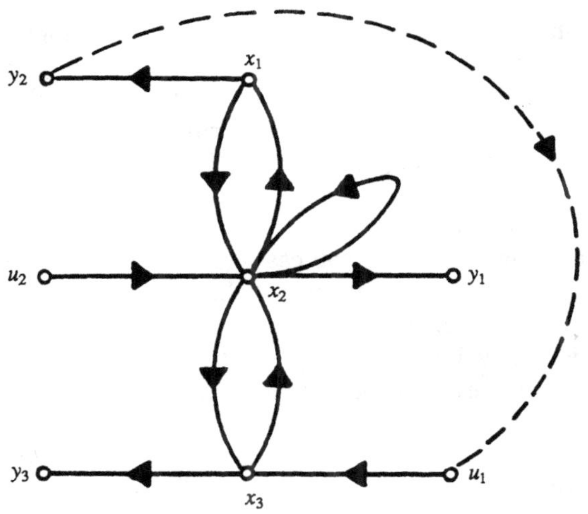

Fig. 1.22. Digraph $\hat{\mathbf{D}}''$.

the graph-theoretic concept of *structural controllability* (Lin, 1974) is an attractive tool which can be used to identify systems where uncontrollability is a parameter independent property. What is even more appealing is the simplicity of structural controllability tests involving binary calculations only (for discussion of numerical problems in testing controllability, see Nour-Eldin, 1987).

Lin's single-input result was generalized to a multi-input case by Shields and Pearson (1976), who introduced an algebraic approach to the problem of structural controllability. This approach has been refined by Glover and Silverman (1976), who produced a recursive algorithm to solve the problem. Further improvements and insights have been provided by many people (see, for example, Hosoe and Matsumoto, 1979; Hosoe, 1980; Mayeda, 1981; Hayakawa *et al.* 1984; Linnemann, 1986).

Among the two parts of structural controllability (Corollary 1.57), *input reachability* was singled out as a topic of special attention (Šiljak, 1977a, b) because of its physical importance and manipulative power in large-scale systems. The concepts of input and output reachability were used to obtain hierarchically ordered subsystems with independent inputs and outputs (Pichai *et al.* 1983). The resulting form of a large-scale system allows for not only a "piece-by-piece" design of decentralized control and estimation, but also breaks down a test for (structural and ordinary) controllability and observability into a number of easier tests involving only low-order subsystems (Chapter 6). Besides being the principal ingredient in theorems on structural controllability and observability, the concept of input and output reachability has been used in a variety of applications (Šiljak, 1978; Burns and Winstead, 1982a, b; Bossermann, 1983; 1988; Maione and Turchiano, 1986, 1988; Evangelatos and Nicholson, 1988; see also Chapters 6 and 7).

The assumption of parameter independence underlying the structural analysis may eliminate some critical cases (Examples 1.33 and 1.34). A linear *parametrization* for structural controllability was considered by Corfmat and Morse (1976), Anderson and Hong (1982), and Anderson (1982). The gain in precision may be outweighed by the amount of additional computational effort required to test parametrized systems. Another way to attack this problem is to look for structures which are controllable despite parameter dependencies—this is *strong structural controllability* (Mayeda and Yamada, 1979). The result, however, is disappointingly conservative. In this context, the most encouraging result is the simple fact that input reachability is independent of parametrization, confirming its importance in the structural analysis of control.

Use of graphs opens up a possibility to consider *vulnerability* of control systems (Šiljak, 1978). Imitating the vulnerability concept in graph theory (Harary *et al.* 1965), we consider a control system vulnerable if a removal of a line, or a set of lines, from the corresponding digraph destroys input reachability (structural controllability). Graph-theoretic procedures have been developed (Pichai *et al.* 1981) to identify the minimal sets of lines which are essential for preserving input reachability and structural controllability of a given system.

More often than not, neither the states, nor their estimates, are available for control. Then the central problem becomes the existence of fixed modes under output or decentralized information constraints. The notion of *structurally fixed modes* was introduced in (Sezer and Šiljak, 1981a) to characterize the existence problem *via* graphs and interconnection matrices. The solution to the problem was given for arbitrary feedback constraints, which resulted in simple binary tests (Pichai *et al.* 1984). Variations of the problem and its solution have appeared in (Linnemann, 1983; Momen and Evans, 1983; Reinschke, 1984; Travé *et al.* 1987; Murota, 1988; Fanti *et al.* 1988).

The next important and interesting step in the development of graph-theoretic concepts in control is the pole assignment problem as formulated by Sezer (1983). He showed how to characterize a minimal set of feedback links that guarantee generic pole assignability for a linear multivariable system using dynamic compensators. Furthermore, Sezer proposed a procedure to identify at least one such feedback pattern. An optimization approach in this context has been presented in (Sezer and Ünyelioğlu, 1988).

Bibliography

Anderson, B. D. O. (1982). Transfer function matrix description of decentralized fixed modes. *IEEE Transactions*, AC-27, 1176–1182.

Anderson, B. D. O., and D. J. Clements (1981). Algebraic characterization of fixed modes in decentralized control. *Automatica*, 17, 703–712.

Anderson, B. D. O., and H. M. Hong (1982). Structural controllability and matrix nets. *International Journal of Control*, 35, 397–416.

Aoki, M. (1972). On feedback stabilizability of decentralized dynamic systems. *Automatica*, 8, 163–173.

Berztiss, A. T. (1975). *Data Structures: Theory and Practice.* Academic Press, New York.

Bosserman, R. W. (1983). Flow analysis sensitivities for models of energy or material flow. *Bulletin of Mathematical Biology*, 45, 807–826.

Bosserman, R. W. (1988). Ecosystem networks: Measures of structure. *Systems and Control Encyclopedia*, M. G. Singh (ed.), Pergamon Press, Oxford, UK, 1352–1356.

Bowie, W. S. (1976). Applications of graph theory in computer systems. *International Journal of Computer and Information Sciences*, 5, 9–31.

Burns, J. R., and W. H. Winstead (1982a). On input–output approach to the structural analysis of digraphs. *IEEE Transactions*, SMC-12, 15–24.

Burns, J. R., and W. H. Winstead (1982b). Input and output redundancy. *IEEE Transactions*, SMC-12, 785–793.

Corfmat, J. P., and A. S. Morse (1976). Structurally controllable and structurally canonical systems. *IEEE Transactions*, AC-21, 129–131.

Davison, E. J. (1977). Connectability and structural controllability. *Automatica*, 13, 109–123.

Davison, E. J., and Ü. Özgüner (1983). Characterizations of decentralized fixed modes for interconnected systems. *Automatica*, 19, 169–182.

Deo, N. (1974). *Graph Theory with Applications to Engineering and Computer Science.* Prentice-Hall, Englewood Cliffs, New Jersey.

Duff, I. S. (1977). A survey of sparse matrix research. *Proceeding of the IEEE*, 65, 500–535.

Duff, I. S. (1981a). On algorithms for obtaining a maximum transversal. *ACM Transactions on Mathematical Software*, 7, 315–330.

Duff, I. S. (1981b). Algorithm 575. Permutations for a zero-free diagonal. *ACM Transactions on Mathematical Software*, 7, 387–390.

Duff, I. S., and J. K. Reid (1978a). An implementation of Tarjan's algorithm for the block triangularization of a matrix. *ACM Journal of Mathematical Software*, 4, 137–147.

Duff, I. S., and J. K. Reid (1978b). Algorithm 529. Permutations to block triangular form. *ACM Transactions on Mathematical Software*, 4, 189–192.

Evangelatos, D. S., and H. Nicholson (1988). Reachability matrix and relation to large-scale systems. *International Journal of Control*, 47, 1163–1177.

Fanti, M. P., B. Maione, and B. Turchiano (1988). Structurally fixed modes of systems described by Rosenbrock's polynomial matrices. *International Journal of Control*, 48, 1947–1965.

Festinger, L., S. Schachter, and K. Back (1950). *Social Pressures in Informal Groups*. Harper, New York.

Franksen, O. I., P. Falster, and F. J. Evans (1979). Structural aspects of controllability and observability. Parts I and II. *Journal of the Franklin Institute*, 308, 79–124.

Glover, K., and L. M. Silverman (1976). Characterization of structural controllability. *IEEE Transactions*, AC-21, 534–537.

Gustavson, F. G. (1976). Finding the block lower triangular form of a sparse matrix. *Sparse Matrix Computations*, J. R. Bunch and D. J. Rose (eds.), Academic Press, New York, 275–289.

Harary, F. (1959). A graph-theoretic method for the complete reduction of a matrix with a view toward finding its eigenvalues. *Journal of Mathematics and Physics*, 38, 104–111.

Harary, F. (1969). *Graph Theory*. Addison-Wesley, Reading, Massachusetts.

Harary, F., R. Z. Norman, and D. Cartwright (1965). *Structural Models: An Introduction to the Theory of Directed Graphs*. Wiley, New York.

Hayakawa, Y., and D. D. Šiljak (1988). On almost invariant subspaces of structured systems and decentralized control. *IEEE Transactions*, 33, 931–939.

Hayakawa, Y., S. Hosoe, M. Hayashi, and M. Ito (1984). On the structural controllability of compartmental systems. *IEEE Transactions*, AC-29, 17–24.

Hopcroft, J. E., and R. Tarjan (1971). Planarity testing in $V \log V$ steps. *Proceedings of the IFIP Congress*, Ljubljana, Yugoslavia, Ta-2, 18–23.

Hosoe, S. (1980). Determination of generic dimensions of controllable subspaces and its application. *IEEE Transactions*, AC-25, 1192–1196.

Hosoe, S., and K. Matsumoto (1979). On the irreducibility condition in the structural controllability theorem. *IEEE Transactions*, AC-24, 963–966.

Kailath, T. (1980). *Linear Systems*. Prentice-Hall, Englewood Cliffs, New Jersey.

Kalman, R. E. (1963). Mathematical description of linear dynamical systems. *SIAM Journal on Control*, 1, 152–192.

Lee, E. B., and L. Markus (1961). *Foundations of Optimal Control Theory*. Wiley, New York.

Lehman, S. K. (1986). Comments on "Fixed modes and local feedbacks in interconnected systems." *International Journal of Control*, 43, 329–333.

Lin, C. T. (1974). Structural controllability. *IEEE Transactions*, AC-19, 201–208.

Linnemann, A. (1983). Fixed modes in parametrized systems. *International Journal of Control*, 38, 319–335.

Linnemann, A. (1986). A further simplification in the proof of the structural controllability theorem. *IEEE Transactions*, AC-31, 638–639.

Maione, B., and B. Turchiano (1986). Input- and output-decoupling structural zeros of linear systems described by Rosenbrock's polynomial matrices. *International Journal of Control*, 44, 1641–1659.

Maione, B., and B. Turchiano (1988). Characterization of decoupling structural zeros of Rosenbrock's polynomial matrices. *International Journal of Control*, 47, 459–476.

Mayeda, H. (1981). On structural controllability theorem. *IEEE Transactions*, AC-26, 795–799.

Mayeda, H., and T. Yamada (1979). Strong structural controllability. *SIAM Journal on Control and Optimization*, 17, 123–138.

McFadden, D. (1967). On the controllability of decentralized microeconomic systems. The assignment Problem. *Mathematical Systems Theory and Economics*, 1, H. W. Kuhn and G. P. Szegö (eds.), Springer, New York, 221–239.

Momen, S., and F. J. Evans (1983). Structurally fixed modes in decentralized systems. Parts I and II. *IEE Proceedings*, 130, 313–327.

Munro, I. (1971). Efficient determination of the transitive closure of a directed graph. *Information Proceedings Letters*, 1, 56–58.

Murota, K. (1987). *Systems Analysis by Graphs and Matroids: Structural Solvability and Controllability*. Springer, Berlin, FRG.

Murota, K. (1988). A matroid-theoretic approach to structurally fixed modes in decentralized control. *Research Memorandum RMI 88-07*, University of Tokyo, Tokyo, Japan.

Nour-Eldin, H. A. (1987). Linear multivariable systems controllability and observability: Numerical aspects. *Systems and Control Encyclopedia*, M. G. Singh (ed.), Pergamon Press, Oxford, UK, 2816–2827.

Pichai, V., M. E. Sezer, and D. D. Šiljak (1981). Vulnerability of dynamic systems. *International Journal of Control*, 34, 1049–1060.

Pichai, V., M. E. Sezer, and D. D. Šiljak (1983). A graphical test for structurally fixed modes. *Mathematical Modeling*, 4, 339–348.

Pichai, V., M. E. Sezer, and D. D. Šiljak (1984). A graph-theoretic characterization of structurally fixed modes. *Automatica*, 20, 247–250.

Purdom, P. (1970). A transitive closure algorithm. *Nordisk Tidskrift for Informationbehandlung*, 10, 76–94.

Reinschke, K. (1984). Graph-theoretic characterization of fixed modes in centralized and decentralized control. *International Journal of Control*, 39, 715–729.

Sezer, M. E. (1983). Minimal essential feedback patterns for pole assignment using dynamic compensation. *Proceedings of the 22nd IEEE Conference on Decision and Control*, San Antonio, Texas, 28–32.

Sezer, M. E., and D. D. Šiljak (1981a). Structurally fixed modes. *Systems and Control Letters*, 1, 60–64.

Sezer, M. E., and D. D. Šiljak (1981b). On structurally fixed modes. *Proceedings of the IEEE International Symposium on Circuits and Systems*, Chicago, Illinois, 558–565.

Sezer, M. E., and K. Ünyelioğlu (1988). On optimum selection of stabilizing feedback structures for multivariable control systems. *Proceedings of the IFAC Symposium on Distributed Intelligence Systems*, Varna, Bulgaria, 145–154.

Shields, R. W., and J. B. Pearson (1976). Structural controllability of multiinput linear systems. *IEEE Transactions*, AC-21, 203–212.

Šiljak, D. D. (1977a). On pure structure of dynamic systems. *Nonlinear Analysis, Theory, Methods, and Applications*, 1, 397–413.

Šiljak, D. D. (1977b). On reachability of dynamic systems. *International Journal of Systems Science*, 8, 321–338.

Šiljak, D. D. (1977c). Vulnerability of dynamic systems. *Proceedings of the IFAC Workshop on Control and Management of Integrated Industrial Processes*, Pergamon Press, London, 133–144.

Šiljak, D. D. (1978). *Large-Scale Dynamic Systems: Stability and Structure*. North-Holland, New York.

Swamy, M. N. S., and K. Thulasiraman (1981). *Graphs, Networks, and Algorithms*. Wiley, New York.

Tarjan, R. (1972). Depth-first search and linear graph algorithms. *SIAM Journal on Computing*, 1, 146–160.

Travé, L., A. M. Tarras, and A. Titli (1987). Minimal feedback structure avoiding structurally fixed modes. *International Journal of Control*, 46, 313–325.

Ulm, M., and H. D. Wend (1986). A concept for parameter independent evaluation of decentralized stabilizability. *Automatica*, 25, 133–136.

Vidyasagar, M. (1980). Decomposition techniques for large-scale systems with nonadditive interactions: Stability and stabilizability. *IEEE Transactions*, AC-25, 773–779.

Vidyasagar, M., and N. Viswanadham (1982). Algebraic characterization of decentralized fixed modes and pole assignment. *Proceedings of the 21st Conference on Decision and Control*, Orlando, Florida, 501–505.

Wang, S. H., and E. J. Davidson (1973). On the stabilization of decentralized control systems. *IEEE Transactions*, AC-18, 473–478.

Warshall, S. A. (1962). A theorem on Boolean matrices. *Journal of the Association of Computing Machinery*, 9, 11–12.

Wend, H. D. (1987). Decomposition and structural properties. *Systems and Control Encyclopedia*, M. G. Singh (ed.), Pergamon Press, Oxford, UK, 948–952.

Wonham, W. M. (1979). *Linear Multivariable Control: A Geometric Approach.* Springer, New York.

Chapter 2 Stabilization

In the stabilization of complex systems, decentralized schemes arise because the centralized designs are either infeasible due to an inherent nonclassical constraint on the information structure (*e.g.*, market systems, where individual agents do not know what other agents are doing), or they are costly because of expensive global information schemes (*e.g.*, power systems spread over distant geographic areas). The case of decentralized control is further strengthened by the fact that imperfect knowledge of the model of interconnections among different parts of a complex system is invariably present in all but trivial problems. It has been known for some time that, unlike centralized control laws, decentralized control structures guarantee robust stability, which tolerates a wide range of nonlinear time-varying uncertainties in the interactions among the subsystems. That is, closed-loop interconnected systems that are stabilized by local feedback laws are *connectively stable.*

A natural way to establish connective stability of a complex system driven by decentralized control is to apply the Matrosov–Bellman concept of vector Liapunov functions. The reason is that each component of a vector function can be used independently to show stability of a locally controlled subsystem. When a vector function establishes stability of the overall system, autonomy of each subsystem is preserved, and, as expected, connective stability follows. Over the years, this simple recipe has achieved a considerable level of sophistication (see Notes and References: Section 2.6). In this chapter, we surrender generality of the concept for design-oriented results, which are useful in decentralized control of complex systems using local state feedback.

Nonexistence of structurally fixed modes is a relatively simple characterization of potentially successful decentralized control structures. It is not sufficient, however, and we use the method of vector Liapunov functions to confirm that a selected control law stabilizes the system. When the method fails, the outcome is inconclusive, which is the case with most of Liapunov-type results. For this reason, there has long been a concerted effort to delineate classes of decentralized systems that can *always* be stabilized by local (state or output) feedback. By describing certain classes of linear interconnected systems, which are decentrally stabilizable by *state feedback*, we include in this chapter the basic part of this effort. We shall rely on graphs to come up with a simple characterization of stabilizability. This is our favorite environment, and similar graph-theoretic conditions for stabilizability of linear systems *via* decentralized *output feedback* will resurface again in Chapter 5.

2.1. Connective Stability

To fix certain ideas before we engage in a presentation of the general stability results available for interconnected systems, let us consider again the two identical penduli of Example 1.36 shown in Figure 1.16. Under the usual simplifying assumptions, the equations of motion were obtained as

$$\begin{aligned} m\ell^2\ddot{\theta}_1 &= mg\ell\theta_1 - ka^2(\theta_1 - \theta_2) + u_1, \\ m\ell^2\ddot{\theta}_2 &= mg\ell\theta_2 - ka^2(\theta_2 - \theta_1) + u_2. \end{aligned} \tag{2.1}$$

Our desire is to keep the penduli in the upright position by using a suitable feedback law *via* the two inputs u_1 and u_2. A special feature of the system is the sliding spring, which can change its position along the length of the penduli in an unpredictable way. All we know about the position a of the spring is that it is bounded, that is, for some constant value $\bar{a} \in [0, \ell]$,

$$0 \leq a \leq \bar{a}. \tag{2.2}$$

The position of the spring determines the degree of coupling among the two penduli, and a feedback control law should stabilize the system despite uncertain variations in the parameter a. At present, we consider a to be unknown but constant.

By choosing the state vector as $x = (\theta_1, \dot{\theta}_1, \theta_2, \dot{\theta}_2)^T$, the interconnected penduli have an input decentralized representation, which can be obtained from (2.1) as

$$\mathbf{S}\colon\ \dot{x} = \left[\begin{array}{cc|cc} 0 & 1 & 0 & 0 \\ \frac{g}{\ell} - \frac{ka^2}{m\ell^2} & 0 & \frac{ka^2}{m\ell^2} & 0 \\ \hline 0 & 0 & 0 & 1 \\ \frac{ka^2}{m\ell^2} & 0 & \frac{g}{\ell} - \frac{ka^2}{m\ell^2} & 0 \end{array}\right] x + \left[\begin{array}{c|c} 0 & 0 \\ \frac{1}{m\ell^2} & 0 \\ \hline 0 & 0 \\ 0 & \frac{1}{m\ell^2} \end{array}\right] u. \tag{2.3}$$

The system **S** can be decomposed along the dashed lines to get

$$\begin{aligned} \mathbf{S}\colon\ \dot{x}_1 &= \begin{bmatrix} 0 & 1 \\ \alpha & 0 \end{bmatrix} x_1 + \begin{bmatrix} 0 \\ \beta \end{bmatrix} u_1 + e \begin{bmatrix} 0 & 0 \\ -\gamma & 0 \end{bmatrix} x_1 + e \begin{bmatrix} 0 & 0 \\ \gamma & 0 \end{bmatrix} x_2, \\ \dot{x}_2 &= \begin{bmatrix} 0 & 1 \\ \alpha & 0 \end{bmatrix} x_2 + \begin{bmatrix} 0 \\ \beta \end{bmatrix} u_2 + e \begin{bmatrix} 0 & 0 \\ \gamma & 0 \end{bmatrix} x_1 + e \begin{bmatrix} 0 & 0 \\ -\gamma & 0 \end{bmatrix} x_2, \end{aligned} \tag{2.4}$$

which is an interconnection of two subsystems

$$\begin{aligned} \mathbf{S}_1\colon\ \dot{x}_1 &= \begin{bmatrix} 0 & 1 \\ \alpha & 0 \end{bmatrix} x_1 + \begin{bmatrix} 0 \\ \beta \end{bmatrix} u_1, \\ \mathbf{S}_2\colon\ \dot{x}_2 &= \begin{bmatrix} 0 & 1 \\ \alpha & 0 \end{bmatrix} x_2 + \begin{bmatrix} 0 \\ \beta \end{bmatrix} u_2, \end{aligned} \tag{2.5}$$

where $x_1 = (\theta_1, \dot{\theta}_1)^T$ and $x_2 = (\theta_2, \dot{\theta}_2)^T$ are states of the subsystems $\mathbf{S}_1$ and $\mathbf{S}_2$, and $\alpha = g/\ell$, $\beta = 1/m\ell^2$, $\gamma = \bar{a}^2 k/m\ell^2$, and $e = (a/\bar{a})^2$.

Since we want stability of **S** when the two subsystems are decoupled as in (2.5), which occurs when $e = 0$, we choose decentralized feedback laws

$$u_1 = -k_1^T x_1, \qquad u_2 = -k_2^T x_2, \tag{2.6}$$

where the feedback gains $k_1 = (k_{11}, k_{12})^T$ and $k_2 = (k_{21}, k_{22})^T$ are selected so that the closed-loop decoupled subsystems

$$\begin{aligned} \hat{\mathbf{S}}_1\colon\ \dot{x}_1 &= \begin{bmatrix} 0 & 1 \\ \alpha - \beta k_{11} & -\beta k_{12} \end{bmatrix} x_1, \\ \hat{\mathbf{S}}_2\colon\ \dot{x}_2 &= \begin{bmatrix} 0 & 1 \\ \alpha - \beta k_{21} & -\beta k_{22} \end{bmatrix} x_2 \end{aligned} \tag{2.7}$$

are stable. Now, the fundamental problem in this context is to show that stability of the closed-loop subsystems $\hat{\mathbf{S}}_1$ and $\hat{\mathbf{S}}_2$ implies stability of the overall closed-loop system

$$\begin{aligned}\hat{\mathbf{S}}\colon\ \dot{x}_1 &= \begin{bmatrix} 0 & 1 \\ \alpha - \beta k_{11} & -\beta k_{12} \end{bmatrix} x_1 + e\begin{bmatrix} 0 & 0 \\ -\gamma & 0 \end{bmatrix} x_1 + e\begin{bmatrix} 0 & 0 \\ \gamma & 0 \end{bmatrix} x_2, \\ \dot{x}_2 &= \begin{bmatrix} 0 & 1 \\ \alpha - \beta k_{21} & -\beta k_{22} \end{bmatrix} x_2 + e\begin{bmatrix} 0 & 0 \\ \gamma & 0 \end{bmatrix} x_1 + e\begin{bmatrix} 0 & 0 \\ -\gamma & 0 \end{bmatrix} x_2\end{aligned} \tag{2.8}$$

for some value $\bar{a}$ corresponding to $\bar{e} = 1$, and then demonstrate the fact that stability achieved by the decentralized control law (2.6) is independent of the value of the coupling coefficient $e \in [0,\ 1]$. In other words, $\hat{\mathbf{S}}_1$, $\hat{\mathbf{S}}_2$, and $\hat{\mathbf{S}}$ are *simultaneously stable* for all $e \in [0,\ 1]$.

At this point, we recognize the fact that e is an *uncertain structural parameter* that determines the degree of coupling between the subsystems $\hat{\mathbf{S}}_1$ and $\hat{\mathbf{S}}_2$. An appropriate framework for stability analysis of systems with uncertain structural parameters is provided by the concept of stability under structural perturbations, that is, *connective stability.* This concept was introduced by the author and further developed by many people, as discussed in Notes and References at the end of this chapter. In this section, we shall present only the core of the concept that is essential for the stabilization by decentralized feedback. After a general outline of the connective stabilization scheme, we shall return to the above example of interconnected penduli to complete the stability analysis. This should provide an illustration of the concept and, it is hoped, motivate applications to a much broader class of control systems.

We start with a general description of a dynamic system $\mathbf{S}$ by a differential equation

$$\mathbf{S}\colon\ \dot{x} = f(t,\ x), \tag{2.9}$$

where $x(t) \in \mathbf{R}^n$ is the state of $\mathbf{S}$ at time $t \in \mathbf{T}$, and the function $f\colon\ \mathbf{T} \times \mathbf{R}^n \to \mathbf{R}^n$ is defined, bounded, and continuous on the domain $\mathbf{T} \times \mathbf{R}^n$, so that solutions $x(t;\ t_0,\ x_0)$ of Equation (2.9) exist for all initial conditions $(t_0,\ x_0) \in \mathbf{T} \times \mathbf{R}^n$ and $t \in \mathbf{T}_0$, where $\mathbf{T} = (\tau,\ +\infty)$, τ is a number or a symbol $-\infty$, and $\mathbf{T}_0 = [t_0, +\infty)$.

We assume that

$$f(t,\ 0) = 0, \qquad \forall t \in \mathbf{T}, \tag{2.10}$$

and $x = 0$ is the unique equilibrium of $\mathbf{S}$. To study stability of the equilibrium under structural perturbations, we assume that the system $\mathbf{S}$ is decomposed as

$$\mathbf{S}\colon\ \dot{x}_i = g_i(t, x_i) + h_i(t, x), \qquad i \in \boldsymbol{N}, \tag{2.11}$$

which is an interconnection of N subsystems

$$\mathbf{S}_i\colon\ \dot{x}_i = g_i(t, x_i), \qquad i \in \boldsymbol{N}, \tag{2.12}$$

where $x_i(t) \in \mathbf{R}^{n_i}$ is the state of the ith subsystem $\mathbf{S}_i$, the function $g_i\colon\ \mathbf{T} \times \mathbf{R}^{n_i} \to \mathbf{R}^{n_i}$ describes the dynamics of $\mathbf{S}_i$, and the function $h_i\colon\ \mathbf{T} \times \mathbf{R}^n \to \mathbf{R}^{n_i}$ represents the interaction of $\mathbf{S}_i$ with the rest of the system $\mathbf{S}$.

To describe the structure of $\mathbf{S}$, we represent the interconnection functions as

$$h_i(t, x) \equiv h_i(t, \bar{e}_{i1}x_1, \bar{e}_{i2}x_2, \ldots, \bar{e}_{iN}x_N), \qquad i \in \boldsymbol{N}, \tag{2.13}$$

where the binary numbers $\bar{e}_{ij}$ are elements of the $N \times N$ *fundamental interconnection matrix* defined by:

2.1. DEFINITION. The $N \times N$ fundamental interconnection matrix $\bar{E} = (\bar{e}_{ij})$ of a system $\mathbf{S}$ has the elements

$$\bar{e}_{ij} = \begin{cases} 1, & x_j \text{ occurs in } h_i(t, x), \\ 0, & x_j \text{ does not occur in } h_i(t, x). \end{cases} \tag{2.14}$$

The matrix $\bar{E}$ is the standard occurrence (interconnection) matrix (Steward, 1962): $\bar{e}_{ij} = 1$ if $\mathbf{S}_j$ acts on $\mathbf{S}_i$, or $\bar{e}_{ij} = 0$ if $\mathbf{S}_j$ does not act on $\mathbf{S}_i$. In applications, $\bar{e}_{ij} = 1$ if there is a physical interconnection from $\mathbf{S}_j$ to $\mathbf{S}_i$, and $\bar{e}_{ij} = 0$ otherwise.

Structural perturbations of $\mathbf{S}$ are described by the interconnection matrix $E\colon\ \mathbf{T} \to \mathbf{R}_+^{N\times N}$, which is defined by:

2.2. DEFINITION. An $N \times N$ interconnection matrix $E = (e_{ij})$ with elements $e_{ij}\colon\ \mathbf{T} \to [0, 1]$ is said to be generated by an $N \times N$ fundamental interconnection matrix $\bar{E} = (\bar{e}_{ij})$ such that $\bar{e}_{ij} = 0$ if and only if $e_{ij}(t) \equiv 0$.

The elements e_{ij} of E represent the strength of coupling between the individual subsystems $\mathbf{S}_i$ of $\mathbf{S}$ at $t \in \mathbf{T}$. When, in particular, an e_{ij} goes to zero, but $\bar{e}_{ij} = 1$, then the subsystem $\mathbf{S}_j$ is decoupled from the subsystem $\mathbf{S}_i$. In this way, the matrix E describes the quantitative as well as qualitative changes in the interconnection structure of $\mathbf{S}$. The fact that E is generated by $\bar{E}$, we denote by $E(t) \in \bar{E}$, or simply, $E \in \bar{E}$ (note the abuse of notation!).

When $E = 0$, the system **S** of (2.11) breaks down into N decoupled subsystems $\mathbf{S}_i$ of (2.12). This means that

$$h_i(t, 0) = 0, \qquad \forall t \in \mathbf{T}. \tag{2.15}$$

Furthermore, we assume that the equilibrium state $x = 0$ of **S** remains invariant under all permissible structural perturbations, that is, for all $E \in \bar{E}$. This implies that when $E = 0$, we have

$$g_i(t, 0) = 0, \qquad \forall t \in \mathbf{T}, \tag{2.16}$$

and $x_i = 0$ is the unique equilibrium state of $\mathbf{S}_i$.

We also need to comment on the decomposition (2.11) regarding the dimensions of the individual state spaces of $\mathbf{S}_i$ in (2.12). We assume at present that

$$\mathbf{R}^n = \mathbf{R}^{n_1} \times \mathbf{R}^{n_2} \times \ldots \times \mathbf{R}^{n_N}, \tag{2.17}$$

and we consider no "overlapping" among the components x_i of x. This means that

$$x = \left(x_1^T, x_2^T, \ldots, x_N^T\right)^T, \tag{2.18}$$

and

$$n = \sum_{i=1}^{N} n_i. \tag{2.19}$$

In Chapter 8, we shall lift this restriction and consider systems composed of overlapping interconnected subsystems.

Now, we associate with **S** a matrix $\bar{E}$ and state:

2.3. DEFINITION. A system **S** is connectively stable if it is stable in the sense of Liapunov for all $E \in \bar{E}$.

Since connective stability is based upon the Liapunov's stability concept, and because there are many types of Liapunov stability, there are many types of connective stability. In this chapter, our interest is in global asymptotic stability under structural perturbations, and if not stated otherwise, when we say "connective stability of **S**," we mean "global asymptotic stability of the equilibrium $x = 0$ of **S** for all $E \in \bar{E}$." For more general treatment of connective stability, see (Šiljak, 1978).

Interconnection matrices $E(t)$, as specified by Definition 2.2, have elements $e_{ij}(t)$ which are continuous function of time such that $e_{ij}(t) \in [0, 1]$ for all $t \in \mathbf{T}$. As far as the (vector) Liapunov method is concerned, this is an unnecessary restriction, and the theory goes through even if the elements e_{ij}: $\mathbf{T} \times \mathbf{R}^n \to \mathbf{R}$ are piece-wise continuous functions of t and continuous functions of x. The elements may have unknown arbitrary forms provided $e_{ij}(t, x) \in [0, 1]$ for all $(t, x) \in \mathbf{T} \times \mathbf{R}^n$. In the following developments, we shall not insist on the dependence of e_{ij}'s on time or state, because the implications will be either obvious or can be readily derived from the basic source (Šiljak, 1978).

2.2. Vector Liapunov Functions

Analysis of connective stability is based upon the concept of vector Liapunov functions, which was introduced independently by Matrosov (1962) and Bellman (1962). Roughly speaking, the concept associates several scalar functions with a given dynamic system in such a way that each function determines a desired stability property in a part of the state space where others do not. These scalar functions are considered as components of a vector Liapunov function, each component being an "aggregate" of the corresponding piece of the state space in much the same way as a single function in the classical Liapunov stability theory aggregates the state space of the entire system. Powerful generalizations of this idea have been obtained by many authors cited in Notes and References (Section 2.6).

We shall present two types of constructions of vector Liapunov functions. The first construction (Šiljak, 1978) is of *linear order* and uses norm-like Liapunov functions for the subsystems. The second approach (Araki, 1978) utilizes quadratic Liapunov functions on the subsystem level and is termed the *quadratic order* construction. These two constructions have been selected because they are easy to use and, under standard conditions, they are the least conservative.

In the *linear construction*, one associates with each decoupled subsystem $\mathbf{S}_i$ of (2.12) a scalar function v_i: $\mathbf{T} \times \mathbf{R}^{n_i} \to \mathbf{R}_+$ such that $v_i(t, x_i)$ is a *continuous* function that satisfies a Lipschitz condition in x_i, that is, for some $\kappa_i > 0$,

$$|v_i(t, x_i') - v_i(t, x_i'')| \le \kappa_i \| x_i' - x_i'' \|, \qquad \forall t \in \mathbf{T}, \qquad \forall x_i', x_i'' \in \mathbf{R}^{n_i}. \tag{2.20}$$

We assume that the function $v_i(t,\, x_i)$ satisfies the inequalities

$$\begin{aligned} \phi_{1i}\left(\|x_i\|\right) &\le v_i\left(t,\, x_i\right) \le \phi_{2i}\left(\|x_i\|\right), \\ D^+ v_i\left(t,\, x_i\right)_{(2.12)} &\le -\phi_{3i}\left(\|x_i\|\right), \qquad \forall\left(t,\, x_i\right) \in \mathbf{T} \times \mathbf{R}^{n_i} \end{aligned} \tag{2.21}$$

where $\phi_{1i},\, \phi_{2i} \in \mathcal{K}_\infty,\, \phi_{3i} \in \mathcal{K}$ are comparison functions (Hahn, 1967; Šiljak, 1978), and $D^+v_i(t,\, x_i)_{(2.12)}$ denotes the Dini derivative of $v_i(t,\, x_i)$ computed with respect to (2.12). Inequalities (2.21) mean that we assume global asymptotic stability of the equilibrium $x_i = 0$ in each subsystem $\mathbf{S}_i$, and that we have a Liapunov function $v_i(t,\, x_i)$ to prove it.

As for the interconnections $h_i(t,\, x)$, we suppose that there are numbers ξ_{ij} such that $\xi_{ij} \ge 0$ for $i \ne j$, and

$$\|h_i(t,\, x)\| \le \sum_{j=1}^{N} \bar{e}_{ij}\xi_{ij}\phi_{3j}\left(\|x_j\|\right), \quad \forall\left(t,\, x\right) \in \mathbf{T} \times \mathbf{R}^n. \tag{2.22}$$

In order to state the condition for connective stability of $\mathbf{S}$, we define the constant $N \times N$ matrix $\bar{W} = (\bar{w}_{ij})$ as

$$\bar{w}_{ij} = \begin{cases} 1 - \bar{e}_{ii}\kappa_i\xi_{ii}, & i = j, \\ -\bar{e}_{ij}\kappa_i\xi_{ij}, & i \ne j. \end{cases} \tag{2.23}$$

Matrix $\bar{W}$ has nonpositive off-diagonal elements, that is, it belongs to the class of matrices $\mathcal{N} = \{W \in \mathbf{R}^{N\times N}\colon\ w_{ij} \le 0,\ i \ne j;\ i,\, j \in \boldsymbol{N}\}$. Under certain conditions, such a matrix belongs also to the important class $\mathcal{M}$ of M-matrices introduced by Ostrowski (1937). Because of our frequent use of M-matrices, we shall give several characterizations of $\mathcal{M}$. For comprehensive treatment of M-matrices, see Fiedler and Pták (1962) and Šiljak (1978), where a more general version of the following theorem is given:

2.4. THEOREM. Let $W \in \mathcal{N}$. Then, the following conditions are equivalent:

(i) There exists a vector $d \in \mathbf{R}_+^N,\ d_i > 0,\ i \in \boldsymbol{N}$, such that the vector $c \in \mathbf{R}_+^N$ defined as

$$c = Wd, \tag{2.24}$$

is positive, that is, $c_i > 0$ for all $i \in \boldsymbol{N}$.

(*ii*) All leading principal minors of W are positive, that is,

$$\begin{vmatrix} w_{11} & w_{12} & \cdots & w_{1k} \\ w_{21} & w_{22} & \cdots & w_{2k} \\ \cdots & \cdots & \cdots & \cdots \\ w_{k1} & w_{k2} & \cdots & w_{kk} \end{vmatrix} > 0, \qquad \forall k \in \boldsymbol{N}. \tag{2.25}$$

(*iii*) W is a positive quasidominant diagonal matrix, that is, $w_{ii} > 0$, $i \in \boldsymbol{N}$, and there exist numbers $d_i > 0$ such that

$$d_i w_{ii} > \sum_{\substack{j=1 \\ j \neq i}}^{N} d_j |w_{ij}|, \qquad \forall i \in \boldsymbol{N}. \tag{2.26}$$

(*iv*) There exists a diagonal matrix $D = \text{diag}\{d_1, d_2, \ldots, d_N\}$ with elements $d_i > 0$, $i \in \boldsymbol{N}$, such that the matrix

$$C = W^T D + DW, \tag{2.27}$$

is positive definite.

(*v*) The real part of each eigenvalue of W is positive, that is, $\text{Re}\{\lambda_i(W)\} > 0$ for all $i \in \boldsymbol{N}$.

(*vi*) The inverse W^{-1} exists and $W^{-1} \geq 0$ element-by-element.

Part (*iv*) was established independently by Tartar (1971) as follows: First, from (*i*) we conclude that the vectors $u = W^{-1}e$ and $v = (W^T)^{-1}e$, $e = (1, 1, \ldots, 1)^T$, are positive element-by-element, and form a positive diagonal matrix $D = \text{diag}\{v_1/u_1, v_2/u_2, \ldots, v_N/u_N\}$. We also recall that $W^T + W$ is positive definite if W is dominant diagonal, that is, satisfies inequalities (2.26) when $d_i = 1$, $i \in \boldsymbol{N}$. Now, choosing $U = \text{diag}\{u_1, u_2, \ldots, u_N\}$ and $V = \text{diag}\{v_1, v_2, \ldots, v_N\}$, we get $W^T D + DW = W^T V U^{-1} + U^{-1} V W = U^{-1}[(VWU)^T + VWU]U^{-1} = U^{-1}(\tilde{W}^T + \tilde{W})U^{-1}$, which is positive definite because $\tilde{W}$ is dominant diagonal. On the other hand, it is easy to see that positive definitness of C implies (*vi*).

At last, we are in a position to prove the basic stability result for composite systems (Šiljak, 1978):

2.5. THEOREM. A system $\mathbf{S}$ is connectively stable if the matrix $\bar{W}$ is an M-matrix.

Proof. We compute

$$
\begin{aligned}
D^+v_i(t,\,x_i)_{(2.11)} &= \limsup_{h\to 0+} \frac{1}{h}\{v_i(t+h\,,\,x_i+h \\
&\quad \times [g_i\,(t,\,x_i) + h_i\,(t,\,x)]) - v_i\,(t,\,x_i)\} \\
&\le D^+v_i\,(t,\,x_i)_{(2.12)} + \kappa_i\|\,h_i\,(t,\,x)\,\| \\
&\le -\,\phi_{3i}\,(\|x_i\|) + \kappa_i \sum_{j=1}^{N} \bar{e}_{ij}\xi_{ij}\phi_{3j}\,(\|x_j\|) \\
&\le -\sum_{j=1}^{N} \bar{w}_{ij}\phi_{3j}\,(\|x_j\|)\,, \\
&\qquad \forall\,(t,\,x) \in \mathbf{T}\times\mathbf{R}^n, \qquad \forall E \in \bar{E}.
\end{aligned}
\tag{2.28}
$$

Next, consider the function ν: $\mathbf{T}\times\mathbf{R}^n \to \mathbf{R}_+$,

$$
\nu(t,\,x) = d^T v(t,\,x), \tag{2.29}
$$

as a candidate for the Liapunov function of system $\mathbf{S}$, where $d \in \mathbf{R}_+^N$ is a constant vector with positive (yet unspecified) components, and v: $\mathbf{T}\times\mathbf{R}^n \to \mathbf{R}_+^N$ is a *vector Liapunov function* $v = (v_1,\,v_2,\,\ldots,\,v_N)^T$. From (2.28), we get

$$
D^+\nu(t,\,x)_{(2.11)} \le -d^T\bar{W}\phi_3(x), \qquad \forall\,(t,\,x) \in \mathbf{T}\times\mathbf{R}^n, \qquad \forall E \in \bar{E}, \tag{2.30}
$$

where $\phi_3(x) = [\phi_{31}\,(\|x_1\|)\,,\ \phi_{32}\,(\|x_2\|)\,,\,\ldots,\,\phi_{3N}\,(\|x_N\|)]^T$. If $\bar{W}$ is an M-matrix, then from (i) of Theorem 2.4, we conclude that there exists a vector $d > 0$ such that (2.29) and (2.30) yield

$$
\begin{aligned}
&\phi_I\,(\|x\|) \le \nu\,(t,\,x) \le \phi_{II}\,(\|x\|)\,, \\
&D^+\nu\,(t,\,x)_{(2.11)} \le -\phi_{III}\,(\|x\|)\,, \qquad \forall\,(t,\,x) \in \mathbf{T}\times\mathbf{R}^n, \qquad \forall E \in \bar{E},
\end{aligned}
\tag{2.31}
$$

where ϕ_I, $\phi_{II} \in \mathcal{K}_\infty$, $\phi_{III} \in \mathcal{K}$ are comparison functions defined as

$$\phi_I(\|x\|) = \sum_{i=1}^{N} d_i \phi_{1i}(\|x_i\|),$$

$$\phi_{II}(\|x\|) = \sum_{i=1}^{N} d_i \phi_{2i}(\|x_i\|), \tag{2.32}$$

$$\phi_{III}(\|x\|) = \sum_{i=1}^{N} c_i \phi_{3i}(\|x_i\|).$$

Inequalities (2.31) are standard conditions for global asymptotic stability of the equilibrium $x = 0$ of **S**. Since the conditions are valid for all $E \in \bar{E}$, stability is also connective. Q.E.D.

2.6. REMARK. This is not the most general formulation of the vector Liapunov functions in the context of large-scale interconnected systems (Šiljak, 1978), but it is all we need in the decentralized control problem. In this formulation, it is clear what advantages Theorem 2.5 offers in numerical simplifications of stability problems: Liapunov functions are constructed for small size subsystems, and stability of the overall system is tested by an M-matrix the size of which is equal to the number of the subsystems. Although numerically attractive, the method is often criticized for its relative conservativeness when compared with the standard single-function approach. Much of the recent work has been devoted to the reduction of conservativeness of the method by more sophisticated decompositions, better choices of subsystem functions, etc., as reported in this chapter and Notes and References in Section 2.6.

2.7. REMARK. From the proof of Theorem 2.5, it follows that the connective property of stability comes for free when vector Liapunov functions are used. That is, whenever stability of the system **S** is established using a specific matrix $\bar{W} \in \mathcal{M}$, stability holds for any $W \in \mathcal{N}$ such that $W \geq \bar{W}$, where the inequality is taken element-by-element. This is obvious from (2.26). By using the interconnection matrices, we define $\bar{W}$ as in (2.23), and have $E \in \bar{E}$ to imply $W \geq \bar{W}$, where W and $\bar{W}$ correspond to E and $\bar{E}$, respectively. In this way, as pointed out by Lakshmikantham (1981), the connective stability concept represents a strengthening of the classical global asymptotic stability property, which aims to extract all that is available in the results obtained by using standard vector Liapunov function methods. As it turns out, this extra information about the connectivity aspect of stability is an important robustness characterization of interconnected systems in a large number of applications (Šiljak, 1978).

2.8. REMARK. We can relax the requirement in (2.22) that the ξ_{ij}'s are nonnegative for $i \neq j$ and still have $D^+v(t, x)_{(2.11)}$ negative. From (2.28), it is obvious that we can set all negative ξ_{ij}'s to zero and test a new matrix $\tilde{W} = (\tilde{w}_{ij})$ for the M-matrix properties, where the elements $\tilde{w}_{ij}$ are defined as (Šiljak, 1980),

$$\tilde{w}_{ij} = \begin{cases} 1 - \bar{e}_{ii}\kappa_i\xi_{ii}, & i = j, \\ -\bar{e}_{ij}\kappa_i \max\{0, \xi_{ij}\}, & i \neq j. \end{cases} \tag{2.33}$$

We note that if any of the ξ_{ii}'s is negative, they should be included in the corresponding diagonal element $\tilde{w}_{ij}$ of $\tilde{W}$, because they enhance the chance for $\tilde{W}$ to be an M-matrix. To see that, one can use the argument of Remark 2.7.

Let us now present the alternative *quadratic construction* of Liapunov functions proposed by Araki (1975, 1978). Due to our special interest in connective stability, we shall modify the original exposition to include structural perturbations.

In the quadratic approach, with each decoupled subsystem $\mathbf{S}_i$ of (2.12) we associate a *continuously differentiable* function v_i: $\mathbf{T} \times \mathbf{R}^{n_i} \to \mathbf{R}_+$, and assume that there exist comparison functions $\phi_{1i}, \phi_{2i} \in \mathcal{K}_\infty$, $\phi_{3i} \in \mathcal{K}$, and a number $\eta_i > 0$, such that

$$\begin{aligned} &\phi_{1i}(\|x_i\|) \leq v_i(t, x_i) \leq \phi_{2i}(\|x_i\|), \\ &\dot{v}_i(t, x_i)_{(2.12)} \leq -\eta_i\phi_{3i}(\|x_i\|), \qquad \forall (t, x_i) \in \mathbf{T} \times \mathbf{R}^{n_i}, \end{aligned} \tag{2.34}$$

where

$$\dot{v}_i(t, x_i)_{(2.12)} = \frac{\partial v_i(t, x_i)}{\partial t} + [\text{grad } v_i(t, x_i)]^T g_i(t, x_i) \tag{2.35}$$

is computed with respect to (2.12) as indicated.

As for the interconnections, we assume that there exist numbers ξ_{ij} such that $\xi_{ij} \geq 0$ for $i \neq j$, and

$$[\text{grad } v_i(t, x_i)]^T h_i(t, x) \leq \phi_{3i}^{1/2}(\|x_i\|) \sum_{j=1}^{N} \bar{e}_{ij}\xi_{ij}\phi_{3j}^{1/2}(\|x_j\|), \quad \forall (t, x) \in \mathbf{T} \times \mathbf{R}^n. \tag{2.36}$$

Finally, we define the $N \times N$ test matrix $\hat{W} = (\hat{w}_{ij})$ as

$$\hat{w}_{ij} = \begin{cases} \eta_i - \bar{e}_{ii}\xi_{ii}, & i = j, \\ -\bar{e}_{ij}\xi_{ij}, & i \neq j, \end{cases} \tag{2.37}$$

and prove the following:

2.9. THEOREM. A system $\mathbf{S}$ is connectively stable if the matrix $\hat{W}$ is an M-matrix.

Proof. Let us select the function ν: $\mathbf{T} \times \mathbf{R}^n \to \mathbf{R}_+$ defined as

$$\nu(t, x) = \sum_{i=1}^{N} d_i v_i(t, x_i) \tag{2.38}$$

to be a candidate for the Liapunov function of $\mathbf{S}$, where the existence of numbers $d_i > 0$, $i \in \boldsymbol{N}$, is yet to be established. Then, we compute $\dot{\nu}(t, x)_{(2.11)}$ as follows:

$$\begin{aligned} \dot{\nu}(t, x)_{(2.11)} &= \frac{\partial \nu(t, x)}{\partial t} + [\text{grad } \nu(t, x)]^T f(t, x) \\ &= \sum_{i=1}^{N} d_i \left\{ \dot{v}_i(t, x_i)_{(2.12)} + [\text{grad } v_i(t, x_i)]^T h_i(t, x) \right\} \\ &\leq \sum_{i=1}^{N} d_i \left\{ -\eta_i\phi_{3i}(\|x_i\|) + \phi_{3i}^{1/2}(\|x_i\|) \sum_{j=1}^{N} \bar{e}_{ij}\xi_{ij}\phi_{3j}^{1/2}(\|x_j\|) \right\} \\ &\leq -\frac{1}{2}\phi_3^T(\|x\|)\left(\hat{W}^T D + D\hat{W}\right)\phi_3(\|x\|), \\ &\quad \forall (t, x) \in \mathbf{T} \times \mathbf{R}^n, \qquad \forall E \in \bar{E}, \end{aligned} \tag{2.39}$$

where $\phi_3(\|x\|) = \left[\phi_{31}^{1/2}(\|x_1\|), \phi_{32}^{1/2}(\|x_2\|), \ldots, \phi_{3N}^{1/2}(\|x_N\|)\right]^T$, and $D = \text{diag}\{d_1, d_2, \ldots, d_N\}$. Now, by (iv) of Theorem 2.4, we conclude from (2.34) and (2.39) that there exist comparison functions $\phi_I, \phi_{II} \in \mathcal{K}_\infty$, $\phi_{III} \in \mathcal{K}$ such that

$$\begin{aligned} &\phi_I(\|x\|) \le \nu(t, x) \le \phi_{II}(\|x\|), \\ &\dot{\nu}(t, x)_{(2.11)} \le -\phi_{III}(\|x\|), \qquad \forall (t, x) \in \mathbf{T} \times \mathbf{R}^n, \qquad \forall E \in \bar{E}, \end{aligned} \tag{2.40}$$

and the system $\mathbf{S}$ is connectively stable. Q. E. D.

2.10. REMARK. Our presentation of the quadratic construction of vector Liapunov functions differs from the original exposition of Araki (1978), not only by inclusion of the connective stability aspect, but also by absence of a number of unused generalities. We also note that the quadratic construction demands the Liapunov functions to be continuously differentiable, whereas in the linear construction only continuity is required.

As an illustration of the application of vector Liapunov functions, we consider stability of *linear systems* composed of interconnected subsystems. This class of systems is of particular interest in the context of control and estimation design. As a by-product of this analysis, we show that, with respect to linear systems, the linear and quadratic constructions of the Liapunov functions are equivalent.

Let us consider a linear system $\mathbf{S}$, which is described as

$$\mathbf{S}\colon\ \dot{x} = Ax, \tag{2.41}$$

where $x(t) \in \mathbf{R}^n$ is the state of $\mathbf{S}$ at time $t \in \mathbf{T}$, and A is a constant $n \times n$ matrix. We assume that the system

$$\mathbf{S}\colon\ \dot{x}_i = A_i x_i + \sum_{j=1}^{N} e_{ij} A_{ij} x_j, \qquad i \in \boldsymbol{N}, \tag{2.42}$$

is an interconnection of N subsystems

$$\mathbf{S}_i\colon\ \dot{x}_i = A_i x_i, \qquad i \in \boldsymbol{N}, \tag{2.43}$$

with $x_i(t) \in \mathbf{R}^{n_i}$ being the state of the subsystem $\mathbf{S}_i$ at time $t \in \mathbf{T}$, and A_i, A_{ij} are constant matrices of appropriate dimensions.

To determine stability of **S** by linear construction, we associate with each free subsystem $\mathbf{S}_i$ a norm-like function $v_i : \mathbf{R}^{n_i} \to \mathbf{R}_+$ defined as

$$v_i(x_i) = \left(x_i^T H_i x_i\right)^{1/2}, \tag{2.44}$$

where H_i is a constant, symmetric, and positive definite $n_i \times n_i$ matrix. The function $v(x_i)$ satisfies the Lipschitz condition (2.20) with $\kappa_i = \lambda_M^{1/2}(H_i)$, where $\lambda_M(H_i)$ is the maximum eigenvalue of the matrix H_i. We assume that the function $v_i(x_i)$ satisfies the inequalities

$$\begin{aligned} &\lambda_m^{1/2}(H_i)\|x_i\| \le v_i(x_i) \le \lambda_M^{1/2}(H_i)\|x_i\|, \\ &D^+ v_i(x_i)_{(2.43)} \le -\frac{1}{2}\lambda_M^{-1/2}(H_i)\lambda_m(G_i)\|x_i\|, \qquad \forall x_i \in \mathbf{R}^{n_i}, \end{aligned} \tag{2.45}$$

where $\lambda_m(H_i)$ and $\lambda_m(G_i)$ are the minimum eigenvalues of the matrices H_i and G_i, which appear in the Liapunov matrix equation

$$A_i^T H_i + H_i A_i = -G_i. \tag{2.46}$$

In fact, we assume that all subsystems $\mathbf{S}_i$ and, thus, the matrices A_i are stable. That is, all eigenvalues of A_i, $i \in \boldsymbol{N}$, have negative real parts. It is a well-known fact (*e.g.*, Hahn, 1967) that A_i is stable if and only if, for any symmetric positive definite G_i, there exists a symmetric positive definite H_i that is the unique solution of Equation (2.46), and the estimates (2.45) of v_i and $\dot{v}_i$ follow.

To bound the interconnections, we define the numbers

$$\xi_{ij} = \lambda_M^{1/2}\left(A_{ij}^T A_{ij}\right), \tag{2.47}$$

and follow the proof of Theorem 2.5 to obtain the test matrix $\bar{W} = (\bar{w}_{ij})$ defined as

$$\bar{w}_{ij} = \begin{cases} 1 - 2\bar{e}_{ii}\lambda_M(H_i)\lambda_m^{-1}(G_i)\lambda_M^{1/2}\left(A_{ii}^T A_{ii}\right), & i = j, \\ -2\bar{e}_{ij}\lambda_M^{1/2}(H_i)\lambda_M^{1/2}(H_j)\lambda_m^{-1}(G_j)\lambda_M^{1/2}\left(A_{ij}^T A_{ij}\right), & i \ne j. \end{cases} \tag{2.48}$$

By Theorem 2.5, the system **S** of (2.42) is connectively stable if the matrix $\bar{W}$ is an M-matrix.

We can simplify the test matrix $\bar{W}$ if we premultiply and postmultiply it by the positive diagonal matrices

$$D_{\mathrm{L}} = \operatorname{diag}\left\{\frac{1}{2}\lambda_M^{-1/2}(H_1),\ \frac{1}{2}\lambda_M^{-1/2}(H_2),\ \ldots,\ \frac{1}{2}\lambda_M^{-1/2}(H_N)\right\},$$

$$D_{\mathrm{R}} = \operatorname{diag}\left\{\lambda_M^{-1/2}(H_1)\lambda_m(G_1),\ \lambda_M^{-1/2}(H_2)\lambda_m(G_2),\ \ldots,\ \lambda_M^{-1/2}(H_N)\lambda_m(G_N)\right\}. \tag{2.49}$$

Then, we get the elements w_{ij} of the new matrix $W = D_{\mathrm{L}}\bar{W}D_{\mathrm{R}}$ as

$$w_{ij} = \begin{cases} \frac{1}{2}\,\frac{\lambda_m(G_i)}{\lambda_M(H_i)} - \bar{e}_{ii}\lambda_M^{1/2}(A_{ii}^T A_{ii}), & i = j, \\ -\bar{e}_{ij}\lambda_M^{1/2}(A_{ij}^T A_{ij}), & i \neq j. \end{cases} \tag{2.50}$$

As is well-known (*e.g.*, Šiljak, 1978), due to the fact that D_{L} and D_{R} of (2.49) have positive elements, we have $\bar{W} \in \mathcal{M}$ if and only if $W \in \mathcal{M}$. In the new test matrix W, however, the subsystem stability matrices H_i and G_i appear in the diagonal elements only, while the off-diagonal elements of W are solely due to the interconnection terms. This separation is convenient for interpretation of the M-matrix conditions in the context of overall system stability.

From Remark 2.7, we know that the larger the diagonal elements w_{ii} and the smaller the off-diagonal elements w_{ij} of W, the better are the chances for W to be an M-matrix. We recall that the ratio $\lambda_m(G_i)/\lambda_M(H_i)$ is an estimate of the degree of stability of the isolated subsystem $\mathbf{S}_i$ (see, for example, Kalman and Bertram, 1960; or more recently, Šiljak, 1978). Clearly, we should choose the stability matrix G_i in (2.46) so as to maximize this ratio. Before we attempt to solve this maximization problem, let us show that the new matrix W is also a test matrix for the quadratic construction of vector Liapunov functions.

We assume again that the subsystems $\mathbf{S}_i$ of (2.43) are stable, but use the quadratic form

$$v_i(x_i) = x_i^T H_i x_i, \tag{2.51}$$

as a Liapunov function for $\mathbf{S}_i$. Inequalities (2.34) are

$$\begin{aligned} &\lambda_m(H_i)\|x_i\|^2 \leq v_i(x_i) \leq \lambda_M(H_i)\|x_i\|^2, \\ &\dot{v}_i(x_i)_{(2.43)} \leq -\lambda_m(G_i)\|x_i\|^2, \qquad \forall x_i \in \mathbf{R}^{n_i}, \end{aligned} \tag{2.52}$$

where the matrices H_i and G_i are those of Equation (2.46).

As for the constraints on interconnections, we use again the numbers ξ_{ij} defined in (2.47), and following the proof of Theorem 2.9, obtain the test matrix $\hat{W} = (\hat{w}_{ij})$ as

$$\hat{w}_{ij} = \begin{cases} \lambda_m(G_i) - 2\bar{e}_{ii}\lambda_M(H_i)\lambda_M^{1/2}\left(A_{ii}^T A_{ii}\right), & i = j, \\ -2\bar{e}_{ij}\lambda_M(H_i)\lambda_M^{1/2}\left(A_{ij}^T A_{ij}\right), & i \neq j. \end{cases} \tag{2.53}$$

We immediately recognize the fact that $W = D_{\mathrm{L}}\hat{W}$, where

$$D_{\mathrm{L}} = \frac{1}{2}\mathrm{diag}\left\{\lambda_M^{-1}(H_1),\, \lambda_M^{-1}(H_2),\, \ldots,\, \lambda_M^{-1}(H_N)\right\}, \tag{2.54}$$

and $\hat{W} \in \mathcal{M}$ if and only if $W \in \mathcal{M}$. Therefore, as far as the linear constant systems are concerned, the linear and quadratic constructions of vector Liapunov functions are equivalent.

Let us now return to the problem of choosing the best subsystem Liapunov functions in the context of the M-matrix conditions. This problem is obviously important, because these conditions are only sufficient for stability of the overall system $\mathbf{S}$. It is reasonable to solve this problem first for linear systems, and we start with the following (Šiljak, 1978):

2.11. PROBLEM.

$$\text{Find: } \max_G \left\{\frac{\lambda_m(G)}{\lambda_M(H)}\right\}, \qquad \text{subject to: } A^T H + HA = -G. \tag{2.55}$$

In order to simplify the notation, the subscript identifying a subsystem is omitted, but we bear in mind that this problem should be solved for each subsystem separately.

To solve Problem 2.11, we assume that all eigenvalues of the matrix A are distinct and are known numerically. This assumption would be prohibitive on the level of the overall system $\mathbf{S}$. Presumably, the subsystems are relatively small, which is not too much to ask. Furthermore, in the stabilization problem, which we consider soon, these eigenvalues are assigned by local feedback, and conventional pole-placement techniques can be used to achieve any desired eigenvalue locations.

We use first the transformation

$$x = T\tilde{x}, \tag{2.56}$$

where T is a nonsingular $n \times n$ matrix such that the equation

$$\mathbf{S}\colon\ \dot{x} = Ax \tag{2.57}$$

becomes

$$\tilde{\mathbf{S}}\colon\ \dot{\tilde{x}} = \Lambda\tilde{x}, \tag{2.58}$$

where the $n \times n$ matrix $\Lambda = T^{-1}AT$ is semisimple, that is, it has the block diagonal form

$$\Lambda = \operatorname{diag}\left\{\begin{bmatrix} -\sigma_1 & \omega_1 \\ -\omega_1 & -\sigma_1 \end{bmatrix}, \ldots, \begin{bmatrix} -\sigma_p & \omega_p \\ -\omega_p & -\sigma_p \end{bmatrix}, \quad -\sigma_{2p+1}, \ldots, -\sigma_n\right\}, \tag{2.59}$$

where the first p blocks correspond to the complex eigenvalues of A, and the rest of the diagonal elements of Λ are the real eigenvalues. By denoting

$$\tilde{G} = T^T G T, \qquad \tilde{H} = T^T H T, \tag{2.60}$$

we replace Problem 2.11 by the following:

2.12. PROBLEM.

$$\text{Find: } \max_{\tilde{G}} \left\{\frac{\lambda_m(\tilde{G})}{\lambda_M(\tilde{H})}\right\}, \tag{2.61}$$
$$\text{subject to: } \Lambda^T\tilde{H} + \tilde{H}\Lambda = \tilde{G}.$$

The solution to Problem 2.12 is known (Šiljak, 1978):

$$\begin{aligned} \tilde{G} &= 2\theta \operatorname{diag}\{\sigma_1, \sigma_1, \sigma_2, \sigma_2, \ldots, \sigma_p, \sigma_p, \sigma_{2p+1}, \ldots, \sigma_n\}, \\ \tilde{H} &= \theta I, \end{aligned} \tag{2.62}$$

where θ is an arbitrary positive number, and I is the $n \times n$ identity matrix. Then,

$$\frac{\lambda_m(\tilde{G})}{\lambda_M(\tilde{H})} = 2\min_q\{\sigma_q\} = 2\sigma_M, \qquad q = 1, 2, \ldots, n, \tag{2.63}$$

where $-\sigma_M < 0$ is the real part of the maximal eigenvalue of the matrix A. Thus, the solution (2.62) yields the largest possible value of the ratio

$\lambda_m(\tilde{G})/\lambda_M(\tilde{H})$, which is $2\sigma_M$, and we can do no better. See (Šiljak, 1978) for a detailed proof of this result.

With the choice (2.62), the test matrix (2.50) becomes a new matrix $W^{\odot} = (w_{ij}^{\odot})$ defined by

$$w_{ij}^{\odot} = \begin{cases} \sigma_M^i - \bar{e}_{ii}\lambda_M^{1/2}\left(\tilde{A}_{ii}^T\tilde{A}_{ii}\right), & i = j, \\ -\bar{e}_{ij}\lambda_M^{1/2}\left(\tilde{A}_{ij}^T\tilde{A}_{ij}\right), & i \neq j, \end{cases} \tag{2.64}$$

where the interconnection matrices A_{ij} are changed to $\tilde{A}_{ij} = T_i^{-1}A_{ij}T_j$, T_i is the transformation matrix which is used to get Λ_i from A_i, and $\sigma_M^i = -\max_k\{\text{Re}\,[\lambda_k(A_i)]\}$, $k \in \boldsymbol{N}_i = \{1, 2, \ldots, n_i\}$.

We should note that in transforming the original subsystems $\mathbf{S}_i$ to $\tilde{\mathbf{S}}_i$ to get the test matrix $W^{\odot}$, it may very well happen that the gain in the size of diagonal elements $-\sigma_M^i$ is overcome by an increase in the off-diagonal elements of $W^{\odot}$ when we go from A_{ij} to $\tilde{A}_{ij}$. In that case, it may be better to solve Problem 2.11 in the original state space (Patel and Toda, 1980). This we consider next.

It is easy to see that the solution to Problem 2.11 is obtained by the choice $G = \theta I$ for any scalar $\theta > 0$. Recall (*e.g.*, Hahn, 1967) that for a given symmetric positive definite G, the symmetric positive definite solution H of the Liapunov matrix equation (2.55) is

$$H = \int_0^\infty e^{A^Tt}\, Ge^{At}dt. \tag{2.65}$$

From (2.65), we have the inequality

$$x^THx \geq \lambda_m(G)x^T\left(\int_0^\infty e^{A^Tt}e^{At}dt\right)x = x^T\hat{H}x, \tag{2.66}$$

which is valid for all x, where $\hat{H}$ is the unique solution of the Liapunov's matrix equation $A^T\hat{H} + \hat{H}A = -\lambda_m(G)I$. By denoting $\hat{G} = \lambda_m(G)I$, from (2.66) and the relation $\lambda_m(G) = \lambda_m(\hat{G})$, we get

$$\frac{\lambda_m(G)}{\lambda_M(H)} \leq \frac{\lambda_m(\hat{G})}{\lambda_M(\hat{H})}. \tag{2.67}$$

Since the solution of Liapunov matrix equation for θG is θH for any G and $\theta > 0$, from (2.67) we conclude that the solution of Problem 2.11 is obtained for $G = \theta I$. Without loss of generality, one can choose $\theta = 1$.

2.13. EXAMPLE. Although we have the largest ratio $\lambda_m(G)/\lambda_M(H)$ for $G = I$ without performing transformations, the resulting value may be far off from σ_M of A. We use a simple example to illustrate this point:

$$\mathbf{S}\colon\ \dot{x} = \begin{bmatrix} 0 & 1 \\ -2 & -3 \end{bmatrix} x + \gamma(t)x, \tag{2.68}$$

where $x(t) \in \mathbf{R}^2$, and $\gamma\colon\ \mathbf{T} \to \mathbf{R}$ is an uncertain piecewise continuous parameter. We should estimate the size of $\gamma(t)$ that can be tolerated by stability of an otherwise linear constant system.

Let us first stay in the original state space and get, from the solution (2.65) of Problem 2.11,

$$G = I, \qquad H = \begin{bmatrix} 5/4 & 1/4 \\ 1/4 & 1/4 \end{bmatrix}. \tag{2.69}$$

Thus,

$$\frac{\lambda_m(G)}{\lambda_M(H)} = \frac{1}{\lambda_M(H)} = 0.763 \tag{2.70}$$

and

$$|\gamma(t)| \leq 0.381, \qquad \forall t \in \mathbf{T}. \tag{2.71}$$

Alternatively, we transform $\mathbf{S}$ applying the transformation

$$x = T\tilde{x}, \qquad T = \begin{bmatrix} 1 & 1 \\ -1 & -2 \end{bmatrix}, \tag{2.72}$$

to get the system

$$\tilde{\mathbf{S}}\colon\ \dot{\tilde{x}} = \begin{bmatrix} -1 & 0 \\ 0 & -2 \end{bmatrix} \tilde{x} + \gamma(t)\tilde{x}. \tag{2.73}$$

Using

$$\tilde{G} = \begin{bmatrix} 1 & 0 \\ 0 & 2 \end{bmatrix}, \qquad \tilde{H} = I, \tag{2.74}$$

we get

$$\frac{\lambda_m(\tilde{G})}{\lambda_M(\tilde{H})} = 2, \tag{2.75}$$

and the resulting bound is

$$|\gamma(t)| < 1, \qquad \forall t \in \mathbf{T}, \tag{2.76}$$

which is more than two and a half times better than (2.71). The reason is that, in the transformed space, the ratio $\lambda_m(G)/2\lambda_M(H) = 1$ is the exact estimate of $\sigma_M = 1$, which is a measure of the degree of stability of **S** when the perturbation term is not present. We should note, however, that in this particular example, the transformation (2.72) leaves the perturbation term unchanged, thus producing the largest bound available in this context. In general, this is not the case, and a transformation such as (2.72) affects the perturbation function, so that the gain in the estimate of the stability degree may be diminished by the increase of the effective size of the transformed perturbation.

2.14. EXAMPLE. At this point, we return to the two penduli of Section 2.1, and apply the method of vector Liapunov functions to establish connective stability of the closed-loop system $\hat{\mathbf{S}}$ of (2.8). For given values of α and β, we choose the feedback gains k_{11}, k_{12}, k_{21}, and k_{22} in (2.6), so that the closed-loop subsystems $\hat{\mathbf{S}}_1$, and $\hat{\mathbf{S}}_2$ of (2.7) are identical and have the eigenvalues at -1 and -2. The system $\hat{\mathbf{S}}$ becomes

$$\hat{\mathbf{S}}\colon\ \dot{x}_1 = \begin{bmatrix} 0 & 1 \\ -2 & -3 \end{bmatrix} x_1 + e \begin{bmatrix} 0 & 0 \\ -\gamma & 0 \end{bmatrix} x_1 + e \begin{bmatrix} 0 & 0 \\ \gamma & 0 \end{bmatrix} x_2,$$
$$\dot{x}_2 = \begin{bmatrix} 0 & 1 \\ -2 & -3 \end{bmatrix} x_2 + e \begin{bmatrix} 0 & 0 \\ \gamma & 0 \end{bmatrix} x_1 + e \begin{bmatrix} 0 & 0 \\ -\gamma & 0 \end{bmatrix} x_2. \tag{2.77}$$

Since the subsystem matrices $\hat{A}_1$ and $\hat{A}_2$ are the same as the system matrix A of **S** in (2.68), we have the ratio $\lambda_m(G)/\lambda_M(H)$ for the diagonal elements w_{ii} of the test matrix W of (2.50), calculated in (2.70). The interconnection bounds are easily computed from (2.47) as $\lambda_M^{1/2}\left(A_{ij}^T A_{ij}\right) = \gamma$. The test matrix W for $\hat{\mathbf{S}}$ is given as

$$W = \begin{bmatrix} 0.381 - \gamma & -\gamma \\ -\gamma & 0.381 - \gamma \end{bmatrix}, \tag{2.78}$$

because the fundamental interconnection matrix is

$$\bar{E} = \begin{bmatrix} 1 & 1 \\ 1 & 1 \end{bmatrix}. \tag{2.79}$$

Applying the M-matrix conditions (2.25) to the matrix W of (2.78), we get

$$0.381 - \gamma > 0, \qquad \begin{vmatrix} 0.381 - \gamma & -\gamma \\ -\gamma & 0.381 - \gamma \end{vmatrix} > 0, \tag{2.80}$$

which produces the bound on the interconnection parameter

$$\gamma < 0.19 \tag{2.81}$$

for connective stability of $\hat{\mathbf{S}}$. This means that the closed-loop system remains stable for all values of the interconnection parameter e such that $e \in [0, 1]$, provided (2.80) holds. It is interesting to note here that a decentralized control law (2.6) can be chosen to stabilize the penduli even if the spring is moved all the way to the bobs, that is, when $\bar{a} = \ell$ and $\gamma = k/m$ (see Example 2.23).

2.3. Stabilization

Armed with the powerful stability conditions of vector Liapunov functions, we are in a position to attack the problem of decentralized stabilization of interconnected systems. The control strategy is simple: stabilize each subsystem when decoupled, and then check stability of the interconnected closed-loop subsystems using M-matrix conditions. This is what we did to control the inverted penduli in the preceding section. Here, we want to provide a general framework for such stabilization practice, which proved to be effective in a variety of applications (Šiljak, 1978). The fact that connective aspect of stability comes for free in this context is most gratifying.

The announced decentralized control strategy will be appended with the global control (Šiljak and Vukčević, 1976a,b), the role of which is to enhance the dominance of subsystems stability by reducing the effect of the interconnections. This two-level control scheme has a considerable flexibility, because, as we shall see soon, the global control may be implemented partially or not at all depending on the physical constraints of the plant as well as the wishes of the designer.

Let us consider the system

$$\mathbf{S}\colon\ \dot{x}_i = A_i x_i + B_i u_i^\ell + \sum_{j=1}^{N} A_{ij} x_j + \Gamma_i u_i^g, \qquad i \in \boldsymbol{N}, \tag{2.82}$$

which is an interconnection of N subsystems

$$\mathbf{S}_i\colon\ \dot{x}_i = A_i x_i + B_i u_i^\ell, \qquad i \in \boldsymbol{N}, \tag{2.83}$$

where $x_i(t) \in \mathbf{R}^{n_i}$ and $u_i^\ell(t) \in \mathbf{R}^{m_i}$ are the state and the local input of $\mathbf{S}_i$, $u_i^g(t) \in \mathbf{R}^{\ell_i}$ is the global input to $\mathbf{S}_i$, and A_i, B_i, A_{ij}, Γ_i are constant matrices of appropriate dimensions.

For local controls we choose the state feedback

$$u_i^\ell = -K_i x_i, \qquad i \in \boldsymbol{N}, \tag{2.84}$$

with constant gain matrices K_i, and the global control laws are

$$u_i^g = -\sum_{j=1}^{N} f_{ij} K_{ij} x_j, \qquad i \in \boldsymbol{N}, \tag{2.85}$$

where the gain matrices K_{ij} are constant and are distributed over the system $\mathbf{S}$ according to a binary $N \times N$ matrix $F = (f_{ij})$.

By applying controls u_i^ℓ and u_i^g to system $\mathbf{S}$, we get the closed-loop system

$$\hat{\mathbf{S}}\colon\ \dot{x}_i = \hat{A}_i x_i + \sum_{j=1}^{N} \hat{A}_{ij} x_j, \qquad i \in \boldsymbol{N}, \tag{2.86}$$

with

$$\hat{A}_i = A_i - B_i K_i, \qquad \hat{A}_{ij} = A_{ij} - f_{ij}\Gamma_i K_{ij}. \tag{2.87}$$

We assume that each pair $(A_i,\, B_i)$ is controllable so that we can always choose the gains K_i to place the eigenvalues of $\hat{A}_i$ at any desired locations (Kailath, 1980). The locations of the eigenvalues $\lambda_k^i(\hat{A}_i)$ are

$$-\sigma_1^i \pm j\omega_1^i,\ \ldots,\ -\sigma_{p_i}^i \pm j\omega_{p_i}^i,\ -\sigma_{2p_i+1}^i,\ \ldots,\ -\sigma_{n_i}^i, \tag{2.88}$$

which are distinct and such that Re $\lambda_k^i(\hat{A}_i) < 0$, $k \in \boldsymbol{N}_i$. With this placement of subsystem eigenvalues, we can use a linear nonsingular transformation

$$x_i = T_i \tilde{x} \tag{2.89}$$

to get the system $\hat{\mathbf{S}}$ in the transformed space as

$$\tilde{\mathbf{S}}\colon\ \dot{\tilde{x}}_i = \Lambda_i \tilde{x}_i + \sum_{j=1}^{N} \Delta_{ij} \tilde{x}_j, \qquad i \in \boldsymbol{N}, \tag{2.90}$$

where

$$\Lambda_i = T_i^{-1} \hat{A}_i T_i, \qquad \Delta_{ij} = T_i^{-1} \hat{A}_{ij} T_j, \tag{2.91}$$

and Λ_i is defined in (2.59).

From the preceding section, we have the norm-like Liapunov function

$$v_i(x_i) = (x_i^T \tilde{H}_i x_i)^{1/2}, \tag{2.51$'$}$$

with H_i replaced by $\tilde{H}_i$, which is the solution of the Liapunov equation

$$\Lambda_i^T \tilde{H}_i + \tilde{H}_i \Lambda_i = -\tilde{G}_i, \tag{2.92}$$

obtained as

$$\tilde{H}_i = I_i, \qquad \tilde{G}_i = \text{diag}\{\sigma_1, \sigma_1, \ldots, \sigma_{p_i}, \sigma_{p_i}, \sigma_{2p_i+1}, \ldots, \sigma_{n_i}\}. \tag{2.93}$$

With this choice of subsystem Liapunov functions, we produce the test matrix $\tilde{W} = (\tilde{w}_{ij})$ defined as

$$\tilde{w}_{ij} = \begin{cases} \sigma_M^i - \lambda_M^{1/2}(\Delta_{ii}^T \Delta_{ii}), & i = j, \\ -\lambda_M^{1/2}(\Delta_{ij}^T \Delta_{ij}), & i \neq j, \end{cases} \tag{2.94}$$

and the system $\tilde{\mathbf{S}}$ and, thus, $\hat{\mathbf{S}}$, are stable if $\tilde{W}$ is an M-matrix.

So far, we did not say how we selected the global gains K_{ij}. From Remark 2.7, we know that the smaller the off-diagonal elements $\tilde{w}_{ij}$ of $\tilde{W}$, the better the chance for $\tilde{W}$ to satisfy the M-matrix conditions. We also recall that this fact holds element-by-element, and each individual gain matrix K_{ij} can be chosen to reduce the effect of the interconnection term A_{ij} independently (Šiljak, 1978).

If $f_{ij} = 1$, then we choose $\tilde{K}_{ij}$ to get

$$\xi_{ij}^* = \min_{\tilde{K}_{ij}} \{\| \tilde{A}_{ij} - \tilde{\Gamma}_i \tilde{K}_{ij} \|\}, \tag{2.95}$$

where

$$\tilde{A}_{ij} = T_i^{-1} A_{ij} T_j, \qquad \tilde{\Gamma}_i = T_i^{-1} \Gamma_i, \qquad \tilde{K}_{ij} = K_{ij} T_j. \tag{2.96}$$

The matrix $\tilde{K}_{ij}^*$ is computed by

$$\tilde{K}_{ij}^* = \tilde{\Gamma}_i^I \tilde{A}_{ij}, \tag{2.97}$$

where $\tilde{\Gamma}_i^I = (\tilde{\Gamma}_i^T \tilde{\Gamma}_i)^{-1} \tilde{\Gamma}_i^T$ is the pseudo-inverse of $\tilde{\Gamma}_i$. In particular, when the rank of the composite matrix $[\tilde{\Gamma}_i \;\; \tilde{A}_{ij}]$ is equal to the rank of the matrix $\tilde{\Gamma}_i$ itself, the choice (2.97) of the matrix $\tilde{K}_{ij}^*$ produces $\Delta_{ij} = 0$. Similar cancellation of nonlinear interconnections by a nonlinear global control law was applied to a model of Large Space Telescope (Šiljak, 1978).

2.15. REMARK. We did not consider the connective aspect of stability, because the emphasis was placed on the feedback structure of the global control. It is easy to involve the interconnection matrix in this context, by rewriting (2.90) as

$$\tilde{\mathbf{S}}\colon\ \dot{\tilde{x}}_i = \Lambda_i \tilde{x}_i + \sum_{j=1}^{N} e_{ij} \Delta_{ij}^* \tilde{x}_j, \qquad i \in \boldsymbol{N}. \tag{2.98}$$

It is now obvious that the concept of connective stability applies directly, save for the fact that any variation in the strength of the effective interconnections Δ_{ij} implies a corresponding variation in the gain $\tilde{K}_{ij}$ of the global control. That is, the interconnection parameters e_{ij} reflect norm-like changes of the effective interconnection matrices $\Delta_{ij}^* = \tilde{A}_{ij} - f_{ij}\tilde{\Gamma}_i \tilde{K}_{ij}^*$.

2.16. EXAMPLE. To illustrate the proposed stabilization procedure, let us consider the interconnected system

$$\begin{aligned} \mathbf{S}\colon\ \dot{x}_1 &= \begin{bmatrix} 0 & 1 & 0 \\ 0 & 0 & 1 \\ -8.86 & 8.50 & 9.36 \end{bmatrix} x_1 + \begin{bmatrix} 0 \\ 0 \\ 1 \end{bmatrix} u_1 + \begin{bmatrix} 3.20 & 1.98 \\ -14.72 & 0.49 \\ -7.92 & 36.01 \end{bmatrix} x_2, \\ \dot{x}_2 &= \begin{bmatrix} 0 & 1 \\ 32.32 & -1.36 \end{bmatrix} x_2 + \begin{bmatrix} 0 \\ 1 \end{bmatrix} u_2 + \begin{bmatrix} 1.69 & 1.26 & 0.08 \\ -7.52 & -5.23 & 0.49 \end{bmatrix} x_1, \end{aligned} \tag{2.99}$$

which has the eigenvalues $\lambda_{1,2} = 0.76 \pm j1.83$, $\lambda_3 = 11.54$, $\lambda_4 = -3.89$, $\lambda_5 = -1.16$, and is unstable. The two subsystems

$$\begin{aligned} \mathbf{S}_1\colon\ \dot{x}_1 &= \begin{bmatrix} 0 & 0 & 0 \\ 0 & 0 & 1 \\ -8.86 & 8.50 & 9.36 \end{bmatrix} x_1 + \begin{bmatrix} 0 \\ 0 \\ 1 \end{bmatrix} u_1, \\ \mathbf{S}_2\colon\ \dot{x}_2 &= \begin{bmatrix} 0 & 1 \\ 32.32 & -1.36 \end{bmatrix} x_2 + \begin{bmatrix} 0 \\ 1 \end{bmatrix} u_2 \end{aligned} \tag{2.100}$$

are in the companion form and are controllable.

Using the local control laws

$$\begin{aligned} u_1^\ell &= -k_1^T x_1, \qquad k_1 = (1791.14,\ 458.50,\ 46.36)^T, \\ u_2^\ell &= -k_2^T x_2, \qquad k_2 = (33.82,\ 1.14)^T, \end{aligned} \tag{2.101}$$

the subsystems eigenvalues

$$\begin{aligned} &\lambda_1^1 = 0.63, \quad \lambda_2^1 = -1.39, \quad \lambda_3^1 = 10.12, \\ &\lambda_1^2 = 5.04, \quad \lambda_2^2 = -6.41 \end{aligned} \tag{2.102}$$

can be relocated to

$$\begin{aligned} &\lambda_1^1 = -10, \quad \lambda_2^1 = -12, \quad \lambda_3^1 = -15, \\ &\lambda_1^2 = -1, \quad \lambda_1^2 = -1.5. \end{aligned} \tag{2.103}$$

Applying transformation (2.89) to each of the closed-loop subsystems $\hat{\mathbf{S}}_1$ and $\hat{\mathbf{S}}_2$, we get the closed-loop overall system $\tilde{\mathbf{S}}$ in the transformed space as

$$\begin{aligned} \tilde{\mathbf{S}}\colon\ \dot{\tilde{x}}_1 &= \begin{bmatrix} -10 & & \bigcirc \\ & -12 & \\ \bigcirc & & -15 \end{bmatrix} \tilde{x}_1 + \begin{bmatrix} -23.52 & -43.78 \\ 40.23 & 68.97 \\ -15.49 & -24.96 \end{bmatrix} \tilde{x}_2, \\ \dot{\tilde{x}}_2 &= \begin{bmatrix} -1 & 0 \\ 0 & -1.5 \end{bmatrix} \tilde{x}_2 + \begin{bmatrix} 179.95 & 247.53 & 367.40 \\ -182.58 & -249.03 & -365.95 \end{bmatrix} \tilde{x}_1. \end{aligned} \tag{2.104}$$

From (2.94) and (2.103), we obtain the aggregate matrix

$$\tilde{W} = \begin{bmatrix} 10 & -98.51 \\ -676.68 & 1 \end{bmatrix}. \tag{2.105}$$

Since $\tilde{W}$ is not an M-matrix, we cannot establish stability of $\tilde{\mathbf{S}}$.

Let us now use the global control as defined in (2.85), with $\Gamma_i = B_i$, the feedback structure matrix

$$F = \begin{bmatrix} 0 & 1 \\ 1 & 0 \end{bmatrix}, \tag{2.106}$$

and the gains

$$\tilde{k}_{12}^* = (-238.95,\ -415.34)^T, \qquad \tilde{k}_{21}^* = (90.63,\ 124.14,\ 183.33)^T, \tag{2.107}$$

which are computed from (2.97). This results in a global closed-loop system

$$\begin{aligned} \tilde{\mathbf{S}}^*\colon\ \dot{\tilde{x}}_1 &= \begin{bmatrix} -10 & & \bigcirc \\ & 12 & \\ \bigcirc & & -15 \end{bmatrix} \tilde{x}_1 + \begin{bmatrix} 0.37 & -2.25 \\ 0.40 & -0.25 \\ 0.44 & 2.73 \end{bmatrix} \tilde{x}_2, \\ \dot{\tilde{x}}_2 &= \begin{bmatrix} -1 & 0 \\ 0 & -1.5 \end{bmatrix} \tilde{x}_2 + \begin{bmatrix} -1.31 & -0.75 & 0.72 \\ -1.31 & -0.75 & 0.72 \end{bmatrix} \tilde{x}_1, \end{aligned} \tag{2.108}$$

and the aggregate matrix

$$\tilde{W}^* = \begin{bmatrix} 10 & -3.55 \\ -2.37 & 1 \end{bmatrix}, \tag{2.109}$$

which is an M-matrix. Therefore, the system $\tilde{\mathbf{S}}$ of (2.104) is stable having the eigenvalues $\lambda_{1,2} = -1.03 \pm j\, 0.16$, $\lambda_3 = -10.27$, $\lambda_4 = -11.99$, $\lambda_5 = -15.17$. It is clear that the system $\tilde{\mathbf{S}}^*$ is connectively stable, for the fundamental interconnection matrix is

$$\bar{E} = \begin{bmatrix} 0 & 1 \\ 1 & 0 \end{bmatrix}, \tag{2.110}$$

and the system $\tilde{\mathbf{S}}^*$ becomes

$$\begin{aligned} \tilde{\mathbf{S}}^*\colon\ \dot{\tilde{x}}_1 &= \begin{bmatrix} -10 & & \bigcirc \\ & 12 & \\ \bigcirc & & -15 \end{bmatrix} \tilde{x}_1 + \bar{e}_{12} \begin{bmatrix} 0.37 & -2.25 \\ 0.40 & -0.25 \\ 0.44 & 2.73 \end{bmatrix} \tilde{x}_2, \\ \dot{\tilde{x}}_2 &= \begin{bmatrix} -1 & 0 \\ 0 & -1.5 \end{bmatrix} \tilde{x}_2 + \bar{e}_{21} \begin{bmatrix} -1.31 & -0.75 & 0.72 \\ -1.31 & -0.75 & 0.72 \end{bmatrix} \tilde{x}_1, \end{aligned} \tag{2.111}$$

leading to the aggregate matrix

$$\tilde{W}^* = \begin{bmatrix} 10 & -3.55\bar{e}_{12} \\ -2.37\bar{e}_{21} & 1 \end{bmatrix}, \tag{2.112}$$

which is an M-matrix for all $E \in \bar{E}$.

It is equally clear that the stabilization procedure, which considers a composite system as a whole, most likely would not produce a connectively stable closed-loop system. The same example treated in this section was considered by Šiljak (1978) to demonstrate this fact.

2.17. REMARK. We must bear in mind that the method of vector Liapunov functions for stabilization of composite systems produces only yes/no answers. No solid evidence is provided for the reasons of success or failure, although it is obvious that the larger the degree of stability of each subsystem the better the chances to "beat" the interconnections. Unfortunately, the degree of stability and the size of interconnections are not independent quantities, and pushing the eigenvalues of the subsystems further to the left may not produce a desired effect. This is best illustrated by the following example due to Sezer and Hüseyin (1981).

2.18. EXAMPLE. The input-decentralized system

$$\mathbf{S}\colon\ \dot{x}_1 = \begin{bmatrix} 0 & 1 \\ 0 & 0 \end{bmatrix} x_1 + \begin{bmatrix} 0 \\ 1 \end{bmatrix} u_1 + \begin{bmatrix} 0 & 2 \\ 0 & 0 \end{bmatrix} x_2,$$
$$\dot{x}_2 = \begin{bmatrix} 0 & 1 \\ 0 & 0 \end{bmatrix} x_2 + \begin{bmatrix} 0 \\ 1 \end{bmatrix} u_2 + \begin{bmatrix} 0 & 2 \\ 0 & 0 \end{bmatrix} x_1 \tag{2.113}$$

has unstable subsystems, but they are controllable and can be stabilized by local feedback control

$$u_1 = -k_1^T x_1, \qquad k_1 = (k_{11},\, k_{12})^T,$$
$$u_2 = -k_2^T x_2, \qquad k_1 = (k_{21},\, k_{22})^T. \tag{2.114}$$

The closed-loop system is obtained as

$$\hat{\mathbf{S}}\colon\ \begin{bmatrix} \dot{x}_1 \\ \dot{x}_2 \end{bmatrix} = \left[\begin{array}{cc|cc} 0 & 1 & 0 & 2 \\ -k_{12} & -k_{11} & 0 & 0 \\ \hline 0 & 2 & 0 & 1 \\ 0 & 0 & -k_{22} & -k_{21} \end{array}\right] \begin{bmatrix} x_1 \\ x_2 \end{bmatrix}. \tag{2.115}$$

A necessary condition for $\hat{\mathbf{S}}$ to be stable is

$$k_{12}k_{22} < 0, \tag{2.116}$$

which is obtained by considering the determinant of the system matrix in (2.115). From this condition it follows that for $\hat{\mathbf{S}}$ to be stable, one of the closed-loop subsystems should be unstable! The vector Liapunov function method would not succeed in this case independent of the choice of local feedback.

To finish this section on a positive note, we should mention the fact that when a system is stabilized *via* the vector Liapunov method, the closed-loop system is robust with respect not only to disconnections of the subsystems, but it can also tolerate a wide range of nonlinear and time-varying uncertainties in the interactions among the subsystems (see Chapter 3).

2.4. Connective Stabilizability

Now, we want to identify a class of complex systems which can *always* be connectively stabilized by local feedback. A way to do this is to characterize the systems for which the method of vector Liapunov functions succeeds (Šiljak and Vukčević, 1977; Ikeda and Šiljak, 1980a,b,c). The basic assumption is that a given system

$$\mathbf{S}\colon\ \dot{x}_i = A_i x_i + b_i u_i + \sum_{j=1}^{N} e_{ij} A_{ij} x_j, \qquad i \in \boldsymbol{N}, \tag{2.117}$$

is composed of controllable single-input subsystems

$$\mathbf{S}_i\colon\ \dot{x}_i = A_i x_i + b_i u_i, \qquad i \in \boldsymbol{N}, \tag{2.118}$$

with the state $x_i(t) \in \mathbf{R}^{n_i}$ and the input $u_i(t) \in \mathbf{R}$. Without loss of generality, we also assume that A_i and b_i are given in the companion form

$$A_i = \begin{bmatrix} 0 & 1 & \dots & 0 \\ 0 & 0 & \dots & 0 \\ \dots & \dots & \dots & \dots \\ 0 & 0 & \dots & 1 \\ -a_1^i & -a_2^i & \dots & -a_{n_i}^i \end{bmatrix}, \qquad b_i = \begin{bmatrix} 0 \\ 0 \\ \vdots \\ 0 \\ 1 \end{bmatrix}. \tag{2.119}$$

To characterize the $n_i \times n_j$ matrices $A_{ij} = (a_{pq}^{ij})$ of the interactions between the subsystems $\mathbf{S}_i$, we follow Ikeda and Šiljak (1980a), and define the integers m_{ij} as

$$m_{ij} = \begin{cases} \displaystyle\max_{(p,q)\colon\, a_{pq}^{ij} \neq 0} \{q - p\}, & A_{ij} \neq 0, \\ -n, & A_{ij} = 0. \end{cases} \tag{2.120}$$

The integers m_{ij} can be interpreted in terms of the structure of A_{ij} as the distance between the main diagonal and a border line of the nonzero elements, which is parallel to the main diagonal. The three characteristic cases are shown in Figure 2.1, with possible nonzero elements a_{pq}^{ij} appearing in shaded areas only.

To stabilize the system $\mathbf{S}$, we apply the decentralized feedback control law

$$u_i = -k_i^T x_i, \qquad i \in \boldsymbol{N}. \tag{2.121}$$

The resulting closed-loop system is

$$\hat{\mathbf{S}}\colon\ \dot{x}_i = (A_i - b_i k_i^T) x_i + \sum_{j=1}^{N} e_{ij} A_{ij} x_j, \qquad i \in \boldsymbol{N}. \tag{2.122}$$

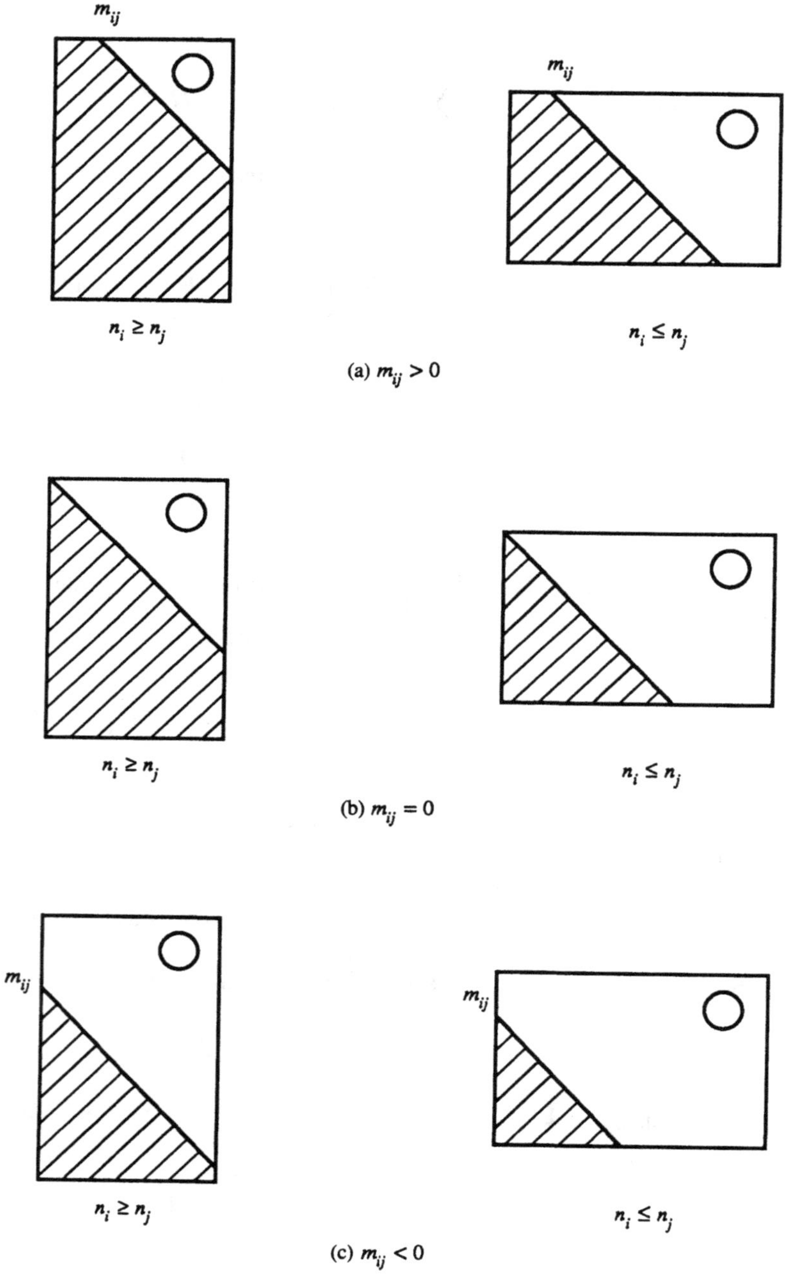

Fig. 2.1. Interconnection matrices.

If we denote by

$$\hat{A}_i = A_i - b_i k_i^T \tag{2.123}$$

the closed-loop matrix of the ith subsystem, we can get $\hat{\mathbf{S}}$ in a compact form

$$\hat{\mathbf{S}}\colon\ \dot{x} = [\hat{A}_D + A_C(E)]x, \tag{2.124}$$

where $x(t) \in \mathbf{R}^n$ is the state of $\hat{\mathbf{S}}$, and

$$\hat{A}_D = \text{diag}\{\hat{A}_1,\, \hat{A}_2,\, \dots,\, \hat{A}_N\}, \qquad A_C(E) = (e_{ij}A_{ij}). \tag{2.125}$$

We are interested in a little more than global connective stability of $\mathbf{S}$:

2.19. DEFINITION. The system $\mathbf{S}$ is said to be decentrally connectively exponentially stabilizable if, for any given positive number π, there exists a decentralized control law such that the solutions $x(t;\, t_0,\, x_0)$ of the system $\hat{\mathbf{S}}$ satisfy the inequality

$$\begin{gathered} \|x(t;\, t_0\, x_0)\| \ \leq \Pi\ \|x_0\|\ \exp\left[-\pi(t - t_0)\right], \\ \forall t \in \mathbf{T}_0, \quad \forall (t_0,\, x_0) \in \mathbf{T} \times \mathbf{R}^n, \quad \forall E \in \bar{E}, \end{gathered} \tag{2.126}$$

where Π and π are positive numbers.

We also need to define a subset $\boldsymbol{I}_r = \{i_1,\, i_2,\, \dots,\, i_r\}$ of the index set $\boldsymbol{N}$, for which $i_1 < i_2 < \dots < i_r$, $r \leq N$. By permuting the elements of $\boldsymbol{I}_r$, we form a set $\boldsymbol{J}_r = \{j_1,\, j_2,\, \dots,\, j_r\}$, and state the following:

2.20. THEOREM. The system $\mathbf{S}$ is decentrally connectively stabilizable if the inequality

$$\sum_{k=1}^{r} (m_{i_k j_k} - 1) < 0 \tag{2.127}$$

holds for all subsets $\boldsymbol{I}_r$ and $\boldsymbol{J}_r$ of $\boldsymbol{N}$.

Proof. This theorem was established in (Ikeda and Šiljak, 1980a) by following the proof of a less general result (Corollary 2.22) obtained by Šiljak and Vukčević (1977).

First, the gain vectors k_i are chosen so that each closed-loop matrix $\hat{A}_i = A_i - b_i k_i^T$ has a set $\mathcal{L}_i$ of distinct real eigenvalues, which is defined as

$$\mathcal{L}_i = \{\lambda_\ell^i\colon\ \lambda_\ell^i = -\rho\sigma_\ell^i\ ,\ \rho \geq 1\ ,\ \sigma_\ell^i > 0\ ,\ \ell \in \boldsymbol{N}_i\}\,. \tag{2.128}$$

The real eigenvalue assumption is not essential, but is convenient to use in the proof. The positive number ρ is to be determined so that the overall system $\hat{\mathbf{S}}$ is exponentially connectively stable.

Then, by the help of transformation

$$x_i = T_i \tilde{x}_i, \tag{2.129}$$

we get the system $\hat{\mathbf{S}}$ in the form

$$\tilde{\mathbf{S}}\colon\ \dot{\tilde{x}}_i = \Lambda_i \tilde{x}_i + \sum_{j=1}^{N} e_{ij} \tilde{A}_{ij} x_j, \qquad i \in \boldsymbol{N}, \tag{2.130}$$

where

$$\Lambda_i = T_i^{-1}(A_i - b_i k_i^T) T_i, \qquad \tilde{A}_{ij} = T_i^{-1} A_{ij} T_j, \tag{2.131}$$

and

$$\Lambda_i = \text{diag}\{-\rho\sigma_1^i,\ -\rho\sigma_2^i,\ \ldots,\ -\rho\sigma_{n_i}^i\}. \tag{2.132}$$

Then, we factorize the matrix T_i as

$$T_i = R_i \tilde{T}_i, \tag{2.133}$$

where the matrix $R_i = R_i(\rho)$ is

$$R_i = \text{diag}\{1,\ \rho,\ \ldots,\ \rho^{n_i - 1}\}, \tag{2.134}$$

and $\tilde{T}_i$ is the Vandermonde matrix

$$\tilde{T}_i = \begin{bmatrix} 1 & 1 & \cdots & 1 \\ -\sigma_1^i & -\sigma_2^i & \cdots & -\sigma_{n_i}^i \\ \cdots & \cdots & \cdots & \cdots \\ (-\sigma_1^i)^{n_i-1} & (-\sigma_2^i)^{n_i-1} & \cdots & (-\sigma_{n_i}^i)^{n_i-1} \end{bmatrix}. \tag{2.135}$$

To measure the magnitude of $\tilde{A}_{ij}$, we use the number

$$\xi_{ij} = \|\tilde{T}_i^{-1}\|\ \|\tilde{T}_i\| \sum_{p=1}^{n_i} \sum_{q=1}^{n_j} |a_{pq}^{ij}|, \tag{2.136}$$

for which we have

$$\|\tilde{A}_{ij}\| \leq \rho^{m_{ij}}\, \xi_{ij}, \tag{2.137}$$

and $\|A_{ij}\| = \lambda_M^{1/2}(A_{ij}^T A_{ij})$. Now, we define two $N \times N$ matrices

$$P = \text{diag}\{\sigma_M^1, \sigma_M^2, \ldots, \sigma_M^N\}, \qquad Q = \left(\rho^{m_{ij}-1}\, \xi_{ij}\right), \tag{2.138}$$

where $\sigma_M^i = \min_k\{\sigma_k^i\}$, and form the aggregate matrix

$$W = P - Q. \tag{2.139}$$

We show that the system $\tilde{\mathbf{S}}$ is stabilizable if $W \in \mathcal{M}$ for some $\rho \geq 1$. This amounts to showing that solutions $x(t;\, t_0,\, x_0)$ satisfy inequality (2.126) for some positive numbers Π and π. For this purpose, we associate with each subsystem $\hat{\mathbf{S}}_i$ a scalar function $\tilde{v}_i : \mathbf{R}^{n_i} \to \mathbf{R}_+$ defined as

$$v_i(\tilde{x}_i) = \tilde{x}_i^T \tilde{x}_i. \tag{2.140}$$

The derivative $\dot{v}_i(\tilde{x}_i)_{(2.130)}$ is majorized as

$$\begin{aligned} \dot{v}_i(\tilde{x}_i)_{(2.130)} &= 2\tilde{x}_i^T \Lambda_i \tilde{x}_i + 2\tilde{x}_i^T \sum_{j=1}^{N} e_{ij} \tilde{A}_{ij} \tilde{x}_j \\ &\leq -2\rho\sigma_m^i v_i(\tilde{x}_i) + v_i^{1/2}(\tilde{x}_i) \sum_{j=1}^{N} e_{ij}\rho^{m_{ij}}\, \xi_{ij} v_j^{1/2}(\tilde{x}_j). \end{aligned} \tag{2.141}$$

As usual, we consider a weighted sum of v_i's,

$$\nu(\tilde{x}) = \sum_{i=1}^{N} d_i v_i(\tilde{x}_i), \tag{2.142}$$

as a candidate of a Liapunov function for $\hat{\mathbf{S}}$, where d_i's are all positive numbers yet to be determined. From (2.140) and (2.142), $\nu(\tilde{x})$ and $\dot{\nu}(\tilde{x})_{(2.130)}$ can be estimated as

$$\begin{aligned} & d_m\|\tilde{x}\|^2 \leq \nu(\tilde{x}) \leq d_M\|\tilde{x}\|^2, \\ & \dot{\nu}(\tilde{x})_{(2.130)} \leq -\rho d_M^{-1} \lambda_m(C)\nu(\tilde{x}), \qquad \forall x \in \mathbf{R}^n, \qquad \forall E \in \bar{E}, \end{aligned} \tag{2.143}$$

where $d_m = \min_i\{d_i\}$, $d_M = \max_i\{d_i\}$,

$$W^T D + DW = C, \tag{2.144}$$

and $D = \text{diag}\{d_1, d_2, \ldots, d_N\}$. Since $W \in \mathcal{M}$ implies the existence of D such that C is positive definite, we conclude from (2.143) that the solutions $x(t; t_0, x_0)$ of $\tilde{\mathbf{S}}$ satisfy inequality (2.126) for $\Pi = d_M^{1/2} d_m^{1/2}$ and $\pi = \frac{1}{2}\rho d_M^{-1}\lambda_m(C)$.

So far we have shown that $W \in \mathcal{M}$ for some $\rho \geq 1$ implies stability of $\hat{\mathbf{S}}$. We should now demonstrate the fact that we can always find a ρ so that $W \in \mathcal{M}$. From the condition of the theorem, we have a possibility of expressing the kth leading principal minor (2.25) of W as $\sigma_M^1 \sigma_M^2 \ldots \sigma_M^k + c_{k1}\rho^{-1} + c_{k2}\rho^{-2} + \ldots + c_{k\ell_k}\rho^{-\ell_k}$, where the $c_{k\ell}$'s are constants and ℓ_k is an integer. This, in turn, implies that every leading principal minor of W is positive for a sufficiently large ρ.

Finally, we should show that the above argument goes through for any prescribed number π. To see that, we consider the system

$$\tilde{\mathbf{S}}_\pi\colon\ \dot{\tilde{x}} = (\Lambda_i + \pi I_{n_i})\,\tilde{x}_i + \sum_{j=1}^{N} e_{ij}\tilde{A}_{ij}\tilde{x}_j, \qquad i \in \boldsymbol{N}, \tag{2.145}$$

instead of $\tilde{\mathbf{S}}$, where I_{n_i} is the $n_i \times n_i$ identity matrix. We note that if the solutions of the system $\tilde{\mathbf{S}}_\pi$ are bounded, then they satisfy inequality (2.126). By redefining the aggregate matrix as

$$W_\pi = P - \rho^{-1}\pi I_N - Q, \tag{2.146}$$

it is easy to show that we can repeat the same argument for W_π that worked for W, and conclude that the solutions of $\tilde{\mathbf{S}}_\pi$ are bounded. This means that, by choosing the gains k_i of the decentralized control law (2.122) appropriately, we can realize any degree π of stability in the closed-loop system $\tilde{\mathbf{S}}$. Q.E.D.

2.21. REMARK. "Sufficiently large ρ" implies that we have to push the eigenvalues of the subsystems far enough to the left. This, however, may result in a local *high-gain feedback* which should cause concern in case noise or modeling errors are significant.

An interesting special case (Šiljak and Vukčević, 1977) of Theorem 2.20, which is given directly in terms of the matrices $A_{ij} = (a_{pq}^{ij})$, is given by:

2.22. COROLLARY. The system $\mathbf{S}$ is decentrally connectively stabilizable if

$$a_{pq}^{ij} = 0, \qquad p < q, \tag{2.147}$$

where $p \in \boldsymbol{N}_i$, $q \in \boldsymbol{N}_j$, and $\boldsymbol{N}_k = \{1, 2, \ldots, n_k\}$.

Restriction (2.147) means that the matrices A_{ij} of the system $\mathbf{S}$ in (2.117) have either of the two structures in Figure 2.1b.

2.23. EXAMPLE. By inspection of the model (2.4) for the inverted penduli of Section 2.1, we conclude that it belongs to the class of systems considered in this section. Let us now apply the proposed stabilization procedure and show that no matter what is the strength of coupling between the two penduli, there is always a set of local feedback gains which stabilize the overall system. In fact, in this simple example we can express explicitly the gains in terms of the coupling.

We first transform the closed-loop system $\hat{\mathbf{S}}$ of (2.8) into the form (2.130) where

$$\Lambda_1 = \Lambda_2 = \Lambda = \text{diag}\{-\rho\sigma_1,\ -\rho\sigma_2\} \tag{2.148}$$

and

$$T_1 = T_2 = \begin{bmatrix} 1 & 1 \\ -\rho\sigma_1 & -\rho\sigma_2 \end{bmatrix}. \tag{2.149}$$

Then, we compute the local feedback gains

$$k_1 = (k_{11},\ k_{12})^T, \qquad k_2 = (k_{21},\ k_{22})^T \tag{2.150}$$

in terms of the eigenvalues $-\rho\sigma_1$ and $-\rho\sigma_2$ of Λ, obtaining

$$\begin{aligned} k_{11} &= k_{21} = \rho\beta^{-1}(\sigma_1 + \sigma_2), \\ k_{12} &= k_{22} = \rho^2\beta^{-1}\sigma_1\sigma_2 + \beta^{-1}\alpha. \end{aligned} \tag{2.151}$$

To calculate the lower bound on ρ for stability, we factor the matrices T_1 and T_2 of (2.149) as in (2.133) to get

$$R_1 = R_2 = \text{diag}\{1,\ \rho\}, \qquad \tilde{T}_1 = \tilde{T}_2 = \begin{bmatrix} 1 & 1 \\ -\sigma_1 & -\sigma_2 \end{bmatrix}. \tag{2.152}$$

Using (2.136) and (2.151), and assuming that $\sigma_1 < \sigma_2$, we compute the aggregate matrix

$$\tilde{W} = \begin{bmatrix} \rho\sigma_1 - \dfrac{2\gamma}{\rho|\sigma_1 - \sigma_2|} & -\dfrac{2\gamma}{\rho|\sigma_1 - \sigma_2|} \\ -\dfrac{2\gamma}{\rho|\sigma_1 - \sigma_2|} & \rho\sigma_1 - \dfrac{2\gamma}{\rho|\sigma_1 - \sigma_2|} \end{bmatrix}. \tag{2.153}$$

For this matrix to satisfy the determinantal conditions (2.25) of Theorem 2.4, that is, to be an M-matrix, we should chose ρ as

$$\rho > 2\left(\frac{\gamma}{\sigma_1|\sigma_1 - \sigma_2|}\right)^{1/2}. \tag{2.154}$$

Finally, in terms of feedback gains (2.151), we interpret the inequality (2.154) as

$$\begin{aligned} k_{11} = k_{21} &> \frac{2(\sigma_1 + \sigma_2)}{\beta\sigma_1^{1/2}|\sigma_1 - \sigma_2|^{1/2}}\,\gamma^{1/2}, \\ k_{12} = k_{22} &> \frac{4\sigma_1\sigma_2}{\beta\sigma_1|\sigma_1 - \sigma_2|}\,\gamma + \frac{\alpha}{\beta}. \end{aligned} \tag{2.155}$$

This choice of the local feedback gains guarantees connective stability of the two interconnected penduli.

2.24. REMARK. Examples 2.18 and 2.23 show the importance of assumptions (2.120) and (2.147) concerning the interconnection structure. Although the assumptions are more general than those of Davison (1974), Ikeda *et al.* (1976), Sezer and Hüseyin (1978), Richter and De Carlo (1984), and many others, it is not the best we can do. A broader class of systems is considered in the next section, *via* graph-theoretic methods (Sezer and Šiljak, 1981). The use of graphs is not surprising in view of the fact that testing of condition (2.127) is a combinatorial problem which involves checking $\Sigma_{r=1}^{N}\, N!/(N-r)!$ inequalities. A further development along these lines can be found in (Shi and Gao, 1986, 1987).

In the rest of this section, we show how Theorem 2.20 can be applied to interconnected systems composed of multi-input subsystems (Ikeda and Šiljak, 1980a). For this purpose, we consider again the system $\mathbf{S}$ of (2.82) in the form

$$\mathbf{S}\colon\ \dot{x}_i = A_i x_i + B_i u_i + \sum_{j=1}^{N} e_{ij} A_{ij} x_j, \qquad i \in \boldsymbol{N}, \tag{2.82$'$}$$

where we ignore the global control. We again assume that all pairs $(A_i,\, B_i)$ are controllable and, without loss of generality, further assume that the decoupled subsystems

$$\mathbf{S}_i\colon\ \dot{x}_i = A_i x_i + B_i u_i, \qquad i \in \boldsymbol{N}, \tag{2.83$'$}$$

are in the Luenberger's canonical form (Luenberger, 1967). Then, each decoupled subsystem $\mathbf{S}_i$ is reducible to a set of single-input components having a companion form, by the use of a preliminary local feedback control

$$u_i = -K_i x_i + G_i v_i, \qquad i \in \boldsymbol{N}, \tag{2.156}$$

where $v(t) \in \mathbf{R}^{\ell_i}$ is the new input vector to the ith subsystem, K_i is an $m_i \times m_i$ gain matrix, G_i is an $m_i \times \ell_i$ constant matrix, and ℓ_i is the number of single-input components contained in $\mathbf{S}_i$. That is, the closed-loop system

$$\bar{\mathbf{S}}\colon\ \dot{x}_i = (A_i - B_i K_i)x_i + B_i G_i v_i + \sum_{j=1}^{N} e_{ij} A_{ij} x_j, \qquad i \in \boldsymbol{N}, \tag{2.157}$$

becomes an input decentralized system with each single-input component having the form (2.117). Now, we can apply Theorem 2.20 to the system $\bar{\mathbf{S}}$ of (2.157). This reduction of a multi-input subsystems to single-input components is not unique. For more details of the reduction process, see Ikeda *et al.* (1976), and Sezer and Hüseyin (1978).

2.25. EXAMPLE. Let us consider a system

$$\mathbf{S}\colon\ \dot{x}_1 = \begin{bmatrix} 0 & 1 & 0 & 0 \\ 2 & 3 & 2 & -1 \\ 0 & 0 & 0 & 1 \\ 2 & 1 & -1 & -2 \end{bmatrix} x_1 + \begin{bmatrix} 0 & 0 \\ 1 & 1 \\ 0 & 0 \\ 0 & 1 \end{bmatrix} u_1 + e_{12} \begin{bmatrix} 1 & 2 \\ 3 & 4 \\ 1 & 0 \\ 2 & 1 \end{bmatrix} x_2,$$

$$\dot{x}_2 = \begin{bmatrix} 0 & 1 \\ 3 & 4 \end{bmatrix} x_2 + \begin{bmatrix} 0 \\ 1 \end{bmatrix} u_2 + e_{21} \begin{bmatrix} 3 & 0 & 1 & 5 \\ 2 & 1 & 4 & 6 \end{bmatrix} x_1, \tag{2.158}$$

where the subsystems are given in the Luenberger canonical form. To transform the first subsystem of (2.158) into a single-input subsystem, we implement the preliminary control

$$u_1 = -\begin{bmatrix} 2 & 3 & 1 & -1 \\ 0 & 0 & 0 & 0 \end{bmatrix} x_1 + \begin{bmatrix} -1 \\ 1 \end{bmatrix} v_1, \tag{2.159}$$

leave the second subsystem as is, and get

$$\bar{\mathbf{S}}\colon\ \dot{x}_1 = \begin{bmatrix} 0 & 1 & 0 & 0 \\ 0 & 0 & 1 & 0 \\ 0 & 0 & 0 & 1 \\ 2 & 1 & -1 & -2 \end{bmatrix} x_1 + \begin{bmatrix} 0 \\ 0 \\ 0 \\ 1 \end{bmatrix} v_1 + e_{12} \begin{bmatrix} 1 & 2 \\ 3 & 4 \\ 1 & 0 \\ 2 & 1 \end{bmatrix} x_2,$$

$$\dot{x}_2 = \begin{bmatrix} 0 & 1 \\ 3 & 4 \end{bmatrix} x_2 + \begin{bmatrix} 0 \\ 1 \end{bmatrix} u_2 + e_{21} \begin{bmatrix} 3 & 0 & 1 & 5 \\ 2 & 1 & 4 & 6 \end{bmatrix} x_1. \tag{2.160}$$

We associate integers m_{ij} with the system $\bar{\mathbf{S}}$ as

$$m_{11} = -6, \qquad m_{12} = 1, \qquad m_{21} = 3, \qquad m_{22} = -6. \tag{2.161}$$

Since

$$(m_{12} - 1) + (m_{21} - 1) = 2 > 0, \tag{2.162}$$

$\bar{\mathbf{S}}$ of (2.160) violates condition (2.127) of Theorem 2.20, and we cannot establish decentral stabilizability of $\mathbf{S}$.

Next, consider the local control law

$$u_1 = -\begin{bmatrix} -2 & -1 & 2 & -1 \\ 2 & 1 & 0 & 0 \end{bmatrix} x_1 + \begin{bmatrix} 1 & -1 \\ 0 & 1 \end{bmatrix} v_1, \tag{2.163}$$

in order to reduce the first subsystem two single-input components while the second subsystem is unchanged. We get

$$\bar{\mathbf{S}}\colon\ \dot{x}_{11} = \begin{bmatrix} 0 & 1 \\ 2 & 3 \end{bmatrix} x_{11} + \begin{bmatrix} 0 \\ 1 \end{bmatrix} v_{11} + e_{12} \begin{bmatrix} 1 & 2 \\ 3 & 4 \end{bmatrix} x_2,$$

$$\dot{x}_{12} = \begin{bmatrix} 0 & 1 \\ -1 & -2 \end{bmatrix} x_{12} + \begin{bmatrix} 0 \\ 1 \end{bmatrix} v_{12} + e_{12} \begin{bmatrix} 1 & 0 \\ 2 & 1 \end{bmatrix} x_2,$$

$$\dot{x}_2 = \begin{bmatrix} 0 & 1 \\ 3 & 4 \end{bmatrix} x_2 + \begin{bmatrix} 0 \\ 1 \end{bmatrix} u_2 + e_{21} \begin{bmatrix} 3 & 0 \\ 2 & 1 \end{bmatrix} x_{11} + e_{21} \begin{bmatrix} 1 & 5 \\ 4 & 6 \end{bmatrix} x_{12}. \tag{2.164}$$

Finally, we rename the states x_{11}, x_{12}, and x_2 as x_1, x_2, and x_3 to make the system fit the standard form (2.117) of the system $\mathbf{S}$. In order to use Theorem 2.20, we compute

$$\begin{aligned} m_{13} &= m_{32} = 1, \qquad m_{23} = m_{31} = 0, \\ m_{ij} &= -6, \qquad \text{for all other pairs } (i,\, j). \end{aligned} \tag{2.165}$$

Since the integers (2.165) satisfy the condition Theorem 2.20, we conclude that the system **S** of (2.164) is decentrally stabilizable by local feedback

$$v_{11} = -k_{11}^T x_{11}, \qquad v_{12} = -k_{12}^T x_{12}, \qquad u_2 = -k_2^T x_2. \tag{2.166}$$

By going back to the original system **S** of (2.158), we further conclude that **S** is decentrally stabilizable by the feedback

$$\begin{aligned} u_1 &= -\left(\begin{bmatrix} -2 & -1 & 2 & -1 \\ 2 & 1 & 0 & 0 \end{bmatrix} + \begin{bmatrix} k_{11}^T & -k_{12}^T \\ 0 & k_{12}^T \end{bmatrix} \right) x_1, \\ u_2 &= -k_2^T x_2. \end{aligned} \tag{2.167}$$

To complete the design, we choose the gains k_{11}, k_{12}, and k_2 by a standard single-input eigenvalue-assignment scheme (*e.g.*, Kailath, 1980).

2.5. Graph-Theoretic Algorithm

Our solution to the decentralized stabilizability problem turned out to be essentially combinatorial, which suggests a use of graphs. This approach is further reinforced by the fact that the outcome of the stabilizability test depends only on the elements of system matrices being equal or different from zero rather than being fixed numerical values. Before we take advantage of these facts using a graph-theoretic formulation, we will present general stabilizability conditions for decentralized control of complex systems (Sezer and Šiljak, 1981). The conditions are not obtained by the method of vector Liapunov functions, but can be interpreted in the context of Small-Gain Theorem (Zames, 1966). The connectivity aspect of decentral stabilizability is not transparent outside of the Liapunov's framework, and we will be content to treat the elements of interconnection matrices as unknown but constant.

Let us again consider the decentrally controlled system

$$\hat{\mathbf{S}}\colon\ \dot{x} = [\hat{A}_D + A_C(E)]\, x, \tag{2.124}$$

where $E = (e_{ij})$ is an $N \times N$ constant matrix that belongs to the class

$$\mathcal{E} = \{E \in \mathbf{R}^{N\times N}\colon\ -1 \le e_{ij} \le 1\}, \tag{2.168}$$

and $\hat{A}_D = \text{diag}\{\hat{A}_1, \hat{A}_2, \dots, \hat{A}_N\}$, $A_C(E) = (e_{ij}A_{ij})$, $A_{ij} = (a_{pq}^{ij})$, as before. Notice that e_{ij}'s are permitted to vary in $[-1, 1]$ instead of $[0, 1]$.

In view of the sign-insensitive constraints (2.22) on the interconnections among the subsystems, the added flexibility concerning e_{ij}'s can be included in all connective stability and stabilization results obtained so far in this chapter.

The system $\hat{\mathbf{S}}$ is exponentially connectively stable with degree π if

$$\operatorname{Re}\{\lambda_M[\hat{A}_D + A_C(E)]\} \leq -\pi, \qquad \forall E \in \mathcal{E}, \tag{2.169}$$

where π is a positive number as in Definition 2.19. Since the subsystems $\mathbf{S}_i$ are all in the companion form (2.119), the gain vectors k_i of (2.122) can always be chosen so that each matrix $\hat{A}_i = A_i - b_i k_i^T$ has a set $\mathcal{L}_i$ of real distinct eigenvalues λ_ℓ^i defined by

$$\mathcal{L}_i = \{\lambda_\ell^i \colon\ \lambda_\ell^i = -\rho_i \sigma_\ell^i,\ \rho_i > 0,\ \sigma_\ell^i > 0,\ \ell \in \boldsymbol{N}_i\}, \tag{2.170}$$

where the σ_ℓ^i are arbitrary positive numbers,

$$\rho_i = \rho^{\nu_i}, \qquad i \in \boldsymbol{N}, \tag{2.171}$$

ρ and ν_i are positive numbers, and $\boldsymbol{N}_i = \{1, 2, \ldots, n_i\}$. To explain the role of the numbers ρ and ν_i, we note first that the underlying idea of the stabilization scheme is to achieve stability of the overall closed-loop system by shifting appropriately the eigenvalues of the decoupled subsystems. The number ρ_i represents the size and ν_i the relative importance of the shift in the ith subsystem.

The result established in (Sezer and Šiljak, 1981) is the following:

2.26. THEOREM. The system $\mathbf{S}$ is decentrally connectively stabilizable if

$$\lim_{\rho \to +\infty} \prod_{\substack{i \in \boldsymbol{I}_r \\ j \in \boldsymbol{J}_r}} a_{pq}^{ij} \frac{\rho_j^{q-1}}{\rho_i^p} = 0 \quad , \tag{2.172}$$

for all $\boldsymbol{I}_r$, $\boldsymbol{J}_r$, $r \in \boldsymbol{N}$, and all (p, q).

Proof. We claim that for a given π there exists a sufficiently large $\bar{\rho} > 0$ such that, whenever $\rho > \bar{\rho}$, the system $\hat{\mathbf{S}}$ exponentially connectively stable with degree π. This is equivalent to saying that there is a $\bar{\rho}$ such that

$$\det\left[(s - \pi)I - \hat{A}_D - A_C(E)\right] \neq 0, \qquad \forall s\colon \operatorname{Re} s \geq 0, \qquad \forall E \in \mathcal{E}. \tag{2.173}$$

Suppose to the contrary that the claim is not true. That is, we assume that the determinant in (2.173) is zero for some s_0 such that Re $s_0 \geq 0$ and some $E_0 = (e_{ij}^o) \in \mathcal{E}$, or, equivalently, there exists a nonzero vector $y \in \mathbf{C}^n$ such that $y = (y_1^T, y_2^T, \ldots, y_N^T)^T$ and

$$\left[(s_0 - \pi)I - \hat{A}_D - A_C(E_0)\right] y = 0, \tag{2.174}$$

which is further equivalent to

$$\left[(s_0 - \pi)I - \hat{A}_i\right] y_i = \sum_{j=1}^{N} e_{ij}^o A_{ij} y_j, \qquad i \in \boldsymbol{N}. \tag{2.175}$$

Using the transformation matrix $\tilde{T}_i$ of (2.135) and

$$R_i(\rho_i) = \text{diag}\{1, \rho_i, \ldots, \rho_i^{n_i - 1}\}, \tag{2.176}$$

we get

$$\tilde{T}_i^{-1} R_i^{-1}(\rho_i) A_i R_i(\rho_i) \tilde{T}_i = -\rho_i \Lambda_i, \tag{2.177}$$

where

$$\Lambda_i = \text{diag}\{\sigma_1^i, \sigma_2^i, \ldots, \sigma_{n_i}^i\}. \tag{2.178}$$

Hence, premultiplying both sides of (2.175) by $\tilde{T}_i^{-1} R_i^{-1}(\rho_i)$ and letting $z_i = \tilde{T}_i^{-1} R_i^{-1}(\rho_i) y_i$, we obtain

$$\left[(s_0 - \pi)I + \rho_i \Lambda_i\right] z_i = \sum_{j=1}^{N} e_{ij}^o \tilde{T}_i^{-1} R_i^{-1}(\rho_i) A_{ij} R_j(\rho_j) \tilde{T}_j z_j, \tag{2.179}$$

or, equivalently,

$$z_i = \left[(s_0 - \pi)I + \rho_i \Lambda_i\right]^{-1} \sum_{j=1}^{N} e_{ij}^o \tilde{T}_i^{-1} R_i^{-1}(\rho_i) A_{ij} R_j(\rho_j) \tilde{T}_j z_j. \tag{2.180}$$

Taking ℓ_∞ norms of both sides of (2.180), and noting that, for sufficiently large ρ_i, we have

$$\| \left[(s_0 - \pi)I + \rho_i \Lambda_i\right]^{-1} \| \leq \frac{1}{\rho_i(\sigma_m^i - \pi/\rho_i)}, \tag{2.181}$$

where $\sigma_M^i = \min\{\sigma_1^i, \sigma_2^i, \ldots, \sigma_{n_i}^i\}$, we obtain

$$\|z_i\| \leq \sum_{j=1}^{N} \ell_{ij}(\rho)\,\|z_j\|, \tag{2.182}$$

where

$$\ell_{ij}(\rho) = \frac{|e_{ij}^o|\,\|T_i^{-1}\|\,\|T_j\|}{\sigma_m^i - \pi/\rho_i}\;\frac{\|R_i^{-1}(\rho_i)A_{ij}R_j(\rho_j)\|}{\rho_i}. \tag{2.183}$$

Defining $L(\rho) = [\ell_{ij}(\rho)]$ and $\bar{z} \in \mathbf{R}_+^N$,

$$\bar{z} = (\|z_1\|,\ \|z_2\|,\ \ldots,\ \|z_N\|)^T, \tag{2.184}$$

(2.182) can be written in a compact form as

$$[I - L(\rho)]\,\bar{z} \leq 0, \tag{2.185}$$

where inequality is taken element-by-element.

Now, condition (2.172) of the theorem, together with (2.176) and (2.183), implies that

$$\lim_{\rho\to+\infty} \prod_{\substack{i\,\in\,\boldsymbol{I}_r\\ j\,\in\,\boldsymbol{J}_r}} \ell_{ij}(\rho) = 0, \tag{2.186}$$

for all $\boldsymbol{I}_r$, $\boldsymbol{J}_r$, $r \in \boldsymbol{N}$. This, in turn, implies that there exists a $\bar{\rho} > 0$ such that, for $\rho > \bar{\rho}$, all principal minors of the matrix $I - L(\rho)$ are positive. Since $I - L(\rho) \in \mathcal{N}$, from (vi) of Theorem 2.4, we have $[I - L(\rho)]^{-1} \geq 0$ for $\rho > \bar{\rho}$. This implies that, for $\rho > \bar{\rho}$, the system $\hat{\mathbf{S}}$ of (2.124) is exponentially connectively stable with degree π, for, otherwise, (2.185) would yield $\bar{z} \leq 0$ contradicting the definition (2.184) of $\bar{z}$. Q.E.D.

Now we will do what we promised at the beginning of this section: to formulate a graph-theoretic test for decentral stabilizability. For this task, we need first to characterize the structure of interconnection matrices A_{ij}. For each A_{ij}, we define the integers

$$\begin{aligned} 1 < p_1^{ij} < p_2^{ij} < \ldots < p_{k_{ij}}^{ij} < n_i, \\ 1 < q_1^{ij} < q_2^{ij} < \ldots < q_{k_{ij}}^{ij} < n_j, \end{aligned} \tag{2.187}$$

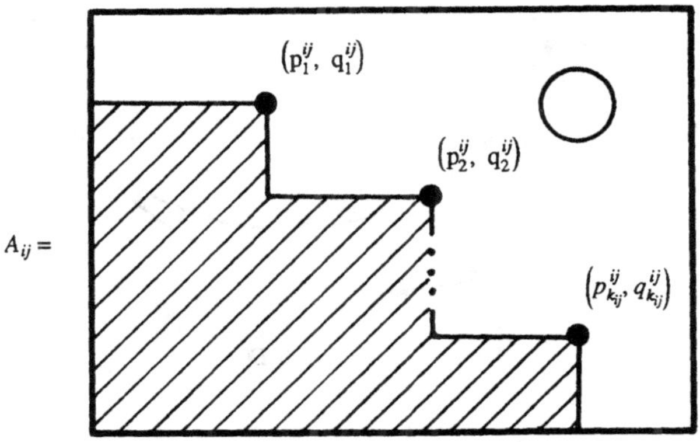

Fig. 2.2. Interconnection matrix.

recursively as

(a) p_ℓ^{ij} is the largest integer such that $a_{pq}^{ij} = 0$ for all $p < p_\ell^{ij}$ and $q > q_{\ell-1}^{ij}$, $\ell = 1, 2, \ldots, k_{ij}$, where $q_0^{ij} = 0$;

(b) q_ℓ^{ij} is the smallest integer such that $a_{pq}^{ij} = 0$ for all $p < p_\ell^{ij} + 1$ and $q > q_\ell^{ij}$, $\ell = 1, 2, \ldots, k_{ij}$.

The integers p_ℓ^{ij}, q_ℓ^{ij}, $\ell = 1, 2, \ldots, k_{ij}$, define a boundary for nonzero elements of the matrix A_{ij} as shown in Figure 2.2.

2.27. LEMMA. Condition (2.172) of Theorem 2.26 holds if and only if it holds when the matrices $A_{ij} = (a_{pq}^{ij})$ are replaced by the corresponding matrices $\bar{A}_{ij} = (\bar{a}_{pq}^{ij})$ defined as

$$\bar{a}_{pq}^{ij} = \begin{cases} 1, & \text{if } (p, q) = (p_\ell^{ij}, q_\ell^{ij}) \text{ for some } \ell = 1, 2, \ldots, k_{ij}, \\ 0, & \text{otherwise.} \end{cases} \tag{2.188}$$

Proof. Necessity is obvious. To prove sufficiency first note that for fixed r, $\boldsymbol{I}_r$, and $\boldsymbol{J}_r$, (2.172) need be checked only for those (p, q) such that $a_{pq}^{ij} \neq 0$, *i.e.*, for which $p \geq p_\ell^{ij}$ and $q \leq q_\ell^{ij}$ for some $\ell = 1, 2, \ldots, k_{ij}$. The proof then follows from the fact that if (2.172) holds for $(p_\ell^{ij}, q_\ell^{ij})$, then it also holds for all (p, q) such that $p \geq p_\ell^{ij}$ and $q \leq q_\ell^{ij}$. Q.E.D.

To construct a digraph $\hat{\mathbf{D}}$ associated with $\hat{\mathbf{S}}$, we first associate a subgraph $\hat{\mathbf{D}}_i = (\mathtt{X}_i, \mathtt{E}_i)$ with the ith closed-loop subsystem $\hat{\mathbf{S}}_i$. Because of the

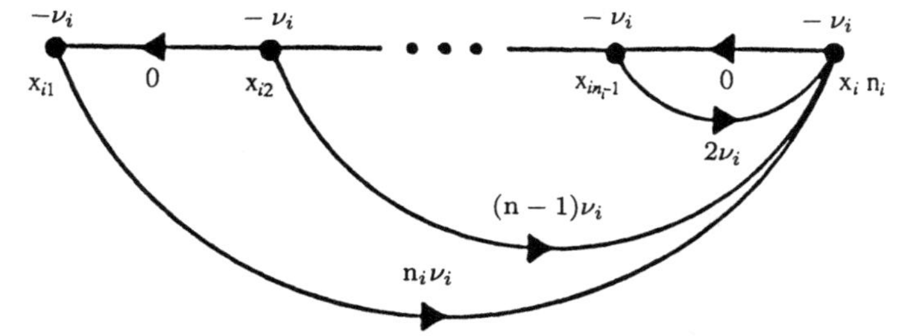

Fig. 2.3. Digraph $\hat{\mathbf{D}}_i$.

special (companion) structure of $\hat{A}_i$, the digraph $\hat{\mathbf{D}}_i$ has the form shown in Figure 2.3. In this analysis, we assign weights to vertices and edges of $\hat{\mathbf{D}}_i$. The weighted digraph $\hat{\mathbf{D}} = (\mathtt{X},\, \mathtt{E}_D \cup \mathtt{E}_C)$ for the system $\hat{\mathbf{S}}$ is defined by

$$\mathtt{X} = \bigcup_{i=1}^{N} \mathtt{X}_i, \qquad \mathtt{E}_D = \bigcup_{i=1}^{N} \mathtt{E}_i, \tag{2.189}$$

and $\mathtt{E}_C$ is a set of edges connecting the vertices belonging to distinct $\mathtt{X}_i$'s, so that there is an edge $x_{jq} \rightarrow x_{ip}$ from vertex x_{jq} of $\hat{\mathbf{D}}_j$ to vertex x_{ip} of $\hat{\mathbf{D}}_i$ if and only if $(p,\, q) = (p_\ell^{ij},\, q_\ell^{ij})$ for some $\ell = 1,\, 2,\, \ldots,\, k_{ij}$; that is, each edge in $\mathtt{E}_C$ corresponds to a corner (nonzero) element in one of A_{ij} in Figure 2.2. Zero weights are assigned to the edges in $\mathtt{E}_C$.

We define the net-weight of a directed path in $\hat{\mathbf{D}}$ as the sum of the weights of the vertices and edges traversed along the path, and prove the following:

2.28. LEMMA. If $\mathbf{P}_i : x_{i\ell} \xrightarrow{i} x_{ik}$ denotes the shortest path in $\hat{\mathbf{D}}_i = (\mathtt{X}_i,\, \mathtt{E}_i)$ from $x_{i\ell} \in \mathtt{X}_i$ to $x_{ik} \in \mathtt{X}_i$, then $\omega\{\mathbf{P}_i\} = (k - \ell - 1)\nu_i$, where $\omega\{\mathbf{P}_i\}$ denotes the weight of $\mathbf{P}_i$.

Proof. The proof follows from the definition of $\omega\{\mathbf{P}_i\}$ and the fact that $\mathbf{P}_i$ is given as

$$\mathbf{P}_i\colon \begin{cases} x_{i\ell} \rightarrow x_{i,\ell-1} \rightarrow \ \ldots \ \rightarrow x_{ik}\ , & \ell > k, \\ \\ x_{i\ell} \rightarrow x_{in_i} \rightarrow x_{i,n_i-1} \rightarrow \ \ldots \ \rightarrow x_{ik}\ , & \ell < k, \end{cases} \tag{2.190}$$

where $\rightarrow$ denotes a single edge in $\mathtt{E}_i$. Q.E.D.

Finally, we arrive at:

2.29. THEOREM. Let $\hat{\mathcal{C}}$ denote the class of all simple cycles $\mathbf{C}$ in $\hat{\mathbf{D}}$ having the properties:

(*i*) $\mathbf{C}$ contains at least one edge in E_C;

(*ii*) for any X_i, $\mathbf{C}$ contains at most one edge in E_C which emerges from X_i; and

(*iii*) if a path $\mathbf{P}_i : x_{i\ell} \xrightarrow{i} x_{ik}$ between two vertices $x_{i\ell}$, $x_{ik} \in \mathrm{X}_i$ is a part of $\mathbf{C}$, then $\mathbf{P}_i$ is the shortest path.

Then, the condition (2.172) of Theorem 2.26 is satisfied if and only if there exists no cycle $\mathbf{C} \in \hat{\mathcal{C}}$ having a nonnegative net-weight.

Proof. By Lemma 2.27, it suffices to show that

$$\lim_{\rho \to +\infty} \prod_{\substack{i \in \boldsymbol{I}_r \\ j \in \boldsymbol{J}_r}} \left(\rho_j^{q-1} / \rho_i^{p} \right) = 0, \tag{2.191}$$

for all r, $\boldsymbol{I}_r$, $\boldsymbol{J}_r$, and all $(p, q) = (p_\ell^{ij}, q_\ell^{ij})$, $\ell = 1, 2, \ldots, k_{ij}$, if and only if there exist no cycles $\mathbf{C} \in \hat{\mathcal{C}}$ in $\hat{\mathbf{D}}$ with $\omega\{\mathbf{C}\} \geq 0$.

To prove necessity, consider a typical cycle $\mathbf{C} \in \hat{\mathcal{C}}$ in $\hat{\mathbf{D}}$, which has the following structure:

$$\mathbf{C}\colon\ x_{i_1 g_1} \to x_{i_2 h_2} \xrightarrow{i_2} x_{i_2 g_2} \to x_{i_3 h_3} \xrightarrow{i_3} \cdots \xrightarrow{i_r} x_{i_r g_r} \to x_{i_1} \xrightarrow{i_1} x_{i_1 g_1}. \tag{2.192}$$

Using Lemma 2.28, the weight of $\mathbf{C}$ can be computed as

$$\omega\{\mathbf{C}\} = \sum_{\ell=1}^{r} (g_\ell - h_\ell - 1)\nu_{i_\ell}. \tag{2.193}$$

Considering

$$\frac{\rho_{i_r}^{g_r - 1}}{\rho_{i_1}^{h_1}} \frac{\rho_{i_1}^{g_1 - 1}}{\rho_{i_2}^{h_2}} \cdots \frac{\rho_{i_{r-1}}^{g_{r-1} - 1}}{\rho_{i_r}^{h_r}} = \rho\omega\{\mathbf{C}\}, \tag{2.194}$$

where the left hand side has the form of the expression in (2.191) with

$$\boldsymbol{I}_r = \{i_1, i_2, \ldots, i_r\}, \qquad \boldsymbol{J}_r = \{j_1, j_2, \ldots, j_{r-1}\}, \tag{2.195}$$

it follows that if $\omega\{\mathbf{C}\} \geq 0$, then (2.191) fails to hold, completing the proof of necessity.

Conversely, if (2.191) fails to hold for some r, $\boldsymbol{I}_r$, $\boldsymbol{J}_r$, and some of $(p, q) = (p_\ell^{ij}, q_\ell^{ij})$, then we can construct a cycle $\mathbf{C}$ in $\hat{\mathbf{D}}$ such that:

(a) $\mathbf{C}$ has the form of (2.192), that is, it contains at least one edge in $\mathbf{E}_C$, and it follows the shortest paths in $\hat{\mathbf{D}}_i$, $i \in \boldsymbol{I}_r$; and

(b) $\omega\{\mathbf{C}\} \geq 0$.

If $\mathbf{C} \in \hat{\mathcal{C}}$, then the proof follows. If, on the other hand, $\mathbf{C} \notin \hat{\mathcal{C}}$ for it contains two or more edges in $\mathbf{E}_C$, which emerge from the same x_i, $i \in \boldsymbol{I}_r$, then breaking $\mathbf{C}$ into two pieces and adding to each piece some edges in $\mathbf{E}_i$, we can construct two cycles $\mathbf{C}_1$ and $\mathbf{C}_2$, at least one of which satisfies the properties (a) and (b) above. Continuing this procedure (if necessary) we finally obtain a cycle $\bar{\mathbf{C}} \in \hat{\mathcal{C}}$ with $\omega\{\bar{\mathbf{C}}\} \geq 0$. This completes the proof of the sufficiency part and, thus, the proof of the theorem. Q.E.D.

2.30. REMARK. We note that conditions (*ii*) and (*iii*) of Theorem 2.29 are not essential, but are useful in minimizing the number of cycles involved in testing condition (2.172). In fact, it is possible to construct the subgraphs $\hat{\mathbf{D}}_i$ associated with $\hat{\mathbf{S}}_i$ in such a way that there is only one path $\mathbf{P}_i$ between any two vertices $x_{i\ell}$, $x_{ik} \in \mathbf{X}_i$, so that any cycle $\mathbf{C}$ satisfying conditions (*i*) and (*ii*) also satisfies (*iii*). Moreover, this reconstruction of the system digraph can be done without effecting the number and weights of the cycles belonging to the class $\hat{\mathcal{C}}$.

2.31. REMARK. For any cycle $\mathbf{C} \in \hat{\mathcal{C}}$, the weight of $\mathbf{C}$ is a linear expression in ν_i, $i \in \boldsymbol{N}$. Therefore, the condition of Theorem 2.29 can be reformulated as a set of linear inequalities

$$\begin{bmatrix} M \\ -I_N \end{bmatrix} \nu < 0, \tag{2.196}$$

where $\nu = (\nu_1, \nu_2, \ldots, \nu_N)$. The matrix M has the elements that are the coefficients of ν_i appearing in the weights of the cycles, with each row corresponding to a cycle in $\hat{\mathcal{C}}$. In (2.196), the $N \times N$ identity matrix I_N is included to ensure positivity of each ν_i. Therefore, stabilizability of $\mathbf{S}$ is reduced to solving linear inequalities (2.196), which can be done using standard techniques (*e.g.*, Zhukhovitskii and Avdeyeva, 1966).

2.32. REMARK. In determining the cycles of $\hat{\mathbf{D}}$, the existing algorithms (Deo, 1974) can be modified to account for only those cycles in $\hat{\mathcal{C}}$ which are required by Theorem 2.29. Certain simplifications can precede the accounting of the cycles in $\hat{\mathcal{C}}$. First, we should eliminate the cycles arising from the self-interconnections, that is, those that are constituted of the edges of $\mathtt{E}_C$ which have the form $x_{ip} \to x_{iq}$. If there exists such an edge with $p < q$, then this edge forms a cycle with a nonnegative weight $(q-p-1)\nu_i$ and (2.196) is violated. On the other hand, if $p \geq q$, the edge can be eliminated without affecting the condition (2.196). Secondly, a decomposition of $\hat{\mathbf{D}}$ into strong components (see Section 1.3) can be used to eliminate all edges between the strong components, thus reducing the accounting of cycles to the level of components. In this way, (2.196) becomes a set of independent inequalities corresponding to the strong components of $\hat{\mathbf{D}}$.

2.33. EXAMPLE. To illustrate Theorem 2.29 let us show that if $A_C = (A_{ij})$ has the structure as shown in Figure 2.4, then the system $\mathbf{S}$ is stabilizable by decentralized feedback. Choose $\nu_N = 1$, and $\nu_{N-1}, \nu_{N-2}, \ldots, \nu_1$ recursively so that

$$\nu_i > \sum_{j=i+1}^{N} (n_j - 2)\nu_j, \qquad i = N-1,\, N-2,\, \ldots,\, 1. \tag{2.197}$$

It is easy to see that the digraph $\hat{\mathbf{D}} = (\mathtt{X},\ \mathtt{E}_D \cup \mathtt{E}_C)$ corresponding to the interconnection matrix A_C in Figure 2.4 has the structure shown in Figure 2.5. Now consider a cycle $\mathbf{C} \in \hat{\mathcal{C}}$ that passes through the subgraphs $\hat{\mathbf{D}}_{i_1}, \hat{\mathbf{D}}_{i_2}, \ldots, \hat{\mathbf{D}}_{i_r}$ in the given order. Letting $i_{r+1} = i_1$ and $i_0 = i_r$ for convenience, the index set $I_r = \{i_1, i_2, \ldots, i_r\}$ can be partitioned into four disjoint subsets:

$$\begin{aligned}
I_{r1} &= \{i_\ell \in I_r\colon\ i_\ell < i_{\ell-1},\, i_{\ell+1}\}, \\
I_{r2} &= \{i_\ell \in I_r\colon\ i_\ell > i_{\ell-1},\, i_{\ell+1}\}, \\
I_{r3} &= \{i_\ell \in I_r\colon\ i_{\ell-1} < i_\ell < i_{\ell+1}\}, \\
I_{r4} &= \{i_\ell \in I_r\colon\ i_{\ell-1} > i_\ell > i_{\ell+1}\}.
\end{aligned} \tag{2.198}$$

A typical cycle and the corresponding partitioning of the index set I_r is illustrated in Figure 2.6. Referring to Figure 2.6 and using Lemma 2.27 the weight of $\mathbf{C}$ can be computed from (2.198) as

$$\omega\{\mathbf{C}\} = \sum_{i\in I_{r1}} (-\nu_i) + \sum_{i\in I_{r2}} (n_i - 2)\nu_i + \sum_{i\in I_{r3}} (\ell_i - 2)\nu_i + \sum_{i\in I_{r4}} (n_i - \ell_i - 2)\nu_i. \tag{2.199}$$

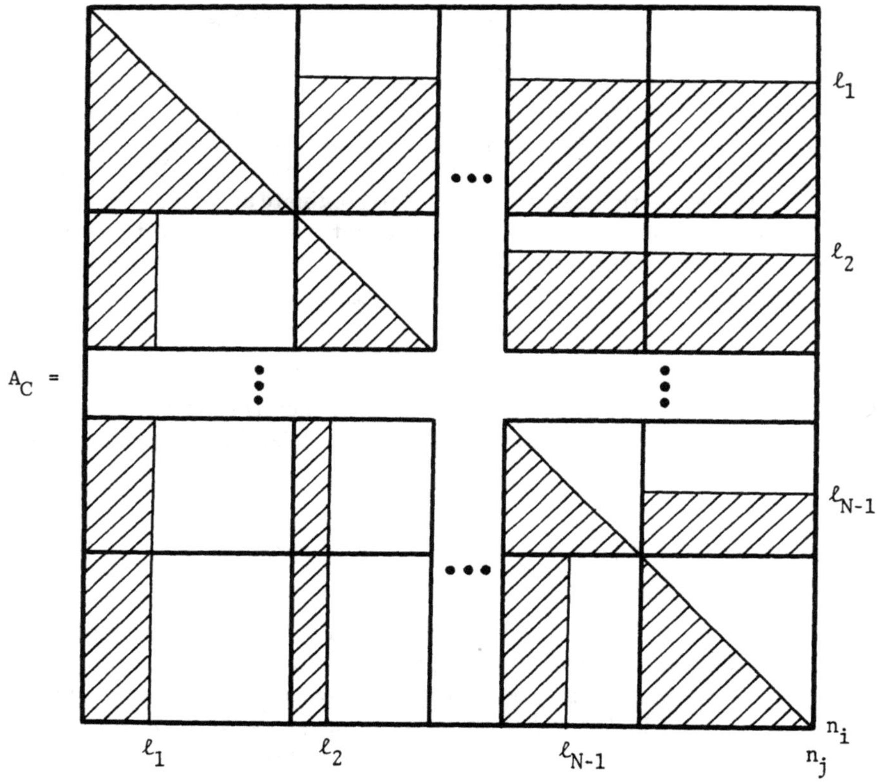

Fig. 2.4. Interconnection matrix.

We also note that if $\min\{i_\ell \in \boldsymbol{I}_r\} = j$, then $j \in \boldsymbol{I}_{r1}$. Since $1 \leq \ell_i \leq n_i$, $i \in \boldsymbol{N}$, (2.199) implies that

$$\omega\{\mathbf{C}\} \leq -\nu_j + \sum_{i \in I_{r2} \cup I_{r3} \cup I_{r4}} (n_i - 2)\nu_i. \tag{2.200}$$

The choice of ν_i, $i \in \boldsymbol{N}$, in (2.197), guarantees that $\omega\{\mathbf{C}\} < 0$, and stabilizability of $\mathbf{S}$ follows from Theorem 2.29.

A logical extension of Theorem 2.29 is to include the output instead of

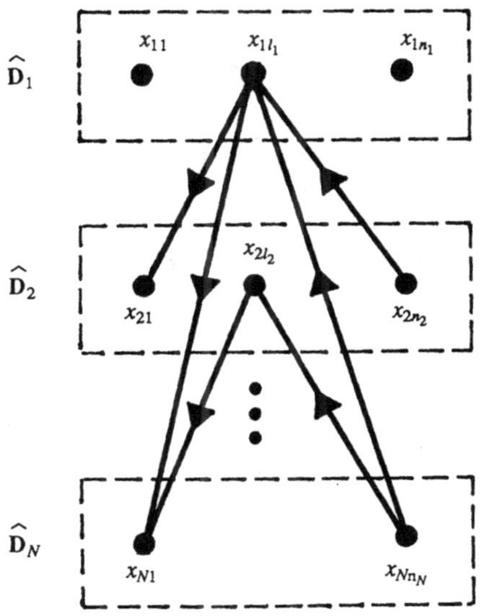

Fig. 2.5. Interconnection structure.

state feedback. This problem is considered in Section 5.4 using the concept of almost invariant subspaces.

2.6. Notes and References

Liapunov's Direct Method is essentially an aggregation process whereby a system of equations involving several state variables is represented by a single scalar function, which contains pertinent information about stability of the entire system. When a system has a large number of variables, the aggregation process based upon a single function may get bogged down in a welter of details with increasing liability in the end result. A crucial contribution to Liapunov's method in the context of complex systems has been made by Matrosov (1962) and Bellman (1962), when they simultaneously, but independently, introduced the concept of *vector Liapunov functions*. By a suitable construction, Bailey (1966) showed how the concept applies to aggregation of interconnected dynamic systems and, thereby, launched the new version of the method in the uncharted area of large-scale dynamic systems.

After a relatively slow start, the vector concept gained momentum and is now a rather well established area of research. A wide variety of models and equally diversified kinds of stability have been considered. The ob-

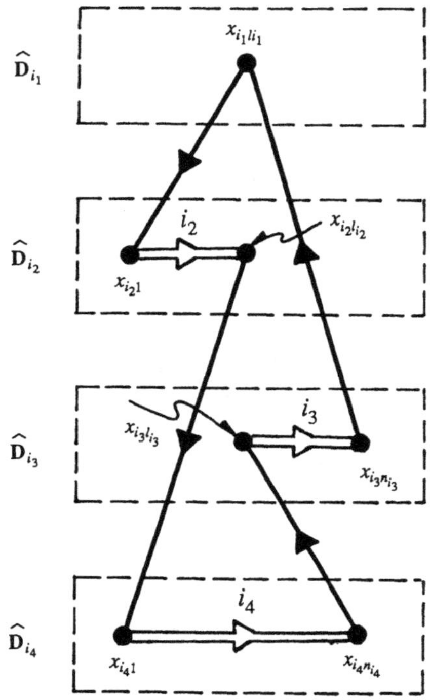

Fig. 2.6. A typical cycle.

tained results have been reported in several recent surveys (Voronov, 1982; Michel, 1983; Šiljak, 1983; Matrosov, 1984; Vidyasagar, 1986; Voronov and Matrosov, 1987) and books (Lakshmikantham and Leela, 1969; Martynyuk, 1975; LaSalle, 1976; Rouche *et al.* 1977; Michel and Miller, 1977; Šiljak, 1978; Matrosov *et al.* 1980; Matrosov and Anapolskii, 1980, 1981; Matrosov, 1981; Jamshidi, 1983; Matrosov and Vasilev, 1984; Voronov, 1985; Matrosov and Malikov, 1986; Grujić *et al.* 1987). We do not plan to review this massive amount of information, but rather concentrate on the areas that are directly related to our interest in robust (connective) stability of decentralized control schemes.

CONNECTIVE STABILITY

To capture the effect on stability of the essential uncertainty residing in interconnections of complex systems, the concept of connective stability was introduced (Šiljak, 1972) as a Liapunov stability under structural (parametric) perturbations. The concept has been generalized considerably over the years (see the surveys of Šiljak, 1978, 1989) to accommodate a wide variety of stability definitions and mathematical models of dynamic systems. Connective stability has been redefined to include input–output models (Moylan, 1980; Vidyasagar, 1981, 1986), time-delay and hereditary systems (Ladde, 1976a, 1977; Anderson, 1979; Lewis and Anderson, 1980; Sinha, 1980, 1981; Shigui, 1988), stochastic models (Ladde and Šiljak, 1976a, b; Šiljak, 1978), the variation of constants method (Aftabizadeh, 1980), evolution systems (Spiteri, 1986), as well as incompletely known systems, simplified models, and symmetric strongly coupled systems (Lunze, 1983, 1984, 1986), bilinear systems (Loparo and Hsu, 1982), and systems under periodic structural perturbations (Burgat and Bernussou, 1978).

Most of the results obtained for connective stability of continuous systems can be carried over to *discrete systems* (Šiljak, 1978). A version of Problem 2.11 has been solved recently (Sezer and Šiljak, 1988) to generate vector Liapunov functions which provide the best estimates of robustness bounds for discrete systems under structural perturbations.

Of our special interest are the *nested structural perturbations* (Section 7.4), because of their relevance to decompositions of complex systems into weakly coupled subsystems. A prime candidate for new research in this direction is the *block diagonal dominance concept* (Ohta and Šiljak, 1985), which is an attractive alternative to vector Liapunov functions in stability analysis of linear systems (Section 7.5).

Applications of connective stability appear in as diverse fields as model ecosystems and arms race, competitive economic analysis, and electric power systems (see Šiljak, 1978). The most extensive reference to the concept has been in biological modeling, as cited in the books by May (1974), Casti (1979), Goh (1980), Svirezhev and Logofet (1983), and Boucher (1985), as well as the papers by Ladde (1976b, c, d), Andretti (1978), Mazanov (1978), Ikeda and Šiljak (1980d, 1982), Olesky and van den Driessche (1982), Solimano and Beretta (1983), Shigesada *et al.* (1984), Spouge (1986), and Casti (1987). Specially interesting are the results in the context of complexity *vs.* stability problem (see Šiljak, 1987a, and references therein).

New avenues for future research on stability under structural perturbations have been opened in a number of areas by new results in Liapunov's stability theory and complex systems. Some of these results are reviewed here. Others can be found in the books and survey papers mentioned above, as well as in the rest of this book (see, in particular, Chapters 7 and 9).

An interesting way to reduce the conservativeness of vector Liapunov functions is to use *cone-valued functions* (Lakshmikantham and Leela, 1977) relative to a cone other than $\mathbf{R}^n_+$. The central object in this context is the *comparison system*, which is considered in the framework of large-scale systems by Bitsoris (1983), Matrosov (1984), and Martynyuk and Krapivnii (1987). An interesting new concept of *majorant Liapunov equations* has been introduced recently by Hyland and Bernstein (1987), which offers a new way to improve flexibility of Liapunov's theory for complex systems.

Matrix Liapunov functions have been introduced by Martynyuk (1975), which offer still another generalization of Liapunov's theory. The concept has been developed in a number of papers to include unstable subsystems (Djordjević, 1983a, b), the Principle of Invariance (Martynyuk, 1984), hybrid systems (Martynyuk, 1985), stochastic systems (Martynyuk, 1987a), and absolute stability (Martynyuk, 1987b). We note that matrix Liapunov functions have not reached the maturity of its vector counterpart, especially in the control applications.

Other relevant contributions to stability theory of complex systems have been made in the areas of partial stability (Bondi *et al.* 1979), absolute stability (Kaszkurewicz and Resende, 1983), limiting equations (Bondi *et al.* 1982), stability regions of multimachine power systems (Saeki *et al.* 1980; Araki *et al.* 1982; Grujić *et al.* 1987; Tsai, 1987), as well as stochastic systems (Socha, 1986; Socha and Popp, 1986; Popp *et al.* 1988; Jumarie, 1988), reaction–diffusion systems (Lakshmikantham and Leela, 1988), and neural networks (Michel and Farell, 1990).

DECENTRALIZED STABILIZATION

Since the size and complexity have been traditionally standard features of mathematical models of economic systems, the fact that a decentralization of control and estimation brings about a considerable reduction of information processing requirements has been long recognized by economists (*e.g.*, Marschak and Radner, 1971; Arrow and Hurwicz, 1977). A similar situation has been present in control of interconnected power areas (*e.g.*, Ćalović, 1986), where each area is controlled locally by variables available in each individual area. Other applications of decentralized control have been in areas of process control, large space structures and traffic control to mention a few, where the modern distributed computer architectures have proven to be effective.

Motivated by the development of decentralized control technology, a considerable number of theoretical results on control with nonclassical (decentralized) information structure constraints were reported in an early survey (Sandell *et al.* 1978). Of special interest has been the problem of decentralized control for interconnected systems, which can tolerate a wide range of uncertainties in the system structure and subsystems dynamics.

Robust control strategies have been invented (Šiljak, 1978, 1989) for systems under structural perturbations, and they are the principal subject of this book.

Stabilization of systems under structural perturbations (*i.e.*, connective stabilization) *via* decentralized feedback has been reported in a number of books (Šiljak, 1978; Singh and Titli, 1979; Singh, 1981; Bernussou and Titli, 1982; Vukobratović and Stokić, 1982; Jamshidi, 1983; Mahmoud *et al.* 1985; Voronov, 1985; Leondes, 1985, 1986), as well as survey papers (Schmidt, 1982; Šiljak, 1983, 1987b; Singh *et al.* 1984). Game-theoretic methods were used by Mageriou and Ho (1977). Stabilization of large systems by variable structure controllers was considered by Khurana *et al.* (1986). Connective stabilizability of systems with time delays was addressed by Ikeda and Šiljak (1980c) and Bourlès (1987).

New decentralized schemes have been developed recently in control of *large space structures* (Kida *et al.* 1984, 1985; Hyland and Collins, 1988), *flight control systems* (Vukobratović and Stojić, 1986, 1988), and *robotics* (Vukobratović and Stokić, 1982; Stokić and Vukobratović, 1984; Vukobratović and Kirćanski, 1985; Mills and Goldenberg, 1987; Gavel, 1988).

UNCERTAIN SYSTEMS

Since robustness is of a primary concern to us, we should pay special attention to the area of uncertain systems. The underlying idea is to design feedback in such a way that stability of the closed-loop system can tolerate uncertain bounded perturbations. A vast literature exists on this subject (for a comprehensive recent survey, see the paper by Corless and Leitmann, 1988).

We recall that Liapunov, in his doctoral dissertation, studied robustness of stability to additive perturbations. Once we realize that interconnections are additive (uncertain) perturbations of the subsystems dynamics, the concept of connective stability appears as a natural outgrowth of Liapunov's original analysis, which is suitable for robust stability studies of complex systems. Then, a decentralized feedback can be conceived as means to build connectively stable systems, that is, systems stable under structural perturbations.

There is an obvious structural similarity between uncertain and interconnected systems (see Šiljak, 1989), which can be exploited to get new results in both fields. By specializing interconnected systems to a single subsystem and treating a self-interconnection as a perturbation, we can reproduce a number of basic results obtained in uncertain systems using the material of this chapter, sometimes with greater generality—especially when "nonmatching conditions" of Sections 2.4 and 2.5 take place. This is not to say that the results on uncertain systems cannot be used to produce improved results in decentralized stabilization; in particular, when we try to reduce the conservativeness of decentralized control strategies (see, for example, Chen, 1987, 1989; Yong, 1988).

Bibliography

Aftabizadeh, A. R. (1980). Variation of constants, vector Lyapunov functions, and comparison theorem. *Applied Mathematics and Computations*, 7, 341–352.

Amemiya, T. (1986). Stability analysis of nonlinearily interconnected systems—application of M-functions. *Journal of Mathematical Analysis and Applications*, 114, 252–277.

Anderson, B. D. O. (1979). Time delays in large-scale systems. *Proceedings of the 24th Conference on Decision and Control*, Fort Lauderdale, Florida, 655–660.

Andretti, F. (1978). Interactions among biological systems: An analysis of asymptotic stability. *Bulletin of Mathematical Biology*, 40, 839–851.

Araki, M. (1975). Application of M-matrices to the stability problems of composite dynamical systems. *Journal of Mathematical Analysis and Applications*, 52, 309–321.

Araki, M. (1978). Stability of large-scale nonlinear systems: Quadratic-order theory of composite-system method. *IEEE Transactions*, AC-23, 129–142.

Araki, M., M. M. Metwally, and D. D. Šiljak (1982). Generalized decompositions for transient stability analysis of multimachine power systems. *Large Scale Systems*, 3, 111–122.

Arrow, K. J., and L. Hurwicz (1977). *Studies in Resource Allocation Processes*. Cambridge University Press, London, UK.

Bailey, F. N. (1966). The application of Lyapunov's second method to interconnected systems. *SIAM Journal of Control*, 3, 443–462.

Bellman, R. (1962). Vector Lyapunov functions. *SIAM Journal of Control*, 1, 32–34.

Bernussou, J., and A. Titli (1982). *Interconnected Dynamical Systems: Stability, Decomposition and Decentralization*. North-Holland, Amsterdam, The Netherlands.

Bitsoris, G. (1983). Stability analysis of non-linear dynamical systems. *International Journal of Control*, 38, 699–711.

Bondi, P., P. Fergola, L. Gambardella, and C. Tenneriello (1979). Partial stability of large-scale systems. *IEEE Transactions*, AC-24, 94–97.

Bondi, P., P. Fergola, L. Gambardella, and C. Tenneriello (1982). Stability of large scale systems *via* limiting equations. *Large Scale Systems*, 3, 27–33.

Boucher, D. H. (ed.) (1985). *The Biology of Mutualism*. Oxford University Press, New York.

Bourlés, H. (1987). α-Stability and robustness of large-scale interconnected systems. *International Journal of Control*, 45, 2221–2232.

Burgat, C., and J. Bernussou (1978). On the stability of interconnected systems with periodic structural perturbations. *International Journal of Systems Science*, 9, 1133–1143.

Ćalović M. S. (1986). Recent developments in decentralized control of generation and power flows. *Proceedings of the 25th Conference on Decision and Control*, Athens, Greece, 1192–1197.

Casti, J. (1979). *Connectivity, Complexity, and Catastrophe in Large-Scale Systems.* Wiley, New York.

Casti, J. (1987). Connective stability. *Systems and Control Encyclopedia*, M. G. Singh (ed.), Pergamon Press, Oxford, UK, 776–777.

Chen, Y. H. (1987). Deterministic control of large-scale uncertain dynamical systems. *Journal of the Franklin Institute*, 323, 135–144.

Chen, Y. H. (1989). Large-scale uncertain systems under insufficient decentralized controllers. *Journal of Dynamic Systems, Measurement, and Control*, 111, 359–363.

Cheres, E., Z. J. Palmor, and S. Gutman (1989). Quantitative measures of robustness for systems including delayed perturbations. *IEEE Transactions*, AC-34, 1203–1204.

Corless, M., and G. Leitmann (1988). Controller design for uncertain systems *via* Lyapunov functions. *Proceedings of the American Control Conference*, Atlanta, Georgia, 2019–2025.

Datta, B. N. (1977). Matrices satisfying Šiljak's conjecture. *IEEE Transactions*, AC-22, 132–133.

Davison, E. J. (1974). The decentralized stabilization and control of a class of unknown nonlinear time-varying systems. *Automatica*, 10, 309–316.

Deo, N. (1974). *Graph Theory with Applications to Engineering and Computer Science.* Prentice-Hall, Englewood Cliffs, New Jersey.

Djordjević, M. (1983a). Stability analysis of interconnected systems with possibly unstable subsystems. *Systems and Control Letters*, 3, 165–169.

Djordjević, M. (1983b). Stability analysis of large scale systems whose subsystems may be unstable. *Large Scale Systems*, 5, 255–262.

Fiedler, M., and V. Pták (1962). On matrices with nonpositive off-diagonal elements and principal minors. *Czechoslovakian Mathematical Journal*, 12, 382–400.

Gavel, D. T., Jr. (1988). *Decentralized Adaptive Control of Large Scale Systems, with Application to Robotics.* Report No. UCRL-53866, Lawrence Livermore National Laboratory, Livermore, California.

Goh, B. S. (1980). *Management and Analysis of Biological Populations.* Elsevier, Amsterdam, The Netherlands.

Grujić, Lj. T., A. A. Martynyuk, and M. Ribbens–Pavella (1987). *Large Scale Systems Stability under Structural and Singular Perturbations.* Springer, New York.

Hahn, W. (1967). *Stability of Motion.* Springer, New York.

Hyland, D. C., and D. S. Bernstein (1987). The majorant Lyapunov equation: A nonnegative matrix equation for robust stability and performance of large scale systems. *IEEE Transactions*, AC-32, 1005–1013.

Hyland, D. C., and E. G. Collins, Jr. (1988). A robust control experiment using an optical structure prototype. *Proceedings of the Automatic Control Conference*, Atlanta, Georgia, 2046–2049.

Ikeda, M., and D. D. Šiljak (1980a). On decentrally stabilizable large-scale systems. *Automatica*, 16, 331–334.

Ikeda, M., and D. D. Šiljak (1980b). Decentralized stabilization of linear time-varying systems. *IEEE Transactions*, AC-25, 106–107.

Ikeda, M., and D. D. Šiljak (1980c). Decentralized stabilization of large scale systems with time delay. *Large Scale Systems*, 1, 273–279.

Ikeda, M., and D. D. Šiljak (1980d). Lotka–Volterra equations: Decompositions, stability, and structure, Part I: Equilibrium analysis. *Journal of Mathematical Biology*, 9, 65–83.

Ikeda, M., and D. D. Šiljak (1982). Lotka–Volterra equations: Decomposition, stability, and structure. Part II. Nonequilibrium analysis. *Nonlinear Analysis, Theory, Methods and Applications*, 6, 487–501.

Ikeda, M., O. Umefuji, and S. Kodama (1976). Stabilization of large-scale linear systems. *Transactions of IECE Japan*, 59-D, 355–362 (also in *Systems, Computers, and Control*, 7, 34–41).

Jamshidi, M. (1983). *Large-Scale Systems: Modeling and Control.* North-Holland, New York.

Jumarie, G. (1988). Simple general method to analyse the moment stability and sensitivity of nonlinear stochastic systems with or without delay. *Journal of Systems Science*, 19, 111–124.

Kailath, T. (1980). *Linear Systems.* Prentice-Hall, Englewood Cliffs, New Jersey.

Kalman, R. E., and J. E. Bertram (1960). Control system analysis and design *via* the "second method" of Lyapunov. Part I: Continuous-time systems. *Transactions of ASME*, Series D, 82, 371–393.

Kaszkurewicz, E., and P. Resende (1983). On the absolute stability of linearly interconnected nonlinear systems. *Proceedings of the IFAC Symposium on Large Scale Systems*, Pergamon Press, Oxford, UK, 309–314.

Khurana, H., S. I. Ahson, and S. S. Lamba (1986). Stabilization of large-scale control systems using variable structure systems theory. *IEEE Transactions*, AC-31, 176–178.

Kida, T., I. Yamaguchi, and Y. Ohkami (1984). Local output-feedback control for large flexible space structures. *Proceedings of the 14th International Symposium on Space Technology and Science*, Tokyo, Japan, 969–974.

Kida, T., I. Yamaguchi, and Y. Ohkami (1985). An approach to local decentralized control for large flexible space structures. *AIAA Guidance, Navigation and Control Conference*, Paper No. 85-1924-C, Snowmass, Colorado.

Ladde, G. S. (1976a). System of functional differential inequalities and functional differential systems. *Pacific Journal of Mathematics*, 66, 161–171.

Ladde, G. S. (1976b). Cellular systems–I. Stability of chemical systems. *Mathematical Biosciences*, 29, 309–330.

Ladde, G. S. (1976c). Cellular systems–II. Stability of compartmental systems. *Mathematical Biosciences*, 30, 1–21.

Ladde, G. S. (1976d). Stability of model ecosystems with time-delay. *Journal of Theoretical Biology*, 61, 1–13.

Ladde, G. S. (1977). Competitive processes–I. Stability of hereditary systems. *Nonlinear Analysis, Theory, Methods and Applications*, 6, 607–631.

Ladde, G. S., and D. D. Šiljak (1976a). Stability of multispecies communities in randomly varying environment. *Journal of Mathematical Biology*, 2, 165–178.

Ladde, G. S., and D. D. Šiljak (1976b). Connective stability of large-scale stochastic systems. *International Journal of Systems Science*, 6, 713–721.

Lakshmikantham, V. (1981). Review of "Large-Scale Dynamic Systems: Stability and Structure" by D. D. Šiljak. *IEEE Transactions*, AC-26, 976–977.

Lakshmikantham, V., and S. Leela (1969). *Differential and Integral Inequalities.* Vols. I and II. Academic Press, New York.

Lakshmikantham, V., and S. Leela (1977). Cone-valued Lyapunov functions. *Nonlinear Analysis, Theory, Methods and Applications*, 1, 215–222.

Lakshmikantham, V., and S. Leela (1988). Reaction–diffusion systems and vector Lyapunov functions. *Differential and Integral Equations*, 1, 41–47.

LaSalle, J. P. (1976). *The Stability of Dynamical Systems.* SIAM, Philadelphia, Pennsylvania.

Leela, S. (1988). Method of vector Lyapunov functions for reaction–diffusion systems. *Proceedings of the International Conference on Theory and Application of Differential Equations*, Ohio University Press, Columbus, Ohio, 120–124.

Leondes, C. T. (ed.) (1985). *Decentralized-Distributed Control and Dynamic Systems.* Part I. Academic Press, New York.

Leondes, C. T. (ed.) (1986). *Decentralized-Distributed Control and Dynamic Systems.* Part II. Academic Press, New York.

Lewis, R. M., and B. D. O. Anderson (1980). Necessary and sufficient conditions for delay-independent stability of linear autonomous systems. *IEEE Transactions*, AC-25, 735–739.

Loparo, K. A., and C. S. Hsu (1982). Analysis of interconnected systems using bilinear models. *Large Scale Systems*, 3, 237–244.

Luenberger, D. G. (1967). Canonical forms for linear multivariable systems. *IEEE Transactions*, AC-12, 290–293.

Lunze, J. (1983). A majorization approach to the quantitative analysis of incompletely known large scale systems. *Zeitschrift für elektrische Infomations und Energietechnik*, 13, 99–117.

Lunze, J. (1984). Stability analysis of large scale interconnected systems by means of simplified models. *System Analysis, Modeling, and Simulation*, 1, 381–398.

Lunze, J. (1986). Dynamics of strongly coupled symmetric composite systems. *International Journal of Control*, 44, 1617–1640.

Mageriou, E. F., and Y. C. Ho (1977). Decentralized stabilization *via* game theoretic methods. *Automatica*, 13, 393–399.

Mahmoud, M. S., M. F. Hassan, and M. G. Darwish (1985). *Large-Scale Control Systems: Theories and Techniques.* Marcel Dekker, New York.

Marschak, J., and R. Radner (1971). *The Economic Theory of Teams.* Yale University Press, New Haven, Connecticut.

Martynyuk, A. A. (1975). *Stability of Motion of Complex Systems* (in Russian). Naukova Dumka, Kiev, USSR.

Martynyuk, A. A. (1984). The Lyapunov matrix functions. *Nonlinear Analysis, Theory, Methods and Applications*, 8, 1223–1226.

Martynyuk, A. A. (1985). Matrix Liapunov functions and stability of hybrid systems. *Prikladnaia Mekhanika*, 21, 89–96.

Martynyuk, A. A. (1987a). Stochastic matrix-valued Lyapunov function and its application. *Stochastic Analysis and Application.* 5, 395–404.

Martynyuk, A. A. (1987b). Absolute stability of singularily-perturbed Lur'e systems and matrix Liapunov functions. *Prikladnaia Mekhanika*, 23, 103–110.

Martynyuk, A. A., and Yu. N. Krapivnii (1987). Nonlinear comparison systems and stability of large-scale systems. *Prikladnaia Mekhanika*, 23, 75–80.

Matrosov, V. M. (1962). On the theory of stability of motion. *Prikladnaia Matematika i Mekhanika*, 26, 992–1000.

Matrosov, V. M. (ed.) (1981). *Algorithms for Theorem Generation in the Method of Vector Liapunov Functions* (in Russian). Nauka, Novosibirsk, USSR.

Matrosov, V. M. (1984). Vector Lyapunov functions method in the analysis of dynamical properties of nonlinear differential equations. *Trends in Theory and Practice of Nonlinear Differential Equations*, V. Lakshmikantham (ed.), Marcel Dekker, New York, 357–374.

Matrosov, V. M., and L. Yu. Anapolskii (eds.) (1980). *Vector-Functions of Liapunov and their Construction* (in Russian). Nauka, Novsibirsk, USSR.

Matrosov, V. M., and L. Yu. Anapolskii (eds.) (1981). *Direct Method in Stability Theory and Its Applications* (in Russian). Nauka, Novosibirsk, USSR.

Matrosov, V. M., and Yu. E. Boiarintsev (eds.) (1986). *Differential Equations and Numerical Methods* (in Russian). Nauka, Novosibirsk, USSR.

Matrosov, V. M., and A. I. Malikov (eds.) (1986). *Liapunov Functions and Their Applications* (in Russian). Nauka, Novosibirsk, USSR.

Matrosov, V. M., and S. N. Vasilev (eds.) (1984). *Method of Liapunov Functions and Its Applications* (in Russian). Nauka, Novosibirsk, USSR.

Matrosov, V. M., L. Yu. Anapolskii, and S. N. Vasilev (1980). *Comparison Method in Mathematical Systems Theory* (in Russian). Nauka, Novosibirsk, USSR.

May, R. M. (1974). *Stability and Complexity in Model Ecosystems.* 2nd ed., Princeton University Press, Princeton, New Jersey.

Mazanov, A. (1978). Acceptor control in model eco-systems. *Journal of Theoretical Biology*, 71, 21–38.

Michel, A. N. (1983). On the status of stability of interconnected systems. *IEEE Transactions*, CAS-30, 326–340.

Michel, A. N., and J. A. Farell (1990). Associative memories via artificial neural networks. *IEEE Control Systems Magazine*, 10, 6–17.

Michel, A. N., and R. K. Miller (1977). *Qualitative Analysis of Large Scale Dynamical Systems.* Academic Press, New York.

Mills, J. K., and A. A. Goldenberg (1987). Global connective stability of a class of robotic manipulators. *Automatica*, 24, 835–839.

Montemayor, J. J., and B. F. Womack (1975). On a conjecture by Šiljak. *IEEE Transactions*, AC-20, 572–573.

Montemayor, J. J., and B. F. Womack (1976). More on a conjecture by Šiljak. *IEEE Transactions*, AC-21, 805–806.

Mori, T., and M. Kuwahara (1981). Comments on Šiljak's conjecture. *IEEE Transactions*, AC-26, 609–611.

Moylan, P. (1980). A connective stability result for interconnected passive systems. *IEEE Transactions*, AC-25, 812–813.

Ohta, Y., and D. D. Šiljak (1985). Overlapping block diagonal dominance and existence of Liapunov functions. *Journal of Mathematical Analysis and Applications*, 112, 396–410.

Olesky, D. D., and P. van den Driessche (1982). Monotone positive stable matrices. *Linear Algebra and Its Applications*, 48, 381–401.

Ostrowski, A. (1937). Über die Derminanten mit überwiegender Hauptdiagonale. *Commentarii Mathematici Helvetici*, 10, 69–76.

Patel, R. V., and M. Toda (1980). Quantitative measures of robustness for multivariable systems. *Proceedings of the JACC*, Paper No. TP8-A, San Francisco, California.

Popp, K., L. Socha, and H. Windrich (1988). Moment stability of linear stochastic nontriangular large-scale systems. *International Journal of Systems Science*, 19, 733–746.

Richter, S., and R. DeCarlo (1984). A homotopy method for eigenvalue assignment using decentralized state feedback. *IEEE Transactions*, AC-29, 149–158.

Rouche, N., P. Habets, and M. Laloy (1977). *Stability Theory by Liapunov's Direct Method.* Springer, New York.

Saeki, M., M. Araki, and B. Kondo (1980). Local stability of composite systems: Frequency-domain condition and estimate of the domain of attraction. *IEEE Transactions*, AC-25, 936–940.

Sandell, N. R., Jr., P. Varaiya, M. Athans, and M. G. Safonov (1978). Survey of decentralized control methods for large scale systems. *IEEE Transactions*, AC-23, 108–128.

Schmidt, G. (1982). What are and what is the origin of complex systems, and which specific problems do they pose to control engineering? (in German). *Regelungstechnik*, 30, 331–339.

Sezer, M. E., and O. Hüseyin (1978). Stabilization of linear interconnected systems using local state feedback. *IEEE Transactions*, SMC-8, 751–756.

Sezer, M. E., and Ö. Hüseyin (1981). Comments on decentralized feedback stabilization. *IEEE Transactions*, AC-26, 547–549.

Sezer, M. E., and D. D. Šiljak (1979). Decentralized stabilization and structure of linear large-scale systems. *Proceedings of the 13th Asilonear Conference*, Pacific Grove, California, 176–181.

Sezer, M. E., and D. D. Šiljak (1981). On decentralized stabilization and structure of linear large scale systems. *Automatica*, 17, 641–644.

Sezer, M. E., and D. D. Šiljak (1988). Robust stability of discrete systems. *International Journal of Control*, 48, 2055–2063.

Sezer, M. E., and D. D. Šiljak (1989). A note on robust stability bounds. *IEEE Transactions*, AC-34, 1212–1215.

Shi, Z. C., and W. B. Gao (1986). Stabilization by decentralized control for large scale interconnected systems. *Large Scale Systems*, 10, 147–155.

Shi, Z. C., and W. B. Gao (1987). On decentralized stabilization of large-scale interconnected systems. *Large Scale Systems*, 13, 187–190.

Shigesada, N., K. Kawasaki, and E. Teramoto (1984). The effects of interference competition on stability, structure and invasion of a multi-species systems. *Journal of Mathematical Biology*, 21, 97–113.

Shigui, R. (1988). Connective stability for large scale systems described by functional differential equations. *IEEE Transactions*, AC-33, 198–200.

Šiljak, D. D. (1972). Stability of large-scale systems under structural perturbations. *IEEE Transactions*, SMC-2, 657–663.

Šiljak, D. D. (1978). *Large-Scale Dynamic Systems: Stability and Structure.* North-Holland, New York.

Šiljak, D. D. (1980). Reliable control using multiple control systems. *International Journal of Control*, 31, 303–329.

Šiljak, D. D. (1983). Complex dynamic systems: Dimensionality, structure, and uncertainty. *Large Scale Systems*, 4, 279–294.

Šiljak, D. D. (1987a). Complex ecosystem stability. *Systems and Control Encyclopedia*, M. G. Singh (ed.), Pergamon Press, Oxford, UK, 672–675.

Šiljak, D. D. (1987b). Interconnected systems: Decentralized control. *Systems and Control Encyclopedia*, M. G. Singh (ed.), Pergamon Press, Oxford, UK, 2557–2560.

Šiljak, D. D. (1989). Parameter space methods for robust control design: A guided tour. *IEEE Transactions*, AC-34, 674–688.

Šiljak, D. D., and M. Vukčević (1976a). Large-scale systems: Stability, complexity, reliability. *Journal of the Franklin Institute*, 301, 49–69.

Šiljak, D. D., and M. Vukčević (1976b). Decentralization, stabilization, and estimation of large-scale systems. *IEEE Transactions*, AC-21, 363–366.

Šiljak, D. D., and M. Vukčević (1977). Decentrally stabilizable linear and bilinear large-scale systems. *International Journal of Control*, 26, 289–305.

Singh, M. G. (1981). *Decentralised Control*. North-Holland, Amsterdam, The Netherlands.

Singh, M. G., and A. Titli (eds.) (1979). *Handbook of Large Scale Systems Engineering Applications*. North-Holland, Amsterdam, The Netherlands.

Singh, M. G., A. Titli, and K. Malinowski (1984). Decentralized control design: An overview. *Proceedings of the IFAC/IFORS Symposium on Large Scale Systems*. Pergamon Press, Oxford, UK, 1–15.

Sinha, A. S. C. (1980). Lyapunov functions for a class of large-scale systems. *IEEE Transactions*, AC-25, 558–560.

Sinha, A. S. C. (1981). Stability of large-scale hereditary systems with infinite retardations. *International Journal of Systems Science*, 12, 111–117.

Socha, L. (1986). The asymptotic stochastic stability in large of the composite stochastic systems. *Automatica*, 22, 605–610.

Socha, L., and K. Popp (1986). The p-moment global exponential stability of linear large scale systems. *Large Scale Systems*, 10, 75–93.

Solimano, F., and E. Beretta (1983). Existence of a globally asymptotically stable equilibrium in Volterra models with continuous time delay. *Journal of Mathematical Biology*, 18, 93–102.

Spiteri, P. (1986). A unified approach to stability conditions of large scale non-linear evolution systems. *Large Scale Systems*, 10, 57–74.

Spouge, J. L. (1986). Increasing stability with complexity in a system composed of unstable subsystems. *Journal of Mathematical Analysis and Applications*, 118, 502–518.

Steward, D. V. (1962). On an approach to techniques for the analysis of the structure of large systems of equations. *SIAM Review*, 4, 321–342.

Stokić, D., and M. Vukobratoviić (1984). Practical stabilization of robotic systems by decentralized control. *Automatica*, 20, 353–358.

Svirezhev, Yu. M., and D. O. Logofet (1983). *Stability of Biological Communities*. Mir, Moscow, USSR.

Tartar, L. (1971). Une nouvelle characterisation des M matrices. *Revue Francaise d'Automatique, Informatique et Recherche Operationelle*, 5, 127–128.

Tsai, W. K. (1987). Minimizing the effects of unknown-but-bounded disturbance in power plants. *Proceedings of the 26th Conference on Decision and Control*, Los Angeles, California, 66–71.

Vidyasagar, M. (1981). *Input–Output Analysis of Large-Scale Interconnected Systems*. Springer, New York.

Vidyasagar, M. (1986). New directions of research in nonlinear system theory. *Proceedings of the IEEE*, 74, 1060–1091.

Voronov, A. A. (1982). Present state and problems of stability theory. *Avtomatika i Telemekhanika*, 42, 5–28.

Voronov, A. A. (1985). *Introduction to Dynamics of Complex Control Systems* (in Russian). Nauka, Moscow, USSR.

Voronov, A. A., and V. M. Matrosov (eds.) (1987). *The Method of Vector Liapunov Functions in Stability Theory* (in Russian). Nauka, Moscow, USSR.

Vukobratović, M., and N. Kirćanski (1985). An approach to adaptive control of robotic manipulators. *Automatica*, 21, 639–647.

Vukobratović, M., and D. Stokić (1982). *Control of Manipulation Robots*. Springer, Berlin, F. R. Germany.

Vukobratović, M. K., and R. D. Stojić (1986). A decentralized approach to integrated flight control synthesis. *Automatica*, 22, 695–704.

Vukobratović, M. K., and R. D. Stojić (1988). *Modern Aircraft Flight Control*. Springer, New York.

Yedavalli, R. K. (1989). Sufficient conditions for the stability of interconnected dynamic systems. *Information and Decision Technologies*, 15, 133–142.

Yong, J. (1988). Stabilization of nonlinear large-scale uncertain dynamical systems. *Journal of Mathematical Analysis and Applications*, 136, 157–177.

Zames, G. (1966). On input-output stability of time-varying nonlinear systems. Part I: Conditions derived using concepts of loop gain, conicity, and positivity. *IEEE Transactions*, AC-11, 228–238.

Zhukhovitskii, S. I., and L. I. Avdeyeva (1966). *Linear and Convex Programming*. Saunders, Philadelphia, Pennsylvania.

Chapter 3 | Optimization

Attempts to formulate decentralized control strategies by extending standard optimization concepts and methods have not been successful; the simple reason is that nonclassical decentralized information structure does not lend itself to a manageable formulation of any kind of optimality principle. This fact is responsible for a relatively large gap between theory and practice of optimal decentralized control. At present, there are a considerable number of *ad hoc* methods and techniques in this context which have few conceptual ingredients in common.

In complex dynamic systems, the notion of optimality is further complicated by the presence of *essential uncertainties* in the interconnections among the subsystems, which cannot be described in either deterministic or stochastic terms. Unlike standard optimization schemes, where robustness is a part of the solution, robustness in complex systems is a part of the problem; it has to be considered from the outset of the design process.

By imitating the decentralized structure of competitive market systems (*e.g.*, Arrow and Hahn, 1971), but otherwise using entirely different techniques, methods have been proposed for optimization of complex interconnected systems (Šiljak, 1978). A market is a place where economic agents meet to exchange commodities and the central question is: Under what conditions in a fully decentralized exchange economy do the agents, who are guided only by their self-interest, reach the state of economy in which their independent decisions are mutually compatible? In dynamic models of market economy, compatibility is demonstrated by establishing the existence of (competitive) equilibrium. Stability of the equilibrium guarantees that compatibility of the decentralized decisions made by different agents can be reached by following a set of rational rules and procedures.

By mimicking the market concept, we shall consider a complex system as constituted of a number of interconnected subsystems with distinct inputs. Each local input is chosen as a linear state feedback to optimize a quadratic performance index with respect to the corresponding subsystem, while ignoring the other subsystems. Then, suitability of the decentralized Linear–Quadratic (LQ) control law is established by showing stability of the overall closed-loop interconnected system. Under certain conditions, the proposed LQ control strategy is *robust* with respect to modeling uncertainties and perturbations in the interconnection structure, that is, the overall system is *connectively stable.*

A locally optimal decentralized LQ control is obviously not optimal (in general) for the interconnected system. When stable, however, it is suboptimal and an *index of suboptimality* is defined, which can be used to measure the cost of robustness to structural perturbations. Furthermore, we will show that suboptimal closed-loop systems tolerate distortions of the LQ control caused by insertion of nonlinearities, gain changes, and phase shifts, as do optimal closed-loop systems.

Inspired by the original paper of Kalman (1964), we will ask: When is an LQ decentralized control optimal? To no one's surprise, we will find that there are no simple answers to this question unless we alter the original set-up and allow modifications of performance indices and feedback control laws. Of course, benefits of solving this type of *inverse optimal problem in decentralized control* are many. First of all, we recover the gain and phase margins of optimal systems, which are usually superior to those of suboptimal LQ controls. Second, we can obtain optimal control by manipulating only subsystems matrices. This can produce a welcome reduction of dimensionality in complex systems, where centralized optimal problems are likely to turn out to be either impossible or uneconomical to solve even with modern computing machinery.

3.1. Suboptimality

We shall introduce in this section the concept of suboptimality relying on the standard optimization theory of Linear–Quadratic (LQ) regulators. This leads to a suitable interpretation of suboptimality in the context of optimal control problems with nonclassical decentralized information structure constraints. Several results in this section are given without proofs, because they appear as straightforward implications of the similar results obtained for complex systems in the subsequent sections of this chapter.

Let us consider a linear system

$$\mathbf{S}\colon\ \dot{x} = Ax + Bu, \tag{3.1}$$

where $x(t) \in \mathbf{R}^n$ is the state and $u(t) \in \mathbf{R}^m$ is the input of $\mathbf{S}$ at time $t \in \mathbf{T}$. The matrices A and B are constant and of appropriate dimensions. We assume that $\mathbf{S}$ is the nominal system with which we associate a quadratic performance index

$$J(x_0, u) = \int_0^{+\infty} (x^T Q x + u^T R u)\, dt, \tag{3.2}$$

where $x_0 = x(0)$ is the initial state, and Q and R are constant symmetric matrices with Q being nonnegative and R positive definite. As is common, we assume that the pair (A, B) is controllable and the pair $(A, Q^{1/2})$ is observable. A system $\mathbf{S}$ and a cost J form the standard optimization problem $(\mathbf{S}, J)$, the solution of which is given by the optimal control law

$$u^{\odot} = -Kx, \tag{3.3}$$

where the gain matrix K is

$$K = R^{-1} B^T P, \tag{3.4}$$

and P is the unique symmetric positive definite solution of the Riccati equation

$$A^T P + PA - PBR^{-1}B^T P + Q = 0. \tag{3.5}$$

The optimal closed-loop system is

$$\hat{\mathbf{S}}^{\odot} : \dot{x} = (A - BK)x, \tag{3.6}$$

and the optimal cost $J^{\odot}(x_0) \equiv J(x_0, u^{\odot})$ is computed as

$$J^{\odot}(x_0) = x_0^T P x_0. \tag{3.7}$$

Now, let us consider a perturbation of the nominal system $\mathbf{S}$, which may, for example, be due to errors committed during the modeling process. We assume that we have the system

$$\mathbf{S}^+ : \dot{x} = (A + A_+)x + Bu, \tag{3.8}$$

where A_+ is a constant matrix representing the uncertainty about $\mathbf{S}$. If we apply the same control (3.3) to $\mathbf{S}^+$, the closed-loop perturbed system

$$\hat{\mathbf{S}}^{\oplus} : \dot{x} = (A + A_+ - BK)x \tag{3.9}$$

would produce a cost $J^{\oplus}(x_0)$ which is, in general, larger than the optimal cost $J^{\odot}(x_0)$. The difference between the costs $J^{\oplus}$ and $J^{\odot}$ measures the effect of modeling uncertainty on the performance of the optimal system $\hat{\mathbf{S}}^{\odot}$. The needed concept is *suboptimality*, which is formalized as follows:

3.1. DEFINITION. A control law $u^{\odot}$ is said to be suboptimal for the pair $(\mathbf{S}^{+}, J)$ if there exists a positive number μ such that

$$J^{\oplus}(x_0) \leq \mu^{-1} J^{\odot}(x_0), \qquad \forall x_0 \in \mathbf{R}^n. \tag{3.10}$$

The number μ is called the degree of suboptimality of $u^{\odot}$.

3.2. REMARK. The cost $J^{\oplus}$ may be infinite if the control $u^{\odot}$ does not stabilize the system $\mathbf{S}^{+}$, that is, if the closed-loop matrix

$$\hat{A}_{+} = A + A_{+} - BK \tag{3.11}$$

is not stable. It is obvious, however, that the value of the performance index $J^{\oplus}$ is

$$J^{\oplus}(x_0) = x_0^T H x_0, \tag{3.12}$$

where

$$\begin{aligned} H &= \int_0^{+\infty} \exp(\hat{A}_{+}^T) G \exp(\hat{A}_{+})\, dt, \\ G &= Q + PBR^{-1}B^T P, \end{aligned} \tag{3.13}$$

and the control $u^{\odot}$ is suboptimal for $\mathbf{S}^{+}$ if and only if the symmetric matrix H exists. It is a well-known fact (*e.g.*, Kalman and Bertram, 1960) that the matrix H is finite whenever the matrix $\hat{A}_{+}$ is stable; that is, it has all eigenvalues with negative real parts. Then, we can compute H as the unique solution of the Liapunov matrix equation

$$\hat{A}_{+}^T H + H\hat{A}_{+} = -G. \tag{3.14}$$

Therefore, stability implies suboptimality. In fact, from Remark 3.2 we can compute the largest (best) value μ^* of μ for which (3.10) holds:

3.3. THEOREM. A control law $u^{\odot}$ is suboptimal with the (largest) degree of suboptimality

$$\mu^* = \lambda_M^{-1}(HP^{-1}) \tag{3.15}$$

for the pair $(\mathbf{S}^{+}, J)$ if and only if the matrix H is finite.

Proof. The expression (3.15) follows directly from inequality (3.10) and the well-known relation

$$\frac{x_0^T H x_0}{x_0^T P x_0} \leq \lambda_M(HP^{-1}), \tag{3.16}$$

which is true if and only if H is finite. Since (3.16) holds for all x_0, and equality in (3.16) takes place for some x_0, the value μ^* is the largest μ for which (3.10) is true. Q.E.D.

While by Remark 3.2 stability implies suboptimality, the converse is not true in general; without additional conditions, suboptimality does not imply stabilizability of the control law $u^{\odot}$. It is not difficult to guess that some kind of observability is needed to establish the converse:

3.4. THEOREM. A control law $u^{\odot}$ is stabilizing for the system $\mathbf{S}^+$ (that is, the matrix $\hat{A}_+$ is stable) if it is suboptimal and the pair of matrices $(\hat{A}_+, G^{1/2})$ is observable.

Proof. Let us assume that the matrix $\hat{A}_+$ has an unstable eigenvalue λ, that is, Re $\lambda > 0$, with the corresponding eigenvector $v \in \mathbf{C}^n$. From (3.13), we have

$$v^H H v = \int_0^{+\infty} \exp(\lambda t) v^H G v \exp(\lambda t) dt, \tag{3.17}$$

where the superscript H denotes the conjugate transpose of v. Since $u^{\odot}$ of (3.3) is suboptimal, H and, thus, $v^H H v$ are finite. But,

$$\exp(\lambda^* t) \exp(\lambda t) \geq 1, \qquad \forall t \geq 0, \tag{3.18}$$

and because $v^H G v$ is nonzero when $(\hat{A}_+, G^{1/2})$ is observable, the right side of (3.17) is not finite. This contradiction establishes the theorem. Q.E.D.

3.5. REMARK. It has been shown (Wonham, 1968) that observability of the pair $(\hat{A}_+, G^{1/2})$ is implied by observability of the pair $(A + A_+, Q^{1/2})$. Therefore, when the system $\mathbf{S}^+$ and the performance index J satisfy the observability condition, the control law $u^{\odot}$ is suboptimal if and only if it is stabilizing.

The formula (3.15) provides the largest value of the degree μ of suboptimality, but it does not offer an explicit characterization of the effect of perturbations on optimality of the control law $u^{\odot}$. Such a characterization is the following:

3.6. THEOREM. A control law $u^{\odot}$ is suboptimal for the pair $(\mathbf{S}^{+}, J)$ if for some positive number μ, the matrix

$$F(\mu) = A_{+}^{T}P + PA_{+} - (1-\mu)(Q + PBR^{-1}B^{T}P) \tag{3.19}$$

is nonpositive definite.

3.7. REMARK. Again, stability of $\hat{\mathbf{S}}$ is not guaranteed by $F(\mu) \leq 0$, and we must have also detectability of the pair $\{\hat{A}_{+}, [-F(0)]^{1/2}\}$, which can be replaced by detectability of the pair $\left(A + A_{+}, Q^{1/2}\right)$.

So far, we established stability of suboptimal regulators in pretty much the same way it was done for optimal regulators (Kalman, 1960). Can we also show that suboptimal control has the classical robustness properties of gain and phase margins, as the optimal control does (*e.g.*, Anderson and Moore, 1971)?

In order to introduce the gain and phase margin and the gain reduction tolerance as measures of robustness of suboptimal systems, we focus our attention on single-input systems and replace the matrices B, K, and R by vectors b, k^{T}, and a scalar ζ. Denoting the open-loop transfer function of $\hat{\mathbf{S}}^{\oplus}$ by

$$g(s) = k^{T}(sI - A - A_{+})^{-1}b, \tag{3.20}$$

we state the following theorem of Sezer and Šiljak (1981a), the proof of which is similar to the proof of a corresponding result for optimal systems (Kalman, 1964):

3.8. THEOREM. Suppose $F(\mu)$ is nonpositive definite for some positive μ. Then,

$$|1 + g(j\omega)|^{2} \leq 1 - (1-\mu)g(j\omega)g(-j\omega), \qquad \forall \omega \in \mathbf{R}. \tag{3.21}$$

Inequality (3.21) implies that the polar plot of $g(j\omega)$ lies outside (inside) the circle with the center at $(\mu-2)^{-1}$ and radius $|\mu-2|^{-1}$ when $\mu < 2$ ($\mu > 2$). These two cases are illustrated in Figure 3.1. For $\mu = 2$, the polar plot lies in the right half of the $g(s)$ plane. When compared with the similar condition for optimality, the inequality (3.21) has an additional term $(1-\mu)g(j\omega)g(-j\omega)$ on the right side. The effect of this term is to expand

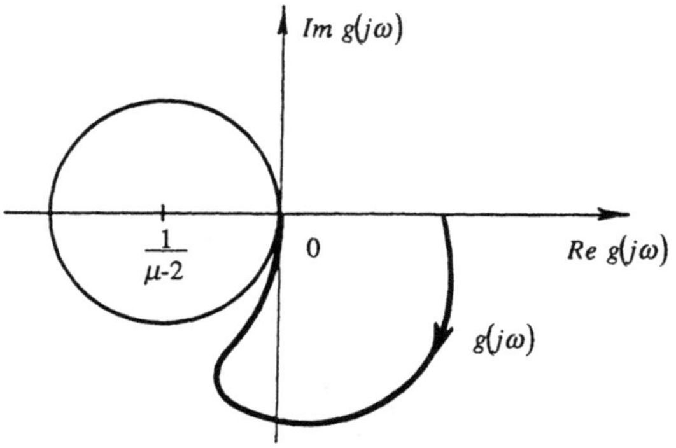

(a) $\mu < 2$

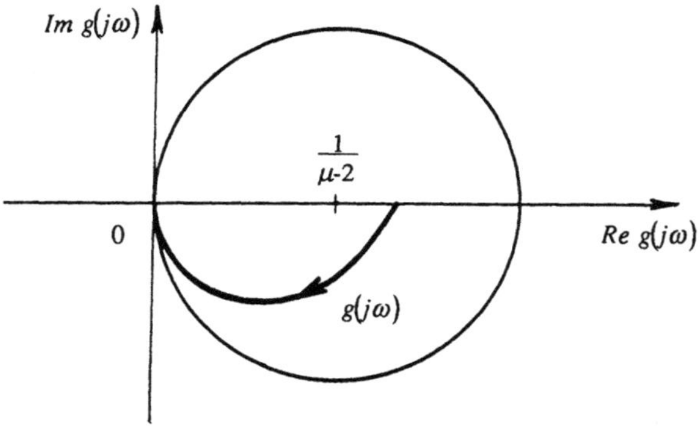

(b) $\mu < 2$

Fig. 3.1. Frequency domain plots.

or shrink the circle, depending on whether $\mu \in (0, 1]$ or $\mu \in [1, +\infty)$, the latter case being interesting in that the performance of the suboptimal regulator is superior to the optimal one!

The role of μ in terms of the gain and phase margin and the gain reduction tolerance of the suboptimal system $\hat{\mathbf{S}}^{\oplus}$ can be formulated as a corollary to Theorem 3.8. We assume that $\hat{\mathbf{S}}^{\oplus}$ is stable, and state the following:

3.9. COROLLARY. Under the condition of Theorem 3.8, the system $\hat{\mathbf{S}}^{\oplus}$ has:

(*i*) infinite gain margin;
(*ii*) at least $\pm\cos^{-1}(1-\frac{1}{2}\mu)$ phase margin;
(*iii*) at least $50\mu\%$ gain reduction tolerance.

3.10. REMARK. When $\mu = 1$, not only is the suboptimal cost at least as good as the optimal, but also the suboptimal system $\hat{\mathbf{S}}^{\oplus}$ has robustness properties (*i*)–(*iii*) no worse than the corresponding optimal system $\hat{\mathbf{S}}^{\odot}$. We also note that, when $\mu \in [1, +\infty)$, which occurs when the perturbation matrix A_+ is "beneficial," the system $\hat{\mathbf{S}}^{\oplus}$ has better robustness properties than $\hat{\mathbf{S}}^{\odot}$. Furthermore, if $\mu \geq 2$, then the gain reduction tolerance exceeds 100%, which corresponds to a "gain reversal."

We are not finished yet! To complete the robustness properties of suboptimal systems, which are germane to LQ regulators, we show that $\hat{\mathbf{S}}^{\oplus}$ can tolerate a large class of nonlinearities in almost the same way as optimal systems (*e.g.*, Anderson and Moore, 1971). To show this, we consider the system

$$\hat{\mathbf{S}}^{\oplus}_{\phi}: \; \dot{x} = (A + A_+)x + b\phi(u), \qquad u = -k^T x, \tag{3.22}$$

where $\phi : \mathbf{R} \to \mathbf{R}$ is a continuous function of the scalar input $u(t)$, and, choosing $v(x) = x^T H x$ as a Liapunov function for $\hat{\mathbf{S}}^{\oplus}_{\phi}$, establish the following:

3.11. THEOREM. Suppose that $F(\mu)$ is nonpositive definite for some positive μ and that the pair $(A + A_+, \, Q^{1/2})$ is observable. Then, the equilibrium $x = 0$ of the system $\hat{\mathbf{S}}^{\oplus}_{\phi}$ is asymptotically stable in the large for all $\phi(u)$ that belong to the class

$$\Phi = \left\{\phi : \; \underline{\kappa} u^2 \leq u\phi(u) \leq \bar{\kappa} u^2, \quad \forall u\right\}, \tag{3.23}$$

where

$$\underline{\kappa} = 1 - \frac{1}{2}(\mu - \Delta\mu), \qquad \bar{\kappa} < +\infty, \tag{3.24}$$

and $\Delta\mu$ is an arbitrarily small positive number.

When we recognize the fact that interconnections in complex systems represent perturbations of subsystems dynamics, then it is logical to extend the results of this section to decentralized control of interconnected systems. The strong motivating factor for this approach is the robustness properties of the resulting control strategies which reinforce the autonomy of individual subsystems. Before we pursue this line of research, we should introduce still another option available in this context.

We recall the role of global control, which was to enhance autonomy of decentrally stabilized subsystems by minimizing the effect of interconnections. This motivates a similar modification of the optimal control law, which can be used to either establish or improve suboptimality in the presence of perturbations. Furthermore, we will delineate the class of perturbations for which simple magnitude adjustments of the optimal control can always be made to ensure a desired suboptimal behavior.

To introduce feedback gain adjustments, we consider the input $u(t)$ to the system $\mathbf{S}^+$ as consisting of two components,

$$u = u^{\odot} + u^{+}, \tag{3.25}$$

where $u^{\odot}$ is the optimal control law (3.3) and u^+ is a corrective input chosen as a linear control

$$u^{+} = -K_{+}x. \tag{3.26}$$

The constant gain matrix K_+ should be determined to account for perturbations. For this purpose, we apply the combined linear control law

$$u^{\oplus} = -(K + K_{+})x \tag{3.27}$$

to the system $\mathbf{S}^+$ and state the following variation of Theorem 3.6:

3.12. THEOREM. A control law $u^{\oplus}$ is suboptimal with degree μ for the pair $(\mathbf{S}^+,\ J)$ if the matrix

$$F_{+}(\mu) = A_{+}^{T}P + PA_{+} + K_{+}^{T}RK_{+} - (1-\mu)\left[Q + (K + K_{+})^{T}R(K + K_{+})\right] \tag{3.28}$$

is nonpositive definite.

3.13. REMARK. The suboptimality test provided by Theorem 3.12 is only a sufficient condition, and it is difficult to see what to do when the test fails. For this reason, we want to delineate the class of perturbation matrices A_+ for which a choice of perturbation gain matrix K_+ can *always* be made to make $F_+(\mu)$ nonpositive definite for some μ. If A_+ is such that it can be factored as

$$A_{+} = BG, \tag{3.29}$$

where G is any constant $m \times n$ matrix, then the system $\mathbf{S}^+$ can always be stabilized by properly choosing K_+. Factorization (3.29) is the so-called "matching condition" (see Remark 3.36), and is equivalent to

$$\operatorname{Im} A_+ \subset \operatorname{Im} B, \tag{3.30}$$

where Im denotes the image (range) of the indicated matrix.

Remembering the stability implications of Theorem 3.6, we proceed to establish the following:

3.14. THEOREM. Suppose that the perturbation matrix A_+ satisfies the matching condition, and the matrix Q is positive definite. Then, there exists a positive number ρ such that, for

$$K_+ = \rho K, \tag{3.31}$$

the matrix $F_+(\mu)$ is nonpositive definite for some $\mu(\rho)$, which implies that the system $\hat{\mathbf{S}}^{\oplus}$ is suboptimal and stable.

Proof. Substituting (3.29) and (3.31) into (3.28), and rearranging the terms, we obtain

$$\begin{aligned} F_+(\mu) = & -\eta^{-1}\left(\eta B^T P - RG\right)^T R^{-1}\left(\eta B^T P - RG\right) \\ & + \eta^{-1} G^T RG - (1-\mu)\, Q, \end{aligned} \tag{3.32}$$

where

$$\eta = (1-\mu)(2\rho + 1) - \mu\rho^2. \tag{3.33}$$

Thus, if there exist a positive μ and a ρ such that $\eta > 0$ and $\eta(1 - \mu) \geq \lambda_M(G^T RG)\lambda_m^{-1}(Q)$, then $F_+(\mu)$ becomes nonpositive definite. These conditions, however, can be satisfied by choosing ρ sufficiently large and $\mu = \rho^{-2}$. The proof follows from Theorem 3.12. Q.E.D.

An interesting feature of the above theorem is that a stabilizing suboptimal control for the perturbed system is obtained by simply increasing the magnitude of the optimal control for the original system. The proof of this result hinges on the fact that by choosing ρ sufficiently large, the Liapunov function, which implies stability of the original system, is at the same time a Liapunov function for the perturbed system—a trick we used in the preceding chapter when we considered connective stabilization of complex systems under arbitrary structural perturbations.

3.2. Complex Systems

Our immediate task now is to adapt the concept of suboptimality to systems composed of interconnected subsystems. We consider the standard model

$$\mathbf{S}\colon\ \dot{x}_i = A_i x_i + B_i u_i + \sum_{j=1}^{N} A_{ij} x_j, \qquad i \in \boldsymbol{N}, \tag{3.34}$$

in the compact form

$$\mathbf{S}\colon\ \dot{x} = A_D x + B_D u + A_C x. \tag{3.35}$$

To generate a decentralized control law, we assume that the decoupled subsystems

$$\mathbf{S}_i\colon\ \dot{x}_i = A_i x_i + B_i u_i, \qquad i \in \boldsymbol{N}, \tag{3.36}$$

or

$$\mathbf{S}_D\colon\ \dot{x} = A_D x + B_D u, \tag{3.37}$$

are controllable, that is, $(A_D,\, B_D)$ is a controllable pair. With $\mathbf{S}_D$, we associate a quadratic cost

$$J_D(x_0,\, u) = \int_0^{+\infty} \left(x^T Q_D x + u^T R_D u\right) dt, \tag{3.38}$$

where

$$\begin{aligned} Q_D &= \operatorname{diag}\{Q_1,\, Q_2,\, \dots,\, Q_N\}, \\ R_D &= \operatorname{diag}\{R_1,\, R_2,\, \dots,\, R_N\}, \end{aligned} \tag{3.39}$$

Q_D is symmetric nonnegative definite, and R_D is a symmetric positive definite matrix. The cost J_D can be considered as a sum of N costs

$$J_i(x_{i0},\, u_i) = \int_0^{+\infty} \left(x_i^T Q_i x_i + u_i^T R_i u_i\right) dt, \qquad i \in \boldsymbol{N}, \tag{3.40}$$

each associated with the corresponding subsystem $\mathbf{S}_i$.

A straightforward strategy to control the complex system **S** is to optimize each subsystem $\mathbf{S}_i$ separately, that is, solve the LQ problem $(\mathbf{S}_D, J_D)$ to get

$$u_D^{\odot} = -K_D x, \tag{3.41}$$

where $K_D = \text{diag}\{K_1, K_2, \ldots, K_N\}$ is given as

$$K_D = R_D^{-1} B_D^T P_D, \tag{3.42}$$

and $P_D = \text{diag}\{P_1, P_2, \ldots, P_N\}$ is the unique symmetric positive definite solution of the Riccati equation

$$A_D^T P_D + P_D A_D - P_D B_D R_D^{-1} B_D^T P_D + Q_D = 0. \tag{3.43}$$

The control $u_D^{\odot}$, when applied to $\mathbf{S}_D$, results in the closed-loop system

$$\hat{\mathbf{S}}_D^{\odot}\colon\ \dot{x} = (A_D - B_D K_D)x, \tag{3.44}$$

which is optimal and produces the optimal cost

$$J_D^{\odot}(x_0) = x_0^T P_D x_0. \tag{3.45}$$

The most important fact about the locally optimal control $u_D^{\odot}$ is that it is decentralized. In other words, the components of $u^{\odot}$ are

$$u_i^{\odot} = -K_i x_i, \qquad i \in \boldsymbol{N}, \tag{3.46}$$

and the system **S** is controlled by driving each individual subsystem $\mathbf{S}_i$ using only the locally available information. Obviously, this strategy is generally not optimal, but under certain conditions it is suboptimal and stable. Most importantly, the final design is robust to a wide range of uncertainties in the interconnections, reinforcing the autonomy of the subsystems.

The closed-loop system

$$\hat{\mathbf{S}}^{\oplus}\colon\ \dot{x} = (A_D - B_D K_D + A_C)x, \tag{3.47}$$

produces a cost $J_D^{\oplus}(x_0)$. The interconnections play the role of perturbations, and suboptimality is formulated as in Definition 3.1:

3.15. DEFINITION. A decentralized control law $u_D^{\odot}$ is said to be suboptimal for the pair $(\mathbf{S}, J_D)$ if there exists a positive number μ such that

$$J_D^{\oplus}(x_0) \le \mu^{-1} J_D^{\odot}(x_0), \qquad \forall x_0 \in \mathbf{R}^n. \tag{3.48}$$

We can take advantage of the fact that the system **S** is considered as an interconnection of subsystems, and obtain a less conservative result than that of Theorem 3.6. We first define an $n \times n$ matrix

$$M_D = \mathrm{diag}\{\mu_1 I_{n_1},\ \mu_2 I_{n_2},\ \ldots,\ \mu_N I_{n_N}\}, \tag{3.49}$$

where the μ_i are positive numbers, and I_{n_i} is the identity matrix of order n_i.

3.16. THEOREM. Suppose there exists a matrix M_D such that the matrix

$$F(M_D) = A_C^T M_D^{-1} P_D + M_D^{-1} P_D A_C + (I - M_D^{-1})(Q_D + K_D^T R_D K_D) \tag{3.50}$$

is nonpositive definite. Then, the control $u_D^{\odot}$ is suboptimal for **S** with degree

$$\mu = \min_{i \in \boldsymbol{N}} \{\mu_i\}. \tag{3.51}$$

Proof. We note that

$$J_D^{\oplus}(x_0) = \lim_{\tau \to \infty} \int_0^{\tau} x^T(t)\left(Q_D + K_D^T R_D K_D\right) x(t)\, dt, \tag{3.52}$$

where $x(t) \equiv x(t,\, x_0)$ is the solution of $\hat{\mathbf{S}}^{\oplus}$ in (3.47), which corresponds to $x(0) = x_0$. From (3.42), (3.43), and (3.50), we get

$$\begin{aligned} Q_D + K_D^T R_D K_D =& F(M_D) - [(A_D - B_D K_D + A_C)^T M_D^{-1} P_D \\ &+ M_D^{-1} P_D (A_D - B_D K_D + A_C)]. \end{aligned} \tag{3.53}$$

Using (3.53) in (3.52), we compute

$$\begin{aligned} J_D^{\oplus}(x_0) =& \lim_{\tau \to \infty} \left[\int_0^{\tau} x^T(t) F(M_D) x(t)\, dt - x^T(\tau) M_D^{-1} P_D x(\tau) \right] \\ &+ x_0^T M_D^{-1} P_D x_0. \end{aligned} \tag{3.54}$$

Now, the nonpositivity of $F(M_D)$ implies that

$$J_D^{\oplus}(x_0) \le x_0^T M_D^{-1} P_D x_0 \le \mu^{-1} x_0^T P_D x_0, \tag{3.55}$$

where μ is given by (3.51), and inequality (3.48) is established. Q.E.D.

3.17. REMARK. When $\mu_i = \mu$ for all $i \in \boldsymbol{N}$, we have the test matrix

$$F(\mu) = A_C^T P_D + P_D A_C - (1-\mu)(Q_D + P_D B_D R_D^{-1} B_D P_D), \tag{3.56}$$

which would have been obtained by application of Theorem 3.6. Allowing different μ_i's for subsystems, we introduce a possibility for scaling of the interconnection matrices and, thus, tailoring of the suboptimality conditions. It can be verified by simple examples that there are cases where $F(\mu)$ is never nonpositive for any μ while, at the same time, a matrix M_D exists such that $F(M_D)$ is nonpositive.

We can strengthen the suboptimality conditions for complex systems as suggested by Theorem 3.3, at the expense of the explicit appearance of the interconnected matrices. We shall do a little more and consider systems that are not input decentralized, that is, have interconnections caused by the input matrix. The overall system is described by

$$\mathbf{S}\colon\ \dot{x} = A_D x + B_D u + A_C x + B_C u, \tag{3.57}$$

where the matrix $B_C = (B_{ij})$ is a block matrix with blocks B_{ij} compatible with the dimensions of the corresponding subsystem state $x_i \in \mathbf{R}^{n_i}$ and input $u_i \in \mathbf{R}^{m_i}$.

With the application of the locally optimal control $u_D^{\odot}$ to $\mathbf{S}$, we get the closed-loop system

$$\hat{\mathbf{S}}^{\oplus}\colon\ \dot{x} = (A_D - B_D K_D)x + A_C x - B_C K_D x, \tag{3.58}$$

and the value of the performance index

$$J_D^{\oplus}(x_0) = x_0^T H x_0, \tag{3.59}$$

where

$$\begin{aligned} H &= \int_0^{+\infty} \exp(\hat{A}^T t)\, G \exp(\hat{A} t)\, dt, \\ G_D &= Q_D + P_D B_D R_D^{-1} B_D^T P_D, \end{aligned} \tag{3.60}$$

and the closed-loop matrix is

$$\hat{A} = A_D - B_D K_D + A_C - B_C K_D. \tag{3.61}$$

We again note that $u_D^{\odot}$ is suboptimal for the system $\mathbf{S}$ of (3.57) if and only if H of (3.60) is finite. Following further the proof of Theorem 3.3, we arrive at:

3.18. THEOREM. A control law $u_D^{\odot}$ is suboptimal with the (largest) degree of suboptimality

$$\mu^* = \lambda_M^{-1}(HP_D^{-1}) \tag{3.62}$$

for the pair $(\mathbf{S},\ J_D)$ if and only if the matrix H is finite.

3.19. REMARK. It is obvious that, when B_C is present in $\mathbf{S}$, the test matrix $F(M_D)$ of Theorem 3.16 becomes

$$\begin{aligned} F(M_D) = {} & (A_C - B_C K_D)^T M_D^{-1} P_D + M_D^{-1} P_D (A_C - B_C K_D) \\ & + (I - M_D^{-1})\left(Q_D + P_D B_D R_D^{-1} B_D^T P_D\right). \end{aligned} \tag{3.63}$$

That is, in the context of Theorem 3.6, the matrix B_C can be interpreted as perturbation of the input matrix B, and included in the test matrix $F(\mu)$ in (3.19).

3.20. EXAMPLE. To show how suboptimality analysis is carried out, we consider a simple system

$$\mathbf{S}\colon\ \begin{bmatrix} \dot{x}_1 \\ \dot{x}_2 \end{bmatrix} = \left[\begin{array}{cc:cc} 1 & -24 & \multicolumn{2}{c}{\multirow{2}{*}{$\bigcirc$}} \\ 1 & -20 & & \\ \hdashline \multicolumn{2}{c:}{\multirow{2}{*}{$\bigcirc$}} & -20 & 1 \\ & & -24 & 1 \end{array}\right] \begin{bmatrix} x_1 \\ x_2 \end{bmatrix}$$

$$+ \left[\begin{array}{c:c} 2 & \multirow{2}{*}{$\bigcirc$} \\ 0 & \\ \hdashline \multirow{2}{*}{$\bigcirc$} & 0 \\ & 2 \end{array}\right] \begin{bmatrix} u_1 \\ u_2 \end{bmatrix} + \left[\begin{array}{cc:cc} \multicolumn{2}{c:}{\multirow{2}{*}{$\bigcirc$}} & 0 & 1 \\ & & 0 & 1 \\ \hdashline 1 & 0 & \multicolumn{2}{c}{\multirow{2}{*}{$\bigcirc$}} \\ 1 & 0 & & \end{array}\right] \begin{bmatrix} x_1 \\ x_2 \end{bmatrix}, \tag{3.64}$$

with the cost

$$J_D(x_0,\, u) = \int_0^{+\infty} \left(x_1^T x_1 + x_2^T x_2 + u_1^2 + u_2^2\right) dt. \tag{3.65}$$

Solving the Riccati equation (3.43), we get

$$P_D = \left[\begin{array}{cc:cc} 0.48 & -0.52 & \multicolumn{2}{c}{\multirow{2}{*}{$\bigcirc$}} \\ -0.52 & 0.62 & & \\ \hdashline \multicolumn{2}{c:}{\multirow{2}{*}{$\bigcirc$}} & 0.62 & -0.52 \\ & & -0.52 & 0.48 \end{array}\right], \tag{3.66}$$

and the decentralized control law becomes

$$\begin{bmatrix} u_1^{\odot} \\ u_2^{\odot} \end{bmatrix} = \left[\begin{array}{cc|cc} 0.96 & -1.04 & \multicolumn{2}{c}{\bigcirc} \\ \hline \multicolumn{2}{c|}{\bigcirc} & -1.04 & 0.96 \end{array}\right] \begin{bmatrix} x_1 \\ x_2 \end{bmatrix}. \tag{3.67}$$

Applying this control to $\mathbf{S}$, we get the closed-loop system

$$\hat{\mathbf{S}}^{\oplus}\colon\ \begin{bmatrix} \dot{x}_1 \\ \dot{x}_2 \end{bmatrix} = \left[\begin{array}{cc|cc} -0.92 & -21.92 & & \\ 1 & -20 & \multicolumn{2}{c}{\bigcirc} \\ \hline & & -20 & 1 \\ \multicolumn{2}{c|}{\bigcirc} & -21.92 & -0.92 \end{array}\right] \begin{bmatrix} x_1 \\ x_2 \end{bmatrix}$$

$$+ \left[\begin{array}{cc|cc} & & 0 & 1 \\ \multicolumn{2}{c|}{\bigcirc} & 0 & 1 \\ \hline 1 & 0 & & \\ 1 & 0 & \multicolumn{2}{c}{\bigcirc} \end{array}\right] \begin{bmatrix} x_1 \\ x_2 \end{bmatrix}. \tag{3.68}$$

To determine suboptimality of decentrally driven system $\hat{\mathbf{S}}^{\oplus}$, we compute the matrix H of (3.60) and get

$$H = \begin{bmatrix} 0.48 & -0.52 & 0.02 & -0.02 \\ -0.52 & 0.63 & -0.03 & 0.03 \\ 0.02 & -0.03 & 0.63 & -0.52 \\ -0.02 & 0.03 & -0.52 & 0.48 \end{bmatrix}. \tag{3.69}$$

This establishes suboptimality of $u_D^{\odot}$ in (3.67). Using Theorem 3.18, we compute the suboptimality index as

$$\mu^* = 0.95, \tag{3.70}$$

and obtain the fundamental inequality of suboptimality:

$$J_D^{\oplus}(x_0) \leq 1.05\ J_D^{\odot}(x_0). \tag{3.71}$$

This means that we are at most 5% off with respect to the performance of optimal decoupled subsystems. This is not surprising because the interconnection matrix A_C in (3.64) or (3.68) is relatively weak and does not disturb significantly the optimal performance of the subsystems.

Theorem 3.4 can be used to show that the computed local control $u_D^{\odot}$ is a stabilizing control as well. We only need to note that, in (3.65), we have $Q_D = \text{diag}\{1, 1, 1, 1\}$ and $(\hat{A}, G_D^{1/2})$ is an observable pair, with $\hat{A}$ and G_D defined by (3.68) and (3.60), respectively.

When the matrix $Q_D + P_D B_D R_D^{-1} B_D^T P_D$ is positive definite, we can use the test matrix $F(\mu)$ from (3.56) to compute the suboptimality index as

$$\begin{aligned}\mu^{\dagger} =& 1 - \lambda_M\{(Q_D + K_D^T R_D K_D)^{-1} \\ & \times \left[(A_C - B_C K_D)^T P_D + P_D (A_C - B_C K_D) \right] \}.\end{aligned} \tag{3.72}$$

If $\mu^{\dagger} > 0$, then it is a suboptimality index for $u_D^{\odot}$, but $\mu^{\dagger} \le \mu^*$. In case of $u_D^{\odot}$ of (3.67),

$$\mu^{\dagger} = 0.93. \tag{3.73}$$

Stabilizability in this case is guaranteed, as in Remark 3.7, by the detectability (observability) of the pair $\{\hat{A}, [-F(0)]^{1/2}\}$, where $\hat{A}$ is given in (3.68) and $F(0)$ is computed from (3.56).

There is still another advantage offered by the concept of suboptimality: a *reduction of dimensionality.* In computing the optimal control $u_D^{\odot}$, we manipulate only subsystem matrices of lower dimensions than the dimension of the overall system. The test matrices, however, involve computations of matrices having dimensions equal to that of the overall system, which may be prohibitive. To remove this difficulty, we can use a condition for suboptimality which closely resembles the stability condition obtained in the preceding chapter by the method of vector Liapunov functions. To state the condition, we define the numbers

$$\begin{aligned}\eta_i &= \lambda_m \left(Q_i + P_i B_i R_i^{-1} B_i^T P_i \right), \\ \xi_{ij} &= \lambda_M^{1/2} \left[(A_{ij} - B_{ij} K_j)^T P_i P_i (A_{ij} - B_{ij} K_j) \right].\end{aligned} \tag{3.74}$$

With these numbers, we specify the two $N \times N$ matrices

$$W_D = \text{diag}\{\eta_1, \eta_2, \ldots, \eta_N\}, \qquad W_C = (\xi_{ij}), \tag{3.75}$$

and prove the following corollary to Theorem 3.16:

3.21. COROLLARY. A control law $u_D^{\odot}$ is suboptimal if for some positive number μ the matrix

$$W(\mu) = W_C^T + W_C - (1 - \mu) W_D \tag{3.76}$$

is nonpositive definite.

Proof. It is easy to see that

$$x^T F(M_D) x \leq \bar{x}^T W(\mu) \bar{x}, \qquad \forall x \in \mathbf{R}^n, \tag{3.77}$$

where $\bar{x} \in \mathbf{R}^N$ is defined as $\bar{x} = (\|x_1\|, \|x_2\|, \ldots, \|x_N\|)^T$, $F(M_D)$ is given in (3.50), and μ in (3.51). The nonpositivity of $W(\mu)$ implies nonpositivity of $F(M_D)$ and, by Theorem 3.16 and Remark 3.19, we have suboptimality of $u_D^{\odot}$. Q.E.D.

Obviously, $W(\mu) \leq 0$ is more restrictive than the condition $F(M_D) \leq 0$, and the additional conservativeness in Corollary 3.21 should be judged satisfactory to the extent it is outweighed by the numerical simplicity in computation of the test matrix $W(\mu)$. In the case of Example 3.20, we can compute the suboptimality index from (3.76) as

$$\mu^{\bullet} = 1 - \lambda_M \left[W_D^{-1} \left(W_C^T + W_C \right) \right], \tag{3.78}$$

which involves only operations on second order matrices. On the other hand, we get

$$\mu^{\bullet} = 0.8, \tag{3.79}$$

which is significantly smaller than $\mu^{\dagger}$ and μ^* of (3.73) and (3.70) obtained by Remark 3.17 and Theorem 3.18.

So far, the reference system for suboptimality has been the collection of optimal decoupled subsystems. This choice promotes the autonomy of subsystems, and can be used to reduce the dimensionality of the control design. If the computation is not a problem and the decentralized information structure is inherent to the system, as is the case with automatic generation control in power systems (*e.g.*, Šiljak, 1978), then it is worthwhile to compare the locally optimal performance with that of the centrally optimal system (Krtolica and Šiljak, 1980).

Let us consider the system $\mathbf{S}$ of (3.57) as

$$\mathbf{S}\colon\ \dot{x} = Ax + Bu, \tag{3.80}$$

where $A = A_D + A_C$ and $B = B_D + B_C$, and the performance index

$$J(x_0, u) = \int_0^{+\infty} (x^T Q x + u^T R u)\, dt, \tag{3.81}$$

from (3.38). We assume again that the pair (A, B) is controllable and the pair $(A, Q^{1/2})$ is observable. It is well known (Kalman, 1960) that the optimal control law is given by

$$u^{\odot} = -R^{-1}B^TPx, \tag{3.82}$$

and the optimal cost by

$$J^{\odot}(x_0) = x_0^TPx_0, \tag{3.83}$$

where P is the unique symmetric positive definite solution of the Riccati equation

$$A^TP + PA - PBR^{-1}B^TP + Q = 0. \tag{3.84}$$

Suboptimality of the decentralized control law

$$u_D^{\odot} = -R_D^{-1}B_D^TP_Dx \tag{3.85}$$

is now defined with respect to the pair $(\mathbf{S}, J)$, and can be evaluated using the costs $J^{\oplus}(x_0)$ of (3.59) and $J^{\odot}(x_0)$ of (3.83), where $J^{\oplus}(x_0)$ is the cost produced by the system $\mathbf{S}$ of (3.80) when driven by $u_D^{\odot}$. That is, to measure the effect of the difference between the control laws $u^{\odot}$ and $u_D^{\odot}$, we consider the positive number $\bar{\mu}$ for which

$$J^{\oplus}(x_0) \leq \bar{\mu}^{-1}J^{\odot}(x_0), \qquad \forall x \in \mathbf{R}^n. \tag{3.86}$$

It is easily seen that the number $\bar{\mu}$ exists if and only if the matrix H of (3.60) is finite. In that case, the largest (best) number $\bar{\mu}^*$ for which (3.86) holds is computed as

$$\bar{\mu}^* = \lambda_M^{-1}(HP^{-1}), \tag{3.87}$$

which is similar to the expression for μ^* in (3.62). This fact implies that, once suboptimality of a control law is established with respect to one reference pair, it is suboptimal to another provided that the corresponding observability conditions are satisfied.

3.3. Robustness of Suboptimal Control

Over the years, control systems evolved from simple servomechanisms to sophisticated control schemes for multivariable systems. One of the essential roles of feedback has been to achieve a satisfactory control of processes having parameters that are either not known exactly due to modeling errors, or are varying in time during operation. In classical control theory

(Bode, 1945; Thaler and Brown, 1960), robustness of feedback control systems to parameter uncertainty has been considered by frequency analysis whereby the gain and phase margins have served to measure the ability of the closed-loop system to withstand gain and phase changes in the open-loop dynamics.

Once it was shown (Kalman, 1964) that optimal LQ regulators satisfy a frequency domain condition, the standard robustness measures of gain and phase margins became available (Anderson and Moore, 1971) for application to optimal control systems designed to meet time domain performance criteria. This opened up a real possibility of establishing these measures as performance characteristics of the optimal multivariable feedback system (Safonov, 1980), thus increasing considerably the applicability of the LQ control design.

A further increase of complexity has appeared in large-scale systems that are characterized by multiplicity of controllers, each associated with one of the interconnected subsystems. This nonclassical information and control structure constraint leads to decentralized control laws that have to be implemented by local feedback loops. The decentralized control schemes have been proven to be effective in handling the modeling uncertainties in the interconnections among the subsystems (Šiljak, 1978), which are invariably present in the control problems of complex systems. An interesting new direction in robustness analysis was taken in a couple of papers (Sezer and Šiljak, 1981a,b), where suboptimal decentralized control schemes were shown to be robust in terms of gain and phase margins and tolerance to nonlinearities inside the subsystems. The basic idea was presented in Section 3.1 for the special case of centralized control system under perturbation. We shall now present the generalized results for the decentralized structures.

First, we consider the question: Can $\hat{\mathbf{S}}^{\oplus}$, for fixed A_C, tolerate distortions in the suboptimal control $u_D^{\odot}$ caused by insertion of nonlinearities into the feedback loops?

Let us consider the system

$$\hat{\mathbf{S}}_{\phi}\colon\ \dot{x} = (A_D + A_C)x + B_D\phi(u), \qquad u = -Kx, \tag{3.88}$$

where the nonlinearity $\phi_i\colon\ \mathbf{R}^m \to \mathbf{R}^n$ is of the form

$$\phi(u) = \left[\phi_1^T(u_1),\ \phi_2^T(u_2),\ \ldots,\ \phi_N^T(u_N)\right]^T, \tag{3.89}$$

with $\phi_i\colon\ \mathbf{R}^{m_i} \to \mathbf{R}^{m_i}$ being a continuous function of u_i.

3.22. THEOREM. Suppose that $F(M_D)$ is nonpositive definite for some M_D, and that the pair $(A_D + A_C,\ Q_D^{1/2})$ is observable. Then, the equilibrium

$x = 0$ of $\hat{\mathbf{S}}_\phi$ is asymptotically stable in the large for all nonlinearities $\phi(u)$ that belong to the class

$$\Phi = \left\{\phi\colon\ \underline{\kappa}_i u_i^T R_i u_i \le u_i^T \phi_i\left(R_i^{-1} u_i\right) \le \bar{\kappa}_i u_i^T u_i, \quad i \in \boldsymbol{N}\right\}, \tag{3.90}$$

where

$$\underline{\kappa}_i = 1 - \frac{1}{2}(\mu_i - \Delta\mu), \qquad \bar{\kappa}_i < +\infty, \tag{3.91}$$

and $\Delta\mu$ is an arbitrarily small positive number.

Proof. Let $v\colon\ \mathbf{R}^n \to \mathbf{R}_+$, which is defined as

$$v(x) = x^T M_D^{-1} P_D x, \tag{3.92}$$

be a candidate for a Liapunov function of $\hat{\mathbf{S}}_\phi$. Using the optimal gain K_D of (3.42), Riccati equation (3.43), and the definition of Φ in (3.90), by straightforward computations we obtain

$$\begin{aligned}
\dot{v}(x)_{(3.88)} =& x^T [(A_D + A_C)^T M_D^{-1} P_D + M_D^{-1} P_D (A_D + A_C)] x \\
&+ 2x^T M_D^{-1} P_D B_D \phi\left(-R_D^{-1} B_D^T P_D\right) \\
=& x^T\left[F(M_D) - Q_D + \left(2M_D^{-1} - I\right) P_D B_D R_D^{-1} B_D^T P_D\right] x \\
&+ 2\sum_{i=1}^{N} \mu_i^{-1} x_i^T P_i B_i \phi_i\left(-R_i^{-1} B_i^T P_i x_i\right) \\
\le& -x^T\left[Q_D + \left(I - 2M_D^{-1}\right) P_D B_D R_D^{-1} B_D^T P_D\right] x \\
&- \sum_{i=1}^{N} 2\mu_i^{-1} \underline{\kappa}_i x_i^T P_i B_i R_i^{-1} B_i^T P_i x_i \\
\le& -x^T\left(Q_D + \Delta\mu P_D B_D R_D^{-1} B_D^T P_D\right) x, \qquad \forall x \in \mathbf{R}^n,
\end{aligned} \tag{3.93}$$

which implies that $\dot{v}(x)_{(3.88)} \le 0$. On the other hand, the observability assumption together with the fact that $\phi(0) = 0$, which is implied by (3.90) and the continuity of $\phi(u)$, ensure that $\dot{v}(x)_{(3.88)}$ is not identically zero along any trajectory of the system $\hat{\mathbf{S}}_\phi$ except the equilibrium $x = 0$. Q.E.D.

Next, we consider the case when the nonlinearities $\phi_i(u_i)$ in (3.89) are replaced by stable linear time-invariant systems $\mathcal{L}_i$ having the states z_i, inputs u_i, and transfer function matrices $L_i(s)$. Denoting the resulting system by $\hat{\mathbf{S}}_L$, we prove:

3.23. THEOREM. Suppose that $F(M)$ is nonpositive definite for some M, and the pair $\left(A_D + A_C,\ Q_D^{1/2}\right)$ is observable. Then, the equilibrium $x = 0,\ z_i = 0,\ i \in \boldsymbol{N}$, of $\mathbf{S}_L$ is asymptotically stable (in the large) provided that

$$L_i(j\omega)R_i^{-1} + R_i^{-1}L_i^H(j\omega) - (2 - \mu_i + \Delta\mu)R_i^{-1} \geq 0, \qquad \forall \omega \in \mathbf{R}, \qquad \forall i \in \boldsymbol{N}, \tag{3.94}$$

where $\Delta\mu$ is an arbitrarily small number.

Proof. Let the collection of the linear systems $\mathcal{L}_i$ inserted into the feedback loops be denoted by $\mathcal{L}$, which has the state $z = \left(z_1^T,\ z_2^T,\ \ldots,\ z_N^T\right)^T$, the input $v = \left(v_1^T,\ v_2^T,\ \ldots,\ v_N^T\right)^T$ with $v_i = -K_i x_i,\ i \in \boldsymbol{N}$, and the block diagonal transfer function $L_D(s) = \text{diag}\{L_1(s),\ L_2(s),\ \ldots,\ L_N(s)\}$. Obviously, $(x = 0,\ z = 0)$ is the equilibrium of $\hat{\mathbf{S}}_L$. For any $x(0) = x_0$, we have

$$x_0^T M_D^{-1} P_D x_0 = x^T(\tau) M_D^{-1} P_D x(\tau) - \int_0^\tau \frac{d}{dt}\left[x^T(t) M_D^{-1} P_D x(t)\right] dt. \tag{3.95}$$

Using (3.42), (3.50), (3.88), and nonpositive definiteness of $F(M_D)$, we obtain, from (3.95),

$$\begin{aligned} x_0^T M_D^{-1} P_D x_0 &- \int_0^\tau x^T(t)\left(Q_D + \Delta\mu M_D^{-1} K_D^T R_D K_D\right) x(t)\, dt \\ &\geq \int_0^\tau \left[x^T(t)\left(I - 2M_D^{-1} - \Delta\mu M_D^{-1}\right) K_D^T R_D K_D x(t)\right. \\ &\qquad \left. - 2x^T(t) M_D^{-1} P_D B_D u(t)\right] dt. \end{aligned} \tag{3.96}$$

Taking the limit as $\tau \to +\infty$, the right hand side of (3.96) can be written, using the Parseval's theorem, as

$$\begin{aligned} \frac{1}{2\pi}\int_{-\infty}^{+\infty} &\left\{x^H(j\omega)\left(I - 2M_D^{-1} - \Delta\mu M_D^{-1}\right) K_D^T R_D K_D x(j\omega)\right. \\ &\left. - x^H(j\omega) M_D^{-1} P_D B_D u(j\omega) - u^H(j\omega) B_D^T M_D^{-1} P_D x(j\omega)\right\} d\omega, \end{aligned} \tag{3.97}$$

where $x(j\omega)$ and $u(j\omega)$ are the Fourier transforms of $x(t)$ and $u(t)$. With $z(0) = 0$, $u(j\omega)$ and $x(j\omega)$ are related by

$$u(j\omega) = L_D(j\omega)v(j\omega) = -L_D(j\omega)K_D x(j\omega). \tag{3.98}$$

Substituting (3.98) into (3.97), we get the expression

$$\begin{aligned} \frac{1}{2\pi}\int_{-\infty}^{+\infty} x^H(j\omega)M_D^{-1}K_D^T R_D[L_D(j\omega)R_D^{-1} + R_D^{-1}L_D^H(j\omega) \\ - (2I - M_D + \Delta\mu I)R_D^{-1}]R_D K_D x(j\omega)\, d\omega, \end{aligned} \tag{3.99}$$

which is nonnegative by (3.94). Thus, from (3.96) we obtain

$$\int_0^{+\infty} x^T(t)\left(Q_D + \Delta\mu M_D^{-1}K_D^T R_D K_D\right)x(t)\, dt \;\le\; x_0^T M_D^{-1}P_D x_0 < \infty, \tag{3.100}$$

which, together with the observability assumption, implies that the solutions of $\hat{\mathbf{S}}_L$ corresponding to the initial condition $x(0) = x_0$, $z(0) = z_0$ tend to zero as $t \to +\infty$. Since x_0 is arbitrary, this is equivalent to saying that the poles of the transfer function $[sI - A_D - A_C + B_D L_D(s)K_D]$ have negative real parts. The proof then follows from the argument in (Safonov, 1980) that all poles of $\hat{\mathbf{S}}_L$ that do not appear as poles of this transfer function (if there are any) must be poles of $\mathcal{L}$, which are assumed to have negative real parts, so that $\hat{\mathbf{S}}_L$ is stable. Q.E.D.

Theorems 3.22 and 3.23 can be viewed as generalizations of the well-known criteria of gain and phase margin to the interconnected systems with multivariable subsystems. To see this, let us consider a special case where the subsystems $\mathbf{S}_i$ have single inputs, that is, $m_i = 1$. Then, the u_i and $\phi_i(u_i)$ in (3.89) as well as the $L_i(s)$ and R_i are scalars, and we obtain the following result from Theorems 3.22 and 3.23:

3.24. COROLLARY. Suppose that each $\mathbf{S}_i$ has a single input, $F(M)$ is nonpositive definite for some M, and the pair $(A_D + A_C,\, Q^{1/2})$ is observable, so that $\hat{\mathbf{S}}$ is stable. Then, the ith feedback loop of $\hat{\mathbf{S}}$ has:

(*i*) infinite gain margin;

(*ii*) at least $\pm\cos^{-1}(1 - \frac{1}{2}\mu_i)$ phase margin;

(*iii*) at least $50\mu_i\%$ gain reduction tolerance.

Proof. By "the ith feedback loop of $\hat{\mathbf{S}}$ has infinite gain margin" we mean as in (Safonov, 1980) that arbitrarily large increases in the gain of the local feedback (3.46) do not destroy stability of $\hat{\mathbf{S}}$. Similar interpretation can be given to loop phase margin and loop gain reduction tolerance.

Under the conditions of the corollary, (3.90) reduces to

$$\underline{\kappa}_i \leq \phi_i(u_i)/u_i \leq \bar{\kappa}_i < +\infty \quad , \qquad i \in \boldsymbol{N}, \tag{3.101}$$

where $u_i \in \mathbf{R}$ and ϕ_i: $\mathbf{R} \to \mathbf{R}$ are scalars, from which (*i*) and (*iii*) follow. Similarly, (3.94) becomes

$$\mathrm{Re}\{\ell_i(j\omega)\} \geq 1 - \frac{1}{2}(\mu_i - \Delta\mu) \qquad \forall \omega \in \mathbf{R}, \qquad i \in \boldsymbol{N}, \tag{3.102}$$

where $\ell_i(s)$ is a scalar transfer function, and (*ii*) is a consequence of (3.102) upon choosing $\ell_i(j\omega) = \exp\{\varphi_i(j\omega)\}$, which implies that $|\varphi_i(j\omega)| \leq \cos^{-1}\left[1 - \frac{1}{2}(\mu_i - \Delta\mu)\right]$. Q.E.D.

The results of this section show that suboptimal control is robust in pretty much the same way as the optimal control. Furthermore, the similarity of the conditions (3.90) and (3.94) with the standard such conditions for optimal multivariable control systems (Safonov, 1980) suggests a possibility that a modification of the suboptimal control might result in an interconnected system that is *optimal* (Sezer and Šiljak, 1981b). Once the interconnections are recognized as perturbation terms disturbing the optimality of the subsystems, the idea of an additional control action in the spirit of the perturbation control is attractive. In fact, this added effort may reshape the locally optimal control into one that is globally optimal, that is, optimal for the overall interconnected system. The reward is the standard robustness properties of the optimal LQ control, which in the case of single input subsystems means that each feedback loop has infinite gain margin, $\pm 60°$ phase margin, and 50% gain reduction tolerance.

Let us assume that $F(M_D)$ is nonpositive definite for some matrix M_D, and modify the weighting matrices $\bar{R}_D$ and $\bar{Q}_D$ as

$$\bar{R}_D = \mathrm{diag}\{\bar{R}_1,\, \bar{R}_2,\, \ldots,\, \bar{R}_N\}, \tag{3.103}$$

where

$$\bar{R}_i = \begin{cases} \left(2\mu_i - \mu_i^2\right)^{-1} R_i, & \mu_i < 2, \\ \mu_i^{-1} R_i, & \mu_i \geq 2, \end{cases} \tag{3.104}$$

with

$$\begin{aligned} \bar{Q}_D = {} & M_D^{-1}\left(Q_D - P_D B_D R_D^{-1} B_D^T P_D\right) + M_D^{-1} P_D B_D R_D^{-1} B_D^T P_D \\ & - A_C^T M_D^{-1} P_D - M_D^{-1} P_D A_C, \end{aligned} \tag{3.105}$$

and prove the following:

3.25. THEOREM. Suppose that $F(M_D)$ is nonpositive definite for some M_D. Then:

(*i*) The matrix $\bar{Q}_D$ is nonnegative definite.

(*ii*) The decentralized control

$$\bar{u}_D^{\odot} = -\bar{K}_D x, \qquad \bar{K}_D = \bar{R}_D^{-1} B_D^T M_D^{-1} P_D, \tag{3.106}$$

is optimal for $\mathbf{S}$ with respect to the modified cost

$$\bar{J}(x_0, u) = \int_0^{+\infty} \left(x^T \bar{Q}_D x + u^T \bar{R}_D u\right) dt, \tag{3.107}$$

and

$$\bar{J}^{\odot}(x_0) = x_0^T M_D^{-1} P_D x_0 \tag{3.108}$$

is the optimal cost.

(*iii*) If, in addition, the pair $(A_D + A_C,\, Q_D^{1/2})$ is observable, then the control $\bar{u}^{\odot}$ stabilizes $\mathbf{S}$.

Proof. Using the expressions (3.42) for K_D and (3.50) for $F(M_D)$ in (3.105), we obtain

$$\begin{aligned} \bar{Q}_D = {} & Q_D - F(M_D) + (I - 2M_D^{-1}) P_D B_D R_D^{-1} B_D^T P_D \\ & + M_D^{-1} P_D B_D \bar{R}_D^{-1} B_D^T P_D M_D^{-1}, \end{aligned} \tag{3.109}$$

and (*i*) is established by the definition of $\bar{R}_i$ in (3.104) and the fact that Q is nonnegative definite and $F(M_D)$ is nonpositive definite by assumption.

Furthermore, it can easily be verified that

$$(A_D+A_C)^T M_D^{-1} P_D + M_D^{-1} P_D (A_D+A_C) - M_D^{-1} P_D B_D \bar{R}_D^{-1} B_D P_D M^{-1} + \bar{Q} = 0, \tag{3.110}$$

proving (*ii*). Finally, observability of the pair $(A_D + A_C,\ Q_D^{1/2})$ implies observability of the pair $\{A_D+A_C-B_D\bar{K}_D,\ (\bar{Q}_D + \bar{K}_D\bar{R}_D\bar{K}_D)^{1/2}\}$ (Wonham, 1979), from which stability of the modified closed-loop system follows, thus (*iii*). Q.E.D.

3.26. REMARK. From (3.104) and (3.106), it follows that the components of the modified optimal control are

$$\bar{u}_i^{\odot} = \begin{cases} (2-\mu_i)u_i^{\odot}, & \mu_i < 2, \\ u_i^{\odot}, & \mu_i \geq 2, \end{cases} \tag{3.111}$$

where the $u_i^{\odot}$ are the local controls that are optimal for the decoupled subsystems $\mathbf{S}_i$ with respect to their individual performance measures J_i. The critical observation is that once suboptimality of the decentralized control $u_D^{\odot}$ is established using $F(M_D)$, by increasing the gains of the feedback loops around the subsystems, we can recover the optimality of the overall system, albeit with some modification of the performance index. The benefit of the gain boost is the improved robustness characteristics, which is really why we went for global optimality in the first place, and why we will return to this problem in the next section.

So far, we had a fixed structure for the complex system $\mathbf{S}$, and we want now to remove that restriction. Our aim is to provide conditions for *connective suboptimality*, which is a natural outgrowth of connective stability in the context of decentralized control.

Our major (but not irremovable) assumption is that the subsystems $\mathbf{S}_i$ have single inputs, that is,

$$B_i = b_i, \qquad i \in \boldsymbol{N}, \tag{3.112}$$

where $b_i \in \mathbf{R}^{n_i}$. The interconnected system is described as

$$\mathbf{S}\colon\ \dot{x} = [A_D + A_C(E)]\,x + Bu, \tag{3.113}$$

which is the model used in Sections 2.4 and 2.5. In fact, the notation and development to the end of this section follows closely those of Section 2.5.

3.27. DEFINITION. A decentralized control law $u_D^{\odot}$ is said to be connectively suboptimal for the pair $(\mathbf{S},\ J_D)$ if there exists a positive number μ such that

$$J_D^{\oplus}(x_0) \leq \mu^{-1} J_D^{\odot}(x_0), \qquad \forall x_0 \in \mathbf{R}^n, \qquad \forall E \in \bar{E}. \tag{3.114}$$

3.28. REMARK. The only difference between this and Definition 3.15 is "for all $E \in \bar{E}$." This is the difference between stability and connective stability, and the conditions for connective suboptimality are obtained by imitating the analysis of connective stabilizability: we delineate a class of interconnection structures for which there *always* exists a connectively suboptimal and stabilizing decentralized control.

Let us assume that the pairs $(A_i,\ b_i)$ of the subsystems $\mathbf{S}_i$ are in the companion (control canonical) form:

$$A_i = \begin{bmatrix} 0 & 1 & \dots & 0 \\ 0 & 0 & \dots & 0 \\ \dots & \dots & \dots & \dots \\ 0 & 0 & \dots & 1 \\ a_1^i & a_2^i & \dots & a_{n_i}^i \end{bmatrix}, \qquad b_i = \begin{bmatrix} 0 \\ 0 \\ \vdots \\ 0 \\ 1 \end{bmatrix}. \tag{3.115}$$

To each interconnection matrix $A_{ij} = \left(a_{pq}^{ij}\right)$ of $A_C = (e_{ij} A_{ij})$ in (3.113), we assign an integer m_{ij} defined as in (2.120) and explained in (2.121). We also recall from Theorem 2.20 that if the integers m_{ij} are such that, for any index set $\boldsymbol{I}_r = \{i_1,\ i_2,\ \dots,\ i_r\} \subset \boldsymbol{N}$ and any permutation $\boldsymbol{J}_r = \{j_1,\ j_2,\ \dots,\ j_r\}$ of $\boldsymbol{I}_r$,

$$\sum_{k=1}^{r} (m_{i_k j_k} - 1) < 0, \tag{3.116}$$

then the interconnected system $\mathbf{S}$ of (3.113) is stabilizable by the decentralized feedback, regardless of what the matrix E is; that is, it is connectively stabilizable. This, however, does not imply automatically that the decentralized control is also suboptimal. In fact, to establish suboptimality *via* Theorem 3.16, we have to choose properly the cost $J_D(x_0,\ u)$. For this, we first need the following:

3.29. LEMMA. Suppose that the integers m_{ij} satisfy the inequality (3.116). Then, there exist numbers ν_i such that

$$\nu_i - \nu_j + 1 - m_{ij} > 0, \qquad \forall\ i,\ j \in \boldsymbol{N}. \tag{3.117}$$

Proof. This is a combinatorial problem that can be solved by the graph-theoretic methods of Section 2.5. Although the proof is straightforward, it is rather lengthy and is, therefore, omitted. Q.E.D.

We now choose the blocks Q_i of the state weighting matrix Q_D as

$$Q_i = D_i(\delta)\bar{Q}_i D_i(\delta), \tag{3.118}$$

where the $\bar{Q}_i$ are arbitrary positive definite matrices, and

$$D_i(\delta) = \text{diag}\{\delta^{\nu_i},\, \delta^{\nu_i+1},\, \ldots,\, \delta^{\nu_i+n_i-1}\}, \tag{3.119}$$

with δ being a positive number to be determined. Also, keeping in mind that, due to the assumption on the dimension of the subsystem inputs, the input weighting matrix R_D in $J_D(x_0,\, u)$ has the form

$$R_D = \text{diag}\{\rho_1,\, \rho_2,\, \ldots,\, \rho_N\}, \tag{3.120}$$

we choose the scalars ρ_i as

$$\rho_i = \delta^{2(\nu_i+n_i)}, \qquad i \in \boldsymbol{N}. \tag{3.121}$$

With this choice of $J_D(x_0,\, u)$, we have:

3.30. THEOREM. Suppose that the numbers m_{ij} satisfy inequality (3.116), and that the matrices Q_D and R_D of the cost $J_D(x_0, u)$ are defined by (3.118)–(3.121). Then, there exists a positive number $\bar{\delta}$ such that, whenever $\delta < \bar{\delta}$, the control law $u_D^{\odot}$ is connectively suboptimal with degree $\mu = \mu(\delta)$ for the system $\mathbf{S}$ of (3.113), such that $\mu(\delta) \to 1$ as $\delta \to 0$, and, at the same time, $u_D^{\odot}$ connectively stabilizes $\mathbf{S}$.

Proof. We first show that with A_i, b_i as in (3.115), Q_i as in (3.118), and ρ_i as in (3.121), the positive definite solution of the Riccati equation

$$A_i^T P_i + P_i A_i - \rho_i^{-1} P_i b_i b_i^T P_i + Q_i = 0 \tag{3.122}$$

has the form

$$P_i = \delta D_i(\delta)\bar{P}_i(\delta)D_i(\delta), \tag{3.123}$$

where $\bar{P}_i(\delta) > 0$ for any $\delta > 0$, and $\lim_{\delta \to 0} \bar{P}_i(\delta) = \bar{P}_{i0} > 0$. For this, we substitute (3.123) into (3.122) and pre- and postmultiply by $D_i^{-1}(\delta)$ to get

$$\bar{A}_i^T(\delta)P_i(\delta) + \bar{P}_i(\delta)\bar{A}_i(\delta) - \bar{P}_i(\delta)b_i b_i^T \bar{P}_i(\delta) + \bar{Q}_i = 0, \tag{3.124}$$

where

$$\bar{A}_i(\delta) = \delta D_i(\delta) A_i D_i^{-1}(\delta) = \begin{bmatrix} 0 & 1 & \dots & 0 \\ 0 & 0 & \dots & 0 \\ \dots & \dots & \dots & \dots \\ 0 & 0 & \dots & 1 \\ \delta a_1^i & \delta a_2^i & \dots & \delta a_{n_i}^i \end{bmatrix}. \tag{3.125}$$

Since the pair $[\bar{A}_i(\delta),\, b_i]$ is controllable, and the pair $[\bar{A}_i(\delta),\, \bar{Q}_i^{1/2}]$ is observable, (3.124) implies that $\bar{P}_i(\delta) > 0$. Positive definiteness of the limit $\bar{P}_{i0}$ follows from continuity of $\bar{A}_i(\delta)$.

Now, to prove Theorem 3.30, we introduce $D = \operatorname{diag}\{D_1,\, D_2,\, \dots,\, D_N\}$, choose $\mu = \mu_i$ for all $i \in \boldsymbol{N}$, define $\bar{F}(\mu) = \mu D^{-1} F D^{-1}$, and use (3.41), (3.50), (3.118), (3.121), and (3.123) to get

$$\bar{F}(\mu) = \delta(D^{-1} A_C^T D \bar{P}_D + \bar{P}_D D A_C D^{-1}) - (1-\mu)(\bar{Q}_D + \bar{P}_D B_D B_D \bar{P}_D), \tag{3.126}$$

where $\bar{Q}_D = \operatorname{diag}\{\bar{Q}_1,\, \bar{Q}_2,\, \dots,\, \bar{Q}_N\}$ and $\bar{P}_D = \bar{P}_D(\delta) = \operatorname{diag}\{\bar{P}_1(\delta),\, \bar{P}_2(\delta),\, \dots,\, \bar{P}_N(\delta)\}$. Let us consider the $N \times N$ matrix

$$\delta \bar{P}_D D A_C D^{-1} = [e_{ij} \delta \bar{P}_i(\delta) D_i(\delta) A_{ij} D_j^{-1}(\delta)]. \tag{3.127}$$

Since, in each block of the matrix (3.127), we have

$$\delta D_i(\delta) A_{ij} D_j^{-1}(\delta) = \left[a_{pq}^{ij} \delta^{\nu_i - \nu_j + 1 - (q-p)} \right], \tag{3.128}$$

from the definition of the numbers m_{ij} in (2.120) and from (3.117), we conclude that the matrix of (3.127) is of the order δ^ν, where

$$\nu = \min_{i,j \in \boldsymbol{N}} \{\nu_i - \nu_j + 1 - m_{ij}\}. \tag{3.129}$$

Thus, as $\delta \to 0$, the first term in $\bar{F}(\mu)$ approaches zero regardless of the values of e_{ij}. Since $\bar{Q}_D + \bar{P}_D B_D B_D^T \bar{P}_D > 0$ for all $\delta > 0$, it follows that for sufficiently small δ, there exists a $\mu > 0$ arbitrarily close to one such that $\bar{F}(\mu) < 0$ for any $E \in \mathcal{E}$. Q.E.D.

We note that the above choice of the cost matrices Q_D and R_D leads to high-gain feedback matrices $k_i^T = \left[k_{i1}\delta^{-2n_i},\, k_{i2}\delta^{-2(n_i-1)}, \ldots, k_{in_i}\delta^{-2}\right]$ which, in turn, result in closed-loop decoupled subsystems having eigenvalues of the order δ^{-2}. It is, in fact, this high-gain characteristic of the suboptimal control, together with the structure of the interconnection matrix A_C, that makes the closed-loop system

$$\hat{\mathbf{S}}: \dot{x} = [A_D - B_D K_D + A_C(E)]x \tag{3.130}$$

connectively suboptimal. Clearly, $\hat{\mathbf{S}}$ also has the sensitivity properties described in Corollary 3.24, that is, for sufficiently small δ, each feedback loop of $\hat{\mathbf{S}}$ has infinite gain margin, approximately 60° phase margin, and 50% gain reduction tolerance.

3.31. EXAMPLE. We consider a special case of interconnected systems, where interconnections enter additively through inputs of the subsystems, which was considered by many people (*e.g.*, Davison, 1974; Ikeda *et al.* 1976; Sezer and Hüseyin, 1978; Saberi and Khalil, 1982). In this case, we have

$$A_{ij} = b_i h_{ij}^T, \qquad \forall\, i, j \in \boldsymbol{N}, \tag{3.131}$$

where $h_{ij} \in \mathbf{R}^{n_j}$ are constant vectors. Clearly, for such systems, A_{ij} matrices have the structure

$$A_{ij} = \begin{bmatrix} & & \bigcirc & & \\ * & * & & \cdots & * \end{bmatrix}, \tag{3.132}$$

when A_i, b_i are brought to the form (3.115), so that the numbers m_{ij} in (2.120) are

$$m_{ij} = n_j - n_i, \qquad i \in \boldsymbol{N}. \tag{3.133}$$

Therefore, (3.116) is satisfied, and Theorem 3.30 ensures the existence of a suboptimal decentralized control law. Furthermore, Theorem 3.25 implies that, by increasing the gains of the local controls, the closed-loop interconnected system can be made optimal. This fact has been recognized by Yasuda *et al.* (1980), and later generalized by Ikeda and Šiljak (1982, 1985) and Ikeda *et al.* (1983).

3.4. Optimality and Robustness

We would like now to broaden the preceding analysis in an essential way and consider optimality of LQ control for nonlinear systems. The LQ control law is generated by optimizing the linear part of the system, and conditions are explored for the control law to remain optimal when a nonlinear time-varying perturbation is present in the system. A strong motivation for this approach is provided by the fact that, as a by-product, we will establish the robustness of the control law for a nonlinear system in terms of the classical sensitivity measures of gain and phase margin used for linear systems.

The approach is inspired by the inverse optimal control problem formulated and solved for linear systems by Kalman (1964), but the inclusion of nonlinear time-varying perturbations is new and is crucial for our consideration of decentralized information structure constraints. The interconnections will be interpreted as perturbations of subsystems that are controlled by decentralized LQ feedback, and robustness in this case is essential because the nonlinearity in modeling of interconnections is almost always a part of the design problem.

We first consider LQ optimality in nonlinear systems without decentralized constraints following the original work of Ikeda and Šiljak (1990). A decentralized version of the results, which were first stated for the linear systems by Ikeda and Šiljak (1982) and Ikeda *et al.* (1983), is postponed until the next section.

Let us consider a nonlinear time-varying system described as

$$\mathbf{S}: \dot{x} = Ax + Bu + f(t, x, u), \tag{3.134}$$

where $x(t) \in \mathbf{R}^n$ is the state, and $u(t) \in \mathbf{R}^m$ is the input of $\mathbf{S}$ at time $t \in \mathbf{R}$. The matrices A and B are constant and constitute a stabilizable pair. The function $f\colon\ \mathbf{R} \times \mathbf{R}^n \times \mathbf{R}^m \to \mathbf{R}^n$ is sufficiently smooth so that (3.134) has the unique solution $x(t) \equiv x(t; t_0, x_0, u)$ for any initial conditions $(t_0, x_0) \in \mathbf{R} \times \mathbf{R}^n$ and any fixed piecewise continuous input $u(\cdot)$. Furthermore, we assume that $f(t, 0, 0) = 0$ for all $t \in \mathbf{R}$, and that $x = 0$ is the unique equilibrium of $\mathbf{S}$ when $u(t) \equiv 0$.

To obtain the LQ control for $\mathbf{S}$, we consider the linear part of $\mathbf{S}$,

$$\mathbf{S}_L\colon\ \dot{x} = Ax + Bu, \tag{3.135}$$

and a quadratic performance index

$$J_L(t_0, x_0, u) = \int_{t_0}^{\infty} (x^T Q x + u^T R u)\, dt, \tag{3.136}$$

where Q and R are positive definite matrices. The optimal control for the pair $(\mathbf{S}_L, J_L)$ is

$$u_L^{\odot} = -Kx, \qquad K = R^{-1} B^T P, \tag{3.137}$$

where P is the positive definite solution of the Riccati equation

$$A^T P + PA - PBR^{-1}B^T P + Q = 0. \tag{3.138}$$

The LQ control law $u_L^{\odot}$ is not optimal for the nonlinear system $\mathbf{S}$ with respect to J_L, but it can be made so (Šiljak, 1980) by introducing another type of performance index that includes the function $f(t, x, u)$:

$$J(t_0, x_0, u) = \int_{t_0}^{\infty} \pi(t, x, u)\, dt, \tag{3.139}$$

where the function π: $\mathbf{R} \times \mathbf{R}^n \times \mathbf{R}^m \to \mathbf{R}$ is defined as

$$\pi(t, x, u) = x^T Q x - 2x^T P f(t, x, u) + u^T R u. \tag{3.140}$$

The basic result is the following:

3.32. THEOREM. If there exist positive numbers α and β such that

$$\pi(t, x, u) \geq \alpha x^T Q x + \beta u^T R u, \qquad \forall (t, x, u) \in \mathbf{R} \times \mathbf{R}^n \times \mathbf{R}^m, \tag{3.141}$$

then the control $u_L^{\odot}$ is optimal for the pair $(\mathbf{S}, J)$.

Proof. Substituting (3.134) and (3.138) into (3.139), we get

$$\begin{aligned} J(t_0, x_0, u) =& x_0^T P x_0 + \lim_{t_1 \to \infty} [-x^T(t_1) P x(t_1) \\ &+ \int_{t_0}^{t_1} (u + R^{-1}B^T P x)^T R (u + R^{-1} B^T P x)\, dt]. \end{aligned} \tag{3.142}$$

If we use $u_L^{\odot}$, we obtain

$$J(t_0, x_0, -R^{-1}B^T P x) = x_0^T P x_0 - \lim_{t_1 \to \infty} x^T(t_1) P x(t_1). \tag{3.143}$$

Because of inequality (3.141), $J(t_0, x_0, u)$ is positive in all arguments, which implies that the infimum of $J(t_0, x_0, u)$ with respect to u exists and satisfies the inequality

$$\inf_u J(t_0, x_0, u) \leq x_0^T P x_0. \tag{3.144}$$

The existence of the infimum of $J(t_0,\ x_0,\ u)$ and inequality (3.141) further imply that $x(t) \to 0$ as $t \to \infty$, which follows from the inequality

$$\inf_u\ J(t_0,\ x_0,\ u) \geq \int_{t_0}^{\infty} (\alpha x^T Q x + \beta u^T R u)\, dt, \tag{3.145}$$

and the positive definiteness of matrices Q and R. Therefore, if we use the LQ control u_L°, we conclude from (3.142) that it is optimal, and

$$\min_u\ J(t_0, x_0, u) = x_0^T P x_0 \tag{3.146}$$

is the optimal cost. Q.E.D.

Because of the presence of the nonlinear time-varying function $f(t,\ x,\ u)$ in $\pi(t,\ x,\ u)$, it is not possible to explain the meaning of the cost functional $J(t_0,\ x_0, u)$ in terms of power, energy, or squares of disturbances from an equilibrium, as in the case of standard LQ control problems. We note, however, that when the condition of Theorem 3.32 is satisfied, J plays the role of an upper bound of the quadratic performance index

$$\tilde{J}_L(t_0,\ x_0,\ u) = \int_{t_0}^{\infty} (\alpha x^T Q x + \beta u^T R u)\, dt, \tag{3.147}$$

so that we expect the optimal control law for J to be suboptimal with respect to $\tilde{J}_L$, which, in turn, justifies the use of J. More importantly, optimality of the pair $(\mathbf{S},\ J)$ implies robustness of the resulting closed-loop system in pretty much the same way as in the case of optimality of the pair $(\mathbf{S}_L,\ J_L)$, which makes Theorem 3.32 potentially useful in applications.

3.33. REMARK. It is obvious that condition (3.141) of Theorem 3.32 is equivalent to the existence of positive numbers α' and β' such that

$$\pi(t,\ x,\ u) \geq \alpha' x^T x + \beta' u^T u, \qquad \forall (t,\ x,\ u) \in \mathbf{R} \times \mathbf{R}^n \times \mathbf{R}^m. \tag{3.148}$$

This inequality is simpler to establish than (3.141), but may appear inferior in the robustness analysis of $\mathbf{S}$.

In applications, knowledge about the nonlinearity $f(t,\ x,\ u)$ is often imprecise, and only sector bounds are available. Then, the following result may be useful:

3.34. COROLLARY. If the nonlinearity $f(t,\ x,\ u)$ satisfies the inequality

$$\|f(t,\ x,\ u)\| \leq \xi \|x\| + \eta \|u\|, \qquad \forall (t,\ x,\ u) \in \mathbf{R} \times \mathbf{R}^n \times \mathbf{R}^m, \tag{3.149}$$

for positive numbers ξ and η such that

$$2\xi\lambda_m(R) + \eta^2\lambda_M(P) < \frac{\lambda_m(Q)\lambda_m(R)}{\lambda_M(P)}, \tag{3.150}$$

then the LQ control law is optimal for the pair $(\mathbf{S},\ J)$.

Proof. Inequality (3.150) follows from the inequality

$$\pi(t,\ x,\ u) \geq [\lambda_m(Q) - 2\xi\lambda_M(P)]\ \|x\|^2 + 2\eta\lambda_M(P)\ \|x\|\ \|u\| + \lambda_m(R)\ \|u\|^2. \tag{3.151}$$

It is easy to see that (3.150) implies (3.148) for some positive numbers α' and β' such that

$$[\lambda_m(Q) - 2\xi\lambda_M(P) - \alpha']\ [\lambda_m(R) - \beta'] = \eta^2\lambda_M^2(P). \tag{3.152}$$

Q.E.D.

Another way to recover optimality of the LQ control law, when applied to $\mathbf{S}$, is to increase the feedback gain (Sezer and Šiljak, 1981b; Ikeda and Šiljak, 1982, 1985). For this approach to work, the function $f(t,\ x,\ u)$ should be separable as

$$f(t,\ x,\ u) = Bg(t,\ x,\ u) + h(t,\ x), \tag{3.153}$$

where the functions g: $\mathbf{R} \times \mathbf{R}^n \times \mathbf{R}^m \to \mathbf{R}^m$ and h: $\mathbf{R} \times \mathbf{R}^n \to \mathbf{R}^n$ are sufficiently smooth, and $g(t,\ x,\ u)$ satisfies the sector condition

$$\|g(t,\ x,\ u)\| \leq \xi'\|x\| + \eta'\|u\|, \qquad \forall (t,\ x,\ u) \in \mathbf{R} \times \mathbf{R}^n \times \mathbf{R}^m, \tag{3.154}$$

with positive numbers ξ' and $\eta' < \lambda_m^{1/2}(R)/\lambda_M^{1/2}(R)$. Then, to reflect the increase of feedback gain, we introduce the number $\rho > 1$ into the optimal LQ control to get

$$u_\rho = -\rho Kx, \qquad K = R^{-1}B^TP. \tag{3.155}$$

At the same time, we modify the cost as

$$J_\rho(t,\, x,\, u) = \int_{t_0}^{\infty} \pi_\rho(t,\, x,\, u)\, dt, \tag{3.156}$$

where π_ρ: $\mathbf{R} \times \mathbf{R}^n \times \mathbf{R}^m \to \mathbf{R}$ is defined as

$$\pi_\rho(t,\, x,\, u) = x^T Q x - 2x^T P f(t,\, x,\, u) + (\rho - 1) x^T P B R^{-1} B^T P x + \rho^{-1} u^T R u. \tag{3.157}$$

3.35. THEOREM. If there exists a positive number γ such that

$$x^T Q x - 2x^T P h(t,\, x) \geq \gamma x^T x, \quad \forall (t,\, x) \in \mathbf{R} \times \mathbf{R}^n, \tag{3.158}$$

then a sufficiently large positive number ρ can be chosen so that the control law u_ρ is optimal for the pair $(\mathbf{S},\, J_\rho)$.

Proof. We will show first that under the condition of the theorem, there exist positive numbers α_ρ and β_ρ such that

$$\pi_\rho(t,\, x,\, u) \geq \alpha_\rho x^T Q x + \beta_\rho u^T R u, \qquad \forall (t,\, x,\, u) \in \mathbf{R} \times \mathbf{R}^n \times \mathbf{R}^m. \tag{3.159}$$

Let us note that the function $\pi_\rho(t,\, x,\, u)$ can be rewritten as

$$\begin{aligned} \pi_\rho(t,\, x,\, u) = {} & x^T Q x - 2x^T P h(t,\, x) - (\rho - 1)^{-1} g(t,\, x,\, u) R g(t,\, x,\, u) \\ & + (\rho - 1)^{-1} \left[(\rho - 1) B^T P x - R g(t,\, x,\, u)\right]^T \\ & \times R^{-1} \left[(\rho - 1) B^T P x - R g(t,\, x,\, u)\right] + \rho^{-1} u^T R u. \end{aligned} \tag{3.160}$$

This expression of $\pi_\rho(t,\, x,\, u)$ and conditions (3.154) and (3.158) imply that for sufficiently large positive ρ, inequality (3.159) holds. Now, we substitute the system description (3.134) and Equation (3.157) into the performance index J_ρ to get

$$\begin{aligned} J_\rho(t_0,\, x_0,\, u) = {} & x_0^T P x_0 + \lim_{t_1 \to \infty} \{[-x^T(t_1) P x(t_1)] \\ & + \rho^{-1} \int_{t_0}^{t_1} \left[u + \rho R^{-1} B^T P x\right]^T \left[u + \rho R^{-1} B^T P x\right] dt\}. \end{aligned} \tag{3.161}$$

Using the same argument as in the proof of Theorem 3.32, we find optimality of the control law u_ρ for the pair $(\mathbf{S}, J_\rho)$. Q.E.D.

3.36. REMARK. Theorem 3.35 implies that increasing the feedback gain, we can recover optimality provided the part of the nonlinearity, which cannot be cancelled by any choice of the input u, is independent of u and is sufficiently small. A special case is the situation when f depends only on t and x, satisfies the so-called "matching condition" (Leitmann, 1979)

$$f(t, x) = Bg(t, x), \tag{3.162}$$

and has bounded gain with respect to x. The specialization of condition (3.162) to linear systems is (3.29), which was used in Section 3.2.

3.37. REMARK. We note that the function $h(t, x)$ in the decomposition (3.153) of $f(t, x, u)$, is not allowed to depend on u, otherwise the proof of Theorem 3.35 cannot go through. This fact implies that although we have introduced a modification of the control u_L as well as the index J_L, the condition for optimality of $(\mathbf{S}, J)$ has not been relaxed enough for Theorem 3.35 to include the result of Theorem 3.32.

At this point the question arises: If we recover LQ optimality in the nonlinear system $\mathbf{S}$ of (3.134), do we get the standard robust stability properties of the resulting closed-loop system? In the rest of this section, we will provide a positive answer to this question and indicate the extent of the robustness of LQ control in this case.

We investigate the robustness of the closed-loop system

$$\hat{\mathbf{S}}\colon\ \dot{x} = (A - BR^{-1}B^TP)x + f(t, x, -R^{-1}B^TPx) \tag{3.163}$$

with respect to distortions of the control law $u_L^{\odot}$. For this purpose, we consider the insertion of a smooth memoryless time-varying nonlinearity $\phi\colon \mathbf{R} \times \mathbf{R}^n \to \mathbf{R}^m$, $\phi(t, 0) \equiv 0$, or a linear time-invariant stable element having an $m \times m$ proper transfer function $L(s)$, in the feedback loop of $\hat{\mathbf{S}}$. The corresponding perturbed versions of $\hat{\mathbf{S}}$ are

$$\hat{\mathbf{S}}_\phi\colon\ \dot{x} = Ax + B\phi(t, -R^{-1}B^TPx) + f\left[t, x, \phi(t, -R^{-1}B^TPx)\right], \tag{3.164}$$

for the nonlinear distortion of $u_L^{\odot}$, and

$$\hat{\mathbf{S}}_{\mathcal{L}}\colon\ \dot{x} = Ax + B[\mathcal{L} * (-R^{-1}B^TPx)] + f\left[t, x, \mathcal{L} * (-R^{-1}B^TPx)\right], \tag{3.165}$$

for the insertion of a linear element, where $*$ denotes the convolution and $\mathcal{L}(t)$ is the inverse Laplace transform of $L(s)$.

By δ, we denote a nonnegative number for which

$$\alpha Q - \delta PBR^{-1}B^T P > 0 \,, \tag{3.166}$$

and prove the following:

3.38. THEOREM. Under the condition of Theorem 3.32, the equilibrium $x = 0$ of the system $\hat{\mathbf{S}}_\phi$ is globally asymptotically stable for any nonlinearity $\phi(t, u)$ such that

$$[\phi(t, u) - u]^T R\, [\phi(t, u) - u] \leq \beta\phi^T(t, u) R\phi(t, u) + \delta u^T Ru, \qquad \forall (t, x) \in \mathbf{R} \times \mathbf{R}^n. \tag{3.167}$$

Proof. The proof is a straightforward application of the Liapunov theory. We consider the quadratic form

$$v(x) = x^T Px, \tag{3.168}$$

and calculate the time derivative $\dot{v}(x)_{(3.164)}$ with respect to $\hat{\mathbf{S}}_\phi$ of (3.164), using the Riccati equation (3.138) and inequality (3.141) of Theorem 3.32, to get

$$\begin{aligned}\dot{v}(x)_{(3.164)} \leq & - x^T(\alpha Q - \delta PBR^{-1}B^T P)x + (1-\beta)\phi^T(t, u_L^\odot)R\phi(t, u_L^\odot) \\ & - 2\phi^T(t, u_L^\odot)Ru_L^\odot + (1-\delta)u_L^{\odot T}Ru_L^\odot, \qquad \forall (t, x) \in \mathbf{R} \times \mathbf{R}^n.\end{aligned} \tag{3.169}$$

Global exponential stability of the equilibrium $x = 0$ of $\hat{\mathbf{S}}_\phi$ follows from $\lambda_m(P)x^Tx \leq v(x) \leq \lambda_M(P)x^Tx$, and $\dot{v}(x)_{(3.164)} \leq -\lambda_m(\alpha Q - \delta PBR^{-1}B^T P)x^Tx$. Q.E.D.

3.39. THEOREM. Under the condition of Theorem 3.32, the equilibrium $x = 0$ of the system $\hat{\mathbf{S}}_\mathcal{L}$ is globally asymptotically stable for any stable linear time-invariant distortion having a transfer function $L(s)$ such that

$$\left[L^H(j\omega) - I\right]^T R[L(j\omega) - I] \leq \beta L^H(j\omega)RL(j\omega) + \delta R, \qquad \forall \omega \in \mathbf{R}. \tag{3.170}$$

Proof. We first recall that superscript H means conjugate transpose. Then, we consider again the quadratic form (3.168) and compute

$$
\begin{aligned}
x^T P x - x_0^T P x_0 = \int_{t_0}^{t} \dot{v}(x)_{(3.165)}\, d\tau \le -\int_{t_0}^{t} x^T(\alpha Q - \delta P B R^{-1} B^T P) x\, d\tau \\
- \int_{t_0}^{t} [(1-\beta)(\mathcal{L} * u_L^{\odot})^T R (\mathcal{L} * u_L^{\odot}) \\
- 2(\mathcal{L} * u_L^{\odot})^T R u_L^{\odot} + (1-\delta) u_L^{\odot T} R u_L]\, d\tau,
\end{aligned}
\tag{3.171}
$$

using inequality (3.148), the Riccati equation (3.138), and the control $u_L^{\odot}$ of (3.137). By representing the second integral in the last inequality of (3.171) using Parseval's formula, we can show that it is nonnegative provided inequality (3.170) holds. Then, $x(t) \to 0$ as $t \to \infty$. This fact and stability of $\mathcal{L}$ implies stability of $\hat{\mathbf{S}}_{\mathcal{L}}$. Q.E.D.

The deeper significance of Theorems 3.38 and 3.39 is that they imply the classical robustness properties of the optimal control design:

3.40. COROLLARY. If the condition (3.141) of Theorem 3.32 is satisfied, then each input channel of the closed-loop system $\hat{\mathbf{S}}$ of (3.163) has:

(*i*) at least

$$20 \log \frac{1 + \sqrt{\theta(\beta - \theta\beta\delta + \delta)}}{1 - \theta\beta} \text{ dB} \tag{3.172}$$

gain margin;

(*ii*) at least

$$20 \log \frac{1 - \theta\delta}{1 + \sqrt{\theta(\beta - \theta\beta\delta + \delta)}} \text{ dB} \tag{3.173}$$

gain reduction tolerance;

(*iii*) at least

$$\pm \cos^{-1}\left[1 - \frac{\theta}{2}(\beta + \delta)\right] \text{ deg} \tag{3.174}$$

phase margin, where $\theta = \lambda_m(R)/\lambda_M(R)$.

Proof. To estimate the gain margin and gain reduction tolerance, we consider as usual the nonlinearity $\phi(t, u)$ channel by channel, that is, $\phi(t, u)$ is described as $\phi(t, u) = [\phi_1(t, u_1), \phi_2(t, u_2), \dots, \phi_m(t, u_m)]^T$ and

ϕ_i: $\mathbf{R} \times \mathbf{R} \to \mathbf{R}$. From Theorem 3.38, we find stability of $\tilde{\mathbf{S}}_\phi$ if each component $\phi_i(t, u_i)$ of $\phi(t, u)$ satisfies the inequality

$$\lambda_M(R)[\phi_i(t, u_i) - u_i]^2 - \beta\lambda_m(R)\phi_i^2(t, u_i) - \delta\lambda_m(R)u_i \le 0,$$

$$\forall (t, u_i) \in \mathbf{R} \times \mathbf{R}, \qquad i \in \boldsymbol{M}, \quad (3.175)$$

where $\boldsymbol{M} = \{1, 2, \ldots, m\}$, or, equivalently,

$$(1 - \theta\beta)\left[\frac{\phi_i(t, u_i)}{u_i}\right]^2 - 2\frac{\phi_i(t, u_i)}{u_i} + (1 - \theta\delta) \le 0,$$

$$\forall (t, u_i) \in \mathbf{R} \times \mathbf{R}, \qquad i \in \boldsymbol{M}. \quad (3.176)$$

This last inequality is solved as

$$\frac{1 - \sqrt{\theta(\beta - \theta\beta\delta + \delta)}}{1 - \theta\beta} \le \frac{\phi_i(t, u_i)}{u_i} \le \frac{1 + \sqrt{\theta(\beta - \theta\beta\delta + \delta)}}{1 - \theta\beta},$$

$$\forall (t, u_i) \in \mathbf{R} \times \mathbf{R}, \qquad i \in \boldsymbol{M}, \quad (3.177)$$

so that the left side provides (*ii*) the gain reduction tolerance, while the right side of the inequality gives (*i*) the gain margin of the ith channel.

For the linear distortion, we assume that the linear element is described by $L(s) = \text{diag}\{L_1(s), L_2(s), \ldots, L_N(s)\}$. Then, inequality (3.170) is implied by

$$\lambda_M(R)|L_i(j\omega) - 1|^2 - \beta\lambda_m(R)|L(j\omega)|^2 - \delta\lambda_m(R) \le 0,$$

$$\forall \omega \in \mathbf{R}, \qquad i \in \boldsymbol{M}. \quad (3.178)$$

To estimate the phase margin, we use $|L_i(j\omega)| = 1$ in (3.178) and get

$$\text{Re}\{L_i(j\omega)\} \ge 1 - \frac{\theta}{2}(\beta + \delta), \qquad \forall \omega \in \mathbf{R}, \qquad i \in \boldsymbol{M}, \qquad (3.179)$$

and (*iii*) follows. Q.E.D.

3.41. REMARK. If the matrix R is diagonal, the stability conditions of Theorems 3.38 and 3.39 can be written channel by channel in the scalar form. Then, inequality (3.148) and, consequently, Corollary 3.40 hold for $\theta = 1$, which results in better estimates (i)–(iii) of the robustness properties.

3.42. REMARK. An interesting special case occurs when the nonlinearity $f(t, x, u)$ is independent of the input u and satisfies the inequality

$$x^T Qx - 2x^T Pf(t, x) \geq \alpha x^T Qx, \qquad \forall (t, x) \in \mathbf{R} \times \mathbf{R}^n, \tag{3.180}$$

for some positive number α. In this case, condition (3.141) holds for $\beta = 1$, and if R is diagonal with positive diagonal elements r_i, we can replace both $\lambda_M(R)$ and $\lambda_m(R)$ by r_i in (3.178) to have $\theta = 1$ in (3.172)–(3.174). Then, $\hat{\mathbf{S}}$ has:

(i) infinite gain margin;

(ii) at least 50% gain reduction tolerance;

(iii) at least $\pm 60°$ phase margin;

which are the standard robustness properties of the optimal LQ control systems.

3.43. REMARK. It is more or less obvious that the results of the robustness part of this section carry over to the control law modifications that were considered in Theorem 3.35. Under the condition (3.158), Theorems 3.38 and 3.39 can be rephrased to accommodate the control law modification with $\beta = \rho\beta_\rho$ and $\delta = \rho^{-1}\delta_\rho$, where δ_ρ is a positive number such that the matrix $\alpha_\rho Q - \delta_\rho PBR^{-1}B^T P$ is positive definite, and ρ, α_ρ, β_ρ are positive numbers defined in the proof of Theorem 3.35.

3.5. Decentrally Optimal Systems

We have demonstrated the robustness of suboptimal decentralized LQ control for linear interconnected plants. What we want to do now is to derive the conditions for optimality of decentralized LQ control when applied to linear subsystems with nonlinear interconnections. A motivation for studying this type of system is the fact that generally the subsystems are known to a great extent by the designer, while the essential uncertainty resides in the interactions among the subsystems. By establishing the global optimality of the decentralized LQ control, we will be able to apply the standard linear robustness measures to the overall nonlinear closed-loop system.

Let us consider a system

$$\mathbf{S}\colon\ \dot{x} = A_D x + B_D u + f_C(t,\, x,\, u), \tag{3.181}$$

which is an interconnection of N linear time-invariant subsystems

$$\mathbf{S}_i\colon\ \dot{x}_i = A_i x_i + B_i u_i, \qquad i \in \boldsymbol{N}, \tag{3.182}$$

where $x_i(t) \in \mathbf{R}^{n_i}$ is the state, and $u_i(t) \in \mathbf{R}^{m_i}$ is the input of $\mathbf{S}_i$ at time $t \in \mathbf{R}$, and A_i and B_i are constant matrices of appropriate dimensions that constitute stabilizable pairs $(A_i,\, B_i)$. In the description of $\mathbf{S}$, a new element is the function $f_C\colon\ \mathbf{R} \times \mathbf{R}^n \times \mathbf{R}^m \to \mathbf{R}^n$, which describes the interconnections among the subsystems. As usual, we assume that $f_C(t,\, x,\, u)$ is sufficiently smooth so that the solutions $x(t) = x(t;\, t_0,\, x_0,\, u)$ of (3.181) exist and are unique for all initial conditions $(t_0,\, x_0) \in \mathbf{R} \times \mathbf{R}^n$ and all fixed inputs $u(\cdot)$. Furthermore, $f_C(t,\, 0,\, 0) \equiv 0$, and $x = 0$ is assumed to be the unique equilibrium of $\mathbf{S}$ when $u(t) \equiv 0$.

An additional assumption in this section is the imposed information structure constraint on the control u, which is compatible with the subsystems $\mathbf{S}_i$. The constraint restricts the feedback control law to

$$u_D = -K_D x, \tag{3.183}$$

where $K_D = \text{diag}\{K_1,\, K_2,\, \ldots,\, K_N\}$, and the submatrices K_i are the gains of the local state feedback.

To stabilize the system $\mathbf{S}$ using the control law (3.183), we adopt the LQ control that is optimal for the decoupled subsystems $\mathbf{S}_i$ represented as

$$\mathbf{S}_D\colon\ \dot{x} = A_D x + B_D u. \tag{3.184}$$

This is our standard strategy for control of interconnected systems. With $\mathbf{S}_D$, we associate a quadratic performance index

$$J_D(t_0,\, x_0,\, u) = \int_{t_0}^{\infty} (x^T Q_D x + u^T R_D u)\, dt, \tag{3.185}$$

where $Q_D = \text{diag}\{Q_1,\, Q_2, \ldots,\, Q_N\}$ and $R_D = \text{diag}\{R_1,\, R_2,\, \ldots,\, R_N\}$ are constant, symmetric, and positive definite matrices, with submatrices Q_i and R_i that are compatible with the dimensions of the corresponding subsystems $\mathbf{S}_i$. The optimal decentralized control law for $\mathbf{S}_D$ is chosen as

$$u_D^{\odot} = -K_D x, \qquad K_D = R_D^{-1} B_D^T P_D, \tag{3.186}$$

where the gain matrix K_D is computed as usual using the Riccati equation (3.43).

Before we consider the optimality of $u_D^{\odot}$ for the interconnected system $\mathbf{S}$, we note that $u_D^{\odot}$ is optimal for $\mathbf{S}_D$ even when J_D is scaled by the weighting matrices

$$\begin{aligned} D_Q &= \mathrm{diag}\{d_1 I_{n_1},\, d_2 I_{n_2},\, \dots,\, d_N I_{n_N}\}, \\ D_R &= \mathrm{diag}\{d_1 I_{m_1},\, d_2 I_{m_2},\, \dots,\, d_N I_{m_N}\}, \end{aligned} \tag{3.187}$$

where the d_i's are positive numbers, and I_{n_i} and I_{m_i} are the identity matrices of order n_i and m_i, respectively. The scaled index $\bar{J}_D$ is

$$\bar{J}_D(t_0,\, x_0,\, u) = \int_{t_0}^{\infty} (x^T \bar{Q}_D x + u^T \bar{R}_D u)\, dt, \tag{3.188}$$

and $\bar{Q}_D = D_Q Q_D$, $\bar{R}_D = D_R R_D$. This is easily seen from the fact that $\bar{J}_D(t_0,\, x_0,\, u)$ is a linear combination

$$\bar{J}_D(t_0,\, x_0,\, u) = \sum_{i=1}^{N} d_i J_i(t_0,\, x_{i0},\, u_i) \tag{3.189}$$

of the subsystem indices

$$J_i(t_0, x_{i0}, u_i) = \int_{t_0}^{\infty} (x_i^T Q_i x_i + u_i^T R_i u_i)\, dt, \tag{3.190}$$

where $x_{i0} \in \mathbf{R}^{n_i}$ is the initial state of the subsystem $\mathbf{S}_i$, and $u_i \in \mathbf{R}^{m_i}$ is its input. The corresponding local feedback is chosen as

$$u_i^{\odot} = -K_i x_i, \qquad K_i = R_i^{-1} B_i^T P_i, \tag{3.191}$$

which is optimal for the pair $(\mathbf{S}_i,\, J_i)$. In fact, the positive definite solution of the Riccati equation

$$A_D^T \bar{P}_D + \bar{P}_D A_D - \bar{P}_D B_D \bar{R}_D^{-1} B_D^T \bar{P}_D + \bar{Q}_D = 0, \tag{3.192}$$

for the pair $(S_D,\, \bar{J}_D)$ is

$$\bar{P}_D = D_Q P_D = \mathrm{diag}\{d_1 P_1,\, d_2 P_2, \dots,\, d_N P_N\}, \tag{3.193}$$

and the optimal feedback control $u_D^{\odot}$ remains the same after the scaling of J_D, that is, $\bar{R}_D^{-1} B_D \bar{P}_D = R_D^{-1} B_D^T P_D$.

Now, to recover optimality of $u_D^{\odot}$ on the level of the overall system $\mathbf{S}$, we imitate the modification of the performance index introduced in the preceding section. We consider the new index

$$\bar{J}_C(t_0, x_0, u) = \int_{t_0}^{\infty} \bar{\pi}_C(t, x, u)\, dt, \tag{3.194}$$

where the function $\bar{\pi}_C$: $\mathbf{R} \times \mathbf{R}^n \times \mathbf{R}^m \to \mathbf{R}$ is defined by

$$\bar{\pi}_C(t, x, u) = x^T \bar{Q}_D x - 2x^T \bar{P}_D f_C(t, x, u) + u^T \bar{R}_D u, \tag{3.195}$$

and by a straightforward application of Theorem 3.32 and Remark 3.33, we establish:

3.44. THEOREM. If there exist positive numbers d_i, $i \in \boldsymbol{N}$, such that the inequality

$$\bar{\pi}_C(t, x, u) \geq \bar{\alpha}' x^T x + \bar{\beta}' u^T u, \qquad \forall (t, x, u) \in \mathbf{R} \times \mathbf{R}^n \times \mathbf{R}^m, \tag{3.196}$$

holds for some positive numbers $\bar{\alpha}'$ and $\bar{\beta}'$, then the decentralized control $u_D^{\odot}$ is optimal for the pair $(\mathbf{S}, \bar{J}_C)$.

Although inequalities (3.141) and (3.148) are equivalent, in Theorem 3.44 we used inequality (3.148) because of its easier testability.

3.45. REMARK. When the system $\mathbf{S}$ is linear and $f_C(t, x, u)$ is replaced by $A_C x$, where $A_C = (A_{ij})$ is a constant block matrix, Theorem 3.44 can be considerably simplified (Šiljak, 1980; Ikeda and Šiljak, 1982). The function $\bar{\pi}_C(t, x, u)$ becomes

$$\bar{\pi}_C(t, x, u) = x^T \bar{Q} x + u^T \bar{R}_D u, \tag{3.197}$$

where

$$\bar{Q} = \bar{Q}_D - A_C^T \bar{P}_D - \bar{P}_D A_C, \tag{3.198}$$

and positivity of $\bar{Q}$ is no longer required. If there exists a positive diagonal matrix D_Q such that the matrix $\bar{Q}$ is *nonnegative* definite and the pair $(A_D + A_C, \bar{Q}^{1/2})$ is detectable, then the decentralized control $u_D^{\odot}$ is optimal for the pair $(\mathbf{S}, \bar{J}_C)$.

In the context of large-scale systems (Šiljak, 1978), it is of interest to restate Theorem 3.44 in terms of the decomposition–aggregation framework of Section 2.3, as suggested by Ikeda and Šiljak (1982). We assume that there are nonnegative numbers ξ_{ij} and η_{ij} such that the components f_i: $\mathbf{R} \times \mathbf{R}^n \times \mathbf{R}^m \to \mathbf{R}^{n_i}$ of the interconnection function $f_C = (f_1^T, f_2^T, \dots, f_N^T)^T$ satisfy the inequalities

$$\| f_i(t, x, u)\| \leq \sum_{j=1}^{N} (\xi_{ij}\|x_j\| + \eta_{ij}\|u_j\|),$$
$$\forall (t, x, u) \in \mathbf{R} \times \mathbf{R}^n \times \mathbf{R}^m, \qquad i \in \boldsymbol{N}, \tag{3.199}$$

where $\|\cdot\|$ denotes the Euclidean norm.

We define three $N \times N$ aggregate matrices $W = (w_{ij})$, $Y = (y_{ij})$, and $Z = \text{diag}\{z_1, z_2, \dots, z_N\}$ as

$$w_{ij} = \begin{cases} \frac{1}{2}\lambda_m(Q_i) - \lambda_M(P_i)\xi_{ii}, & i = j, \\ -\lambda_M(P_i)\xi_{ij}, & i \neq j, \end{cases}$$
$$y_{ij} = \lambda_M(P_i)\eta_{ij}, \tag{3.200}$$
$$z_i = \lambda_m(R_i).$$

Using these matrices, we can get the inequality

$$\bar{\pi}_C(t, x, u) \geq (\bar{x}^T, \bar{u}^T)W_C(\bar{x}^T, \bar{u}^T)^T, \qquad \forall (t, x, u) \in \mathbf{R} \times \mathbf{R}^n \times \mathbf{R}^m, \tag{3.201}$$

where

$$W_C = \begin{bmatrix} W^T D + DW & -DY \\ -Y^T D & DZ \end{bmatrix}, \tag{3.202}$$

and $\bar{x} = (\|x_1\|, \|x_2\|, \dots, \|x_N\|)^T$, $\bar{u} = (\|u_1\|, \|u_2\|, \dots, \|u_N\|)^T$. Obviously, $\bar{x}^T\bar{x} = x^Tx$ and $\bar{u}^T\bar{u} = u^Tu$. We also note that the matrix W of (3.200) has nonpositive diagonal elements, and the M-matrix conditions of Theorem 2.4 can be used to establish positive definiteness of the matrix W_C. An immediate corollary to Theorem 3.44 is the following:

3.46. COROLLARY. If there exists a positive diagonal matrix D such that the matrix W_C is positive definite, then the decentralized control law $u_D^{\odot}$ is optimal for the pair $(\mathbf{S}, \bar{J}_C)$.

3.47. REMARK. The necessary and sufficient condition for the matrix W_C to be positive definite is

$$W^T D + DW - DY(DZ)^{-1}Y^T D > 0. \tag{3.203}$$

For (3.203) to hold, $W^T D + DW$ should be positive definite, which in turn implies that $W \in \mathcal{M}$. Therefore, to test (3.203) we should first check positivity of the leading principal minors of W (see Theorem 2.4) and, if they are positive, we should proceed to compute a positive D and then test positive definiteness of W_C by checking (3.202). It is more or less obvious from (3.202) that if the interconnections $f_i(t, x, u)$ are weak and ξ_{ij} and η_{ij} are sufficiently small, then Corollary 3.46 would be satisfied.

There are several interesting special cases of the interconnection function $f_C(t, x, u)$. An easy case is when the function is independent of the control u, that is, we have $f_C(t, x)$. Then, $Y = 0$ and optimality of the control law $u_D^{\odot}$ for the pair $(\mathbf{S}, \bar{J}_C)$ is guaranteed by $W \in \mathcal{M}$. Another case occurs when $\mathbf{S}$ is an input-decentralized system (Section 1.5), that is, the ith component $f_i(t, x, u_i)$ of the interconnection function f_C depends only on the ith component u_i of the control u, for all $i \in \boldsymbol{N}$. In this case, Y is diagonal and the left side of (3.203) can be written as

$$\begin{aligned} &W^T D + DW - DY(DZ)^{-1}DY \\ &\quad = \left(W - \frac{1}{2}YZ^{-1}Y\right)^T D + D\left(W - \frac{1}{2}YZ^{-1}Y\right). \end{aligned} \tag{3.204}$$

Using this expression, we get:

3.48. COROLLARY. If $\mathbf{S}$ is an input-decentralized system and $W - \frac{1}{2}YZ^{-1}Y \in \mathcal{M}$, then the decentralized control law $u_D^{\odot}$ is optimal for the pair $(\mathbf{S}, \bar{J}_C)$.

Finally, we consider again the linear case when $f_C(t, x, u) \equiv A_C x$. Then, in (3.199), we only need to compute $\xi_{ij} = \lambda_M^{1/2}(A_{ij}^T A_{ij})$ for the matrix $W = (w_{ij})$ defined in (3.200). As shown by Ikeda and Šiljak (1982), optimality of $u_D^{\odot}$ is implied by $W \in \mathcal{M}$, because then $\bar{Q}$ of (3.198) is positive definite, which establishes the required optimality (see Remark 3.45).

In the context of decentralized control, the LQ optimality of local feedback control loops implies robustness to variations in open-loop dynamics of the subsystems. This fact was first established for linear interconnections (Ikeda and Šiljak, 1982). The linear restriction has been removed in (Ikeda and Šiljak, 1985), but the interconnection functions were required to depend on time and state only. Here, we remove this last restriction, and consider the closed-loop system

$$\hat{\mathbf{S}}\colon\ \dot{x} = \left(A_D - B_D R_D^{-1} B_D^T P_D\right) x + f_C\left(t,\, x,\, -R_D^{-1} B_D^T P_D x\right), \qquad (3.205)$$

and assume that the nonlinear distortion of $u_D^{\odot}$ has occurred. This fact is described by a nonlinear function $\phi_D\colon\ \mathbf{R} \times \mathbf{R}^n \to \mathbf{R}^N$, which is defined as $\phi_D(t,\, u) = [\phi_1^T(t,\, u_1),\, \phi_2^T(t,\, u_2),\, \ldots,\, \phi_N^T(t,\, u_N)]^T$, and $\phi_D(t,\, 0) \equiv 0$. The system $\hat{\mathbf{S}}$ becomes

$$\hat{\mathbf{S}}_\phi\colon\ \dot{x} = A_D x + B_D \phi_D(t,\, -R_D^{-1} B_D^T P_D x) + f_C\left[t,\, x,\, \phi_D(t,\, -R_D^{-1} B_D^T P_D x)\right]. \qquad (3.206)$$

To show robustness, let us introduce the diagonal matrices

$$\begin{aligned} D_\alpha &= \text{diag}\{\alpha_1 I_{n_1},\, \alpha_2 I_{n_2},\, \ldots,\, \alpha_N I_{n_N}\}, \\ D_\beta &= \text{diag}\{\beta_1 I_{n_1},\, \beta_2 I_{n_2},\, \ldots,\, \beta_N I_{n_N}\}, \qquad (3.207) \\ D_\delta &= \text{diag}\{\delta_1 I_{n_1},\, \delta_2 I_{n_2},\, \ldots,\, \delta_N I_{n_N}\}, \end{aligned}$$

where the α_i's and β_i's are positive numbers and the δ_i's are nonnegative numbers, and assume that these matrices can be taken so that

$$\bar{\pi}_C(t,\, x,\, u) \geq x^T D_\alpha \bar{Q}_D x + u^T D_\beta \bar{R}_D u, \qquad \forall (t,\, x,\, u) \in \mathbf{R} \times \mathbf{R}^n \times \mathbf{R}^m,$$

$$D_\alpha \bar{Q}_D - P_D B_D D_\delta R_D^{-1} B_D^T P_D > 0. \qquad (3.208)$$

The existence of such matrices is obvious by (3.196) of Theorem 3.44. Similarly, as in Theorem 3.38, we obtain:

3.49. THEOREM. Under the condition (3.208), the equilibrium $x = 0$ of the system $\hat{\mathbf{S}}_\phi$ is globally asymptotically stable for any nonlinear function $\phi(t,\, u)$ such that

$$[\phi_i(t,\, u_i) - u_i]^T R_i\, [\phi_i(t,\, u_i) - u_i] \leq \beta_i \phi_i^T(t,\, u_i) R_i \phi_i(t,\, u_i) + \delta_i u_i^T R_i u_i,$$

$$\forall (t,\, u_i) \in \mathbf{R} \times \mathbf{R}^{m_i}. \qquad (3.209)$$

Linear distortions are handled similarly. As in Section 3.4, we introduce the matrix $L_D(s) = \text{diag}\{L_1(s), L_2(s), \ldots, L_N(s)\}$, where $L_i(s)$ is an $m_i \times m_i$ stable transfer function, with the inverse Laplace transform $\mathcal{L}_D(t) = \text{diag}\{\mathcal{L}_1(t), \mathcal{L}_2(t), \ldots, \mathcal{L}_N(t)\}$. The linear closed-loop system is described by

$$\hat{\mathbf{S}}_{\mathcal{L}}\colon\ \dot{x} = A_D x + B_D\left[\mathcal{L}_D * \left(-R_D^{-1}B_D^T P_D x\right)\right] + f_C\left[t,\, x,\, \mathcal{L}_D * \left(-R_D^{-1}B_D^T P_D x\right)\right]. \tag{3.210}$$

Imitating the proof of Theorem 3.39, we can establish the following:

3.50. THEOREM. Under the condition (3.208), the equilibrium $x = 0$ of the system $\hat{\mathbf{S}}_{\mathcal{L}}$ is globally asymptotically stable for any stable linear time-invariant distortion having a transfer function $L(s)$ such that

$$[L_i^H(j\omega) - I_{m_i}]^T R_i [L_i(j\omega) - I_{m_i}] \leq \beta_i L_i^H(j\omega) R_i L_i(j\omega) + \delta_i R_i, \qquad \forall\, \omega \in \mathbf{R}. \tag{3.211}$$

With Theorems 3.49 and 3.50 in hand, we can calculate directly the gain and phase margin of each input channel using the formulas of Corollary 3.40 as suggested by Remark 3.41. The numbers β, δ, and θ should be replaced by β_i, δ_i, and θ_i, with β_i and δ_i being defined by (3.209) and (3.211), and $\theta_i = \lambda_m(R_i)/\lambda_M(R_i)$.

The numbers β_i and δ_i can also be computed in the decomposition–aggregation framework of Section 2.3, but the results may be conservative. If the computational simplicity of the framework is a deciding factor, then we modify the $N \times N$ matrices W and Z to get $\tilde{W} = (\tilde{w}_{ij})$ and $\tilde{Z} = \text{diag}\{z_1, z_2, \ldots, z_N\}$, which are defined as

$$\tilde{w}_{ij} = \begin{cases} \frac{1}{2}(1-\alpha_i)\lambda_m(Q_i) - \lambda_M(P_i)\xi_{ij}, & i = j, \\ -\lambda_M(P_i)\xi_{ij}, & i \neq j, \end{cases} \tag{3.212}$$

$$z_i = (1-\beta_i)\lambda_m(R_i),$$

and form the matrix $\tilde{W}_C$ as W_C of (3.202):

$$\tilde{W}_C = \begin{bmatrix} \tilde{W}^T D + D\tilde{W} & -DY \\ -Y^T D & D\tilde{Z} \end{bmatrix}. \tag{3.213}$$

Now, stability conditions (3.209) and (3.211) are satisfied by the positive numbers α_i, β_i, and the nonnegative numbers δ_i such that $\tilde{W}_C$ is nonnegative definite and

$$\alpha_i Q_i - \delta_i P_i B_i R_i^{-1} B_i^T P_i > 0. \tag{3.214}$$

An extension of these facts to control law modifications is straightforward. We present the results without proofs. Following the approach of Ikeda and Šiljak (1982), we split the interconnection function as

$$f_C(t, x, u) = B_D g_C(t, x, u) + h_C(t, x), \tag{3.215}$$

where $g_C = \left(g_1^T, g_2^T, \dots, g_N^T\right)$, g_i: $\mathbf{R} \times \mathbf{R}^n \times \mathbf{R}^m \to \mathbf{R}^{m_i}$, and $h_C = \left(h_1^T, h_2^T, \dots, h_N^T\right)^T$, h_i: $\mathbf{R} \times \mathbf{R}^n \to \mathbf{R}^{n_i}$. We assume that $g_i(t, x, u)$ satisfies the inequality

$$\|g_i(t, x, u)\| \le \xi_i' \|x\| + \eta_i' \|u\|, \qquad \forall (t, x, u) \in \mathbf{R} \times \mathbf{R}^n \times \mathbf{R}^m, \tag{3.216}$$

for some positive numbers ξ_i' and η_i' such that $\sum_{i=1}^N \lambda_M(R_i)(\eta_i')^2 < \sum_{i=1}^N \lambda_m(R_i)$. The modification of the decentralized control law is similar in form to that of the preceding section:

$$u_D^\rho = -\rho_D R_D^{-1} B_D^T P_D x, \tag{3.217}$$

where $\rho_D = \operatorname{diag}\{\rho_1 I_{m_1}, \rho_2 I_{m_2}, \dots, \rho_N I_{m_N}\}$, and $\rho_i > 1$ for all $i \in \boldsymbol{N}$. The performance index is modified accordingly as

$$\bar{J}_\rho(t_0, x_0, u) = \int_{t_0}^{\infty} \bar{\pi}_\rho(t, x, u)\, dt, \tag{3.218}$$

where $\bar{\pi}_\rho$: $\mathbf{R} \times \mathbf{R}^n \times \mathbf{R}^m \to \mathbf{R}$ is defined by

$$\begin{aligned} \bar{\pi}_\rho(t, x, u) = {} & x^T \bar{Q}_D x - 2x^T \bar{P}_D f_C(t, x, u) \\ & + x^T \bar{P}_D B_D(\rho_D - I_m) \bar{R}_D^{-1} B_D^T \bar{P}_D x + u^T \rho_D^{-1} \bar{R}_D u. \end{aligned} \tag{3.219}$$

3.51. THEOREM. If there exist positive numbers d_i, $i \in \boldsymbol{N}$, and a positive number γ' such that

$$x^T \bar{Q}_D x - 2x^T \bar{P}_D h_C(t, x) \ge \gamma' x^T x, \qquad \forall (t, x) \in \mathbf{R} \times \mathbf{R}^n, \tag{3.220}$$

then positive numbers ρ_i can be chosen so that the modified control law u_D^ρ is optimal for the pair $(\mathbf{S}, \bar{J}_\rho)$.

As in the proof of Theorem 3.35, we first establish the inequality

$$\bar{\pi}_\rho(t, x, u) \geq x^T D_\alpha \bar{Q}_D x + u^T D_\beta \bar{R}_D u, \qquad \forall (t, x, u) \in \mathbf{R} \times \mathbf{R}^n \times \mathbf{R}^m, \tag{3.221}$$

and then show optimality of u_D^ρ.

Again, the interesting case is that of linear interconnections (Ikeda and Šiljak, 1982) when $f_C(t, x, u)$ is replaced by $A_C x$, where $A_C = (A_{ij})$ is a block matrix with blocks represented as

$$A_{ij} = B_i G_{ij} + H_{ij}, \tag{3.222}$$

where G_{ij} and H_{ij} are constant matrices of the appropriate dimensions. If the splitting is done so that

$$\operatorname{Im} B_i \perp \operatorname{Im} H_{ij}, \tag{3.223}$$

which is always possible, then the term $B_i G_{ij}$ can be neutralized by high-gain feedback (ρ_i sufficiently large) leaving the matrices H_{ij} (hopefully, sufficiently small) for optimality considerations.

In the context of the system $\mathbf{S}$ of (3.181) introduced at the beginning of this section, the splitting (3.222) is

$$A_C = B_D G_C + H_C, \tag{3.224}$$

where $G_C = (G_{ij})$ and $H_C = (H_{ij})$ are block matrices. In this case, a sufficiently large $\rho > 1$ can be chosen for u_D^ρ to be optimal with respect to the pair $(\mathbf{S}, J_D)$ if there exists a positive D_Q such that the matrix

$$\bar{Q}_\rho = D_Q Q_D - H_C^T P_D D_Q - D_Q P_D H_C \tag{3.225}$$

is positive definite.

Positive definiteness of $\bar{Q}_\rho$ can easily be tested by the decomposition–aggregation method (Section 2.2), which we show in the general context of nonlinear interactions. We assume that each component h_i of the function h_C satisfies the inequality

$$\|h_i(t, x)\| \leq \sum_{j=1}^{N} \gamma'_{ij} \|x_j\|, \qquad \forall (t, x) \in \mathbf{R} \times \mathbf{R}^n, \tag{3.226}$$

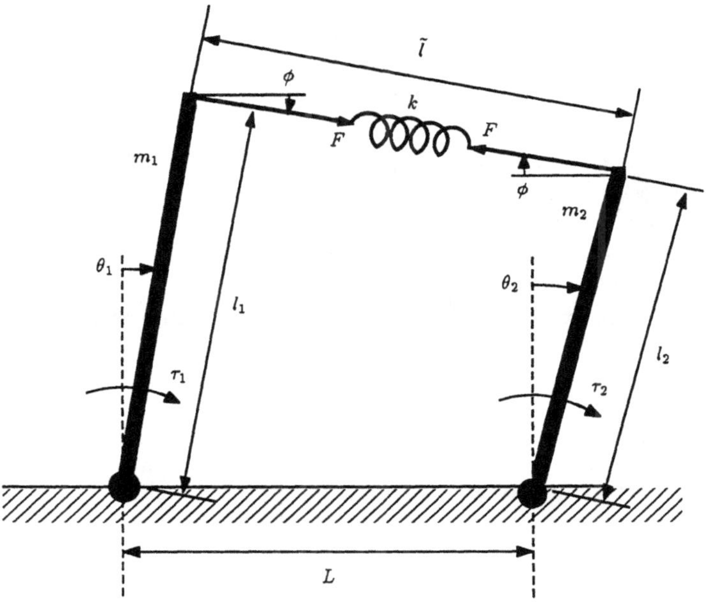

Fig. 3.2. Inverted penduli.

where the γ'_{ij} are nonnegative numbers. We define the $N \times N$ matrix $W' = (w'_{ij})$ by

$$w'_{ij} = \begin{cases} \frac{1}{2}\lambda_m(Q_i) - \lambda_M(P_i)\gamma'_{ii}, & i = j, \\ -\lambda_M(P_i)\gamma'_{ij}, & i \neq j. \end{cases} \tag{3.227}$$

Then, condition (3.220) holds if $W' \in \mathcal{M}$, that is, W' is an M-matrix, which by Theorem 3.51 implies optimality of u_D^ρ for the pair $(\mathbf{S}, J_D)$ when the gain constant ρ is chosen sufficiently large.

3.52. EXAMPLE. To demonstrate the proposed optimization scheme, we consider a system of two inverted penduli coupled by a spring. The system is shown in Figure 3.2, where the variables are:

θ_i — angular displacement of pendulum i, $i = 1, 2$;
τ_i — torque input generated by the actuator for pendulum i, $i = 1, 2$;
F — spring force; (3.228)
$\tilde{\ell}$ — spring length;
ϕ — slope of the spring to the earth;

and the constants are:

$$\begin{aligned} \ell_i &\text{ — length of pendulum } i,\ i = 1,\ 2; \\ m_i &\text{ — mass of pendulum } i,\ i = 1,\ 2; \\ L &\text{ — distance of two penduli;} \\ k &\text{ — spring constant.} \end{aligned} \tag{3.229}$$

The mass of each pendulum is uniformly distributed. The length of the spring is chosen so that $F = 0$ when $\theta_1 = \theta_2 = 0$, which implies that $(\theta_1,\ \dot{\theta}_1,\ \theta_2,\ \dot{\theta}_2)^T = 0$ is an equilibrium of the system if $\tau_1 = \tau_2 = 0$. For simplicity, we assume that the mass of the spring is zero, and restrict the movement of the penduli as $|\theta_i| < \pi/6$, $i = 1,\ 2$.

The equations of motion of the coupled penduli are written as

$$\begin{aligned} m_1(\ell_1/2)\ddot{\theta}_1 &= \tau_1 + m_1 g(\ell_1/2)\ \sin\theta_1 + \ell_1 F\ \cos(\theta_1 - \phi), \\ m_2(\ell_2/2)\ddot{\theta}_2 &= \tau_2 + m_2 g(\ell_2/2)\ \sin\theta_2 - \ell_2 F\ \cos(\theta_1 - \phi), \end{aligned} \tag{3.230}$$

where g is the constant of gravity, and

$$\begin{aligned} F &= k\left(\tilde{\ell} - [L^2 + (\ell_2 - \ell_1)^2]^{1/2}\right), \\ \tilde{\ell} &= [(L + \ell_2\ \sin\theta_2 - \ell_1\ \sin\theta_1)^2 + (\ell_2\ \cos\theta_2 - \ell_1\ \cos\theta_1)^2]^{1/2}, \\ \phi &= \tan^{-1}\frac{\ell_1\ \cos\theta_1 - \ell_2\ \sin\theta_2}{L + \ell_2\ \sin\theta_2 - \ell_1\ \sin\theta_1}. \end{aligned} \tag{3.231}$$

We consider each pendulum as one component, and rewrite (3.230) as

$$\mathbf{S}\text{:}\quad \frac{d}{dt}\begin{bmatrix} \theta_1 \\ \dot{\theta}_1 \\ \theta_2 \\ \dot{\theta}_2 \end{bmatrix} = \left[\begin{array}{cc|cc} 0 & 1 & 0 & 0 \\ g & 0 & 0 & 0 \\ \hline 0 & 0 & 0 & 1 \\ 0 & 0 & g & 0 \end{array}\right]\begin{bmatrix} \theta_1 \\ \dot{\theta}_1 \\ \theta_2 \\ \dot{\theta}_2 \end{bmatrix} + \left[\begin{array}{c|c} 0 & 0 \\ 2/m_1\ell_1 & 0 \\ \hline 0 & 0 \\ 0 & 2/m_2\ell_2 \end{array}\right]\begin{bmatrix} \tau_1 \\ \tau_2 \end{bmatrix}$$

$$+ \left[\begin{array}{c} 0 \\ g(\sin\theta_1 - \theta_1) + 2(1/m_1)F\ \cos(\theta_1 - \phi) \\ \hline 0 \\ g(\sin\theta_2 - \theta_2) - 2(1/m_2)F\ \cos(\theta_2 - \phi) \end{array}\right]. \tag{3.232}$$

For the ith decoupled subsystem in (3.232) and the performance index

$$J_i(\theta_{i0},\, \dot{\theta}_{i0},\, \tau_i) = \int_0^{\infty} (m_i g \theta_i^2 + m_i \dot{\theta}_i^2 + \tau_i^2)\, dt, \tag{3.233}$$

we compute the local optimal control, where $m_i g \theta_i^2$ and $m_i \dot{\theta}_i^2$ in (3.233) are chosen to give a physical meaning to J_i. That is, $(1/2)(\ell_i/2)^2 m_i g \theta_i^2$ represents the potential energy and $(1/2)(\ell_i/2)^2 m_i \dot{\theta}_i^2$ is the kinetic energy, for small θ_i and $\dot{\theta}_i$. Obviously, the proposed optimization scheme is not restricted to this choice of J_i.

Using the matrices

$$\begin{aligned} A_i &= \begin{bmatrix} 0 & 1 \\ g & 0 \end{bmatrix}, & B_i &= \begin{bmatrix} 0 \\ 2/m_i \ell_i \end{bmatrix}, \\ Q_i &= \begin{bmatrix} m_i g & 0 \\ 0 & m_i \end{bmatrix}, & R_i &= 1, \end{aligned} \tag{3.234}$$

the positive definite solution of the Riccati equation

$$A_i^T P_i + P_i A_i - P_i B_i R_i^{-1} B_i^T P_i + Q_i = 0, \tag{3.235}$$

is calculated as

$$P_i = \frac{\sqrt{g}\, m_i \ell_i + \sqrt{g m_i^2 \ell_i^2 + 4 m_i}}{4} \begin{bmatrix} \sqrt{g}\,\sqrt{g m_i^2 \ell_i^2 + 4 m_i} & \sqrt{g}\, m_i \ell_i \\ \sqrt{g}\, m_i \ell_i & m_i \ell_i \end{bmatrix}, \tag{3.236}$$

and we obtain the optimal control law for the ith subsystem as

$$\tau_i = -\frac{\sqrt{g}\, m_i \ell_i + \sqrt{g m_i^2 \ell_i^2 + 4 m_i}}{2} \begin{bmatrix} \sqrt{g} & 1 \end{bmatrix} \begin{bmatrix} \theta_i \\ \dot{\theta}_i \end{bmatrix}. \tag{3.237}$$

Now, we consider optimality of the decentralized control strategy defined by (3.237) for the overall system **S** of (3.232). We assume the following numerical values:

$$\begin{aligned} \ell_1 &= 1(\text{m}), & \ell_2 &= 0.8(\text{m}), \\ m_1 &= 1(\text{Kg}), & m_2 &= 0.8(\text{Kg}), \\ L &= 1.2(\text{m}), & g &= 9.8(\text{m}/\text{sec}^2), \\ k &= 0.02(\text{N}/\text{m}) \text{ and } 2(\text{N}/\text{m}); \end{aligned} \tag{3.238}$$

and compute the decentralized control law

$$\begin{bmatrix} \tau_1 \\ \tau_2 \end{bmatrix} = -\left[\begin{array}{cc|cc} 10.715 & 3.423 & 0 & 0 \\ \hline 0 & 0 & 7.340 & 2.345 \end{array}\right] \begin{bmatrix} \theta_1 \\ \dot{\theta}_1 \\ \theta_2 \\ \dot{\theta}_2 \end{bmatrix}. \tag{3.239}$$

The matrices Q_i and P_i, $i = 1, 2$, are

$$Q_1 = \begin{bmatrix} 9.8 & 0 \\ 0 & 1 \end{bmatrix}, \qquad Q_2 = \begin{bmatrix} 7.84 & 0 \\ 0 & 0.8 \end{bmatrix},$$

$$P_1 = \begin{bmatrix} 19.901 & 5.357 \\ 5.357 & 1.711 \end{bmatrix}, \qquad P_2 = \begin{bmatrix} 9.857 & 2.349 \\ 2.349 & 0.750 \end{bmatrix}, \tag{3.240}$$

and the nonlinear term in (3.232) is written as

$$f_C(\theta_1, \theta_2) = \begin{bmatrix} 0 \\ 9.8(\sin\theta_1 - \theta_1) + 2F\cos(\theta_1 - \phi) \\ 0 \\ 9.8(\sin\theta_2 - \theta_2) - 2.5F\cos(\theta_2 - \phi) \end{bmatrix}, \tag{3.241}$$

with

$$\begin{aligned} F &= k\{[3.08 - 2.4\sin\theta_1 + 1.92\sin\theta_2 - 1.6\cos(\theta_1 - \theta_2)]^{1/2} - 1.217\}, \\ \phi &= \tan^{-1}\frac{\cos\theta_1 - 0.8\cos\theta_2}{1.2 - \sin\theta_1 + 0.8\sin\theta_2}. \end{aligned} \tag{3.242}$$

We note here that when the nonlinear term f_C in (3.181) is independent of the input u, the condition (3.196) of Theorem 3.44 is reduced to

$$x^T \bar{Q}_D x - 2x^T \bar{P}_D f_C(t, x) \geq \bar{\alpha}' x^T x. \tag{3.243}$$

We use this fact to show optimality of the decentralized control law of (3.239). In the case $k = 0.02$, we set $d_1 = d_2 = 1$ and compute

$$[\theta_1 \quad \dot{\theta}_1 \quad \theta_2 \quad \dot{\theta}_2] \left[\begin{array}{cc|cc} 9.8 & 0 & 0 & 0 \\ 0 & 1 & 0 & 0 \\ \hline 0 & 0 & 7.84 & 0 \\ 0 & 0 & 0 & 0.8 \end{array}\right] \begin{bmatrix} \theta_1 \\ \dot{\theta}_1 \\ \theta_2 \\ \dot{\theta}_2 \end{bmatrix}$$

$$- 2[\theta_1 \quad \dot{\theta}_1 \quad \theta_2 \quad \dot{\theta}_2] \left[\begin{array}{cc|cc} 19.901 & 5.357 & 0 & 0 \\ 5.357 & 1.711 & 0 & 0 \\ \hline 0 & 0 & 9.857 & 2.349 \\ 0 & 0 & 2.349 & 0.750 \end{array}\right]$$

$$\times \left[\begin{array}{c} 0 \\ 9.8(\sin\theta_1 - \theta_1) + 2F\cos(\theta_1 - \phi) \\ \hline 0 \\ 9.8(\sin\theta_2 - \theta_2) - 2.5F\cos(\theta_2 - \phi) \end{array}\right]$$

$$\geq [\theta_1 \quad \dot{\theta}_1 \quad \theta_2 \quad \dot{\theta}_2] \left[\begin{array}{cc|cc} 9.8 & 0 & 0 & 0 \\ 0 & 1 & 0 & 0 \\ \hline 0 & 0 & 7.84 & 0 \\ 0 & 0 & 0 & 0.8 \end{array}\right] \begin{bmatrix} \theta_1 \\ \dot{\theta}_1 \\ \theta_2 \\ \dot{\theta}_2 \end{bmatrix}$$

$$- 2[\,|\theta_1| \quad |\dot{\theta}_1| \quad |\theta_2| \quad |\dot{\theta}_2|\,] \left[\begin{array}{c} 5.357(9.81|\sin\theta_1 - \theta_1| + 2|F|) \\ 1.711(9.8|\sin\theta_1 - \theta_1| + 2|F|) \\ \hline 2.349(9.8|\sin\theta_2 - \theta_2| + 2.5|F|) \\ 0.750(9.8|\sin\theta_2 - \theta_2| + 2.5|F|) \end{array}\right]$$

$$\geq [\,|\theta_1| \quad |\dot{\theta}_1| \quad |\theta_2| \quad |\dot{\theta}_2|\,] \left[\begin{array}{cc|cc} 3.118 & -1.067 & -1.366 & -0.171 \\ -1.067 & 1 & -0.265 & 0 \\ \hline -1.366 & -0.265 & 4.858 & -0.476 \\ -0.171 & 0 & -0.476 & 0.8 \end{array}\right] \begin{bmatrix} |\theta_1| \\ |\dot{\theta}_1| \\ |\theta_2| \\ |\dot{\theta}_2| \end{bmatrix}$$

$$\geq 0.05(\theta_1^2 + \dot{\theta}_1^2 + \theta_2^2 + \dot{\theta}_2^2), \tag{3.244}$$

where we used the inequalities

$$\begin{aligned} |\sin\theta_i - \theta_i| &\leq 0.045|\theta_i|, \\ |F| &\leq k(4.566|\theta_1| + 3.873|\theta_2|), \end{aligned} \tag{3.245}$$

which hold for $|\theta_i| < \pi/6$. Thus, the condition (3.243) is satisfied, and the decentralized state feedback of (3.239) is the optimal control law for the overall system $\mathbf{S}$ of (3.232) and the performance index (3.139). In this case, as mentioned in Remark (3.42), the resultant closed-loop nonlinear system has in each channel: (*i*) infinite gain margin, (*ii*) at least 50% gain reduction tolerance, and (*iii*) at least $\pm 60°$ phase margin.

In the case $k = 2(\mathrm{N/m})$, it can be readily shown that when $x = (\theta_1, \dot{\theta}_1, \theta_2, \dot{\theta}_2)^T = (-\pi/9, 1.6, \pi/9, -2.1)^T$, the condition (3.243) does not hold for any $d_1, d_2 > 0$. This implies that optimality of the decentralized control law of (3.239) cannot be restored by modifying only the performance index, and an increase of the feedback gains is necessary. Fortunately, in this case, the gain increases can eventually restore optimality because the nonlinearity $f_C(\theta_1, \theta_2)$ of (3.241) satisfies the matching condition. To show this fact, we write

$$f_C(\theta_1, \theta_2) = \begin{bmatrix} 0 & 0 \\ 2 & 0 \\ 0 & 0 \\ 0 & 3.125 \end{bmatrix} \begin{bmatrix} 4.9(\sin\theta_1 - \theta_1) + F\cos(\theta_1 - \phi) \\ 3.316(\sin\theta_2 - \theta_2) - 0.8F\cos(\theta_2 - \phi) \end{bmatrix}, \tag{3.246}$$

and note that the nonlinear functions on the right side of (3.246) have finite gains with respect to θ_1 and θ_2, which are implied by (3.245). Then, the condition (3.220) of Theorem 3.51 always holds with $h_c(t, x) = 0$, and by increasing the gains ρ_1, ρ_2, the decentralized control

$$\begin{bmatrix} \tau_1 \\ \tau_2 \end{bmatrix} = -\begin{bmatrix} \rho_1 & 0 \\ 0 & \rho_2 \end{bmatrix} \left[\begin{array}{cc|cc} 10.715 & 3.423 & 0 & 0 \\ \hline 0 & 0 & 7.340 & 2.345 \end{array}\right] \begin{bmatrix} \theta_1 \\ \dot{\theta}_1 \\ \theta_2 \\ \dot{\theta}_2 \end{bmatrix}, \tag{3.247}$$

becomes optimal for the system $\mathbf{S}$ of (3.232) and the performance index $\bar{J}_\rho$ of (3.218).

3.6. Notes and References

The suboptimality concept was introduced by Popov (1960) as a performance deterioration analysis of control systems. Wide classes of control systems were subsequently considered by many people (Dorato, 1963;

Durbeck, 1965; Rissanen, 1966; McClamroch and Rissanen, 1967; McClamroch, 1969; Burghart, 1969; Chang and Peng, 1972). Control structure constraints of the output type were introduced in the suboptimality context by Kosut (1970). Decentralized constraints in linear autonomous systems were considered by Bailey and Ramapriyan (1973).

A general approach to suboptimality of nonlinear time-varying systems was introduced in (Šiljak and Sundareshan, 1976), where the Hamilton–Jacobi theory was used to get decentralized connective suboptimality results (see also, Šiljak, 1978). A stochastic version of the suboptimality concept, which generalized a number of results and the definition of the concept as well, was obtained for the full-blown LQG problem (Krtolica and Šiljak, 1980). Overlapping information structure constraints were solved (Ikeda *et al.* 1981; Hodžić *et al.* 1983; Hodžić and Šiljak, 1986) within the framework of the Inclusion Principle (Chapter 8).

For us, one of the most important implications of suboptimality is the robustness in terms of the classical measures of *gain and phase margin* (Sezer and Šiljak, 1981a,b). This implication is somewhat surprising since the meaning of decentralized suboptimality is quite different from that of centralized optimality. Nonlinear distortions of suboptimal control law are also tolerated by suboptimal closed-loop systems. This type of results, which were presented in Sections 3.1 and 3.3, increase considerably our confidence in suboptimal decentralized control.

If we are willing to modify the performance index, then we can recover global optimality from locally optimal LQ control subsystems (Özgüner, 1975). This we can do even if the interconnections are nonlinear (Šiljak, 1979, 1980). To broaden the class of potentially optimal decentralized control systems, we have to allow for modifications of the control laws as well (Ikeda and Šiljak, 1982; Ikeda *et al.* 1983; Yasuda, 1986). This is *the inverse optimal problem of decentralized control* which we solved in Sections 3.4 and 3.5 following (Ikeda and Šiljak, 1985, 1990). Other solutions can be found in (Ikeda and Šiljak, 1982; Young, 1985; Zheng, 1986; Saberi, 1988; Yang *et al.* 1989).

A very important and useful area of research is the *parametric optimization* framework for design of decentralized control laws. Initial results were obtained by Levine and Athans (1970) for the output control constraints. Similar methods were used by Geromel and Bernussou (1982) to get the best available decentralized control laws. An interesting application of these results to interconnected power systems was described by Gopal and Ghodekar (1985). Of special value is the work of Friedlander (1985), because he proposed to compute the parameters of decentralized controllers using only the locally available information; the communication requirements between subsystems are minimized in a significant way.

A comprehensive survey of parameter optimization methods for design of decentralized controllers is available in (Geromel and Bernussou, 1987). In the context of this chapter, these methods can be used to improve suboptimality of locally optimal LQ control laws; feedback gains can be adjusted by gradient algorithms to get the best results.

Finally, we should comment on the *hierarchical control* of large-scale systems initiated by Mesarović *et al.* (1970), and developed by many people (see the books by Findeisen *et al.* 1980; Singh, 1980; and Jamshidi, 1983; and the recent surveys by Haimes and Li, 1988; and Malinowski, 1989). The concept is a type of decentralized optimization scheme, where a coordinating (centralized) control is introduced to ensure that the local controls are properly modified to result in an overall optimal system. Defined in this way, the concept is a mixture of decentralized and centralized optimal control, which is hard to justify on both accounts. No one knows if the limitations of the hierarchical schemes can be removed by developing better design procedures, or if these limitations can be attributed to the lack of a suitable axiomatic definition of the optimization problem. Simply speaking, it has never been clearly demonstrated under reasonable assumptions that hierarchical control is superior to centralized or decentralized alternatives. We regard hierarchical control as a problem for the future and shall not consider it here.

Bibliography

Anderson, B. D. O., and J. B. Moore (1971). *Linear Optimal Control*, Prentice-Hall, Englewood Cliffs, New Jersey.

Arrow, K. J., and F. H. Hahn (1971). *General Competitive Analysis.* Holden-Day, San Francisco, California.

Bailey, F. N., and H. K. Ramapriyan (1973). Bounds on suboptimality in the control of linear dynamic systems. *IEEE Transactions*, AC-18, 532–534.

Bode, H. W. (1945). *Network Analysis and Feedback Amplifier Design.* Van Nostrand, New York.

Burghart, J. H. (1969). A technique for suboptimal feedback control of nonlinear systems. *IEEE Transactions*, AC-14, 530–533.

Chang, S. S. L., and T. K. C. Peng (1972). Adaptive guaranteed cost control of systems with uncertain parameters. *IEEE Transactions*, AC-17, 474–483.

Davison, E. J. (1974). The decentralized stabilization and control of a class of unknown nonlinear time-varying systems. *Automatica*, 10, 309–316.

Dorato, P. (1963). On sensitivity of optimal control systems. *IEEE Transactions*, AC-8, 256–257.

Durbeck, R. (1965). An approximation technique for suboptimal control. *IEEE Transactions*, AC-10, 144–149.

Findeisen, W., F. N. Bailey, M. Brdyś, K. Malinowski, P. Tatjewski, and A. Woźniak (1980). *Control and Coordination in Hierarchical Systems.* John Wiley, New York.

Friedlander, B. (1985). Decentralized design of decentralized controllers. *Control and Dynamic Systems*, C. T. Leondes (ed.), Academic Press, New York, 22, 165–194.

Geromel, J. C., and J. Bernussou (1982). Optimal decentralized control of dynamic systems. *Automatica*, 18, 545–557.

Geromel, J. C., and J. Bernussou (1987). Structure-constrained control: Parametric optimization. *Systems and Control Encyclopedia*, M. G. Singh (ed.), Pergamon Press, Oxford, UK, 4678–4685.

Gopal, M., and J. G. Ghodekar (1985). On decentralized controllers for interconnected power systems. *International Journal of Systems Science*, 16, 1391–1407.

Haimes, Y. Y., and D. Li (1988). Hierarchical multiobjective analysis for large-scale systems: Review and current status. *Automatica*, 24, 53–69.

Hodžić, M., and D. D. Šiljak (1986). Decentralized estimation and control with overlapping information sets. *IEEE Transactions*, AC-31, 83–86.

Hodžić, M., R. Krtolica, and D. D. Šiljak (1983). A stochastic inclusion principle. *Proceedings of the* 22nd *IEEE Conference on Decision and Control*, San Antonio, Texas, 17–22.

Ikeda, M., and D. D. Šiljak (1982). When is a linear decentralized control optimal? *Analysis and Optimization of Systems*, A. Bensoussan and J. L. Lions (eds.), Springer, Berlin, FRG, 419–431.

Ikeda, M., and D. D. Šiljak (1985). On optimality and robustness of LQ regulators for nonlinear and interconnected systems. *Proceedings of the IFAC Workshop on Model Error Concepts and Compensations*, Boston, Massachusetts, 77–82.

Ikeda, M., and D. D. Šiljak (1990). Optimality and robustness of linear–quadratic control for nonlinear systems. *Automatica*, 26, 499-511.

Ikeda, M., O. Umefuji, and S. Kodama (1976). Stabilization of large-scale linear systems. *Systems, Computers, and Control*, 7, 34–41.

Ikeda, M., D. D. Šiljak, and D. E. White (1981). Decentralized control with overlapping information sets. *Journal of Optimization Theory and Applications*, 34, 279–310.

Ikeda, M., D. D. Šiljak, and K. Yasuda (1983). Optimality of decentralized control for large-scale systems. *Automatica*, 19, 309–316.

Jamshidi, M. (1983). *Large-Scale Systems: Modeling and Control.* North-Holland, Amsterdam, The Nederlands.

Kalman, R. E. (1960). Contributions to the theory of optimal control. *Boletin de la Sociedad Matematica Mexicana*, 5, 102–119.

Kalman, R. E. (1964). When is a linear control system optimal? *Transactions of ASME, Journal of Basic Engineering*, 86, 1–10.

Kalman, R. E., and J. E. Bertram (1960). Control system analysis and design via the "Second Method" of Lyapunov. *Transactions of ASME, Journal of Basic Engineering*, 82, Part I: 371–393; Part II: 394–400.

Kosut, R. L. (1970). Suboptimal control of linear time-invariant systems subject to control structure constraints. *IEEE Transactions*, AC-15, 557–563.

Krtolica, R., and D. D. Šiljak (1980). Suboptimality of decentralized stochastic control and estimation. *IEEE Transactions*, AC-25, 76–83.

Leitmann, G. (1979). Guaranteed asymptotic stability for some linear systems with bounded uncertainties. *Transactions of ASME, Journal of Dynamic Systems, Measurements, and Control*, 101, 212–216.

Levine, W. S., and M. Athans (1970). On the determination of the optimal constant output feedback gains for linear multivariable systems. *IEEE Transactions*, AC-15, 44–48.

Malinowski, K. (1989). Hierarchical control under uncertainty. *Preprints of the 5th IFAC/IFORS Symposium on Large Scale Systems*, Berlin, DDR.

Mao, C. J., and W. S. Lin (1990). Decentralized control of interconnected systems with unmodeled nonlinearity and interaction. *Automatica*, 26, 263–268.

McClamroch, N. H. (1969). Evaluation of suboptimality and sensitivity of control and filtering processes. *IEEE Transactions*, AC-13, 282–285.

McClamroch, N. H., and J. Rissanen (1967). A result on the performance deterioration of optimum systems. *IEEE Transactions*, AC-12, 209–210.

Mesarović, M. D., D. Macko, and Y. Takahara (1970). *Theory of Hierarchical, Multilevel, Systems.* Academic Press, New York.

Özgüner, Ü. (1975). Local optimization in large scale composite dynamic systems. *Proceedings of the 9th Asilomar Conference on Circuits, Systems, and Computers*, Pacific Grove, California, 87–91.

Popov, V. M. (1960). Criterion of quality for nonlinear systems. *Proceedings of the 5th IFAC World Congress*, Moscow, USSR, 173–176.

Rissanen, J. J. (1966). Performance deterioration of optimum systems. *IEEE Transactions*, AC-11, 530–532.

Saberi, A. (1988). On optimality of decentralized control for a class of nonlinear interconnected systems. *Automatica*, 24, 101–104.

Saberi, A., and H. Khalil (1982). Decentralized stabilization of a class of non-linear interconnected systems. *International Journal of Control*, 36, 803–818.

Safonov, M. G. (1980). *Stability and Robustness of Multivariable Feedback Systems*, MIT Press, Boston, Massachusetts.

Sezer, M. E., and Ö. Hüseyin (1978). Stabilization of linear interconnected systems using local state feedback. *IEEE Transactions*, SMC-8, 751–756.

Sezer, M. E., and D. D. Šiljak (1981a). Robustness of suboptimal control: Gain and phase margin. *IEEE Transactions*, AC-26, 907–911.

Sezer, M. E., and D. D. Šiljak (1981b). Sensitivity of large-scale control systems. *Journal of the Franklin Institute*, 312, 179-197.

Šiljak, D. D. (1978). *Large-Scale Dynamic Systems: Stability and Structure.* North-Holland, New York.

Šiljak, D. D. (1979). Overlapping decentralized control. *Large Scale Systems Engineering Applications*, M. G. Singh and A. Titli (eds.), North-Holland, Amsterdam, Holland, 145–166.

Šiljak, D. D. (1980). Reliable control using multiple control systems. *International Journal of Control*, 31, 309–339.

Šiljak, D. D. (1987). Interconnected systems: Decentralized control. *Systems and Control Encyclopedia*, M. G. Singh (ed.), Pergamon Press, Oxford, UK, 2557–2560.

Šiljak, D. D. (1988). Parameter space methods for robust control design: A guided tour. *IEEE Transactions*, AC-34, 674–688.

Šiljak, D. D., and M. K. Sundareshan (1976). A multilevel optimization of large-scale dynamic systems. *IEEE Transactions*, AC-21, 79–84.

Sinai, M. (1986). Suboptimality bounds on decentralized control and estimation of large-scale discrete-time linear systems. *Control and Dynamic Systems*, C. T. Leondes (ed.), 24, 67–103.

Singh, M. G. (1980). *Dynamical Hierarchical Control.* North-Holland, Amsterdam, The Netherlands.

Thaler, G. J., and R. G. Brown (1960). *Analysis and Design of Feedback Control Systems.* McGraw-Hill, New York.

Wonham, W. M. (1968). On a matrix Riccati equation of stochastic control. *SIAM Journal of Control*, 6, 681–697.

Wonham, W. M. (1979). *Linear Multivariable Control: A Geometric Approach.* Springer, New York.

Yang, T. C., N. Munro, and A. Brameller (1989). Improved condition for the optimality of decentralized control for large-scale systems. *IEE Proceedings*, 136-D, 44–46.

Yasuda, K. (1986). Decentralized optimal control for large-scale interconnected systems. *Control and Dynamic Systems*, C. T. Leondes (ed.), Academic Press, New York, 23, 139–163.

Yasuda, K., T. Hikata, and K. Hirai (1980). On decentrally optimizable interconnected systems. *Proceedings of the 19th IEEE Conference on Decision and Control*, Albuquerque, New Mexico, 536–537.

Young, K. D. (1985). On near optimal decentralized control. *Automatica*, 21, 607–610.

Zheng, D. Z. (1986). Optimization of linear–quadratic regulator systems in the presence of parameter perturbations. *IEEE Transactions*, AC-31, 667–670.

Chapter 4 Estimation and Control

Our ability to control an interconnected system by decentralized state feedback depends crucially upon the availability of states at each subsystem. In most practical cases, the states of subsystems are not accessible as outputs, and we have to solve the problem of state determination in complex systems. The important part of the problem is the constraint that the state determination be carried out decentrally; we want to build a *decentralized asymptotic observer.* In this way, we can provide an observer for each subsystem, which makes states available locally to each individual controller.

For a union of subsystem observers to represent an asymptotic observer for the overall system, the interconnections among the subsystems should appear as inputs to the observers. It is comforting to note that this does not mean that the interconnections need to be read as outputs. Such an assumption would be unrealistic, because the interconnections are almost always inaccessible variables inside the plant. We can use the state estimates instead and construct the necessary inputs. Unfortunately, this means that the observers have to exchange their estimates, thus requiring an excessive data communication between the subsystems. This is the first problem we must come to grips with in building decentralized regulators for complex plants.

An observer is acceptable only if it is asymptotically stable. At this point in our exposition, we do not expect stabilizability of a decentralized observer to be straightforward. We are glad to find, however, that the stabilization problem can be solved *via* "duality" by the methods of Chapter 2. The decentrally stabilizable structures will reappear in a dual form, and high-gain feedback will again be used to make sure that the overall observer is asymptotically stable. Most importantly, the decentralized observer and controller can be designed independently of each other. This is the so-called *separation property*, which has an added significance in solving difficult control problems involving decentralized information structure constraints.

In the second part of this chapter, we shall consider the stochastic decentralized control and present a decentralized version of the well-known Linear–Quadratic–Gaussian (LQG) design. The optimal control strategies in this context have been much talked about, but very little has been accomplished. We take a pragmatic approach to this problem, and develop *suboptimal control laws*, which are stochastic versions of the laws obtained in the preceding chapter. Actually, this is a rewarding path to take, because we are trading the futile optimality quest for the *robustness* of suboptimal control laws to a wide variety of parametric (structural) perturbations.

4.1. An Interconnected Observer

Our first task is to show how to determine the state of an interconnected system by building only local observers for the subsystems. Each observer produces a state estimate of the subsystem it is attached to, and the question arises: When does the union of the local estimates constitute an estimate of the overall system? At the time this problem was introduced (Šiljak and Vukčević, 1976), it was expected that the design of decentralized observers is a natural counterpart to the design of the controllers using decentralized state feedback. It turned out, however, that to build local observers one has to allow for an exchange of state estimates among the observers. The exchange is necessary for the separation principle to hold, so that one can design the observers independently of the controllers that use the estimated states for stabilization purposes. The obvious concern is that the exchange of estimates among the subsystems may be costly when, for example, the controllers are distributed geographically in mutually distant areas. On the other hand, when decentralized observers are used to simplify the computations or speed up the observation process using a parallel processing scheme, the exchange is not a dominating factor. Furthermore, the exchange can be removed altogether by giving up the separation property. We start our exposition of decentralized state determination with the basic result of Šiljak and Vukčević (1976), which requires only the standard observer theory (*e.g.*, O'Reilly, 1983).

Let us consider a linear time invariant system

$$\mathbf{S}\colon\ \begin{aligned} \dot{x}_i &= A_i x_i + B_i u + \sum_{j=1}^{N} A_{ij} x_j, \\ y_i &= C_i x_i, \qquad\qquad i \in \boldsymbol{N}, \end{aligned} \tag{4.1}$$

which is an interconnection of N subsystems

$$\mathbf{S}_i\colon\ \dot{x}_i = A_i x_i + B_i u, \qquad\qquad (4.2)$$
$$y_i = C_i x_i, \qquad i \in \boldsymbol{N},$$

where, as usual, $x_i(t) \in \mathbf{R}^{n_i}$ is the state, $y_i(t) \in \mathbf{R}^{\ell_i}$ is the output of $\mathbf{S}_i$ and $u(t) \in \mathbf{R}^m$ is the input of $\mathbf{S}$ at time $t \in \mathbf{T}$.

In order to estimate the states $x_i(t)$, we assume that all pairs (A_i, C_i) are observable and build an observer

$$\hat{\mathbf{S}}\colon\ \dot{\hat{x}}_i = \tilde{A}_i\hat{x}_i + L_i y_i + B_i u + \sum_{j=1}^{N} \tilde{A}_{ij}\hat{x}_j + \sum_{j=1}^{N} L_{ij} y_j, \qquad i \in \boldsymbol{N}, \qquad (4.3)$$

where

$$\tilde{A}_i = A_i - L_i C_i, \qquad \tilde{A}_{ij} = A_{ij} - L_{ij} C_j, \qquad (4.4)$$

and the gain matrices L_i and L_{ij} have appropriate dimensions.

For the errors in estimation

$$\tilde{x}_i = x_i - \hat{x}_i, \qquad i \in \boldsymbol{N}, \qquad (4.5)$$

we get the equations

$$\tilde{\mathbf{S}}\colon\ \dot{\tilde{x}}_i = \tilde{A}_i\tilde{x}_i + \sum_{j=1}^{N} \tilde{A}_{ij}\tilde{x}_j, \qquad i \in \boldsymbol{N}. \qquad (4.6)$$

To obtain an asymptotic observer from (4.6), we stabilize the dual of $\tilde{\mathbf{S}}$, which is described as

$$\tilde{\mathbf{S}}^*\colon\ \dot{\tilde{x}}_i = \tilde{A}_i^T\tilde{x}_i + \sum_{j=1}^{N} \tilde{A}_{ji}^T\tilde{x}_j, \qquad i \in \boldsymbol{N}. \qquad (4.7)$$

Using (4.4), we get $\tilde{\mathbf{S}}^*$ as

$$\tilde{\mathbf{S}}^*\colon\ \dot{\tilde{x}}_i = (A_i^T - C_i^T L_i^T)\tilde{x}_i + \sum_{j=1}^{N} (A_{ji}^T - C_i^T L_{ji}^T)\tilde{x}_j, \qquad i \in \boldsymbol{N}. \qquad (4.8)$$

Stabilization of $\tilde{\mathbf{S}}^*$ can proceed by applying the decentralized scheme of Section 2.4. We assume that the pairs (A_i, C_i) are observable, which implies that the pairs (A_i^T, C_i^T) are controllable. Then, we can place the eigenvalues of $\tilde{A}_i^T = A_i^T - C_i^T L_i^T$ at desired locations by a choice of local gain L_i. The global gains L_{ij} are selected to minimize the influence of the interactions A_{ij} wherever appropriate. Finally, stability of the overall closed-loop system $\tilde{\mathbf{S}}^*$ is tested using the method of vector Liapunov functions. This recipe is described first.

For ease of presentation, we use our standard compact form for $\mathbf{S}$:

$$\mathbf{S}\colon\ \dot{x} = A_D x + Bu + A_C x,\qquad y = C_D x, \tag{4.9}$$

which represents an output-decentralized system, that is, $C_D = \mathrm{diag}\{C_1, C_2, \ldots, C_N\}$. The form of the matrix B is not essential. The interconnected observer is

$$\hat{\mathbf{S}}\colon\ \dot{\hat{x}} = A_D\hat{x} + L_D(y - C_D\hat{x}) + Bu + A_C\hat{x} + L_C(y - C_D\hat{x}), \tag{4.10}$$

where

$$L_D = \mathrm{diag}\{L_1, L_2, \ldots, L_N\}, \qquad L_C = (L_{ij}), \tag{4.11}$$

are constant feedback matrices with blocks L_i and L_{ij} having appropriate dimensions.

The observation error

$$\tilde{x} = x - \hat{x} \tag{4.12}$$

is described by the error system

$$\tilde{\mathbf{S}}\colon\ \dot{\tilde{x}} = (A_D - L_D C_D)\tilde{x} + (A_C - L_C C_D)\tilde{x}. \tag{4.13}$$

For $\hat{\mathbf{S}}$ to be an asymptotic observer, we require that $\tilde{\mathbf{S}}$ is globally asymptotically stable, that is,

$$\tilde{x}(t) \to 0, \qquad \text{as}\quad t \to +\infty. \tag{4.14}$$

This may not be completely satisfactory, and we want to say something about the speed of (4.14).

4.1. DEFINITION. A system $\hat{\mathbf{S}}$ is an exponential observer with degree π for a system $\mathbf{S}$ if there exist two positive numbers Π and π such that

$$\|\tilde{x}(t;\, t_0,\, x_0)\| \le \Pi\|x_0\|\ \exp\left[-\pi(t - t_0)\right], \qquad \forall t \in \mathbf{T}_0,\quad \forall(t_0, x_0) \in \mathbf{T} \times \mathbf{R}^n, \tag{4.15}$$

where $\tilde{x}(t;\, t_0,\, x_0)$ is a solution of the system $\tilde{\mathbf{S}}$.

What we want is to find the feedback gains L_D and L_C so that (4.15) holds for a sufficiently large value of π. We show first how to do this when $L_C = 0$ and only local feedback gains L_D are used. That is, we consider stabilization of

$$\tilde{\mathbf{S}}^*\colon\ \dot{\tilde{x}} = (A_D^T - C_D^T L_D^T)\tilde{x} + A_C^T \tilde{x}. \tag{4.16}$$

For L_D, we choose

$$L_D = S_D C_D^T, \tag{4.17}$$

where $S_D = \text{diag}\{S_1, S_2, \ldots, S_N\}$ is the symmetric positive definite solution of the Riccati equation

$$(A_D + \pi I)S_D + S_D(A_D + \pi I)^T - S_D C_D^T C_D S_D + Q_D = 0, \tag{4.18}$$

and $Q_D = \text{diag}\{Q_1, Q_2, \ldots, Q_N\}$ is a symmetric nonnegative definite matrix.

4.2. THEOREM. A system

$$\hat{\mathbf{S}}\colon\ \dot{\hat{x}} = A_D\hat{x} + L_D(y - C_D\hat{x}) + Bu + A_C\hat{x} \tag{4.19}$$

is an exponential observer with degree π for the system $\mathbf{S}$ if

$$\lambda_M^{1/2}(A_C^T A_C) \le \frac{1}{2}\frac{\lambda_m(Q_D + S_D C_D^T C_D S_D)}{\lambda_M(S_D)}. \tag{4.20}$$

Proof. We select the function $v\colon\ \mathbf{R}^n \to \mathbf{R}_+$ defined as

$$v(\tilde{x}) = \tilde{x}^T S_D \tilde{x} \tag{4.21}$$

to be a candidate for the Liapunov function of the system $\tilde{\mathbf{S}}^*$. As usual, we calculate

$$\begin{aligned} \dot{v}(\tilde{x})_{(4.16)} &= \tilde{x}^T\left[(A_D - L_D C_D)S_D + S_D(A_D - L_D C_D)^T\right]\tilde{x} + 2\tilde{x}^T A_C S_D \tilde{x} \\ &= -2\pi\tilde{x}^T S_D\tilde{x} - \tilde{x}^T(Q_D + S_D C_D^T C_D S_D)\tilde{x} + 2\tilde{x}^T A_C S_D\tilde{x} \\ &\le -2\pi\tilde{x}^T S_D\tilde{x} \\ &\le -2\pi v(\tilde{x}), \qquad \forall (t, \tilde{x}) \in \mathbf{T} \times \mathbf{R}^n, \end{aligned} \tag{4.22}$$

using condition (4.20) of the theorem. The last inequality in (4.22) implies (4.15) with $\Pi = \lambda_M^{1/2}(S_D)\lambda^{-1/2}(S_D)$. Q.E.D.

If the interconnections are not sufficiently small, the condition (4.20) is not satisfied, and the observer $\hat{\mathbf{S}}$ of (4.19) is useless. A way to make the observer $\hat{\mathbf{S}}$ work in this case is to use the global control concept of Šiljak and Vukčević (1976). Then, instead of the size limitation, the interconnection matrix A_C and the output matrix C_D are required to satisfy a rank condition that makes the perfect neutralization of the interconnections possible.

4.3. THEOREM. Suppose that

$$\begin{aligned} &(i)\ \text{Im}\ C_D = \mathbf{R}^\ell; \\ &(i)\ \text{Im}\ C_D^T \supset \text{Im}\ A_C^T\ . \end{aligned} \tag{4.23}$$

Then the system

$$\hat{\mathbf{S}}\colon\ \dot{\hat{x}} = A_D\hat{x} + L_D(y - C_D\hat{x}) + Bu + A_C\hat{x} + L_C(y - C_D\hat{x}), \tag{4.24}$$

with $L_C = A_C C_D^T (C_D C_D^T)^{-1}$, is an exponential observer with degree π for the system $\mathbf{S}$.

Proof. The error system is $\tilde{\mathbf{S}}$ of (4.13). From the proof of Theorem 4.2, it is clear that $\tilde{\mathbf{S}}$ is exponentially stable with degree π if the second term $(A_C - L_C C_D)\tilde{x}$ in (4.13) is zero. From the condition (i) of the theorem, it follows that C_D has full row rank and has the right inverse $C_D^{\#} = C_D^T(C_D C_D^T)^{-1}$. This fact and (ii) imply that A_C can be factored as $A_C = L_C C_D$, with L_C as defined in the theorem. Q.E.D.

4.4. REMARK. Although global feedback in the observation scheme (4.23) neutralized the interconnection terms, the price may be too high because the local observers have to exchange the observed state $\hat{x}$. The price would be reduced if we could use the output y instead of the state $\hat{x}$. This possibility is presented by the observation scheme

$$\hat{\mathbf{S}}\colon\ \dot{\hat{x}} = A_D\hat{x} + Bu + L_D(y - C_D\hat{x}) + L_C y \tag{4.25}$$

for the system $\mathbf{S}$ of (4.9). Then, the observation error $\tilde{x} = x - \hat{x}$ is described by

$$\tilde{\mathbf{S}}\colon\ \dot{\tilde{x}} = (A_D - L_D C_D)\tilde{x} + (A_C - L_C C_D)x. \tag{4.26}$$

Under the conditions (*i*)–(*ii*) of Theorem 4.3, the term $(A_C - L_C C_D)x$ would vanish, and $\hat{\mathbf{S}}$ of (4.25) is an exponential observer for **S**. It is crucial, however, to recognize the fact that the observer $\hat{\mathbf{S}}$ of (4.25) is not robust if the state $x(t)$ of **S** is not properly bounded, since an imperfection in the cancellation of the term $A_C x$ by the global output feedback $L_C y$ could cause the error system $\tilde{\mathbf{S}}$ of (4.26) to blow up. One way to fix this is to consider the estimation and control design as a joint problem, as we show in Section 4.4.

4.2. Decentralized Feedback

Even if global feedback is applied to reduce the size of interconnections, the union of local observers may not be an asymptotic observer for the overall system, unless the interconnections are completely neutralized. It is, therefore, natural to delineate a class of systems for which interconnected observers can *always* be built using only decentralized feedback. This is a dual to the problem of decentrally stabilizable composite systems treated in Section 2.4, and we are prepared to surrender the generality of the system description for the assured success in building an interconnected observer, even though it may involve a high-gain feedback (Šiljak and Vukčević, 1978).

The interconnected system

$$\mathbf{S}\colon\ \dot{x}_i = A_i x_i + B_i u_i + \sum_{j=1}^{N} A_{ij} x_j,$$
$$y_i = c_i^T x_i, \qquad i \in \boldsymbol{N}, \tag{4.27}$$

is a special case of **S** described by (4.1), because each subsystem $\mathbf{S}_i$ has a single output $y_i \in \mathbf{R}$, and $c_i \in \mathbf{R}^{n_i}$ is a constant vector. We assume that all the pairs (A_i, c_i) are observable, which allows us to further assume without loss of generality that the subsystem matrices A_i and c_i have the following form:

$$A_i = \begin{bmatrix} 0 & 0 & \dots & 0 & -a_1^i \\ 1 & 0 & \dots & 0 & -a_2^i \\ \dots & \dots & \dots & \dots & \dots \\ 0 & 0 & \dots & 1 & -a_{n_i}^i \end{bmatrix}, \qquad c_i = \begin{bmatrix} 0 \\ 0 \\ \vdots \\ 1 \end{bmatrix}. \tag{4.28}$$

The crucial assumption is that the interconnection matrices $A_{ij} = (a_{pq}^{ij})$ are such that

$$a_{pq}^{ij} = 0, \qquad p > q, \qquad p \in \boldsymbol{N}_i, \qquad q \in \boldsymbol{N}_j, \tag{4.29}$$

where $\boldsymbol{N}_i = \{1, 2, \dots, n_i\}$ and $\boldsymbol{N}_j = \{1, 2, \dots, n_j\}$. These are the structural constraints on the interconnections, which make the local stabilization possible.

To construct an interconnected estimator for the system $\mathbf{S}$ of (4.27) that uses decentralized feedback only, we need the following:

4.5. THEOREM. Given a system $\mathbf{S}$, the local feedback gains ℓ_i can always be selected so that the system

$$\hat{\mathbf{S}}\colon\ \dot{\hat{x}}_i = A_i \hat{x}_i + \ell_i y_i + B_i u + \sum_{j=1}^{N} A_{ij} \hat{x}_j, \qquad i \in \boldsymbol{N}, \tag{4.30}$$

is an asymptotic observer for $\mathbf{S}$.

Proof. The observation error is described by the dual system

$$\tilde{\mathbf{S}}^*\colon\ \dot{\tilde{x}}_i = \left(A_i^T - c_i \ell_i^T\right) \tilde{x}_i + \sum_{j=1}^{N} A_{ji}^T \tilde{x}_j, \qquad i \in \boldsymbol{N}. \tag{4.31}$$

Each gain ℓ_i can be chosen for the matrix $A_i^T - c_i \ell_i^T$ to have a set of distinct real eigenvalues

$$\mathcal{L}_i = \{\lambda_k^i\colon\ \lambda_k^i = \rho \sigma_k^i, \quad \rho \geq 1, \quad \sigma_k^i > 0, \quad \forall k \in \boldsymbol{N}\}, \tag{4.32}$$

for any specified value of ρ. What we want to show is that by choosing a sufficiently large ρ, we can make the error system $\tilde{\mathbf{S}}$ exponentially stable for a given π.

Applying the nonsingular transformation

$$\tilde{x}_i = \bar{T}_i \bar{x}_i, \qquad i \in \boldsymbol{N}, \tag{4.33}$$

we get $\tilde{\mathbf{S}}^*$ as

$$\bar{\mathbf{S}}^*\colon\ \dot{\bar{x}}_i = \Lambda_i \bar{x}_i + \sum_{j=1}^{N} \bar{A}_{ji}^T \bar{x}_j, \qquad i \in \boldsymbol{N}, \tag{4.34}$$

where

$$\Lambda_i = \bar{T}_i^{-1} \left(A_i^T - c_i \ell_i^T\right) \bar{T}_i, \qquad \bar{A}_{ji}^T = \bar{T}_i^{-1} A_{ji}^T \bar{T}_j, \tag{4.35}$$

with Λ_i in the diagonal form

$$\Lambda_i = \text{diag}\{-\rho\sigma_1^i, -\rho\sigma_2^i, \ldots, -\rho\sigma_{n_i}^i\}. \tag{4.36}$$

The transformation matrix T_i can be factored as

$$\bar{T}_i = R_i T_i, \tag{4.37}$$

with the factors

$$R_i = \text{diag}\{1, \rho, \ldots, \rho^{n_i-1}\} \tag{4.38}$$

and the Vandermonde matrix

$$T_i = \begin{bmatrix} 1 & 1 & \ldots & 1 \\ -\sigma_1^i & -\sigma_2^i & \ldots & -\sigma_{n_i}^i \\ \ldots & \ldots & \ldots & \ldots \\ (-\sigma_1^i)^{n_i-1} & (-\sigma_2^i)^{n_i-1} & \ldots & (-\sigma_{n_i}^i)^{n_i-1} \end{bmatrix}. \tag{4.39}$$

With this modification, each error subsystem

$$\bar{\mathbf{S}}_i^*\colon\ \dot{\bar{x}}_i = \bar{\Lambda}_i \bar{x}_i \tag{4.40}$$

has a degree of exponential stability

$$\bar{\pi}_i = \rho\pi_i, \tag{4.41}$$

where

$$\pi_i = \min_{k \in \boldsymbol{N}_i} \{\sigma_k^i\}. \tag{4.42}$$

As in Section 2.5, we choose the Liapunov function

$$v_i(\bar{x}_i) = \bar{x}_i^T \bar{H}_i \bar{x}_i, \tag{4.43}$$

such that

$$\Lambda_i \bar{H}_i + \bar{H}_i \Lambda_i = -\bar{G}_i \tag{4.44}$$

and

$$\begin{aligned} \bar{G}_i &= 2\ \text{diag}\{\rho\sigma_1^i, \rho\sigma_2^i, \ldots, \rho\sigma_{n_i}^i\}, \\ \bar{H}_i &= I_i. \end{aligned} \tag{4.45}$$

Using the function

$$\nu(\bar{x}) = \sum_{i=1}^{N} d_i v_i(\bar{x}_i), \tag{4.46}$$

as a Liapunov function for the system $\bar{\mathbf{S}}^*$ with the state $\bar{x} = (\bar{x}_1^T, \bar{x}_2^T, \ldots, \bar{x}_N^T)^T$, we get the $N \times N$ aggregate matrix $\bar{W} = (\bar{w}_{ij})$ defined by

$$\bar{w}_{ij} = \begin{cases} \bar{\pi}_i - \bar{\xi}_{ij}, & i = j, \\ -\bar{\xi}_{ij}, & i \neq j, \end{cases} \tag{4.47}$$

where

$$\bar{\xi}_{ij} = \lambda_M^{1/2}(\bar{A}_{ji}\bar{A}_{ji}^T) \tag{4.48}$$

and

$$\bar{A}_{ji} = T_i^{-1} R_i^{-1} A_{ji}^T R_j T_j. \tag{4.49}$$

Now, we note that the diagonal elements $\bar{w}_{ii} = \frac{1}{2}\rho\pi_i$ depend linearly on the adjustable parameter ρ. The off-diagonal elements $\bar{w}_{ij} = \bar{\xi}_{ij}(\rho)$, however, are bounded functions of ρ. The elements $\rho^{p-q}a_{pq}^{ji}$ of the matrices $R_i A_{ji}^T R_j$ are either zero for $p > q$ due to the assumption (4.29), or they are bounded for $p \leq q$ due to nonpositive powers of ρ. Therefore, we have

$$\lim_{\rho \to +\infty} R_i^{-1} A_{ji}^T R_j = D_{ij}, \tag{4.50}$$

where the $n_i \times n_j$ matrix $D_{ij} = (d_{pq}^{ij})$ is defined by

$$d_{pq}^{ij} = \begin{cases} a_{qp}^{ij}, & p = q, \\ 0, & p \neq q. \end{cases} \tag{4.51}$$

From (4.49) and (4.50), we define

$$\bar{D}_{ij} = T_i^{-1} D_{ij} T_j, \tag{4.52}$$

and conclude that

$$\lim_{\rho \to \infty} \bar{\xi}_{ij}(\rho) = \lambda_M^{1/2}(\bar{D}_{ij}^T \bar{D}_{ij}), \tag{4.53}$$

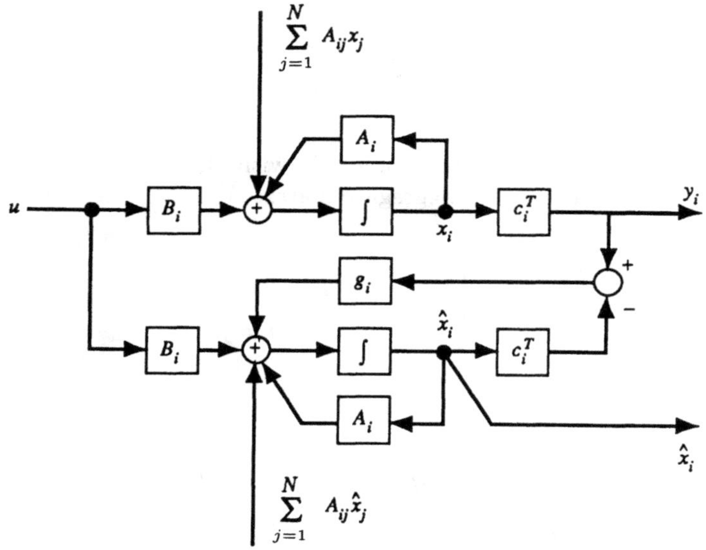

Fig. 4.1. Subsystem with observer.

which implies that the off-diagonal elements $\bar{w}_{ij}$ of the matrix $\bar{W}$ are bounded in ρ. Therefore, we can always choose local gain vectors ℓ_i to make $\bar{W}$ an M-matrix. This implies stability of $\bar{\mathbf{S}}^*$ and, thus, $\tilde{\mathbf{S}}^*$, *via* Theorem 2.5. Q.E.D.

The final form of the interconnected observer corresponding to the system $\mathbf{S}$ of (4.27) is given as

$$\hat{\mathbf{S}}\colon\ \dot{\hat{x}}_i = A_i\hat{x}_i + \ell_i(y_i - c_i^T\hat{x}_i) + B_iu + \sum_{j=1}^{N} A_{ij}\hat{x}_j, \qquad i \in \boldsymbol{N}. \tag{4.54}$$

The block diagram of the ith subsystem $\mathbf{S}_i$ and its observer $\hat{\mathbf{S}}_i$ are shown in Figure 4.1. We note that $\hat{\mathbf{S}}_i$ has as input the observed states from all other local observers, which prevents $\hat{\mathbf{S}}$ from being a fully decentralized observer for $\mathbf{S}$. We shall address this question in Section 4.4. At present, we illustrate the design of interconnected observers using the following example.

4.6. EXAMPLE. Let us consider the interconnected system

$$\mathbf{S}\colon\ \dot{x}_1 = \begin{bmatrix} 0 & 0 & -2 \\ 1 & 0 & -1 \\ 0 & 1 & -1 \end{bmatrix} x_1 + B_1 u + \begin{bmatrix} 4 & 5 \\ 0 & 6 \\ 0 & 0 \end{bmatrix} x_2,$$

$$\dot{x}_2 = \begin{bmatrix} 0 & -3 \\ 1 & -2 \end{bmatrix} x_2 + B_2 u + \begin{bmatrix} 2 & 3 & 2 \\ 0 & 4 & 1 \end{bmatrix} x_1, \tag{4.55}$$

$$y_1 = [0 \quad 0 \quad 1]\, x_1,$$

$$y_2 = [0 \quad 1]\, x_2.$$

The open-loop (dual) error system ($\ell_1 = 0,\ \ell_2 = 0$) is described by the equations

$$\dot{\tilde{x}}_1 = \begin{bmatrix} 0 & 1 & 0 \\ 0 & 0 & 1 \\ -2 & -1 & -1 \end{bmatrix} \tilde{x}_1 + \begin{bmatrix} 2 & 0 \\ 3 & 4 \\ 2 & 1 \end{bmatrix} \tilde{x}_2,$$

$$\dot{\tilde{x}}_2 = \begin{bmatrix} 0 & 1 \\ -3 & -2 \end{bmatrix} \tilde{x}_2 + \begin{bmatrix} 4 & 0 & 0 \\ 5 & 6 & 0 \end{bmatrix} \tilde{x}_1. \tag{4.56}$$

The eigenvalues of the two decoupled subsystems are

$$\lambda_1^1 = 1.3532, \qquad \lambda_{2,3}^1 = 0.1766 \pm j1.2028,$$
$$\lambda_{1,2}^2 = -1 \pm j1.4142. \tag{4.57}$$

By using the feedback gains

$$\ell_1^T = (4,\ 10,\ 5), \qquad \ell_2^T = (-1,\ 1), \tag{4.58}$$

we relocate the eigenvalues (4.57) to the new locations

$$\lambda_1^1 = -\sigma_1^1 = -1, \qquad \lambda_2^1 = -\sigma_2^1 = -2, \qquad \lambda_3^1 = -\sigma_3^1 = -3,$$
$$\lambda_1^2 = -\sigma_1^2 = -1, \qquad \lambda_2^2 = -\sigma_2^2 = -2, \tag{4.59}$$

where we have used $\rho = 1$.

We construct the transformation matrices R_1, R_2, T_1, and T_2 for $\rho > 1$ as

$$R_1 = \begin{bmatrix} 1 & 0 & 0 \\ 0 & \rho & 0 \\ 0 & 0 & \rho^2 \end{bmatrix}, \qquad T_1 = \begin{bmatrix} 1 & 1 & 1 \\ -1 & -2 & -3 \\ 1 & 4 & 9 \end{bmatrix},$$
$$R_2 = \begin{bmatrix} 1 & 0 \\ 0 & \rho \end{bmatrix}, \qquad T_2 = \begin{bmatrix} 1 & 1 \\ -1 & -2 \end{bmatrix}. \tag{4.60}$$

The numbers π_1 and π_2 are both set at 1. Then, the aggregate matrix $\bar{W}$ defined by (4.47) is

$$\bar{W} = \begin{bmatrix} \rho & -\bar{\xi}_{12}(\rho) \\ -\bar{\xi}_{21}(\rho) & \rho \end{bmatrix}, \tag{4.61}$$

which for $\rho = 1$ takes the form

$$\bar{W} = \begin{bmatrix} 1 & -17.0011 \\ -12.2936 & 1 \end{bmatrix}, \tag{4.62}$$

where $\bar{\xi}_{12}$ and $\bar{\xi}_{21}$ are computed from

$$\bar{\xi}_{12} = \lambda_M^{1/2}(T_1^{-1} A_{21}^T T_2), \qquad \bar{\xi}_{21} = \lambda_M^{1/2}(T_2^{-1} A_{12}^T T_1), \tag{4.63}$$

and A_{12}^T and A_{21}^T are defined in (4.56).

It is obvious that the test matrix $\bar{W}$ of (4.62) is not an M-matrix, and we have to try a higher value of ρ. From (4.50) and (4.56), we find that

$$D_{12} = \begin{bmatrix} 2 & 0 \\ 0 & 4 \\ 0 & 0 \end{bmatrix}, \qquad D_{21} = \begin{bmatrix} 4 & 0 & 0 \\ 0 & 6 & 0 \end{bmatrix}, \tag{4.64}$$

and, for $\rho > 15$, we get $\bar{\xi}_{12} \simeq 32.55$, $\bar{\xi}_{21} \simeq 18.98$. Thus, for $\rho = 25$, we get

$$\bar{W} = \begin{bmatrix} 25 & -32.55 \\ -18.98 & 25 \end{bmatrix}, \tag{4.65}$$

which is an M-matrix, and the observer is stable. The eigenvalues of the overall closed-loop system $\tilde{\mathbf{S}}$ are

$$\lambda_1^1 = -36.0364, \qquad \lambda_{2,3} = -25.9599 \pm j3.5219, \qquad \lambda_{4,5} = -68.5213 \pm j6.0474. \tag{4.66}$$

For the chosen value of $\rho = 25$, we have the subsystem eigenvalue sets $\mathcal{L}_1$ and $\mathcal{L}_2$ of (4.32) as

$$\mathcal{L}_1 = \{\lambda_1^1 = -25,\ \lambda_2^1 = -50,\ \lambda_3^1 = -75\}, \qquad \mathcal{L}_2 = \{\lambda_1^2 = -25,\ \lambda_2^2 = -50\}. \tag{4.67}$$

The local gains which produce $\mathcal{L}_1$ and $\mathcal{L}_2$ are

$$\ell_1^T = (93748,\ 6874,\ 149), \qquad \ell_2^T = (1247,\ 73), \tag{4.68}$$

which completes the design of the interconnected observer for the system $\mathbf{S}$ of (4.55).

4.7. REMARK. We note that the local gains (4.68) are relatively high. This is expected, as the local feedback needs to stabilize the observer and, at the same time, neutralize the effect of interconnections. By using the global feedback, the local gains could be reduced considerably.

4.8. REMARK. The single-output assumption in system $\mathbf{S}$ of (4.27) can be removed along the lines outlined in Example 2.25. Furthermore, we presented only full-order observation schemes. It is relatively straightforward to broaden the schemes to include subsystem observers of minimal dimensions using the standard observer theory (*e.g.*, O'Reilly, 1983).

4.3. Separation Property

When an observer is used to provide the state for the control laws, the basic question is how to combine a solution of the estimation problem with a solution of the control problem that assumes that the states (not estimates of the states) are known. In the classical control theory this fact does not cause any difficulty, because it is well-known (*e.g.*, Kailath, 1980) that the observer and the controller can be designed independently of each other. This is the so-called *separation property* of the observer–controller (or regulator) design recognized first by Luenberger (1964).

In the context of complex systems, the separation property has been established for the decentralized observers–controllers design under the condition that observers exchange the observed states (Šiljak and Vukčević, 1976). As a final product, we obtain a decentralized regulator for each subsystem separately. The significance of this fact is that we can build decentralized regulators "piece-by-piece" for complex systems such as electric power systems, chemical plants, *etc.*, using multiprocessors in a parallel computing scheme with great advantages of size, speed, robustness, and cost.

In controlling a plant represented by the system **S** of (4.9), we want to use a decentralized feedback. We assume that the system is in the input decentralized form, that is, the input matrix B is a block diagonal matrix $B_D = \text{diag}\{B_1, B_2, \ldots, B_N\}$. Thus,

$$\mathbf{S}\colon\ \begin{aligned} \dot{x} &= A_D x + B_D u + A_C x, \\ y &= C_D x. \end{aligned} \tag{4.69}$$

The crucial fact, however, is that in controlling the plant **S**, instead of the control law $u(t) \in \mathbf{R}^m$ defined as

$$u = r - K_D x, \tag{4.70}$$

we want to use the control law

$$u = r - K_D \hat{x}, \tag{4.71}$$

where the true (unavailable) state $x(t) \in \mathbf{R}^n$ of **S** is replaced by the estimated state $\hat{x}(t) \in \mathbf{R}^n$ generated by the observer

$$\hat{\mathbf{S}}\colon\ \dot{\hat{x}} = (A_D - L_D C_D)\hat{x} - B_D K_D \hat{x} + L_D y + B_D r + A_C \hat{x}, \tag{4.72}$$

and $r(t) \in \mathbf{R}^m$ is the reference input of **S**. For simplicity, we decide not to use the global feedback in the systems **S** and $\hat{\mathbf{S}}$; it does not affect the separation property anyway. Applying the control (4.71) to **S**, we get

$$\check{\mathbf{S}}\colon\ \begin{aligned} \dot{x} &= A_D x - B_D K_D \hat{x} + A_C x + B_D r, \\ y &= C_D x. \end{aligned} \tag{4.73}$$

The joint closed-loop system is

$$\check{\mathbf{S}} \,\&\, \hat{\mathbf{S}}\colon \quad \begin{bmatrix} \dot{x} \\ \dot{\hat{x}} \end{bmatrix} = \begin{bmatrix} A_D + A_C & -B_D K_D \\ L_D C_D & A_D - L_D C_D - B_D K_D + A_C \end{bmatrix} \begin{bmatrix} x \\ \hat{x} \end{bmatrix} + \begin{bmatrix} B_D \\ B_D \end{bmatrix} r. \tag{4.74}$$

By applying the nonsingular transformation

$$\begin{bmatrix} x \\ \tilde{x} \end{bmatrix} = \begin{bmatrix} I & 0 \\ I & -I \end{bmatrix} \begin{bmatrix} x \\ \hat{x} \end{bmatrix}, \tag{4.75}$$

we get the joint system, which involves the closed-loop system $\check{\mathbf{S}}$ and the error system $\tilde{\mathbf{S}}$,

$$\check{\mathbf{S}} \,\&\, \tilde{\mathbf{S}}\colon \quad \begin{bmatrix} \dot{x} \\ \dot{\tilde{x}} \end{bmatrix} = \begin{bmatrix} A_D - B_D K_D + A_C & B_D K_D \\ 0 & A_D - L_D C_D + A_C \end{bmatrix} \begin{bmatrix} x \\ \tilde{x} \end{bmatrix} + \begin{bmatrix} B_D \\ 0 \end{bmatrix} r, \tag{4.76}$$

which can be split into two independent systems. Therefore, the controller and observer can be designed independently of each other, and the separation property is established. Both designs can be carried out using the stabilization methods of Chapters 2 and 3, which is the most important consequence of the separation property in the context of decentralized regulator theory.

4.4. Decentralized Observer

In certain applications, the exchange of the observed states among the subsystem observers may cause prohibitive communication costs as well as a degradation of the system performance due to transmission delays of relevant data between various parts of the system. As pointed out in Remark 4.4, it is possible to remove the exchange altogether if we are willing to surrender the separation property and build jointly the controller and the observer for the overall system. In this section, we consider the no-exchange alternative and show how the results in decentralized stabilization of Chapter 2 can be carried over to the decentralized observer problem.

We consider again the output-decentralized system $\mathbf{S}$ of (4.9) and use the observer $\hat{\mathbf{S}}$ of (4.19) without the term $A_C\hat{x}$, that is,

$$\hat{\mathbf{S}}_D\colon \quad \dot{\hat{x}} = A_D\hat{x} + L_D(y - C_D\hat{x}) + B_D u. \tag{4.77}$$

The observed state $\hat{x}$ obtained locally by $\hat{\mathbf{S}}_D$ is used in the control law (4.71) to control the system $\mathbf{S}$ of (4.9). This produces the closed-loop system $\check{\mathbf{S}}$ and the observer $\hat{\mathbf{S}}_D$ as

$$\begin{aligned} \check{\mathbf{S}}\colon\ \dot{x} &= A_D x - B_D K_D \hat{x} + A_C x + B_D r, \\ y &= C_D x, \\ \hat{\mathbf{S}}_D\colon\ \dot{\hat{x}} &= A_D \hat{x} - L_D C_D \hat{x} - B_D K_D \hat{x} + L_D y + B_D r, \end{aligned} \tag{4.78}$$

which can be rewritten as a single joint system

$$\check{\mathbf{S}}\,\&\,\hat{\mathbf{S}}_D\colon\ \begin{bmatrix} \dot{x} \\ \dot{\hat{x}} \end{bmatrix} = \begin{bmatrix} A_D + A_C & -B_D K_D \\ L_D C_D & A_D - L_D C_D - B_D K_D \end{bmatrix} \begin{bmatrix} x \\ \hat{x} \end{bmatrix} + \begin{bmatrix} B_D \\ B_D \end{bmatrix} r. \tag{4.79}$$

Using the same transformation (4.77), we get

$$\check{\mathbf{S}}\,\&\,\tilde{\mathbf{S}}\colon\ \begin{bmatrix} \dot{x} \\ \dot{\tilde{x}} \end{bmatrix} = \begin{bmatrix} A_D - B_D K_D + A_C & -B_D K_D \\ A_C & A_D - L_D C_D \end{bmatrix} \begin{bmatrix} x \\ \tilde{x} \end{bmatrix}, \tag{4.80}$$

which cannot be split into two independent systems $\check{\mathbf{S}}$ and $\tilde{\mathbf{S}}$, and the separation property does not hold. For the system $\hat{\mathbf{S}}_D$ to be an asymptotic observer for $\mathbf{S}$, the joint system $\check{\mathbf{S}}\,\&\,\tilde{\mathbf{S}}$ should be asymptotically stable. Stability of $\check{\mathbf{S}}\,\&\,\tilde{\mathbf{S}}$, however, can be determined using the standard methods of vector Liapunov functions (see Chapter 2).

Although application of the vector Liapunov function is straightforward, there is a disturbing fact that the interconnection term $-B_D K_D$ in (4.80) contains the gain matrix K_D, which makes the problem nonstandard: increasing the gain K_D to stabilize the closed-loop matrix $A_D - B_D K_D$ increases the interconnection term $-B_D K_D$, thus making the vector Liapunov function method less effective. We explore this problem further by rewriting (4.79) as

$$\check{\mathbf{S}}\,\&\,\tilde{\mathbf{S}}\colon\ \begin{bmatrix} \dot{x} \\ \dot{\tilde{x}} \end{bmatrix} = \begin{bmatrix} A_D - B_D K_D & -B_D K_D \\ 0 & A_D - L_D C_D \end{bmatrix} \begin{bmatrix} x \\ \tilde{x} \end{bmatrix} + \begin{bmatrix} A_C & 0 \\ A_C & 0 \end{bmatrix} \begin{bmatrix} x \\ \tilde{x} \end{bmatrix}. \tag{4.81}$$

This new form suggests that on the subsystem level the separation property still holds, but global stability involves the joint interconnection problem of the system–observer combination.

Let us study the problem in some detail. Obviously, the system $\check{\mathbf{S}}\,\&\,\tilde{\mathbf{S}}$ can be viewed as an interconnection of N subsystems, which can be described as

$$\check{\mathbf{S}}\,\&\,\tilde{\mathbf{S}}\colon\quad \begin{bmatrix} \dot{x}_i \\ \dot{\tilde{x}}_i \end{bmatrix} = \begin{bmatrix} A_i - B_iK_i & -B_iK_i \\ 0 & A_i - L_iC_i \end{bmatrix} \begin{bmatrix} x_i \\ \tilde{x}_i \end{bmatrix} + \sum_{j=1}^{N} \begin{bmatrix} A_{ij} & 0 \\ A_{ij} & 0 \end{bmatrix} \begin{bmatrix} x_j \\ \tilde{x}_j \end{bmatrix}, \quad i \in \boldsymbol{N}. \tag{4.82}$$

Stabilization can proceed as in Section 2.4. We start with

$$(\check{\mathbf{S}}\,\&\,\tilde{\mathbf{S}})_i\colon\quad \begin{bmatrix} \dot{x}_i \\ \dot{\tilde{x}}_i \end{bmatrix} = \begin{bmatrix} A_i - B_iK_i & -B_iK_i \\ 0 & A_i - L_iC_i \end{bmatrix} \begin{bmatrix} x_i \\ \tilde{x}_i \end{bmatrix}, \tag{4.83}$$

and note that on the level of subsystems, the separation property remains valid. Since the pair $(A_i,\, B_i)$ is controllable, we can choose K_i to produce the optimal pair $(H_i^c,\, G_i^c)$ of Liapunov matrices for the controller. Observability of $(A_i,\, C_i)$ implies that a selection of L_i can be made to obtain the optimal pair $(H_i^e,\, G_i^e)$ for the observer part of $(\check{\mathbf{S}}\,\&\,\tilde{\mathbf{S}})_i$. In the transformed space defined by the nonsingular matrices T_{ci} and T_{ei}, the individual Liapunov functions are

$$v_i^c(\bar{x}_i) = \|\bar{x}_i\|^2, \qquad v_i^e(\bar{\tilde{x}}_i) = \|\bar{\tilde{x}}_i\|^2. \tag{4.84}$$

Following the construction of Section 2.2 , we use the function $v_i\colon\ \mathbf{R}^{2n_i} \to \mathbf{R}_+$, defined as

$$v_i(z_i) = d_{i1}v_i^c(\bar{x}_i) + d_{i2}v_i^e(\bar{\tilde{x}}_i), \tag{4.85}$$

for a Liapunov function of $(\check{\mathbf{S}}\,\&\,\tilde{\mathbf{S}})_i$, where $z_i \in \mathbf{R}^{2n_i}$ is the vector $z_i = (\bar{x}_i^T,\, \bar{\tilde{x}}_i^T)^T$, and the positive numbers d_{i1} and d_{i2} are yet to be determined. We define

$$d_{im} = \min(d_{i1},\, d_{i2}), \quad d_{iM} = \max(d_{i1},\, d_{i2}), \tag{4.86}$$

and produce the estimates

$$\begin{aligned} d_{im}\|z_i\|^2 &\le v_i(z_i) \le d_{iM}\|z_i\|^2, \\ \dot{v}_i(z_i)_{(4.83)} &\le -\lambda_m(U_i)z_i^{+}, \qquad \forall z_i \in \mathbf{R}^{2n_i}, \end{aligned} \tag{4.87}$$

where $z_i^+ = (\|\bar{x}_i\|,\ \|\tilde{\bar{x}}_i\|)^T$ and

$$
\begin{aligned}
&U_i = W_i^T D_i + D_i W_i, \\
&W_i = \begin{bmatrix} \sigma_{cM}^i & -\xi_{ce}^i \\ 0 & \sigma_{eM}^i \end{bmatrix}, \qquad D_i = \operatorname{diag}\{d_{i1},\ d_{i2}\}, \\
&\sigma_{cM}^i = \sigma_M^i(\Lambda_i^c), \qquad \sigma_{eM}^i = \sigma_M^i(\Lambda_i^e), \\
&\xi_{ce}^i = \lambda_M^{1/2}\left[(T_{ci}^{-1} B_i K_i T_{ei})^T (T_{ci}^{-1} B_i K_i T_{ei})\right].
\end{aligned}
\tag{4.88}
$$

The interconnections in (4.82) among the subsystems $(\check{\mathbf{S}}\,\&\,\tilde{\mathbf{S}})_i$ can be bounded using the numbers

$$
\xi_{ij} = \lambda_M^{1/2}(\bar{A}_{ij}^T \bar{A}_{ij}), \tag{4.89}
$$

where the interconnection matrix $\bar{A}_{ij}$ is defined in the transformed space as

$$
\bar{A}_{ij} = \begin{bmatrix} T_{ci}^{-1} A_{ij} T_{cj} & 0 \\ T_{ei}^{-1} A_{ij} T_{ej} & 0 \end{bmatrix}. \tag{4.90}
$$

Finally, we formulate the $N \times N$ aggregate matrix $W = (w_{ij})$ for the overall system $\check{\mathbf{S}}\,\&\,\tilde{\mathbf{S}}$ of (4.82):

$$
w_{ij} = \begin{cases} \lambda_m(U_i) - 2d_{iM}\xi_{ii}, & i = j, \\ -2d_{iM}\xi_{ij}, & i \neq j, \end{cases} \tag{4.91}
$$

4.9. THEOREM. A system

$$
\hat{\mathbf{S}}_D\colon\ \dot{\hat{x}} = A_D\hat{x} + L_D(y - C_D\hat{x}) + B_D u, \tag{4.77}
$$

is a decentralized asymptotic observer for a system

$$
\begin{aligned}
\mathbf{S}\colon\ \dot{x} &= A_D x + B_D u + A_C x, \\
y &= C_D x,
\end{aligned}
\tag{4.69}
$$

and the closed-loop system $\check{\mathbf{S}}\,\&\,\tilde{\mathbf{S}}$ is stable if $W \in \mathcal{M}$.

Proof. Straightforward application of the method of vector Liapunov functions outlined in Section 2.2. Q.E.D.

4.10. REMARK. Stability of the joint system $\check{\mathbf{S}}\,\&\,\tilde{\mathbf{S}}$ was established in a hierarchical fashion because the subsystems themselves were treated as composite subsystems involving the closed-loop subsystems and the local error subsystems as components. A considerably more general hierarchical stability analysis is available in Section 7.4, which can be applied to nonlinear systems as well.

Of course, when W of Theorem 4.9 fails the M-matrix test, one has to redesign the controller and observer on the subsystem level and test the new W matrix to see if $W \in \mathcal{M}$. No assurance is given that the process would succeed even if the poles of the closed-loop system are pushed further and further to the left in the complex plane. It is possible, however, to delineate classes of systems for which decentralized observers can *always* be built by high-gain feedback to produce a global asymptotic observer. As expected, these classes are duals to those that are decentrally stabilizable in the sense of Section 2.4. In the rest of this section, we present the decentralized observer scheme of Bachmann (1983), which is based upon the stabilization result of Ikeda and Šiljak (1980) contained in Theorem 2.20.

We first specify the interconnected system

$$\begin{aligned} \mathbf{S}\colon\ \dot{x}_i &= A_i x_i + b_i u_i + \sum_{j=1}^{N} A_{ij} x_j, \\ y_i &= c_i^T x_i, \qquad i \in \boldsymbol{N}, \end{aligned} \tag{4.92}$$

which has the single-input–single-output subsystems

$$\begin{aligned} \mathbf{S}_i\colon\ \dot{x}_i &= A_i x_i + b_i u_i, \\ y_i &= c_i^T x_i, \qquad\qquad i \in \boldsymbol{N}, \end{aligned} \tag{4.93}$$

and $x_i(t) \in \mathbf{R}^{n_i}$, which is standard. The matrices A_i, b_i, c_i are

$$A_i = \begin{bmatrix} 0 & 1 & \dots & 0 \\ 0 & 0 & \dots & 0 \\ \dots & \dots & \dots & \dots \\ 0 & 0 & \dots & 1 \\ 0 & 0 & \dots & 0 \end{bmatrix}, \qquad b_i \begin{bmatrix} 0 \\ 0 \\ \vdots \\ 0 \\ 1 \end{bmatrix}, \qquad c_i = \begin{bmatrix} c_{i1} \\ 0 \\ \vdots \\ 0 \\ 0 \end{bmatrix}. \tag{4.94}$$

A nonstandard feature is the fact that the last row of A_i is zero because it is included in A_{ii}. A restrictive assumption, however, is the new requirement that in c_i only $c_{i1} \neq 0$, which makes (4.94) very special, indeed.

The overall closed-loop system $\check{\mathbf{S}}\,\&\,\tilde{\mathbf{S}}$ of (4.80) has the following specifications:

$$
\begin{aligned}
A_D &= \operatorname{diag}\{A_{11},\, A_{22},\, \dots,\, A_{NN}\}, & B_D &= \operatorname{diag}\{b_1,\, b_2,\, \dots,\, b_N\},\\
A_C &= (A_{ij}), & C_D &= \operatorname{diag}\{c_1^T,\, c_2^T,\, \dots,\, c_N^T\},\\
K_D &= \operatorname{diag}\{k_1^T,\, k_2^T,\, \dots,\, k_N^T\}, & L_D &= \operatorname{diag}\{\ell_1,\, \ell_2,\, \dots,\, \ell_N\},
\end{aligned}
\tag{4.95}
$$

where the vectors $k_i,\, \ell_i \in \mathbf{R}^{n_i}$ are

$$
k_i = (k_1^i,\, k_2^i,\, \dots,\, k_{n_i}^i)^T, \qquad \ell_i = (\ell_1^i,\, \ell_2^i,\, \dots,\, \ell_{n_i}^i), \tag{4.96}
$$

with the components chosen as

$$
k_j^i = \rho^{n_i-j+1}\, \Sigma_j^{ci}, \qquad \ell_j^i = \rho^{n_i-j+1}\, \Sigma_j^{ei}, \tag{4.97}
$$

where Σ_j^{ci} and Σ_j^{ei} are the corresponding sums of the products of σ_j^{ci} and σ_j^{ei}, which are the real and distinct eigenvalues, so that in the transformed spaces the subsystems closed-loop matrices are

$$
\begin{aligned}
\Lambda_i^c &= \operatorname{diag}\{-\rho\sigma_1^{ci},\, -\rho\sigma_2^{ci},\, \dots,\, -\rho\sigma_{n_i}^{ci}\},\\
\Lambda_i^e &= \operatorname{diag}\{-\rho\sigma_1^{ei},\, -\rho\sigma_2^{ei},\, \dots,\, -\rho\sigma_{n_i}^{ei}\}.
\end{aligned}
\tag{4.98}
$$

The transformations are defined as

$$
x_i = R_i T_{ci} \bar{x}_i, \qquad \tilde{x}_i = R_i T_{ei} \bar{\tilde{x}}_i, \tag{4.99}
$$

where $R = \operatorname{diag}\{1,\, \rho,\, \dots,\, \rho^{n_i-1}\}$ depends on ρ, but T_{ci} and T_{ei} are nonsingular matrices with elements depending on σ_j^{ci} and σ_j^{ei} only. The transformed matrices are

$$
\begin{aligned}
\Lambda_i^c &= T_{ci}^{-1} R_i^{-1} (A_i - b_i k_i^T) R_i T_{ci},\\
\bar{A}_{ij}^c &= T_{ci}^{-1} R_i^{-1} A_{ij} R_j T_{cj},\\
\bar{b}_i \bar{k}_i^T &= T_{ci}^{-1} R_i^{-1} b_i k_i^T R_i T_{ei},\\
\Lambda_i^e &= T_{ei}^{-1} R_i^{-1} (A_i - \ell_i c_i^T) R_i T_i T_{ei},\\
\bar{A}_{ij}^e &= T_{ei}^{-1} R_i^{-1} A_{ij} R_j T_{cj}.
\end{aligned}
\tag{4.100}
$$

At last, by relying on Theorem 2.20, we establish the following result (Bachmann, 1983):

4.11. THEOREM. A union of disconnected subsystems

$$\hat{\mathbf{S}}_D\colon\ \dot{\hat{x}}_i = A_i\hat{x}_i + \ell_i(y_i - c_i^T\hat{x}_i) + b_iu_i, \qquad i \in \boldsymbol{N}, \tag{4.101}$$

is a decentralized observer for a system

$$\begin{aligned} \mathbf{S}\colon\ \dot{x}_i &= A_ix_i + b_iu_i + \sum_{j=1}^{N} A_{ij}x_j, \\ y_i &= c_i^T x_i, \qquad\qquad i \in \boldsymbol{N}, \end{aligned} \tag{4.102}$$

and the closed-loop system $\check{\mathbf{S}}\,\&\,\tilde{\mathbf{S}}$ is stable if the matrices A_{ij} are such that the inequality

$$\sum_{k=1}^{r} (m_{i_k j_k} - 1) < 0 \tag{4.103}$$

holds for all subsets $\boldsymbol{I}_r$ and $\boldsymbol{J}_r$ of $\boldsymbol{N}$, where m_{ij} is defined in (2.120).

Proof. We have all we need to define the $2N \times 2N$ aggregate matrix

$$W' = \begin{bmatrix} \rho D_c - Z_c & -D_s \\ -\rho Z_e & \rho D_e \end{bmatrix}, \tag{4.104}$$

where the $N \times N$ submatrices are

$$\begin{aligned} D_c &= \mathrm{diag}\{\sigma_{cM}^1,\ \sigma_{cM}^2,\ \ldots,\ \sigma_{cM}^N\}, \qquad Z_c = (\rho^{m_{ij}}\xi_{ij}^c), \\ D_s &= \mathrm{diag}\{\xi_1,\ \xi_2,\ \ldots,\ \xi_N\}, \\ D_e &= \mathrm{diag}\{\sigma_{eM}^1,\ \sigma_{eM}^2,\ \ldots,\ \sigma_{eM}^N\}, \qquad Z_e = (\rho^{m_{ij}}\xi_{ij}^e), \end{aligned} \tag{4.105}$$

and

$$\xi_{ij}^c = \|T_{ci}^{-1}\| \; \|T_{cj}\| \sum_{p=1}^{n_j} \sum_{q=1}^{n_j} |a_{pq}^{ij}|,$$

$$\xi_{ij}^e = \|T_{ei}^{-1}\| \; \|T_{cj}\| \sum_{p=1}^{n_i} \sum_{q=1}^{n_j} |a_{pq}^{ij}|, \tag{4.106}$$

$$\xi_i = \|T_{ci}^{-1}\| \; \|T_{ei}\| \sum_{p=1}^{n_i} \binom{n_i}{p-1} (\sigma_{cM}^i)^{n_i-p+1}.$$

Using W' of (4.104) we define a new matrix

$$W'' = \rho^{-1} \begin{bmatrix} I & \rho^{-1} D_s D_e^{-1} \\ 0 & I \end{bmatrix} W', \tag{4.107}$$

or

$$W'' = \begin{bmatrix} W & 0 \\ -Z_e & D_e \end{bmatrix}, \qquad W = D_c - \rho^{-1}(Z_c - D_s D_e^{-1} Z_e). \tag{4.108}$$

Now, $W' \in \mathcal{M}$ if and only if $W'' \in \mathcal{M}$, that is, $W \in \mathcal{M}$. On the other hand, we notice that in the split $W = D_c - W_C$, $W_C = \rho^{-1}(Z_c - D_s D_e^{-1} Z_e)$, the $N \times N$ matrix $W_C = (\rho^{m_{ij}-1} \xi_{ij}^C)$ has the nonnegative numbers ξ_{ij}^C independent of ρ, and the theorem follows from the proof of Theorem 2.20. Q.E.D.

4.5. Stochastic Control

Despite a relatively long history, the optimization methods for decentralized control and estimation of stochastic systems are unsatisfactory. Due to nonclassical constraints on information and control structure, the standard optimization procedures are ineffective even for simple systems of small dimensions. The situation is further complicated by the fact that imperfect knowledge of interactions among various components of the system is invariably a part of the problem. No one knows if the limitations of the standard methods can be removed by developing better procedures, or if these limitations can be attributed to the lack of a precise axiomatic definition of the optimization problem.

To bypass these difficulties, the problem of decentralized stochastic control is formulated as a problem of interconnecting subsystems that are optimized by local feedback. The recipe was introduced in Section 3.1: treat interconnections as perturbation terms, and determine suboptimality of locally optimal controls when applied to the overall interconnected system. The measure of suboptimality is the size of the performance bound of the interconnected system relative to the performance of the nominal system, be it the disconnected locally optimal subsystems or the overall globally optimal system. The most appealing feature of this design is the robustness characteristics of the closed-loop system.

We start our exposition of the suboptimal LQG design with a linear stochastic differential equation of the Itô type:

$$\mathbf{S}\colon\ dx = Ax\,dt + Bu\,dt + dv, \tag{4.109}$$

where $x(t) \in \mathbf{R}^n$ is the state of $\mathbf{S}$, $u(t) \in \mathbf{R}^m$ is its input, and $A(t)$ and $B(t)$ are $n \times n$ and $n \times m$ matrices which are continuous in $t \in \mathbf{R}$. In (4.109), $v(t) \in \mathbf{R}^n$ is the input noise, which is a Wiener process independent of the initial state $x(t_0) = x_0$ with $\mathcal{E}(dv\,dv^T) = R_v\,dt$, where $\mathcal{E}$ denotes mathematical expectation, $R_v(t)$ is the incremental covariance matrix, which is continuous and uniformly bounded in time, and $\mathcal{E}dv = 0$. It is also assumed that the Gaussian statistics $\mathcal{E}x_0 = m_0$, $\mathrm{cov}(x_0, x_0) = R_0$, and $R_v(t)$ for the system $\mathbf{S}$ are given *a priori.*

The system $\mathbf{S}$ is decomposed to get

$$\mathbf{S}\colon\ dx_i = A_i x_i\,dt + B_i u_i\,dt + \sum_{j=1}^{N} A_{ij} x_j\,dt + dv_i, \qquad i \in \boldsymbol{N}, \tag{4.110}$$

which represent an interconnection of N subsystems

$$\mathbf{S}_i\colon\ dx_i = A_i x_i\,dt + B_i u_i\,dt + dv_i, \qquad i \in \boldsymbol{N}, \tag{4.111}$$

where $x_i(t) \in \mathbf{R}^{n_i}$, $u_i(t) \in \mathbf{R}^{m_i}$, $v_i(t) \in \mathbf{R}^{n_i}$ are state, input, and noise of $\mathbf{S}_i$, such that $\sum_{i=1}^N n_i = n$, $\sum_{i=1}^N m_i = m$; $A_i(t)$, $B_i(t)$, $A_{ij}(t)$ are matrices of appropriate dimensions and continuous in time.

The fundamental assumption concerning the system $\mathbf{S}$ is that the input matrix B is a block diagonal matrix, that is, $B_D = \mathrm{diag}\{B_1, B_2, \ldots, B_N\}$, and $\mathbf{S}$ is an input-decentralized system. We assume, however, that the noise inputs of the subsystems are mutually correlated.

We consider the system $\mathbf{S}$ in the compact form

$$\mathbf{S}\colon\ dx = A_D x\,dt + B_D u\,dt + A_C x\,dt + dv. \tag{4.112}$$

With **S** we associate a quadratic cost

$$J_D = x_f^T P_f x_f + \int_{t_0}^{t_f} (x^T Q_x x + u^T Q_u u)\, dt, \tag{4.113}$$

and consider the expected value $\mathcal{E}J$ of the cost J as a measure of the system performance, where the matrices $Q_x(t)$ and $Q_u(t)$ are uniformly bounded and continuous functions of t, and

$$\begin{aligned} Q_x &= \text{diag}\{Q_{x1}, Q_{x2}, \ldots, Q_{xN}\}, \\ Q_u &= \text{diag}\{Q_{u1}, Q_{u2}, \ldots, Q_{uN}\}, \end{aligned} \tag{4.114}$$

where each Q_{xi} is symmetric nonnegative definite and has dimension $n_i \times n_i$, and Q_{ui} is symmetric positive definite and has dimension $m_i \times m_i$. Furthermore, we assume that $Q_u^{-1}(t)$ is uniformly bounded and continuous. In J above, $x_f^T P_f x_f$ is the terminal cost, where $x(t_f) = x_f$, $t_f > t_0$, is the terminal state, and

$$P_f = \text{diag}\{P_{f1}, P_{f2}, \ldots, P_{fN}\} \tag{4.115}$$

is a symmetric positive definite matrix. The block diagonal choice of Q_x, Q_u, and P_f is motivated by a decentralized strategy to be used in controlling the system **S**. In fact, this choice implies that to each decoupled subsystem $\mathbf{S}_i$ of (4.111) there corresponds a cost

$$J_i = x_{fi}^T P_{fi} x_{fi} + \int_{t_0}^{t_f} (x_i^T Q_{xi} x_i + u_i^T Q_{ui} u_i)\, dt. \tag{4.116}$$

It is a well-known result of Kalman (*e.g.*, Åström, 1970) that the optimal control law for the system $\mathbf{S}_D$ with $A_C(t) \equiv 0$,

$$\mathbf{S}_D\colon\ dx = A_D x\, dt + B_D u\, dt + dv, \tag{4.117}$$

is

$$u_D^{\odot} = -K_D x, \tag{4.118}$$

where the matrix $K_D(t, t_f)$ is given by

$$K_D = Q_u^{-1} B_D^T P_D. \tag{4.119}$$

The matrix $P_D(t;\, t_f,\, P_f)$ is the symmetric positive definite solution of the Riccati equation

$$\dot{P}_D = -A_D^T P_D - P_D A_D + P_D B_D Q_u^{-1} B_D^T P_D - Q_x, \tag{4.120}$$

with $t \in (t_0, t_f)$ and $P_D(t_f; t_f, P_f) = P_f$, provided the pair $(A_D, Q_x^{1/2})$ is differentially observable (Weiss, 1966; Anderson and Moore, 1969).

Due to the fact that the matrices A_D, B_D, Q_x, Q_u, and P_f are block diagonal, we have also $K_D = \text{diag}\{K_1, K_2, \ldots, K_N\}$ and $P_D = \text{diag}\{P_1, P_2, \ldots, P_N\}$ block diagonal, so that the optimal control is given by a set of local controls

$$u_i^{\odot} = -K_i x_i, \qquad i \in \boldsymbol{N}, \tag{4.121}$$

which result in an optimal decoupled closed-loop system $\check{\mathbf{S}}_D^{\odot}$.

When the decentralized control (4.121) is applied to the interconnected system $\mathbf{S}$, the value of the expected cost $\mathcal{E}J_D^{\oplus}$ is generally different from $\mathcal{E}J_D^{\odot}$ which is computed for decoupled subsystems, that is, when $A_C(t) \equiv 0$. We propose to bound $\mathcal{E}J_D^{\oplus}$ from above by $\mu^{-1}\mathcal{E}J_D^{\odot}$, where μ is a positive number. In this way, we provide a measure of suboptimality of the local decentralized control when it is applied to the interconnected system $\mathbf{S}$. This design policy may be justified by the fact that the uniqueness of the optimal decentralized control law that solves the complete LQG problem is not guaranteed (Witsenhausen, 1968), and a search for such a control may be unreasonable. Furthermore, the suboptimal policy based upon the decoupled subsystems used as reference can preserve autonomy of the subsystems when interconnected, thus leading to control laws that are suboptimal despite structural perturbations (Section 4.8).

A stochastic version of suboptimality (Definition 3.15) is:

4.12. DEFINITION. A decentralized control

$$u_D^{\odot} = -Q_u^{-1} B_D^T P_D x \tag{4.122}$$

is said to be uniformly suboptimal for the system $\mathbf{S}$ and with respect to the pair $(\check{\mathbf{S}}_D^{\odot}, J_D^{\odot})$ if there exists a positive number μ such that

$$\mathcal{E}[J_D^{\oplus} | x_0] \leq \mu^{-1} \mathcal{E}[J_D^{\odot} | x_0] \tag{4.123}$$

holds for all $t_0 \in (0, +\infty)$, $t_f \in (0, +\infty)$, and $x_0 \in \mathbf{R}^n (x_0 \neq 0)$.

Our immediate interest is to characterize suboptimality in terms of the matrix A_C:

4.13. THEOREM. A decentralized control $u_D^{\odot}$ is uniformly suboptimal if the matrix

$$\begin{aligned} F(t_0, t_f; \mu) = {} & A_C^T(t_0) P_D(t_0; t_f, P_f) + P_D(t_0; t_f, P_f) A_C(t_0) \\ & - (1-\mu) P_D(t_0; t_f, P_f) B_D(t_0) Q_u^{-1}(t_0) P_D(t_0; t_f, P_f) + Q_x(t_0) \end{aligned} \tag{4.124}$$

is nonpositive definite for some $\mu \leq 1$, and all $t_0 \in [0, +\infty)$, $t_f \in (t_0, +\infty)$.

Proof. We represent the performance index $\mathcal{E}J^{\oplus}$ as

$$\mathcal{E}J^{\oplus}_D = \mathcal{E}x_0^T P^{\oplus}_D(t_0)x_0 + \operatorname{tr}\int_0^{t_f} P^{\oplus}_D(t)R_v(t)\,dt, \tag{4.125}$$

where the symmetric positive definite $n \times n$ matrix $P^{\oplus}_D(t;\, t_f,\, P_f)$ satisfies the differential equation

$$\frac{d}{dt}P^{\oplus}_D + A_K^T P^{\oplus}_D + P^{\oplus}_D A_K + K_D^T Q_u K_D + Q_x = 0, \tag{4.126}$$

with $P^{\oplus}_D(t_f;\, t_f,\, P_f) = P_f$, and $A_K = A_K(t,\, t_f)$ is given as

$$A_K(t,\, t_f) = A(t) - B_D(t)K_D(t,\, t_f). \tag{4.127}$$

Using the Riccati equation (4.120) with (4.126), it is easy to obtain

$$\frac{d}{dt}(\mu P^{\oplus}_D - P_D) + A_K^T(\mu P^{\oplus}_D - P_D) + (\mu P^{\oplus}_D - P_D)A_K + F = 0, \tag{4.128}$$

where $\mu P^{\oplus}_D(t_f;\, t_f,\, P_f) - P_D(t_f;\, t_f,\, P_f) = -(1-\mu)P_f$. It is obvious that the matrix $\mu P^{\oplus}_D - P_D$ is always nonpositive definite under the conditions of the theorem. Since (4.125) implies

$$\mu\mathcal{E}J^{\oplus}_D - \mathcal{E}J^{\odot}_D = \mathcal{E}x_0^T[\mu P^{\oplus}_D(t_0) - P_D(t_0)]x_0 + \operatorname{tr}\int_{t_0}^{t_f} [\mu P^{\oplus}_D(t) - P_D(t)]R_v(t)\,dt, \tag{4.129}$$

the theorem follows. Q.E.D.

4.14. REMARK. The integrability conditions for the mean-square integrals $J^{\odot}_D$, $J^{\oplus}_D$, and $\int_{t_0}^{t_f} x^T F x\,dt$ are obviously satisfied because $x(t)$ is a Gaussian process. These integrals are random variables that are Lebesque-integrable, and they are defined on the Riesz space which includes the case $t_f = +\infty$ (see, for example, Gikhman and Skorokhod, 1969). In this case, the basic inequality $\mathcal{E}J^{\oplus} \leq \mu^{-1}\mathcal{E}J^{\odot}$ implies either $\mathcal{E}J^{\oplus} < +\infty$ or $\mathcal{E}J^{\oplus} = +\infty$ and $\mathcal{E}J^{\odot} = +\infty$.

Remark 4.14 and continuity of $F(t_0, t_f; \mu)$ in t_f imply that the conditions of Theorem 4.13 are meaningful for $t_f = +\infty$. The additional condition is that the pairs $(A_D, Q_x^{1/2})$ and (A_D, B_D) are uniformly observable and controllable, respectively.

It is now of interest to find out whether an analogous statement to Theorem 4.13 could be made for a criterion defined as expectation of the time average of a quadratic cost function over the infinite time interval. This kind of criterion is most often used to obtain a finite performance measure of time-invariant control systems over the infinite time interval. Therefore, we consider the criterion

$$\mathcal{E}\bar{J}_D = \lim_{t_f \to +\infty} \frac{1}{t_f - t_0} \mathcal{E} \int_{t_0}^{t_f} (x^T Q_x x + u^T Q_u u)\, dt, \tag{4.130}$$

and establish the following:

4.15. COROLLARY. A decentralized control $u_D^{\odot}$ is suboptimal if the matrix $F(t_0, t_f; u)$ is nonpositive definite for some $\mu \leq 1$ and all $t_0 \in [0, +\infty)$, $t_f \in (t_0, +\infty]$.

Proof. It is clear that Theorem 4.13 applies to

$$\frac{1}{t_f - t_0} \mathcal{E} \int_{t_0}^{t_f} (x^T Q_x x + u^T Q_u u)\, dt, \tag{4.131}$$

for $t_0 < t_f < +\infty$. Thus, when F is nonpositive definite on (t_0, t_f), we have

$$\mathcal{E} \int_{t_0}^{t_f} (x^T Q_x + u^T Q_u u)\, dt \leq \mu^{-1} \mathcal{E} \left[x_0^T P_D(t_0) x_0 + \operatorname{tr} \int_{t_0}^{t_f} P_D(t) R_v(t)\, dt \right], \tag{4.132}$$

and, consequently,

$$\begin{aligned} &\frac{1}{t_f - t_0} \mathcal{E} \int_{t_0}^{t_f} (x^T Q_x x + u^T Q_u u)\, dt \\ &\leq \mu^{-1} \frac{1}{t_f - t_0} \mathcal{E} \left[x_0^T P_D(t_0) x_0 + \operatorname{tr} \int_{t_0}^{t_f} P_D(t) R_v(t)\, dt \right]. \end{aligned} \tag{4.133}$$

In view of Remark 4.14, expectations in (4.133) are well-defined for

$t \to +\infty$. Therefore, by continuity,

$$\lim_{t_f \to +\infty} \frac{1}{t_f - t_0} \mathcal{E} \int_{t_0}^{t_f} (x^T Q_x x + u^T Q_u u)\, dt$$

$$\leq \mu^{-1} \lim_{t_f \to +\infty} \frac{1}{t_f - t_0} \mathcal{E} \left[x_0^T P_D(t_0) x_0 + \operatorname{tr} \int_{t_0}^{t_f} P_D(t) R_v(t)\, dt \right], \quad (4.134)$$

and, thus,

$$\mathcal{E} \bar{J}_D^{\oplus} \leq \mu^{-1} \mathcal{E} \bar{J}_D^{\odot}, \quad (4.135)$$

which completes the proof. Q.E.D.

Suboptimality when established by Corollary 4.15 implies exponential stability of the unforced closed-loop system

$$\mathbf{S}_K\colon\ \dot{x} = A_K(t) x, \quad (4.136)$$

where

$$A_K(t) = A(t) - B_D(t) K_D(t), \quad (4.137)$$

and

$$K_D(t) = Q_u^{-1}(t) B_D^T(t) \bar{P}_D(t). \quad (4.138)$$

In the expression for $K_D(t)$, $\bar{P}_D(t) = \lim_{t \to +\infty} P_D(t;\, t_f,\, 0)$.

Now, we define

$$\bar{F}(t;\, \mu) = \lim_{t_f \to +\infty} F(t,\, t_f;\, \mu) \quad (4.139)$$

and

$$\bar{Q}_D(t) = \bar{P}_D(t) B_D(t) Q_u^{-1}(t) B_D^T(t) \bar{P}_D(t) + Q_x(t), \quad (4.140)$$

to prove the following:

4.16. COROLLARY. The equilibrium $x = 0$ of the system $\mathbf{S}_K$ is exponentially stable in the large if the matrix $\bar{Q}_D(t)$ is bounded from below by a matrix $\alpha_0 I_n$, where α_0 is a positive number, and for some $\mu > 0$, the matrix

$$\bar{F}(t;\, \mu) = A_C^T(t) \bar{P}_D(t) + \bar{P}_D(t) A_C(t) - (1 - \mu) \bar{Q}_D(t) \quad (4.141)$$

is nonpositive definite for all $t \in [0, +\infty)$.

Proof. To prove uniform asymptotic (exponential) stability of the equilibrium state of $\mathbf{S}_K$, we use the quadratic form

$$V(t,\, x) = x^T \bar{P}_D(t) x, \tag{4.142}$$

as a candidate for a Liapunov function of $\mathbf{S}_K$. From uniform and differential observability of the pair $(A_D, Q_x^{1/2})$, and uniform controllability of the pair $(A_D,\, B_D)$, it follows that there are positive numbers α_1 and α_2 such that

$$\alpha_1 I_n \leq \bar{P}_D(t) \leq \alpha_2 I_n, \qquad \forall t \in [0,\, +\infty), \tag{4.143}$$

where $\bar{P}_D(t) \geq \alpha I_n$ denotes, as usual, nonpositive definiteness of the matrix $\bar{P}_D(t) - \alpha I_n$. Hence,

$$\alpha_1 \|x\|^2 \leq V(t,\, x) \leq \alpha_2 \|x\|^2, \qquad \forall t \in [0,\, +\infty)\ ,\ \forall x \in \mathbf{R}^n. \tag{4.144}$$

Moreover, by computing the derivative of $V(t,\, x)$ with respect to (4.136), we obtain

$$\dot{V}(t,\, x)_{(4.136)} = x^T \bar{F}(t;\, 0) x. \tag{4.145}$$

Since $\bar{F}(t;\, \mu) = \bar{F}(t;\, 0) + \mu \bar{Q}_D(t)$, we have

$$\bar{F}(t;\, 0) \leq -\mu \bar{Q}_D(t), \qquad \forall t \in [0, +\infty). \tag{4.146}$$

Furthermore,

$$x^T \bar{F}(t;\, 0) x \leq -\mu \alpha_0 \|x\|^2. \tag{4.147}$$

As the upper bound in (4.147) does not depend on the initial time t_0, and the solutions to (4.136) are continuous in the initial state x_0, (4.147) concludes the proof. Q.E.D.

In case of a time-invariant system $\mathbf{S}$, we have $\bar{F}(t;\, \mu) = \bar{F}(\mu)$ and $\bar{Q}_D(t) = \bar{Q}_D$. For this case, it is convenient to compute the largest degree of suboptimality μ^* that is provided by Corollary 4.16,

$$\mu^* = 1 - \lambda_M[\bar{F}(1)\bar{Q}_D^{-1}], \tag{4.148}$$

where $\lambda_M(\,\bullet\,)$ is the maximum eigenvalue of the indicated matrix.

4.17. REMARK. The restriction $\mu \leq 1$ in Theorem 4.13 and Corollary 4.15 can be removed when $P_f(t) \equiv 0$, which is the case analyzed by

Krtolica and Šiljak (1980). When $\mu > 1$, the interconnection matrix $A_C(t)$ can be considered as beneficial because it provides an improvement rather than deterioration of the overall system performance when compared with the performance of the optimal decoupled subsystems for $A_C(t) \equiv 0$.

4.6. Estimation

By duality, the results obtained for suboptimality of decentralized control can be applied to the problem of decentralized estimation when the subsystem statistics is uncorrelated. To show this, let us consider the state estimation problem for the system $\mathbf{S}$ with noisy observations:

$$\begin{aligned} \mathbf{S}\colon\ dx &= A_D x\, dt + B_D u\, dt + A_C x\, dt + dv, \\ dy &= C_D x\, dt + dw, \end{aligned} \tag{4.149}$$

where $C_D(t) = \text{diag}\{C_1(t),\, C_2(t),\, \ldots,\, C_N(t)\}$ is a continuous $\ell \times n$ matrix, and $C_i(t)$ is a $\ell_i \times n_i$ submatrix of $C_D(t)$. In (4.149), $w(t) \in \mathbf{R}^\ell$ is the observation noise, which is a Wiener process independent of the initial state x_0 and the input noise $v(t)$. Furthermore, $\mathcal{E}\, dw = 0$, $\mathcal{E}(dw\, dw^T) = R_w\, dt$, the matrix $R_w(t)$ is continuous positive definite and uniformly bounded, and the matrices $R_w^{-1}(t)$ and $R_v(t)$ are uniformly bounded on the time interval $[0, +\infty)$.

For $A_C(t) \equiv 0$, the system $\mathbf{S}$ becomes

$$\begin{aligned} \mathbf{S}_D\colon\ dx &= A_D x\, dt + B_D u\, dt + dv, \\ dy &= C_D x\, dt + dw, \end{aligned} \tag{4.150}$$

which is the union of the decoupled subsystems

$$\begin{aligned} \mathbf{S}_i\colon\ dx_i &= A_i x_i\, dt + B_i u_i\, dt + dv_i, \\ dy_i &= C_i x_i\, dt + dw_i, \qquad i \in \boldsymbol{N}. \end{aligned} \tag{4.151}$$

It is well known (*e.g.*, Åström, 1970) that the minimum mean-square estimation error for the decoupled system $\mathbf{S}_D$ is achieved by the state estimate $\hat{x}^\odot$ defined by the Kalman–Bucy estimator (filter)

$$\hat{\mathbf{S}}_D^\odot\colon\ d\hat{x}^\odot = A_D \hat{x}^\odot\, dt + B_D u\, dt + L_D(dy - C_D \hat{x}^\odot\, dt), \tag{4.152}$$

and $\hat{x}^{\odot}(t_0) = x_0$, where $L_D(t, t_0)$ is given by

$$L_D = S_D C_D^T R_w^{-1}. \tag{4.153}$$

The matrix $S_D(t;\, t_0,\, R_0)$ is the symmetric positive definite solution of the Riccati equation

$$\dot{S}_D = A_D S_D + S_D A_D^T - S_D C_D^T R_w^{-1} C_D S_D + R_v, \tag{4.154}$$

where $t \in (t_0, +\infty)$ and $S_D(t_0; t_0, R_0) = R_0$, provided the pair $(A_D,\, R_v^{1/2})$ is differentially controllable (Bucy and Joseph, 1968). The minimum estimation error

$$(\sigma^{\odot})^2 = \mathcal{E}[a^T x(t) - a^T \hat{x}^{\odot}(t)]^2 \tag{4.155}$$

is achieved for all weighting vectors $a \in \mathbf{R}^n$, where $x(t) = x(t;\, t_0,\, x_0)$. In terms of the covariance matrix S_D,

$$(\sigma^{\odot})^2 = a^T S_D a. \tag{4.156}$$

When we apply the interconnected filter

$$\hat{\mathbf{S}}^{\oplus}\colon\; d\hat{x} = A\hat{x}\, dt + B_D u\, dt + L_D(dy - C\hat{x}\, dt) \tag{4.157}$$

to the interconnected system $\mathbf{S}$, the resulting state estimate $\hat{x}^{\oplus}(t;\, t_0,\, x_0)$ produces, in general, a different value of the mean-square error

$$(\sigma^{\oplus})^2 = \mathcal{E}[a^T x(t) - a^T \hat{x}^{\oplus}(t)]^2. \tag{4.158}$$

In analogy to the matrix $S_D(t;\, t_0,\, R_0)$ for $A_C \equiv 0$, we can define a matrix $S_D^{\oplus}(t) = S_D^{\oplus}(t;\, t_0,\, R_0)$ for $A_C \not\equiv 0$ as

$$S_D^{\oplus}(t) = \mathcal{E}\{[x(t) - \hat{x}^{\oplus}(t)]\, [x(t) - \hat{x}^{\oplus}(t)]^2, \tag{4.159}$$

which is the error correlation matrix for the interconnected system, and for which we have

$$(\sigma^{\oplus})^2 = a^T P_D^{\oplus} a. \tag{4.160}$$

In the spirit of Definition 4.12, we bound from above $\sigma^{\oplus}$ by a multiple of a $\sigma^{\odot}$, and state the following:

4.18. DEFINITION. An interconnected estimator $\hat{\mathbf{S}}^{\oplus}$ for the system $\mathbf{S}$ and with respect to the pair $(\hat{\mathbf{S}}_D^{\odot},\, J_D^{\odot})$, is said to be uniformly suboptimal if there exists a positive number ζ such that

$$(\sigma^{\oplus})^2 \le \zeta^{-1} (\sigma^{\odot})^2 \tag{4.161}$$

holds for all $t_0 \in (-\infty, +\infty)$, $t \in (t_0, +\infty)$, and all $a \in \mathbf{R}^n$ $(a \neq 0)$.

We consider the case when $R_v = \text{diag}\{R_{v1}, R_{v2}, \ldots, R_{vN}\}$ and $R_w = \text{diag}\{R_{w1}, R_{w2}, \ldots, R_{wN}\}$. This means that the input noise and observation noise of each subsystem are independent. We prove the following corollary to Theorem 4.13:

4.19. COROLLARY. An interconnected estimator $\hat{\mathbf{S}}^{\oplus}$ is uniformly suboptimal if the matrix

$$G(t, t_0; \zeta) = A_C(t)S_D(t; t_0, R_0) + S_D(t; t_0, R_0)A_C^T(t) \\ - (1-\zeta)[S_D(t; t_0, R_0)C_D^T(t)R_w^{-1}(t)C_D(t)S_D(t; t_0, R_0) + R_v(t)] \tag{4.162}$$

is nonpositive definite for some $\zeta \leq 1$ and all $t_0 \in (-\infty, +\infty), t \in (t_0, +\infty)$.

Proof. The dual of the estimation problem defined by $\mathbf{S}$ of (4.149) and $(\sigma^{\odot})^2$ of (4.155) is to find a deterministic vector function $\ell(-t; -t_0)$ that minimizes the performance index

$$z^T(-t_0)R_0z(-t_0) + \int_{-t}^{-t_0} (z^T R_v z + \ell^T R_w \ell)\, d(-\tau) = \mathcal{E}[a^T x(t) - a^T \hat{x}^{\odot}(t)]^2, \tag{4.163}$$

constrained by the equation

$$\frac{dz}{d(-\tau)} = A_D^T z + A_C^T z + C_D^T \ell, \tag{4.164}$$

where $-\tau \in (-t, -t_0)$ and $z(-t; -t, a) = a$ (*e.g.*, Åström, 1970). This problem has a well-known solution for $A_C(t) \equiv 0$:

$$\ell = -L_D^T z, \tag{4.165}$$

where L_D is given by (4.153). Theorem 4.13, when applied to the deterministic system (4.164), implies that inequality (4.161) of Definition 4.18 holds for an arbitrary vector $a \in \mathbf{R}^n$ on any time interval (t_0, t) if the matrix $G(t, t_0; \zeta)$ of the corollary is nonpositive definite for all $t_0 \in (-\infty, +\infty)$, $t \in (t_0, +\infty)$. Q.E.D.

An obvious dual of Corollary 4.16 can be formulated by using Corollary 4.19 to establish exponential stability of the unforced closed-loop system

$$\tilde{\mathbf{S}}_L\colon\ \dot{\tilde{x}} = A_L(t)\tilde{x}, \tag{4.166}$$

where $\tilde{x}(t) = x(t) - \hat{x}(t)$ is the expected estimation error,

$$A_L(t) = A(t) - L_D(t)C_D(t), \tag{4.167}$$

and

$$L_D(t) = \bar{S}_D(t)C_D^T(t)R_w^{-1}(t), \tag{4.168}$$

with $\bar{S}_D(t) = \lim_{t_0 \to -\infty} S_D(t;\, t_0,\, 0)$. In case of a time-invariant version of the system $\mathbf{S}$ in (4.149) and infinite observation interval, we have a constant matrix

$$\bar{G}(\zeta) = A_C\bar{S}_D + \bar{S}_D A_C - (1-\zeta)(\bar{S}_D C_D^T R_w^{-1} C_D \bar{S}_D + R_v). \tag{4.169}$$

Again, nonpositive definiteness of $\bar{G}(\zeta)$ guarantees suboptimality of the estimator $\hat{\mathbf{S}}^{\oplus}$. If, in addition, $\bar{S}_D C_D^T R_w^{-1} C_D \bar{S}_D + R_v$ is positive definite, then the unforced suboptimal estimator is also exponentially stable.

4.7. Incomplete State Information

By joining the results obtained separately for control and estimation, we can provide a suboptimal solution to the decentralized control problem with incomplete state information. We consider again the system

$$\mathbf{S}\colon\; \begin{aligned} dx &= A_D x\,dt + B_D u\,dt + A_C x\,dt + dv, \\ dy &= C_D x\,dt + dw, \end{aligned} \tag{4.149}$$

with the cost

$$J_D = x_f^T P_f x_f + \int_{t_0}^{t_f} (x^T Q_x x + u^T Q_u u)\,dt, \tag{4.113}$$

and use

$$\mathcal{E}J_D = \mathcal{E}\{\mathcal{E}[J_D|\mathcal{Y}_t]\} \tag{4.170}$$

as a measure of system performance, where $\mathcal{Y}_t = \{y(\tau)\colon\; \tau \in (t_0,\, t)\}$. We remark that $\mathcal{E}[J_D|\mathcal{Y}_t]$ depends on the control law, which is a mapping from the space of observations $\mathcal{Y} = \{\mathcal{Y}_t\}$ into the control space $\mathcal{U}$, that is, $u\colon\; (t_0,\, t_f) \times \mathcal{Y} \to \mathcal{U}$.

It is well known (*e.g.*, Åström, 1970) that the optimal control law for the system **S** with $A_C(t) \equiv 0$ is

$$u_D^{\odot} = -K_D \hat{x}^{\odot}, \tag{4.171}$$

where $K_D = Q_u^{-1} B_D^T P_D$, and $P_D(t;\, t_0,\, P_f)$ is the symmetric positive definite solution of the Riccati equation

$$\dot{P}_D = A_D^T P_D - P_D A_D + P_D B_D Q_u^{-1} B_D^T P_D - Q_x, \tag{4.120}$$

with $t \in (t_0,\, t_f)$ and $P_D(t_f;\, t_f,\, P_f) = P_f$. The optimal state estimate $\hat{x}^{\odot}$ is defined by

$$\hat{\mathbf{S}}_D^{\odot}\colon\; d\hat{x}^{\odot} = A_D \hat{x}^{\odot}\, dt + B_D u^{\odot}\, dt + L_D(dy - C_D \hat{x}^{\odot}\, dt), \tag{4.172}$$

and $\hat{x}^{\odot}(t_0) = m_0$, where the matrix $L_D(t,\, t_0)$ is specified by $L_D = S_D C_D^T R_w^{-1}$, and $S_D(t;\, t_0,\, R_0)$ is the covariance matrix of the estimation error, which is governed by the equation

$$\dot{S}_D = S_D A_D^T + A_D S_D - S_D C_D^T R_w^{-1} C_D S_D + R_w, \tag{4.173}$$

where $S_D(t_0;\, t_0,\, R_0) = R_0$. We denote by $\mathcal{E} J_D^{\odot}$ the optimal value of $\mathcal{E} J_D$ with respect to $u \in \mathcal{U}$, which is obtained as

$$\min_u \mathcal{E} J_D = \mathcal{E} \min_u \mathcal{E}[J_D | \mathcal{Y}_t], \tag{4.174}$$

and by $\mathcal{E} J_D^{\oplus}$ the value of $\mathcal{E} J_D$ when the feedback of the same form as $u_D^{\odot}$ is applied to the interconnected system **S**. In the latter case, the suboptimal decentralized control is determined by

$$u_D = -K_D \hat{x}, \tag{4.175}$$

where the state estimate $\hat{x}(t;\, t_0,\, m_0)$ is provided by the estimator $\hat{\mathbf{S}}^{\oplus}$.

In order to formulate an analog to Theorem 4.13, we observe that the combined closed-loop system $\check{\mathbf{S}}$ & $\tilde{\mathbf{S}}$ can be described as

$$\check{\mathbf{S}} \,\&\, \tilde{\mathbf{S}}\colon\; dz = \bar{A} z\, dt + \Gamma\, d\bar{v}, \tag{4.176}$$

where the augmented state vector $z \in \mathbf{R}^{2n}$ is defined as $z = (x^T,\, \tilde{x}^T)^T$, $\tilde{x} = x - \hat{x}$, and $z(t_0;\, t_0,\, z_0) = z_0$. In (4.176), we have the matrices $\bar{A} = \bar{A}(t;\, t_0,\, t_f)$ and $\Gamma = \Gamma(t;\, t_0)$ as

$$\bar{A} = \begin{bmatrix} A - B_D K_D & B_D K_D \\ 0 & A - L_D C_D \end{bmatrix}, \qquad \Gamma = \begin{bmatrix} I_n & 0 \\ 0 & -K_D \end{bmatrix}. \tag{4.177}$$

The Wiener process $\bar{v}(t) \in \mathbf{R}^{2n}$ and the initial state z_0 are defined as $\bar{v} = (v^T, w^T)^T$ and $z_0 = (x_0^T, \tilde{x}_0^T)^T$, $\tilde{x}_0 = x_0^T - m_0$, where $\mathcal{E}\bar{v} = 0$, $\mathcal{E}z_0 = (m_0^T, 0)^T$, and covariance matrices $\bar{R}_v(t) = \text{cov}[\bar{v}(t), \bar{v}(t)]$, $\bar{R}_0 = \mathbf{COV}(z_0, z_0)$ are given by

$$\bar{R}_v = \begin{bmatrix} R_v & 0 \\ 0 & R_w \end{bmatrix}, \qquad \bar{R}_0 = \begin{bmatrix} R_0 & R_0 \\ R_0 & R_0 \end{bmatrix}. \tag{4.178}$$

When $A_C(t) \equiv 0$, we denote the augmented state z of the optimal closed-loop system $(\check{\mathbf{S}} \ \& \ \tilde{\mathbf{S}})_D$ by $\mathcal{E}z^{\odot}$, and the corresponding matrix $\bar{A}$ by $\bar{A}_D$. We introduce the matrix $\bar{A}_C = \bar{A} - \bar{A}_D = \text{diag}\{A_C, A_C\}$, and thus reduce the problem of suboptimality characterization to the one formulated in Section 4.5. In order to state an analog to Theorem 4.13, we note that the optimal performance index $\mathcal{E}J^{\odot}$ may be represented as

$$\mathcal{E}J_D^{\odot} = \mathcal{E}z_0^T\Phi^{\odot}(t_0)z_0 + \text{tr} \int_{t_0}^{t_f} \Phi^{\odot}(t)\Gamma(t)\bar{R}_v(t)\Gamma^T(t)\, dt, \tag{4.179}$$

where the symmetric nonnegative $2n \times 2n$ matrix $\Phi^{\odot}(t;\, t_f,\, \Phi_f)$ satisfies the differential equation

$$\frac{d}{dt}\Phi^{\odot} + \bar{A}_D^T\Phi^{\odot} + \Phi^{\odot}\bar{A}_D + \Theta = 0 \tag{4.180}$$

with $\Phi^{\odot}(t_f;\, t_f,\, \Phi_f) = \Phi_f$, and the symmetric nonnegative matrices $\Theta(t;\, t_f)$, Phi_f defined by

$$\Theta = \begin{bmatrix} Q_x + K_D^TQ_uK_D & -K_D^TQ_uK_D \\ -K_D^TQ_uK_D & K_D^TQ_uK_D \end{bmatrix}, \qquad \Phi_f = \begin{bmatrix} P_f & 0 \\ 0 & 0 \end{bmatrix}. \tag{4.181}$$

Now, we consider the control law $u_D^{\odot}$ of (4.171), the system $\mathbf{S}$ of (4.149) and the cost J_D of (4.113), and prove the following:

4.20. THEOREM. A decentralized control law $u_D^{\odot}$ is suboptimal if the matrix

$$H(t_0,\, t_f;\, \mu) = \bar{A}_C^T(t_0)\Phi^{\odot}(t;\, t_f,\, \Phi_f) + \Phi^{\odot}(t_0;\, t_f,\, \Phi_f)\bar{A}_C(t_0) - (1-\mu)\Theta(t_0,\, t_f) \tag{4.182}$$

is nonpositive definite for some $\mu \leq 1$, and all $t_0 \in [0, +\infty)$, $t_f \in (t_0, +\infty)$.

Proof. We observe that in analogy to (4.116), and using (4.180), we can write

$$\mathcal{E}J_D^{\oplus} = \mathcal{E}z_0^T\Phi^{\oplus}(t_0)z_0 + \operatorname{tr}\int_{t_0}^{t_f} \Phi^{\oplus}(t)\Gamma(t)\bar{R}_v(t)\Gamma^T(t)\,dt, \tag{4.183}$$

where the symmetric nonnegative definite $2n \times 2n$ matrix $\Phi^{\oplus}(t;\, t_f,\, \Phi_f)$ satisfies the differential equation

$$\frac{d}{dt}\Phi^{\oplus} + \bar{A}^T\Phi^{\oplus} + \Phi^{\oplus}\bar{A} + \Theta = 0, \tag{4.184}$$

with $\Phi^{\oplus}(t_f;\, t_f,\, \Phi_f) = \Phi_f$. Therefore,

$$\frac{d}{dt}(\mu\Phi^{\oplus} - \Phi^{\odot}) + \bar{A}^T(\mu\Phi^{\oplus} - \Phi^{\odot}) + (\mu\Phi^{\oplus} - \Phi^{\odot})\bar{A} + H = 0, \tag{4.185}$$

where

$$\mu\Phi^{\oplus}(t_f;\, t_f,\, \Phi_f) - \Phi^{\odot}(t_f;\, t_f,\, \Phi_f) = -(1-\mu)\Phi_f, \tag{4.186}$$

and

$$\begin{aligned}\mu\mathcal{E}J_D^{\oplus} - \mathcal{E}J_D^{\odot} =& \mathcal{E}z_0^T[\mu\Phi^{\oplus}(t_0) - \Phi^{\odot}(t_0)]z_0 \\ &+ \operatorname{tr}\int_{t_0}^{t_f} [\mu\Phi^{\oplus}(t) - \Phi^{\odot}(t)]\bar{R}_v(t)\,dt\end{aligned} \tag{4.187}$$

which proves the theorem. Q.E.D.

4.21. REMARK. Since the matrix F is a principal minor of H, Theorem 4.20 implies the same degree of suboptimality μ of the control law $u_D^{\odot}$ of Theorem 4.13 in the case of complete state estimation.

4.8. Structural Perturbations

We return to our favorite subject of structural perturbations and consider performance of an interconnected dynamic system driven by locally optimal decentralized control and subject to unpredictable changes of the interconnection structure. Our objective is to present the conditions under which suboptimality of a stochastic decentralized control strategy is invariant to structural perturbations.

To incorporate changes of the interconnection structure, we modify the system description (4.110) to include the elements $e_{ij}(t)$ of the $N \times N$ interconnection matrix $E = (e_{ij})$:

$$\mathbf{S}\colon\ dx_i = A_i x_i\,dt + B_i u_i\,dt + \sum_{j=1}^{N} e_{ij} A_{ij} x_j\,dt + dv_i, \qquad i \in \boldsymbol{N}, \tag{4.188}$$

where the elements $e_{ij}(t)$ of the matrix $E(t)$ are continuous functions of time such that $e_{ij}(t) \in [0,\ 1]$ for all $t \in [0,\ +\infty)$. We recall the definition of the $N \times N$ fundamental interconnection matrix $\bar{E} = (\bar{e}_{ij})$:

$$\bar{e}_{ij} = \begin{cases} 1\,, & A_{ij} \not\equiv 0, \\ 0\,, & A_{ij} \equiv 0. \end{cases} \tag{4.189}$$

We state the following:

4.22. DEFINITION. A decentralized control

$$u_D^{\odot} = -Q_u^{-1} B_D^T P_D x \tag{4.122}$$

is connectively suboptimal with a degree μ for the system $\mathbf{S}$ and with respect to the pair $(\check{\mathbf{S}}_D^{\odot}, J_D^{\odot})$, if it is suboptimal with the degree μ for all $E \in \bar{E}$.

In order to derive conditions for connective suboptimality, it is convenient to introduce also the $n \times n$ matrices $\tilde{A}_{k\ell}$ obtained from the matrix $P_D A_C = (e_{ij} P_i A_{ij})$, where $A_C = (e_{ij} A_{ij})$, by setting $e_{ij} \equiv 1$ when $i = k$, $j = \ell$, and $e_{ij} \equiv 0$ otherwise. Furthermore, we define the numbers ξ_{ij} and η as

$$\xi_{ij} = \inf_{\substack{t_f \in (0,+\infty) \\ t \in [0,t_f)}} \lambda_M \left\{ \frac{1}{2} [\tilde{A}_{ij}^T(t,\ t_f) + \tilde{A}_{ij}^T(t,\ t_f)] \right\}, \qquad \forall i, j \in \boldsymbol{N},$$

$$\eta = \inf_{\substack{t_f \in (0,+\infty) \\ t \in [0,t_f)}} \min_{i \in \boldsymbol{N}} \lambda_m [W_i(t,\ t_f)], \tag{4.190}$$

where

$$W_i(t,\ t_f) = P_i(t,\ t_f,\ P_f) B_i(t) Q_{ui}^{-1}(t) B_i^T(t) P_i(t,\ t_f;\ P_f) + Q_{xi}(t). \tag{4.191}$$

We assume $\eta > 0$, and establish the following result:

4.23. THEOREM. A decentralized control law $u_D^{\odot}$ is connectively suboptimal if for some $\mu \leq 1$,

$$\sum_{i,j=1}^{N} \bar{e}_{ij}\xi_{ij} \leq \frac{1}{2}(1-\mu)\eta. \tag{4.192}$$

Proof. We note that from the definition of the matrices $\tilde{A}_{ij}$ and the numbers ξ_{ij}, we have

$$\sum_{i,j=1}^{N} \bar{e}_{ij}\xi_{ij} \geq \sum_{i,j=1}^{N} e_{ij}\frac{x^T\tilde{A}_{ij}(t,\, t_f)x}{x^Tx} = \frac{x^TP_D(t,\, t_f,\, P_f)A_C(t)x}{x^Tx} \tag{4.193}$$

$$\forall t \in [0,\, t_f), \quad \forall t_f \in (0,\, +\infty), \quad \forall E \in \bar{E},$$

where we recall $A_C = (e_{ij}A_{ij})$. Then, from the condition of the theorem and (4.193), we conclude that $F(t,\, t_f;\, \mu)$ of Theorem 4.13 is nonpositive definite for all $t \in [0,\, t_f)$, $t_f \in (0,\, +\infty)$. Hence, the theorem follows from Theorem 4.13. Q.E.D.

Besides suboptimality under structural perturbations, Theorem 4.23 provides also a simple test for suboptimality of decentralized control strategies. Although the test is more restrictive than that of Theorem 4.13, it involves only subsystem matrices and is a useful computational simplification when the interconnection matrix A_C is large.

4.24. REMARK. According to Corollary 4.15 the inequality (4.192) of Theorem 4.23 implies exponential connective stability of the closed-loop deterministic system $\mathbf{S}_K$ of (4.136), where $A_K(t)$ is appropriately redefined to include the elements $e_{ij}(t)$. It follows from the proof of the corollary that

$$\|x(t;\, t_0,\, x_0)\| \leq \Pi\|x_0\| \exp[-\pi(t-t_0)], \tag{4.194}$$

where $x(t;\, t_0,\, x_0)$ is a solution to (4.136), $\Pi = \lambda_{\text{sup}}^{1/2}[\bar{P}_D(t)]\lambda_{\text{inf}}^{-1/2}[\bar{P}_D(t)]$, $\pi = -\frac{1}{2}\lambda_{\text{inf}}[\bar{F}(t;\, 0)]\, \lambda_{\text{sup}}^{-1}[\bar{P}_D(t)]$ are both positive numbers, and $\lambda_{\text{sup}}(\,\bullet\,)$ and $\lambda_{\text{inf}}(\,\bullet\,)$ denote the least upper bound and the greatest lower bound of the maximal and minimal eigenvalues. The fact that $\lambda_{\text{sup}}[\bar{P}_D(t)]$, $\lambda_{\text{inf}}[\bar{P}_D(t)]$, and $\lambda_{\text{inf}}[\bar{F}(t;\, 0)]$ are all positive numbers follows from the uniform boundedness of the matrix $\bar{P}_D(t)$ and inequality (4.192). From this inequality and

(a)

$$\bar{E} = \begin{bmatrix} 0 & 1 \\ 1 & 0 \end{bmatrix}$$

$$\begin{bmatrix} 0 & 0 \\ 1 & 0 \end{bmatrix} \qquad \begin{bmatrix} 0 & 1 \\ 0 & 0 \end{bmatrix} \qquad \begin{bmatrix} 0 & 0 \\ 0 & 0 \end{bmatrix}$$

(b)

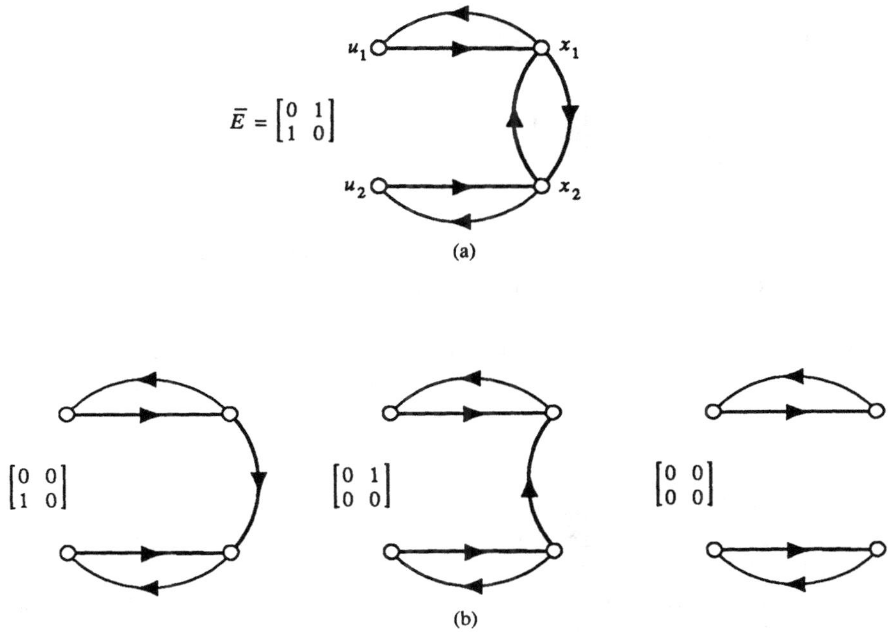

Fig. 4.2.   Structural perturbations: interconnections; (a) basic structure; (b) perturbed structures.

the proof of Theorem 4.13, it further follows that inequality (4.194) holds for all $E(t) \in \bar{E}$, which implies connective exponential stability in the large of the equilibrium $x = 0$ of the closed-loop system $\mathbf{S}_K$.

In Figure 4.2, we illustrate the meaning of structural perturbations which are characterized by Theorem 4.13 for a case of two interconnected subsystems. Following Šiljak (1978), we describe the basic structure of the overall system by the digraph on Figure 4.2a, which corresponds to the fundamental interconnection matrix $\bar{E}$ (feedback loops involving inputs are excluded). The structural perturbations with binary matrices $E \in \bar{E}$ are shown in Figure 4.2b.

4.25. REMARK.   Finally, we want to comment on the fact that the closed-loop system $\mathbf{S}_K$ driven by a suboptimal decentralized control is robust; it can tolerate a wide range of nonlinearities in the interactions among the subsystems. We introduce a modification of $\mathbf{S}_K$ whereby the time-varying matrices $A_{ij}(t)$ are replaced by nonlinear time-varying matrices $A_{ij}(t, x)$. The elements $a_{ij}(t, x)$ of $A_{ij}(t, x)$ are sufficiently smooth so that the solutions $x(t;\, t_0,\, x_0)$ of (4.188) exist and are unique for all initial conditions $(t_0,\, x_0) \in \mathbf{T} \times \mathbf{R}^n$ and all time $t \in \mathbf{T}_0$. The ξ_{ij} are now redefined as

$$\xi_{ij} = \sup_{\substack{t,\, t_f \in \mathbf{T}_0 \\ x \in \mathbf{R}^n}} \lambda_M \left\{ \frac{1}{2} [\tilde{A}_{ij}^T(t,\, t_f,\, x) + \tilde{A}_{ij}(t,\, t_f,\, x)] \right\}, \qquad i \in N. \tag{4.195}$$

Under the condition (4.192) of Theorem 4.13, where the ξ_{ij}'s of (4.195) are used, the decentralized control law $u_D^{\odot}$ is connectively suboptimal with degree μ for the nonlinear version of **S** in (4.188). It is obvious that by duality this conclusion carries over to decentralized estimators.

4.9. Degenerate Control

Structural perturbations in the feedback structure are of special interest because they can take place not only as a consequence of operational conditions, but as a means of control design. To come up with a decentralized control law, one may design a centralized control strategy and disconnect the feedback links to fit a prescribed information structure constraint—this is termed the *degenerate control* (Krtolica and Šiljak, 1980). Of course, there is no guarantee that the control obtained in this way would be stable, let alone optimal in any prescribed way. Our objective here is to present the sufficient conditions for suboptimality and stability of degenerate control laws.

We consider again the system **S** of (4.109), but, in the equation

$$\mathbf{S}\colon\; dx = Ax\,dt + Bu\,dt + dv, \tag{4.196}$$

we do not assume that the matrix B is block diagonal. With **S**, we associate the same cost J_D of (4.113). The globally optimal control law of Kalman is

$$u^{\odot} = -Kx, \tag{4.197}$$

where the matrix $K(t,\, t_f)$ is given as

$$K = Q_u^{-1} B^T P, \tag{4.198}$$

and $P(t;\, t_f,\, P_f)$ is the symmetric positive definite solution of the Riccati equation

$$\dot{P} = -A^T P - PA + PBQ_u^{-1} B^T P - Q_x, \tag{4.199}$$

with $t \in (t_0, t_f)$ and $P(t_f; t_f, P_f) = P_f$, provided the pair $(A, Q_x^{1/2})$ is differentially observable. The matrices K and P are no longer necessarily block-diagonal. The closed-loop system driven by the optimal control law $u^\odot$ is

$$\check{\mathbf{S}}^\odot\colon\ dx = (A - BK)x\,dt + dv, \tag{4.200}$$

and the corresponding expected value of the optimal performance index is

$$\mathcal{E}J_D^\odot = \mathcal{E}\left\{x_0^T P(t_0;\, t_f,\, P_f)x_0 + \int_{t_0}^{t_f} \mathrm{tr}[P(t;\, t_f,\, P_f)R_v(t)]\,dt\right\}. \tag{4.201}$$

Now, in order to consider disturbances in the control structure, we introduce the notation

$$B = B_D + B_C, \qquad K = K_D + K_C, \tag{4.202}$$

where $B_D = \mathrm{diag}\{B_1, B_2, \ldots, B_N\}$, $K_D = \mathrm{diag}\{K_1, K_2, \ldots, K_N\}$, $B_C = (B_{ij})$, and $K_C = (K_{ij})$ are matrices with blocks $B_i(t)$, $K_i(t)$, $B_{ij}(t)$, and $K_{ij}(t)$ that are continuous in time and of appropriate dimensions. Additive disturbances $-B_C(t)$ and $-K_C(t)$ of the input and feedback gain matrices $B(t)$ and $K(t)$ result in $B(t) \equiv B_D(t)$ and $K(t) \equiv K_D(t)$, as indicated by thin lines in Figure 4.3. The resulting (now decentralized) control law is

$$u_D = -K_D x, \tag{4.203}$$

and the perturbed closed-loop system becomes

$$\check{\mathbf{S}}^\oplus\colon\ dx = (A - B_D K_D)x\,dt + dv. \tag{4.204}$$

After the disturbance, the expected value of the performance index is

$$\mathcal{E}J_D^\oplus = \mathcal{E}\left\{x_f^T P_f x_f + \int_{t_0}^{t_f} x^T(Q_x x + K_D^T Q_u K_D)x\,dt\right\}. \tag{4.205}$$

A sufficient condition for the basic inequality

$$\mathcal{E}J_D^\oplus \le \mu^{-1}\mathcal{E}J_D^\odot \tag{4.206}$$

is given by the following:

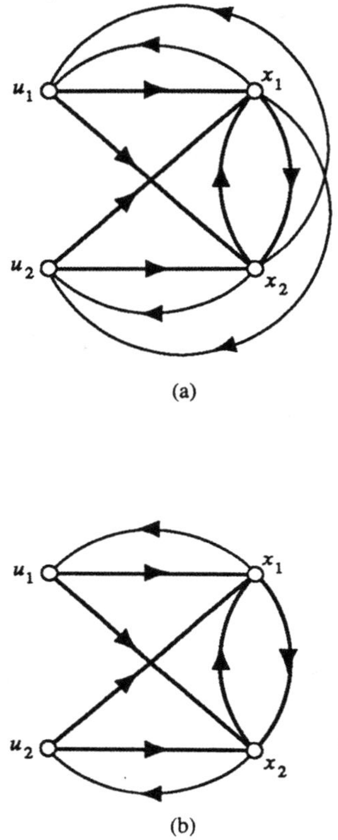

Fig. 4.3. Structural perturbations: control; (a) basic structure; (b) perturbed structure.

4.26. THEOREM. A decentralized control law $u_D^{\odot}$ is suboptimal for the system **S** with respect to the pair $(\check{\mathbf{S}}^{\odot},\, J_D^{\odot})$ if, for some $\mu \leq 1$, the matrix

$$
\begin{aligned}
F_{BK}(t,\, t_f;\, \mu) = &- K_D^T(t_0,\, t_f) B_D^T(t_0) P(t_0;\, t_f,\, P_f) \\
&- P(t_0;\, t_f,\, P_f) B_D(t_0) K_D(t_0,\, t_f) - Q_x(t_0) \\
&+ P(t_0;\, t_f,\, P_f) B(t_0) Q_u^{-1}(t_0) B^T(t_0) P(t_0;\, t_f,\, P_f) \\
&+ \mu [Q_x(t_0) + K_D^T(t_0,\, t_f) Q_u(t_0) K_D(t_0,\, t_f)]
\end{aligned}
\tag{4.207}
$$

is nonpositive definite for all $t_0 \in [0,\, +\infty)$ and $t_f \in (t_0,\, +\infty)$.

Proof. In analogy to the proof of Theorem 4.13, we use the representation (4.125) of $\mathcal{E}J_D^{\oplus}$ to obtain (4.151), where the matrix $\mu P^{\oplus}(t;\, t_f,\, P_f) - P(t;\, t_f,\, P_f)$ is the solution of the differential equation

$$\frac{d}{dt}(\mu P^{\oplus} - P) + A_K^T(\mu P^{\oplus} - P) + (\mu P^{\oplus} - P)A_K + F_{BK} = 0, \tag{4.208}$$

with $\mu P^{\oplus}(t_f;\, t_f,\, P_f) - P(t_f;\, t_f,\, P_f) = -(1-\mu)P_f$, and $A_K = A - BK$. Q.E.D.

Remark 4.14, which is concerned with the case $t = +\infty$, applies to Theorem 4.26 directly. Furthermore, a statement analogous to Corollary 4.15 can also be established.

An interesting special case arises when the original matrix B is block diagonal, and perturbations affect only the feedback control law, that is, the gain matrix K. In this case, $B \equiv B_D$, and after perturbation the centralized optimal control law $u^{\odot}$ is reduced to a decentralized control law u_D that is suboptimal with a degree $\mu \leq 1$. This fact is confirmed by the form of the matrix F_{BK}, which becomes

$$F_K = K_C^T Q_u K_C - (1-\mu)(Q_x + K_D^T Q_u K_D). \tag{4.209}$$

From this expression, it is obvious that F_K can be nonpositive definite only if $\mu \leq 1$.

We now introduce the matrix

$$W_K(t,\, t_f) = Q_x(t) + K_D^T(t,\, t_f)Q_u(t)K_D(t,\, t_f), \tag{4.210}$$

and state the following result:

4.27. THEOREM. A decentralized control law u_D is suboptimal for the system $\mathbf{S}$ and with respect to the pair $(\check{\mathbf{S}}^{\odot},\, J_D^{\odot})$ if, for some $\mu \leq 1$, the matrix

$$G_K(t,\, t_f;\, \mu) = K_C^T(t,\, t_f)K_C(t,\, t_f) - (1-\mu)\lambda_M^{-1}[Q_u(t)]W_K(t,\, t_f) \tag{4.211}$$

is nonpositive definite for all $t \in [0,\, t_f)$ and $t_f \in (0,\, +\infty)$.

Proof. The theorem follows from the fact that the matrix $K_C^T Q_u K_C - \lambda_M(Q_u)K_C^T K_C$ is always nonpositive definite. Q.E.D.

The suboptimality condition of Theorem 4.27 can be restated as the inequality

$$\| K_C(t,\, t_f)\|^2 \leq (1-\mu)\lambda_m[W_K(t,\, t_f)]/\lambda_M[Q_u(t)], \tag{4.212}$$

which is easy to test. Furthermore, from this inequality we see that suboptimality of the degenerate decentralized control requires that the perturbation of the control structure produce "small effects" on the centralized optimal control system. This fact leads naturally to connective stability investigations of degenerate forms of decentralized control.

Let us introduce the elements $\delta_{ij}(t)$ of the $N \times N$ feedback interconnection matrix $\Delta = (\delta_{ij})$ into the description of the optimal control law as

$$u^{\odot}(x, t; t_f) = -K_i(t, t_f) - \sum_{j=1}^{N} \delta_{ij}(t) K_{ij}(t, t_f) x, \qquad i \in \boldsymbol{N}. \tag{4.213}$$

In fact, we redefine the matrix K_C as $K_C = (\delta_{ij} K_{ij})$. The elements $\delta_{ij}(t) \in [0, 1]$ of the matrix $\Delta(t)$ are continuous functions of time, and $\delta_{ij}(t) \leq \bar{\delta}_{ij}$, where $\bar{\Delta} = (\delta_{ij})$ described the fundamental control structure as

$$\bar{\delta}_{ij} = \begin{cases} 1\,, & K_{ij}(t, t_f) \not\equiv 0, \\ 0\,, & K_{ij}(t, t_f) \equiv 0. \end{cases} \tag{4.214}$$

It is evident that continuity in t of the elements of the matrix blocks $\delta_{ij}(t) K_{ij}(t, t_f)$ is preserved. We also redefine the numbers

$$\begin{aligned} \tilde{\xi}_{ijk} &= \sup_{\substack{t_f \in (0, +\infty) \\ t \in [0, t_f)}} \| K_{ki}^T(t, t_f) K_{kj}(t, t_f) \|, \qquad \forall i, j, k \in \boldsymbol{N}, \\ \tilde{\eta} &= \inf_{\substack{t_f \in (0, +\infty) \\ t \in [0, t_f)}} \lambda_m[W_K(t, t_f)] / \sup_{\substack{t_f \in (0, +\infty) \\ t \in [0, t_f)}} \lambda_M[Q_u(t)]. \end{aligned} \tag{4.215}$$

We prove the following:

4.28. THEOREM. A decentralized control law u_D is connectively suboptimal with a degree μ for the system $\mathbf{S}$ with respect to the pair $(\check{\mathbf{S}}^{\odot}, J_D^{\odot})$ if

$$\sum_{i,j,k=1}^{N} \bar{\delta}_{ki} \bar{\delta}_{kj} \tilde{\xi}_{ijk} \leq (1 - \mu) \tilde{\eta}. \tag{4.216}$$

Proof. From the condition (4.216) of the theorem, it follows that

$$\sum_{i,j,k=1}^{N} \bar{\delta}_{ki}\bar{\delta}_{kj} \| K_{ki}^T(t,\, t_f) K_{kj}(t,\, t_f) \| \leq (1-\mu) \frac{\lambda_m[W_K(t,\, t_f)}{\lambda_M[Q_u(t)]},$$

$$\forall t \in [0,\, t_f)\ ,\ \forall t_f \in (0, +\infty). \tag{4.217}$$

We remark that

$$x^T K_C^T K_C x = \sum_{i,j,k=1}^{N} \delta_{ki}\delta_{kj} x_i^T K_{ki}^T K_{kj} x_j, \tag{4.218}$$

and

$$\sum_{i,j,k=1}^{N} \delta_{ki}\delta_{kj} x_i^T K_{ki}^T K_{kj} x_j \leq \sum_{i,j,k=1}^{N} \bar{\delta}_{ki}\bar{\delta}_{kj} \| K_{ki}^T K_{kj} \| \ \|x\|^2,$$

$$\forall t \in [0,\, t_f), \qquad \forall t_f \in (0,\, +\infty), \qquad \forall x \in \mathbf{R}^n. \tag{4.219}$$

Inequalities (4.217) and (4.219) imply that the conditions of Theorem 4.26 are satisfied for all $\Delta(t) \in \bar{\Delta}$. Q.E.D.

In analogy to the arguments of Remark 4.24, we conclude that when $\tilde{\eta} > 0$, the inequality (4.216) of the above theorem implies exponential connective stability of the closed-loop deterministic system

$$\mathbf{S}_K\colon\ \dot{x} = A_K x, \tag{4.220}$$

which corresponds to the use of the optimal control law $u^{\odot}$ and the absence of the input noise on an infinite time interval. That is,

$$A_K(t) = A(t) - B(t) Q_u^{-1}(t) B^T(t) \bar{P}(t), \tag{4.221}$$

and $\bar{P}(t) = \lim_{t_f \to +\infty} P(t;\, t_f,\, 0)$, where $P(t;\, t_f,\, P_f)$ is the solution of the Riccati equation (4.199). Furthermore, we note that the connectivity aspect of stability of the system $\mathbf{S}$ involves only the perturbations of feedback links as illustrated in Figure 4.4. Connective suboptimality corresponding to the optimal control law $u^{\odot}$ includes the suboptimal control u_D as a special case when $\Delta(t) \equiv 0$. Indeed, the third digraph of Figure 4.4b is the last digraph

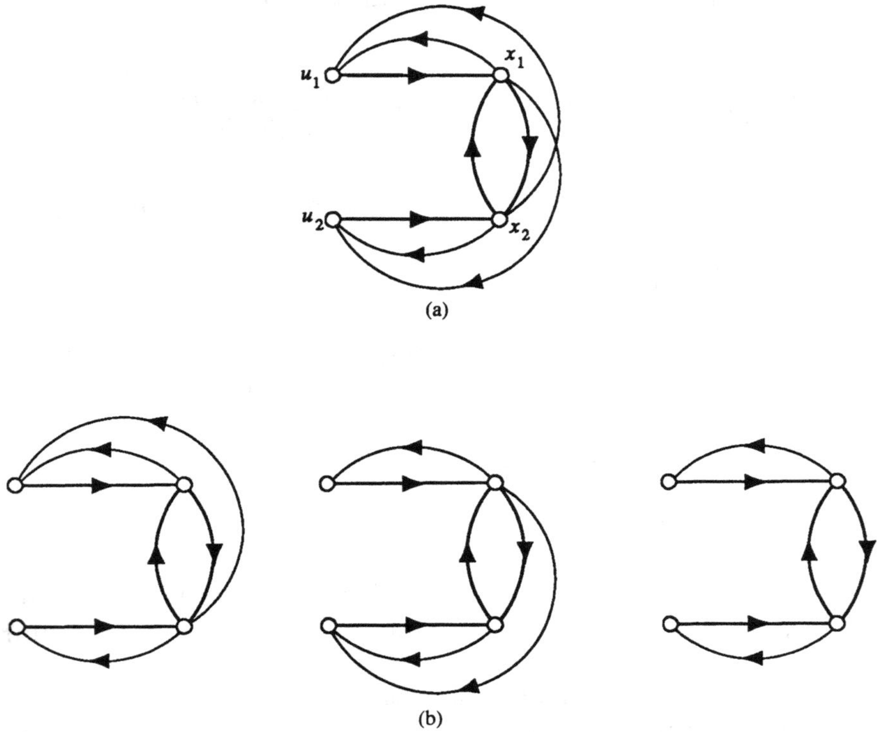

Fig. 4.4. Structural perturbations: feedback; (a) basic structure; (b) perturbed structures.

of Figure 4.2b. However, the condition (4.216) of connective suboptimality is more restrictive than nonpositivity of the matrix F_K in (4.209), which is equivalent to suboptimality of the perturbed control law u_D.

4.10. Notes and References

The early development of decentralized state determination (Šiljak, 1978) indicated that building observers for complex linear plants requires much more effort and ingenuity than solving a decentralized stabilization problem *via* state feedback. One of the first decentralized observers (Weissenberger, 1974) was constructed together with the corresponding control law. The separation property (Section 4.3) was established in (Šiljak and Vukčević, 1976) at the price of the exchange of state estimates among the local observers (see also Sundareshan, 1977; Sundareshan and Huang, 1984).

Unlike centralized observers, observability is not enough of a guarantee that a decentralized observer can be made asymptotically stable using feedback. We had to apply the concept of vector Liapunov functions and the M-matrix test to establish stability. One way to assure success of the test is to restrict the class of interconnections so that a decentralized observer can *always* be built by high-gain feedback (Šiljak and Vukčević, 1978: Section 4.2; Bachmann, 1983: Section 4.4; Bachmann and Konik, 1984).

If we want to avoid the exchange of state estimates among the subsystem observers, we should consider the interconnections as unknown inputs and reject their influence on the estimation error. This approach was developed by Viswanadham and Ramakrishna (1982) using the theory of *unknown-input observers* (Basile and Marro, 1969; see the survey by Viswanadham, 1987).

For an unknown-input (or disturbance decoupling) observer to work, the interconnections have to satisfy a rather restrictive condition. A remedy is to bring back into action the high-gain feedback (Section 4.2) and apply almost disturbance decoupling observers (Willems, 1982) as proposed by Ikeda and Willems (1986).

Observers for discrete-time systems were developed by reformulating the continuous-time results (Mahmoud and Singh, 1981; Hassan *et al.* 1983). Applications of decentralized observers to mechanical systems (robotics) were presented by Stokić and Vukobratović (1983). A comprehensive survey on decentralized state reconstruction can be found in (Kuhn and Schmidt, 1984). Suboptimality *vs.* robustness issues in decentralized control using observers were considered recently by Shahian (1986).

In the context of *stochastic decentralized problems* of estimation and control, it has been long recognized that solutions to LQG problems with nonclassical (decentralized) information patterns may be nonlinear. Using a simple discrete-time LQG problem, Witsenhausen (1968) showed that a nonlinear decentralized control law outperforms the best law of the linear variety. This important discovery predicted difficulties in searching for optimal solutions to decentralized stochastic control problems. Over the years, many authors have more or less successfully addressed these difficulties, but a satisfactory solution to the optimal problem has never been found.

A suboptimal solution to the decentralized LQG control problem (Krtolica and Šiljak, 1980: Sections 4.5–9) is useful, not only because it produces linear control laws, but because it is *robust*. In controlling complex plants, it is natural to assume that the interconnections are uncertain. Suboptimal control can cope with unknown perturbations in the interconnections of a complex plant as well as controller connections. It has also been shown (Hodžić *et al.* 1983; Hodžić and Šiljak, 1986a) that the suboptimal scheme can accomodate decentralized estimation and control laws with *overlapping information sets* (see Chapter 8).

Suboptimal LQG control laws provide a considerable autonomy to the subsystems, which opens up a real possibility for distributed parallel computations of the subsystems estimates (Hodžić and Šiljak, 1986b). A study of new iterative methods for parallel estimation was initiated (Hodžić and Šiljak, 1986c), which laid the groundwork for multirate schemes implementable by flexible multiprocessor systems. Jacobi and Gauss–Seidel decentralized filters have been introduced, which can exploit a considerable wealth of experience present in corresponding numerical methods (Hageman and Young, 1981; Bertsekas and Tsitsiklis, 1989; Kaszkurewicz *et al.* 1989). The iterative estimation schemes are ideally suited for realizations *via* multiprocessor systems with their inherent advantages in response time, reliability, program modularity, incremental redundancy, *etc.*, over the mainframe computer implementations (Liebowitz and Carson, 1985).

Bibliography

Anderson, B. D. O., and J. B. Moore (1969). New results in linear system stability. *SIAM Journal of Control*, 7, 398–414.

Åström, K. J. (1970). *Introduction to Stochastic Control Theory*. Academic Press, New York.

Bachmann, W. (1983). On decentralized output feedback stabilizability. *Large Scale Systems*, 4, 165–176.

Bachmann, W., and D. Konik (1984). On stabilization of decentralized dynamic output feedback systems. *Systems and Control Letters*, 5, 89–95.

Basile, G., and G. Marro (1969). On the observability of linear time-invariant systems with unknown inputs. *Journal of Optimization Theory and Applications*, 3, 410–415.

Bertsekas, D. P., and J. N. Tsitsiklis (1989). *Parallel and Distributed Computation*. Prentice-Hall, Englewood Cliffs, New Jersey.

Bucy, R. S., and P. D. Joseph (1968). *Filtering for Stochastic Processes with Applications to Guidance*. Wiley, New York.

Gikhman, J. J., and A. V. Skorokhod (1969). *Introduction to the Theory of Random Processes*. Saunders, Philadelphia, Pennsylvania.

Hageman, L. A., and D. M. Young (1981). *Applied Iterative Methods*, Academic Press, New York.

Hassan, M. F., M. I. Younis, and M. S. Mahmoud (1983). Linear multilevel estimators for stabilization of discrete systems. *International Journal of Systems Science*, 14, 731–743.

Hodžić, M., and D. D. Šiljak (1986a). Decentralized estimation and control with overlapping information sets. *IEEE Transactions*, AC-31, 83–86.

Hodžić, M., and D. D. Šiljak (1986b). A parallel estimation algorithm. *Parallel Processing Techniques for Simulation*, M. G. Singh, A. Y. Allidina, and B. K. Daniels (eds.). Plenum Press, New York, 113–121.

Hodžić, M., and D. D. Šiljak (1986c). Iterative methods for parallel-multirate estimation. *Preprints of the 4th IFAC/IFORS Symposium on Large Scale Systems*, Zurich, Switerland, 169–175.

Hodžić, M., R. Krtolica, and D. D. Šiljak (1983). A stochastic inclusion principle. *Proceedings of the* 22nd IEEE Control and Decision Conference, San Antonio, Texas, 17–22.

Ikeda, M., and D. D. Šiljak (1980). On decentrally stabilizable large-scale systems. *Automatica*, 16, 331–334.

Ikeda, M., and J. C. Willems (1986). An observer theory for decentralized control of large-scale interconnected systems (unpublished paper).

Kailath, T. (1980). *Linear Systems*. Prentice-Hall, Englewood Cliffs, New Jersey.

Kaszkurewicz, E., A. Bhaya, and D. D. Šiljak (1989). On the convergence of parallel asynchronous block-iterative computations. *Linear Algebra and Its Applications* (to appear).

Krtolica, R., and D. D. Šiljak (1980). Suboptimality of decentralized stochastic control and estimation. *IEEE Transactions*, AC-25, 76–83.

Kuhn, U., and G. Schmidt (1984). Decentralized observation: A unifying presentation of five basic schemes. *Large Scale Systems*, 7, 17–31.

Liebowitz, B. H., and J. H. Carson (1985). *Multiple Processor Systems for Real-Time Applications*, Prentice-Hall, Englewood Cliffs, New Jersey.

Luenberger, D. G. (1964). Observing the state of a linear system. *IEEE Transactions*, MIL-8, 74–80.

Mahmoud, M. S., and M. G. Singh (1981). Decentralized state reconstruction of interconnected discrete systems. *Large Scale Systems*, 2, 151–158.

O'Reilly, J. (1983). *Observers for Linear Systems*. Academic Press, New York.

Shahian, B. (1986). Decentralized control using observers. *Control and Dynamic Systems*, C. T. Leondes (ed.), 24, 105–124.

Šiljak, D. D. (1978). *Large-Scale Dynamic Systems: Stability and Structure*, North-Holland, New York.

Šiljak, D. D., and M. Vukčević (1976). Decentralization, stabilization, and estimation of large-scale linear systems. *IEEE Transactions*, AC-21, 363–366.

Šiljak, D. D., and M. Vukčević (1978). On decentralized estimation. *International Journal of Control*, 27, 113–131.

Stokić, D., and M. Vukobratović (1983). Decentralized regulator and observer for a class of large scale nonlinear mechanical systems. *Large Scale Systems*, 5, 189–206.

Sundareshan, M. K. (1977). Decentralized observation in large-scale systems. *IEEE Transactions*, SMC-7, 863–867.

Sundareshan, M. K., and P. C. K. Huang (1984). On the design of a decentralized observation scheme for large-scale systems. *IEEE Transactions*, AC-29, 274–276.

Viswanadham, N. (1987). Large-Scale Systems: Observers. *Systems and Control Encyclopedia*, M. G. Singh (ed.). Pergamon Press, Oxford, UK, 2697–2702.

Viswanadham, N., and A. Ramakrishna (1982). Decentralized estimation and control for interconnected systems. *Large Scale Systems*, 3, 255–266.

Weiss, L. (1966). Correction and addendum: The concepts of differential controllability and differential observability. *Journal of Mathematical Analysis and Applications*, 13, 577–578.

Weissenberger, S. (1974). Tolerance of decentrally optimal controllers to nonlinearity and coupling. *Proceedings of the 12th Allerton Conference on Circuit and System Theory*, University of Illinois, Monticello, Illinois, 87–95.

Willems, J. C. (1982). Almost invariant subspaces: An approach to high-gain feedback design; Part II: Almost conditionally invariant subspaces. *IEEE Transactions*, AC-27, 1071–1085.

Witsenhausen, H. S. (1968). A counter example in stochastic optimal control. *SIAM Journal of Control*, 6, 131–147.

Chapter 5 Output Control

Decentralized control is an appealing concept, because it offers an essential reduction in communication requirements without significant, if any, loss of performance. Each subsystem is controlled on the basis of locally available information, with states of one subsystem being seldom used in control of another subsystem. This attractive property disappears when states of subsystems are not accessible at the local level. In building observers or estimators to reconstruct the states, we may be forced to exchange states between the subsystems, and thus violate the information structure constraints of decentralized control laws. A logical way out is to consider static or dynamic output feedback satisfying simultaneously the global (decentralized) and local (output) information constraints.

We shall first identify a class of interconnected systems that are always stabilizable by decentralized output feedback. We hasten to note that the conditions do not require that the subsystems are in any kind of canonical form. The control strategy is to stabilize each decoupled subsystem using *dynamic controllers* and, at the same time, make sure that the gains of the loops in the overall system are sufficiently small, so that stability is retained in the presence of interconnections. The same strategy was used in the static version of decentralized state feedback for linear plants (Section 2.5), where it produced robustly stable closed-loop systems. We shall achieve here the same kind of robustness by using output feedback. Furthermore the compensated system is nonlinear, consisting of linear plants interconnected by memoryless nonlinearities.

Our recipe for building connectively stable systems is to use decentralized feedback and treat interconnections as a disturbance of subsystems dynamics. If static feedback is our first choice, the geometric concept of almost invariant subspaces for disturbance decoupling with stability is an appealing alternative to the small gain approach *via* dynamic controllers.

We shall present graph-theoretic conditions under which an interconnected system is connectively stabilizable by static output feedback. Again, the subsystems need not be in a canonical form. Furthermore, the conditions are *generic.* Although they are conservative, they should be considered satisfactory to the extent that the conservativeness is outweighed by easy testability requiring only binary (combinatorial) calculations.

In the last part of this chapter, we shall give an *adaptive* version of decentralized output feedback. The local feedback gains are adaptively adjusted to the levels necessary to neutralize the interconnections and, at the same time, drive the subsystems with unknown parameters toward the performance of the locally chosen reference models, as close as desired. It is important to note that the resulting adaptive schemes retain the connective stability property and robustness to nonlinearities that are enjoyed by the fixed-gain decentralized controllers. Furthermore, the adaptive schemes are likely to use lower gains, since adaptation extends only to the level that is necessary for the required performance, instead of to the conservatively high level that may have been determined through a fixed-gain design procedure.

5.1. Dynamic Output Feedback

We consider an interconnected system $\mathbf{S}$ composed of N linear time-invariant subsystems $\mathbf{S}_i$ described as

$$\begin{aligned} \mathbf{S}_i:\ \dot{x}_i &= A_i x_i + b_i u_i + P_i v_i, \\ y_i &= c_i^T x_i, \\ w_i &= Q_i x_i, \qquad i \in \boldsymbol{N}, \end{aligned} \tag{5.1}$$

where $x_i(t) \in \mathbf{R}^{n_i}$ is the state, $u_i(t) \in \mathbf{R}$ is the control input, and $y_i(t) \in \mathbf{R}$ is the measured output of $\mathbf{S}_i$ at time $t \in \mathbf{R}$. The matrices A_i, P_i, Q_i and vectors b_i, c_i are constant and of appropriate dimensions. In Equation (5.1), $v_i(t) \in \mathbf{R}^{m_i}$ and $w_i(t) \in \mathbf{R}^{\ell_i}$ are the interaction inputs and outputs at time $t \in \mathbf{R}$, which are associated with subsystem $\mathbf{S}_i$, and which represent, respectively, the effect of other subsystems on $\mathbf{S}_i$ and the effect of $\mathbf{S}_i$ on the other subsystems. The interaction inputs and outputs are related by

$$v_i = f_i(w), \qquad i \in \boldsymbol{N}, \tag{5.2}$$

where the nonlinear function $f_i\colon \mathbf{R}^\ell \to \mathbf{R}^{m_i}$ is continuous in $w = (w_1^T, w_2^T, \ldots, w_N^T)^T$, $\ell = \sum_{i=1}^N \ell_i$, and satisfies the "conical" condition

$$\|f_i(w)\| \le \kappa_i \|w\|, \qquad \forall w \in \mathbf{R}^\ell, \qquad \forall i \in \boldsymbol{N}, \tag{5.3}$$

with $\kappa_i > 0$.

We assume that there exists a subset $\boldsymbol{I} \subset \boldsymbol{N}$ such that the following factorization holds:

$$\begin{aligned} P_i &= b_i p_i^T, \qquad i \in \boldsymbol{I}, \\ Q_i &= q_i c_i^T, \qquad i \in \boldsymbol{N} - \boldsymbol{I}. \end{aligned} \tag{5.4}$$

In other words, we assume that for each subsystem $\mathbf{S}_i$ either the interaction input v_i has the same effect on $\mathbf{S}_i$ as the control input $u_i (i \in \boldsymbol{I})$, or the interaction output w_i is a reproduction of the measured output $y_i (i \in \boldsymbol{N} - \boldsymbol{I})$. It should be emphasized that this assumption about the interconnection structure is essential in the following development.

We apply to the interconnected system $\mathbf{S}$ a decentralized dynamic controller

$$\begin{aligned} \mathbf{C}_i\colon\ \dot{z}_i &= F_i z_i + g_i y_i, \\ u_i &= -h_i^T z_i - k_i y_i, \qquad i \in \boldsymbol{N}, \end{aligned} \tag{5.5}$$

where $z_i(t) \in \mathbf{R}^{r_i}$ is the state of $\mathbf{C}_i$, and F_i, g_i, h_i, and k_i are constant and of appropriate dimensions. The compensated subsystem $\hat{\mathbf{S}}_i = \mathbf{S}_i \,\&\, \mathbf{C}_i$ can be represented as

$$\begin{aligned} \hat{\mathbf{S}}_i\colon\ \dot{\hat{x}}_i &= \hat{A}_i \hat{x}_i + \hat{P}_i v_i, \\ w_i &= \hat{Q}_i \hat{x}_i, \qquad i \in \boldsymbol{N}, \end{aligned} \tag{5.6}$$

where $\hat{x}_i = (x_i^T, z_i^T)^T$ is the state of $\hat{\mathbf{S}}_i$, and

$$\hat{A}_i = \begin{bmatrix} A_i - b_i k_i c_i^T & -b_i h_i^T \\ g_i c_i^T & F_i \end{bmatrix}, \quad \hat{P}_i = \begin{bmatrix} P_i \\ 0 \end{bmatrix}, \quad \hat{Q}_i = [\,Q_i \quad 0\,], \qquad i \in \boldsymbol{N}. \tag{5.7}$$

Our goal is to choose the order r_i and the parameters F_i, g_i, h_i, and k_i of the controllers $\mathbf{C}_i$, such that the null solution $\hat{x}_i = 0$, $i \in \boldsymbol{N}$, of the compensated interconnected system $\hat{\mathbf{S}}$, consisting of the closed-loop subsystems $\hat{\mathbf{S}}_i$ of (5.6) and the interconnections in (5.2), is asymptotically stable in the large for the nonlinearities f_i satisfying (5.3); that is, $\hat{\mathbf{S}}$ is absolutely stable in the sense of Lur'e–Postnikov (*e.g.*, Šiljak, 1969). For this purpose, we first investigate the effect of the proper choice of each individual controller $\mathbf{C}_i$ on the subsystem $\mathbf{S}_i$.

We begin with the analysis of decoupled subsystems. To simplify the notation, we drop the subscript i and describe $\mathbf{S}_i$ of (5.1) as

$$\begin{aligned} \mathbf{S}_i\colon\ \dot{x} &= Ax + bu + pv, \\ y &= c^T x, \\ w &= q^T x, \end{aligned} \tag{5.8}$$

where $x(t) \in \mathbf{R}^n$ is the state of $\mathbf{S}_i$ at $t \in \mathbf{R}$. Furthermore, we assume that the interaction input $v(t)$ and interaction output $w(t)$ of $\mathbf{S}_i$ are scalars; thus, $u(t)$, $v(t)$, $y(t)$, $w(t) \in \mathbf{R}$ at $t \in \mathbf{R}$. We also assume that $\mathbf{S}_i$ is controllable by the control input $u(t)$, and observable from the measured output $y(t)$; that is, the triple (A, b, c^T) is controllable and observable.

Similarly, we represent a typical controller $\mathbf{C}_i$ of (5.5) as

$$\begin{aligned} \mathbf{C}_i\colon\ \dot{z} &= Fz + gy, \\ u &= -h^T z - ky, \end{aligned} \tag{5.9}$$

where $z(t) \in \mathbf{R}^r$ is the state of $\mathbf{C}_i$ at $t \in \mathbf{R}$. The closed-loop subsystem $\hat{\mathbf{S}}_i = \mathbf{S}_i \,\&\, \mathbf{C}_i$ is described by

$$\begin{aligned} \hat{\mathbf{S}}_i\colon\ \dot{\hat{x}} &= \hat{A}\hat{x} + \hat{p}v, \\ w &= \hat{q}^T \hat{x}, \end{aligned} \tag{5.10}$$

where $\hat{x} = (x^T,\, z^T)^T$, and

$$\hat{A} = \begin{bmatrix} A - bkc^T & -bh^T \\ gc^T & F \end{bmatrix}, \qquad \hat{p} = \begin{bmatrix} p \\ 0 \end{bmatrix}, \qquad \hat{q} = \begin{bmatrix} q \\ 0 \end{bmatrix}. \tag{5.11}$$

We define now the following transfer functions:

$$\begin{aligned} G_{yu}(s) &= c^T(sI - A)^{-1}b, \\ G_{yv}(s) &= c^T(sI - A)^{-1}p, \\ G_{wu}(s) &= q^T(sI - A)^{-1}b, \\ H(s) &= k + h^T(sI - F)^{-1}g, \\ W(s) &= \hat{q}^T(sI - \hat{A})^{-1}\hat{p}. \end{aligned} \tag{5.12}$$

If $p = b$ in (5.8), then, by direct computation, we obtain

$$W(s) = \frac{G_{wu}(s)}{1 + G_{yu}(s)H(s)}. \tag{5.13}$$

Similarly, if $q = c$ in (5.8), then $W(s)$ can be computed as

$$W(s) = \frac{G_{yu}(s)}{1 + G_{yu}(s)H(s)}. \tag{5.14}$$

Since the transfer functions in (5.13) and (5.14) are of the same form (in fact, they have the same poles), the following argument applies to both of them equally well. Therefore, from now on, we shall assume that $p = b$ in (5.8), so that $W(s)$ is as given in (5.13).

We begin our development by defining $G_{yu}(s)$ and $G_{wu}(s)$ in (5.12) explicitly as

$$G_{yu}(s) = \beta_0 \frac{\beta(s)}{\alpha(s)} = \beta_0 \frac{s^{n-\nu-1} + \beta_1 s^{n-\nu-2} + \cdots + \beta_{n-\nu-1}}{s^n + \alpha_1 s^{n-1} + \cdots + \alpha_n}, \tag{5.15}$$

where $0 \le \nu \le n-1$, and

$$G_{wu}(s) = \frac{\gamma(s)}{\alpha(s)} = \frac{\gamma_1 s^{n-1} + \gamma_2 s^{n-2} + \cdots + \gamma_n}{s^n + \alpha_1 s^{n-1} + \cdots + \alpha_n}. \tag{5.16}$$

We assume that $G_{yu}(s)$ has $n - \nu - 1$ zeros. We shall further assume that these zeros lie in the open left-half complex plane $\mathbf{C}_-$. We choose the order of $\mathbf{C}$ to be ν; that is, $r = \nu$. Then, $H(s)$ of (5.12) has the form

$$H(s) = \eta_0 \frac{\eta(s)}{\xi(s)} = \eta_0 \frac{s^\nu + \eta_1 s^{\nu-1} + \cdots + \eta_\nu}{s^\nu + \xi_1 s^{\nu-1} + \cdots + \xi_\nu}. \tag{5.17}$$

Finally, using (5.15)–(5.17), we can express $W(s)$ in (5.13) as

$$W(s) = \frac{\gamma(s)\xi(s)}{\alpha(s)\xi(s) + \beta_0\eta_0\beta(s)\eta(s)}. \tag{5.18}$$

We now choose the parameters of the controller **C** as follows:

(*a*) The coefficients η_j, $j = 1, 2, \ldots, \nu$, in (5.17) are chosen arbitrarily, subject to the condition that $\eta(s)$ has all its zeros in $\mathbf{C}_-$.

(*b*) The coefficients ξ_j, $j = 1, 2, \ldots, \nu$, and η_0 in (5.17) are chosen as

$$\begin{aligned} \xi_j &= \tilde{\xi}_j\rho^j, \qquad j = 1, 2, \ldots, \nu, \\ \beta_0\eta_0 &= \tilde{\xi}_{\nu+1}\rho^{\nu+1}, \end{aligned} \tag{5.19}$$

where $\rho > 0$ is a parameter to be specified, and $\tilde{\xi}_j$, $j = 1, 2, \ldots, \nu + 1$, are such that the polynomial

$$\tilde{\xi}(s) = s^{\nu+1} + \tilde{\xi}_1 s^{\nu} + \tilde{\xi}_2 s^{\nu-1} + \cdots + \tilde{\xi}_{\nu+1} \tag{5.20}$$

has only distinct zeros and all in $\mathbf{C}_-$.

Denoting by $\mathbf{C}_+$ the closed right-half complex plane, we now prove the following result (Hüseyin, *et al.* 1982):

5.1. LEMMA. Under the conditions (*a*) and (*b*) above, for any $\epsilon > 0$ there exists a $\bar{\rho} > 0$ such that, whenever $\rho > \bar{\rho}$, the following conditions are satisfied:

(*i*) $\hat{A}$ has all eigenvalues in $\mathbf{C}_-$;

(*ii*) $|W(s)| < \epsilon, \qquad \forall s \in \mathbf{C}_+$.

Proof. Let $W(s)$ of (5.18) be written as

$$W(s) = \frac{\psi(s)}{\theta(s)} = \frac{\psi_1 s^{n+\nu-1} + \psi_2 s^{n+\nu-2} + \cdots + \psi_{N+\nu}}{s^{n+\nu} + \theta_1 s^{n+\nu-1} + \cdots + \theta_{n+\nu}}, \tag{5.21}$$

where ψ_j and θ_j, $j = 1, 2, \ldots, n + \nu$, are functions of the parameter ρ. Comparing (5.18) and (5.21), and using (5.19), we have

$$\theta(s) = s^{\nu}\alpha(s) + \rho\tilde{\xi}_1 s^{\nu-1}\alpha(s) + \cdots + \rho^{\nu}\tilde{\xi}_{\nu}\alpha(s) + \rho^{\nu+1}\tilde{\xi}_{\nu+1}\beta(s)\eta(s). \tag{5.22}$$

The proof of the lemma is based upon the following variation of a result obtained by Inan (1978):

As $\rho \to +\infty$, $k+1$ zeros of $\theta(s)$ approach ρ times the zeros of $\tilde{\xi}(s)$ in (5.20), while the remaining $n-1$ zeros of $\theta(s)$ approach the zeros of $\beta(s)\eta(s)$ in the following sense:

(a) Let μ be a zero of $\tilde{\xi}(s)$, which is by assumption simple. Then, there exists $M_1 > 0$ and $\bar{\rho}_1 > 0$ such that an exact zero $\mu(\rho)$ of $\theta(s)$ satisfies

$$|\mu(\rho) - \rho\mu| \leq M_1, \qquad \forall \rho > \bar{\rho}_1. \tag{5.23}$$

(b) Let λ be a zero of $\beta(s)\eta(s)$ with multiplicity k. Then, there exists $M_2 > 0$ and $\bar{\rho}_2 > 0$ such that exactly k zeros, $\lambda_j(\rho)$, $j = 1, 2, \ldots, k$, of $\theta(s)$ satisfy

$$|\lambda_j(\rho) - \lambda| \leq M_2\rho^{-1/k}, \qquad \forall \rho > \bar{\rho}_2. \tag{5.24}$$

Now, let us consider $\theta(s)$ of (5.22), where $\beta(s)$ has its zeros in $\mathbf{C}_-$ by assumption, and $\eta(s)$ and $\tilde{\xi}(s)$ have their zeros in $\mathbf{C}_-$ by the choice of the parameters of the controller $\mathbf{C}$. Then, the above result of Inan (1978) ensures that there exists $\bar{\rho}_3 > 0$ such that $\theta(s)$ has all of its zeros in $\mathbf{C}_-$ for all $\rho > \bar{\rho}_3$. Since the triple (A, b, c^T) is controllable and observable, the proof of the first part of the lemma is complete.

To prove the second part of the lemma, let us denote the zeros of $\tilde{\xi}(s)$ by μ_j, $j = 1, 2, \ldots, \nu+1$, and those of $\beta(s)\eta(s)$ by λ_j, $j = 1, 2, \ldots, n-1$. Also, let $\mu_j(\rho)$, $j = 1, 2, \ldots, \nu+1$, and $\lambda_j(\rho)$, $j = 1, 2, \ldots, n-1$, denote the exact zeros of $\theta(s)$ which approach $\rho\mu_j$ and λ_j, respectively, as $\rho \to +\infty$. Let us rewrite $W(s)$ in (5.21) as

$$\begin{aligned} W(s) &= \frac{\psi(s)}{\prod_{j=1}^{\nu+1}[s-\mu_j(\rho)]\,\prod_{j=1}^{n-1}[s-\lambda_j(\rho)]} \\ &= \frac{\psi(s)}{\prod_{j=1}^{\nu+1}(s-\rho\mu_j)\,\prod_{j=1}^{n-1}(s-\lambda_j)} \prod_{j=1}^{\nu+1}\frac{s-\rho\mu_j}{s-\mu_j(\rho)} \prod_{j=1}^{n-1}\frac{s-\lambda_j}{s-\lambda_j(\rho)} \\ &= \frac{\xi(s)}{\prod_{j=1}^{\nu+1}(s-\rho\mu_j)}\,\frac{\gamma(s)}{\prod_{j=1}^{n-1}(s-\lambda_j)} \prod_{j=1}^{\nu+1}\frac{s-\rho\mu_j}{s-\mu_j(\rho)} \prod_{j=1}^{n-1}\frac{s-\lambda_j}{s-\lambda_j(\rho)}. \end{aligned} \tag{5.25}$$

Since $\gamma(s)$ has degree $n-1$, and $\lambda_j \in \mathbf{C}_-$, $j = 1, 2, \ldots, n-1$, we have, for some $M_3 > 0$,

$$\left| \frac{\gamma(s)}{\Pi_{j=1}^{n-1}(s-\lambda_j)} \right| < M_3, \qquad \forall s \in \mathbf{C}_+. \tag{5.26}$$

On the other hand, using the above result of Inan (1978), it is easy to show that there exists $\bar{\rho}_4 > 0$ such that, for $\rho > \bar{\rho}_4$,

$$\left| \frac{s-\rho\mu_j}{s-\mu_j(\rho)} \right| < 2, \qquad \forall s \in \mathbf{C}_+, \qquad \forall j = 1, 2, \ldots, \nu+1, \tag{5.27a}$$

and

$$\left| \frac{s-\lambda_j}{s-\lambda_j(\rho)} \right| < 2, \qquad \forall s \in \mathbf{C}_+, \qquad \forall j = 1, 2, \ldots, n-1. \tag{5.27b}$$

Now, consider the first term on the right-hand side of (5.25), which can be expanded into partial fractions as

$$\frac{\xi(s)}{\Pi_{j=1}^{\nu+1}(s-\rho\mu_j)} = \sum_{j=1}^{\nu+1} \frac{R_j(\rho)}{s-\rho\mu_j}, \tag{5.28}$$

where

$$R_j(\rho) = \frac{\xi(\rho\mu_j)}{\Pi_{\substack{k=1\\k\neq j}}^{\nu+1}(\rho\mu_j-\rho\mu_k)}, \qquad \forall j = 1, 2, \ldots, \nu+1. \tag{5.29}$$

Using (5.19), $R_j(\rho)$ in (5.28) can be computed as

$$\begin{aligned} R_j(\rho) &= \frac{(\rho\mu_j)^\nu + \tilde{\xi}_1\rho(\rho\mu_j)^{\nu-1} + \cdots + \tilde{\xi}_\nu\rho^\nu}{\Pi_{\substack{k=1\\k\neq j}}^{\nu+1}(\rho\mu_j-\rho\mu_k)} \\ &= \frac{\mu_j^\nu + \tilde{\xi}_i\mu_j^{\nu-1} + \cdots + \tilde{\xi}_\nu}{\Pi_{\substack{k=1\\k\neq j}}^{\nu+1}(\mu_j-\mu_k)}, \end{aligned} \tag{5.30}$$

so that

$$|R_j(\rho)| = M_{j4} < +\infty, \qquad \forall j = 1, 2, \ldots, \nu+1. \tag{5.31}$$

Since $\mu_j \in \mathbf{C}_-$, (5.31) implies that

$$\left|\frac{R_j(\rho)}{s-\rho\mu_j}\right| \leq \frac{M_{j4}}{\rho|\mathrm{Re}\mu_j|}, \qquad \forall s \in \mathbf{C}_+, \qquad \forall j = 1, 2, \ldots, \nu+1, \tag{5.32}$$

so that from (5.28) we get

$$\left|\frac{\xi(s)}{\Pi_{j=1}^{\nu+1}(s-\rho\mu_j)}\right| \leq \rho^{-1}\sum_{j=1}^{\nu+1}\frac{M_{j4}}{|\mathrm{Re}\mu_j|} = \rho^{-1}M_4. \tag{5.33}$$

From (5.25) and inequalities (5.26), (5.27), and (5.33), it follows that if $\rho > \max\{\bar{\rho}_3, \bar{\rho}_4\}$, then

$$|W(s)| < 2^{n+\nu}\rho^{-1}M_3M_4, \qquad \forall s \in \mathbf{C}_+. \tag{5.34}$$

Now, for any $\epsilon > 0$, letting $\bar{\rho}_5 = 2^{n+\nu}\epsilon^{-1}M_3M_4$ and $\bar{\rho} = \max\{\bar{\rho}_3, \bar{\rho}_4, \bar{\rho}_5\}$, it follows that

$$|W(s)| < \epsilon, \qquad \forall s \in \mathbf{C}_+, \qquad \forall \rho > \bar{\rho}. \tag{5.35}$$

Q.E.D.

The properties of the Lemma 5.1 have already given a clue about the mechanism of achieving stability of the closed-loop interconnected system: a small-gain version of the circle criterion (Zames, 1966), which we summarize below.

Let us consider the compensated interconnected system $\hat{\mathbf{S}}$, which can be described in compact form as

$$\hat{\mathbf{S}}\colon\ \dot{\hat{x}} = \hat{A}\hat{x} + \hat{P}f(w), \qquad w = \hat{Q}\hat{x}, \tag{5.36}$$

where $\hat{x}(t) \in \mathbf{R}^n$, $\hat{x} = (\hat{x}_1^T, \hat{x}_2^T, \ldots, \hat{x}_N^T)^T$, is the state of $\hat{\mathbf{S}}$, $f\colon \mathbf{R}^\ell \to \mathbf{R}^m$, $f(w) = [f_1^T(w), f_2^T(w), \ldots, f_N^T(w)]^T$, $m = \sum_i^N m_i$, is the interconnection function, and

$$\hat{A} = \mathrm{diag}\{\hat{A}_1, \hat{A}_2, \ldots, \hat{A}_N\}, \qquad \hat{P} = \mathrm{diag}\{\hat{P}_1, \hat{P}_2, \ldots, \hat{P}_N\}, \qquad \hat{Q} = \mathrm{diag}\{\hat{Q}_1, \hat{Q}_2, \ldots, \hat{Q}_N\}, \tag{5.37}$$

where

$$\hat{P}_i = [\,P_i^T \quad 0^T\,]^T, \qquad \hat{Q}_i = [\,Q_i \quad 0\,] \tag{5.38}$$

are constant matrices of appropriate dimensions. The linear part of $\hat{\mathbf{S}}$ has the transfer function

$$W(s) = \hat{Q}(sI - A)^{-1}\hat{P}, \tag{5.39}$$

which is in block diagonal form, that is, $W(s) = \text{diag}\{W_1(s), W_2(s), \dots, W_N(s)\}$, with the diagonal blocks

$$W_i(s) = \hat{Q}_i(sI - A_i)^{-1}\hat{P}_i, \qquad i \in \boldsymbol{N}, \tag{5.40}$$

corresponding to the compensated subsystem $\hat{\mathbf{S}}_i$ of (5.6).

As far as the interconnection function $f(w)$ is concerned, we require that it belongs to the class

$$\mathcal{F} = \{f\colon\ \|f(w)\| \le \kappa\|w\|, \quad \forall w \in \mathbf{R}^\ell\}, \tag{5.41}$$

for some positive number κ.

To prove the main result of this section, we need the following small-gain version of the circle criterion (*e.g.*, Narendra and Taylor, 1973):

5.2. THEOREM. Suppose the matrix $A = \text{diag}\{A_1, A_2, \dots, A_N\}$ has all of its eigenvalues in $\mathbf{C}_-$. Then, the equilibrium $\hat{x} = 0$ of the system $\hat{\mathbf{S}}$ is globally asymptotically stable for all $f(w) \in \mathcal{F}$, that is, the system $\hat{\mathbf{S}}$ is absolutely stable, if

$$\sup_{\omega \in \mathbf{R}} \ \max_k |\lambda_k[W^T(-j\omega)W(j\omega)]| < \frac{1}{\kappa^2}, \tag{5.42}$$

where $\lambda_k(\bullet)$ denotes an eigenvalue of the indicated matrix.

Using Lemma 5.1 and Theorem 5.2, we arrive at the following (Hüseyin *et al.* 1982):

5.3. THEOREM. Suppose that the subsystem $\mathbf{S}_i = (A_i, b_i, c_i^T)$, $i \in \boldsymbol{N}$, are controllable and observable and the matrices A_i, $i \in \boldsymbol{N}$, have all their eigenvalues in $\mathbf{C}_-$, and the pairs (P_i, Q_i) can be factored as in (5.4). Then, there exists a set of local controllers $\mathbf{G}_i$, $i \in \boldsymbol{N}$, defined by (5.5), such that the closed-loop interconnected system $\hat{\mathbf{S}}$ is absolutely stable.

Proof. From the condition (5.4), it follows that the subsystem transfer functions have the forms

$$
\begin{aligned}
W_i(s) &= \hat{Q}_i(sI - \hat{A}_i)^{-1}\hat{b}_i p_i^T, \qquad i \in \boldsymbol{I}, \\
W_i(s) &= q_i \hat{c}_i^T(sI - \hat{A}_i)^{-1}\hat{P}_i, \qquad i \in \boldsymbol{N} - \boldsymbol{I},
\end{aligned}
\tag{5.43}
$$

where $\hat{A}_i$ is $\hat{A}$ of (5.11), and $\hat{b}_i$ and $\hat{c}_i^T$ are obtained from b_i and c_i^T by augmenting with zeros. Then, from Lemma 5.1 and conditions of the theorem, we conclude that the local controllers $\mathbf{C}_i$ can be designed so that the matrices $\hat{A}_i$ have all their eigenvalues in C_-, and for any set of positive numbers ϵ_i, $i \in \boldsymbol{N}$,

$$
\|W_i(s)\| < \epsilon_i, \qquad \forall s \in \mathsf{C}_+. \tag{5.44}
$$

In particular, (5.44) holds for all $s = j\omega$, $\omega \in \mathsf{R}$. Thus, the matrix $\hat{A}$ of (5.36) has all its eigenvalues in C_-, and letting $\epsilon_i = 1/\kappa^2$, the transfer function $W(s)$ in (5.39) satisfies condition (5.42). Q.E.D.

We note that assumptions (5.4) concerning the pairs (P_i, Q_i) are the same as those used in decentralized stabilization by static state feedback discussed in Sections 2.4 and 2.5. These assumptions make it possible to design local controllers that can stabilize the interconnected system independently of the size of the interconnections, so long they are bounded. They may be too restrictive, however, and one should try to obtain weaker constraints by a more detailed analysis of the effect of local controllers on the subsystems and their interconnections, as it was done for the case of state feedback in Section 2.5. We should also note that these assumptions can be a promising starting point for approaching the problem of suboptimal design of local controllers, the state feedback version of which has been considered in Section 3.4.

5.2. Structured Almost Invariant Subspaces

By recognizing the fact that interconnections among the subsystems can be considered as perturbations, Linneman (1985) and Willems and Ikeda (1984) derived conditions for decentralized stabilization using the concept of almost disturbance decoupling and small gain theorems in the context of almost invariant subspaces formulated by Willems (1981, 1982). Although the conditions are of connective stability type, they are parameter dependent because the relevant subspaces vary as functions of system parameters. One way to overcome this dependency is to select the cases in which the relevant subspaces are invariant under parametric perturbations. This leads to investigation of the subspaces in the framework of structured systems (Hayakawa and Šiljak, 1988).

Our aim here is to present conditions for a structured system to have the property that the supremal $\mathcal{L}_p$-almost invariant (controllability) subspace is generically the entire state space and that the infimal $\mathcal{L}_p$-almost conditional invariant (complementary observability) subspace is generically the zero space. In the next section, the conditions are reformulated in terms of directed graphs, which is consistent with our emphasis on the structural approach to decentralized control.

Let us review first certain well-known results concerning a linear constant system

$$\mathbf{S}\colon\ \dot{x} = Ax + Bu,$$
$$y = Cx, \tag{5.45}$$

where $x(t) \in \mathcal{X} := \mathbf{R}^n$, $u(t) \in \mathcal{U} := \mathbf{R}^m$, and $y(t) \in \mathcal{Y} := \mathbf{R}^\ell$ are the state, input, and output of $\mathbf{S}$ at time $t \in \mathbf{R}$, and $\mathcal{X}$, $\mathcal{U}$, and $\mathcal{Y}$ are normed vector spaces. With $\mathbf{S}$ we associate the following almost invariant subspaces (Willems, 1981):

$\mathcal{V}^*_{b,\mathrm{Ker}\ C}$ — the supremal $\mathcal{L}_p$-almost invariant subspace "contained" in Ker C;

$\mathcal{R}^*_{b,\mathrm{Ker}\ C}$ — the supremal $\mathcal{L}_p$-almost controllability subspace "contained" in Ker C;

$\mathcal{S}^*_{b,\mathrm{Im}\ B}$ — the infimal $\mathcal{L}_p$-almost conditional invariant subspace "containing" Im B;

$\mathcal{N}^*_{b,\mathrm{Im}\ B}$ — the infimal $\mathcal{L}_p$-almost complementary observability subspace "containing" Im B.

These spaces can be computed by the well-known algorithms

$$\begin{aligned} \mathcal{V}_{k+1} &= \mathrm{Ker}\, C \cap A^{-1}(\mathcal{V}_k + \mathrm{Im}\, B), \qquad \mathcal{V}_0 = \mathcal{X}, \\ \mathcal{S}_{k+1} &= \mathrm{Im}\, B + A(\mathrm{Ker}\, C \cap \mathcal{S}_k), \qquad \mathcal{S}_0 = \{0\}, \end{aligned} \tag{5.46}$$

and

$$\begin{aligned} \mathcal{V}^*_{b,\mathrm{Ker}\ C} &= \mathcal{V}_n + \mathcal{S}_n, \\ \mathcal{R}^*_{b,\mathrm{Ker}\ C} &= \mathcal{S}_n, \\ \mathcal{S}^*_{b,\mathrm{Im}\ B} &= \mathcal{V}_n \cap \mathcal{S}_n, \\ \mathcal{N}^*_{b,\mathrm{Im}\ B} &= \mathcal{V}_n. \end{aligned} \tag{5.47}$$

We recall that the triple (A, B, C), or the system **S**, is called standard when the matrix B has full column rank and C has full row rank, and state the following result (Morse, 1973):

5.4. LEMMA. Let a triple (A, B, C) be standard. Then, the state-space $\mathcal{X}$ can be decomposed into four independent subspaces $\mathcal{X}_1$, $\mathcal{X}_2$, $\mathcal{X}_3$, and $\mathcal{X}_4$ (*i.e.*, $\mathcal{X} = \mathcal{X}_1 \oplus \mathcal{X}_2 \oplus \mathcal{X}_3 \oplus \mathcal{X}_4$) such that

$$\begin{aligned} \mathcal{X}_1 \oplus \mathcal{X}_2 &= \mathcal{V}_n, \\ \mathcal{X}_2 \oplus \mathcal{X}_4 &= \mathcal{S}_n, \end{aligned} \tag{5.48}$$

where $\mathcal{V}_n$ and $\mathcal{S}_n$ are defined in (5.46). Furthermore, there exist three nonsingular matrices T, H, G and two matrices F, J such that

$$\begin{bmatrix} T^{-1} & T^{-1}J \\ 0 & H \end{bmatrix} \begin{bmatrix} sI_n - A & B \\ C & 0 \end{bmatrix} \begin{bmatrix} T & 0 \\ FT & G \end{bmatrix}$$

$$= \left[\begin{array}{cccc|cc} sI_{n_1} - A_1 & & & \bigcirc & 0 & 0 \\ & sI_{n_2} - A_2 & & & B_2 & 0 \\ & & sI_{n_3} - A_3 & & 0 & 0 \\ \bigcirc & & & sI_{n_4} - A_4 & 0 & B_4 \\ \hline 0 & 0 & C_3 & 0 & 0 & 0 \\ 0 & 0 & 0 & C_4 & 0 & 0 \end{array}\right], \tag{5.49}$$

where the $n_i \times n_i$ matrix A_i corresponds to $T^{-1}A|_{\mathcal{X}_i}T$, $i = 1, 2, 3, 4$, and B_2, B_4, C_3, and C_4 have dimensions $n_2 \times m_c$, $n_4 \times m_p$, $\ell_0 \times n_3$, and $\ell_p \times n_4$. In (5.49), the matrix A_1 is in the rational form, the pairs (A_2, B_2) and (A_3^T, C_3^T) are in Brunovsky canonical form, and the triple (C_4, A_4, B_4) is in the prime canonical form.

5.5. REMARK. We note that $m_p = \ell_p$ because (C_4, A_4, B_4) is prime.

From the above facts and, in particular, Equations (5.47) and (5.48), we can get the relations

$$\begin{aligned} \mathcal{V}^*_{b,\text{Ker } C} &= \mathcal{X}_1 \oplus \mathcal{X}_2 \oplus \mathcal{X}_4, \\ \mathcal{R}^*_{b,\text{Ker } C} &= \mathcal{X}_2 \oplus \mathcal{X}_4, \\ \mathcal{S}^*_{b,\text{Im } B} &= \mathcal{X}_2, \\ \mathcal{N}^*_{b,\text{Im } B} &= \mathcal{X}_1 \oplus \mathcal{X}_2. \end{aligned} \tag{5.50}$$

By $\rho(\bullet)$, we denote the rank of the indicated matrix, and we prove the following:

5.6. THEOREM. Let the system **S** be standard. Then,

(*i*) $\mathcal{V}^*_{b,\text{Ker } C} = \mathcal{X}$ if and only if

$$\rho\left(\begin{bmatrix} sI - A & B \\ C & 0 \end{bmatrix}\right) = n + \ell, \qquad \text{for almost all } s \in \mathbf{C}; \tag{5.51}$$

(*ii*) $\mathcal{R}^*_{b,\text{Ker } C} = \mathcal{X}$ if and only if

$$\rho\left(\begin{bmatrix} sI - A & B \\ C & 0 \end{bmatrix}\right) = n + \ell, \qquad \text{for all } s \in \mathbf{C}; \tag{5.52}$$

(*iii*) $\mathcal{S}^*_{b,\text{Im } B} = \{0\}$ if and only if

$$\rho\left(\begin{bmatrix} sI - A & B \\ C & 0 \end{bmatrix}\right) = n + m, \qquad \text{for almost all } s \in \mathbf{C}; \tag{5.53}$$

(*iv*) $\mathcal{N}^*_{b,\text{Im } B} = \{0\}$ if and only if

$$\rho\left(\begin{bmatrix} sI - A & B \\ C & 0 \end{bmatrix}\right) = n + m, \qquad \text{for all } s \in \mathbf{C}. \tag{5.54}$$

Proof. (*i*) From (5.50), $\mathcal{V}^*_{b,\text{Ker } C} = \mathcal{X}$ is equivalent to saying that $\mathcal{X}_3 = \{0\}$ holds in Lemma 5.4, *i.e.*, $n_3 = 0$ in (5.49). Notice that the submatrices $[sI_{n_2} - A_2 \;\; B_2]$ and $\begin{bmatrix} sI_{n_4} - A_4 & B_4 \\ C_4 & 0 \end{bmatrix}$ in (5.49) have full row rank for all $s \in \mathbf{C}$ because (A_2, B_2) is controllable and (C_4, A_4, B_4) is prime. Therefore, a necessary and sufficient condition for $\mathcal{V}^*_{b,\text{Ker } C} = \mathcal{X}$ to hold is that the matrix in the right hand side of (5.49) have full row rank for all $s \in \mathbf{C}$ except at the eigenvalues of A_1. From the above fact, and also from the nonsingularity of matrices $\begin{bmatrix} T^{-1} & T^{-1}J \\ 0 & H \end{bmatrix}$ and $\begin{bmatrix} T & 0 \\ FT & G \end{bmatrix}$, the assertion (i) is automatic.

(ii) From (5.50), $\mathcal{R}^*_{b,\text{Ker } C} = \mathcal{X}$ is equivalent to saying that $\mathcal{X}_1 = \mathcal{X}_3 = \{0\}$ holds in Lemma 5.4, *i.e.*, $n_1 = n_3 = 0$ in (5.49). Therefore the assertion (ii) is easy to see by relying on the full row rank of $[\, sI_{n_2} - A_2 \quad B_2 \,]$ and $\begin{bmatrix} sI_{n_4} - A_4 & B_4 \\ C_4 & 0 \end{bmatrix}$.

(iii) From (5.50), $\mathcal{S}^*_{b,\text{Im } B} = \{0\}$ is equivalent to saying that $\mathcal{X}_2 = \{0\}$ holds in Lemma 5.4, *i.e.*, $n_2 = 0$ in (5.49). Therefore, the assertion (iii) follows from the full column rank of $\begin{bmatrix} sI_{n_3} - A_3 \\ C_3 \end{bmatrix}$ and $\begin{bmatrix} sI_{n_4} - A_4 & B_4 \\ C_4 & 0 \end{bmatrix}$, and by using the same argument as used in (i).

(iv) From (5.50), $\mathcal{N}^*_{b,\text{Im } B} = \{0\}$ is equivalent to saying that $\mathcal{X}_1 = \mathcal{X}_2 = \{0\}$ holds in Lemma 5.4, *i.e.*, $n_1 = n_2 = 0$ in (5.49). Therefore, the assertion is trivial. Q.E.D.

5.7. REMARK. Even if $\mathbf{S}$ is not standard, parts (i) and (ii) of Theorem 5.6 hold when $C = 0$. In fact, if $C = 0$, (i) is trivial because $\mathcal{V}^*_{b,\text{Ker } 0} = \mathcal{X}$ and (5.51) holds for $\ell = 0$. Furthermore, (ii) becomes the well-known rank condition for controllability because $\mathcal{R}^*_{b,\text{Ker } 0}$ is the controllable subspace. Similarly, (iii) and (iv) hold when $B = 0$, because (iii) becomes a trivial statement and (iv) becomes the rank condition for observability, since $\mathcal{N}^*_{b,\text{Im } 0}$ is the unobservable subspace.

Theorem 5.6 establishes the relationships between the almost invariant subspaces and system matrices for an important special case that is of interest in this chapter. In particular, the relationships are useful in formulating the generic properties of these subspaces in the graph-theoretic framework.

In the context of structural analysis, instead of a system $\mathbf{S}$ and a triple (A, B, C), we consider a structured system $\tilde{\mathbf{S}}$ described by a triple of structured matrices $(\tilde{A}, \tilde{B}, \tilde{C})$, each having a number of fixed zero elements while the rest of the elements are independent free parameters. The parameter space $\mathbf{R}^\nu$ has the dimension ν equal to the total number of the indeterminate entries of $\tilde{A}$, $\tilde{B}$, and $\tilde{C}$. When A, B and C are numeric matrices obtained from $\tilde{A}$, $\tilde{B}$, and $\tilde{C}$ by fixing their indeterminates at some specific values in a set $\mathcal{P} \subset \mathbf{R}^\nu$, we write $(A, B, C) \in \mathcal{P}$. We need the following:

5.8. DEFINITION. Given a structured triple $(\tilde{A}, \tilde{B}, \tilde{C})$, we define:

(i) The supremal $\mathcal{L}_p$-almost invariant (controllability) subspace contained in Ker C is said to be generically equal to the whole state-space $\mathcal{X}$, which is denoted by $\mathcal{V}^*_{b,\mathrm{Ker}\ C} \overset{*}{=} \mathcal{X}$ $(\mathcal{R}^*_{b,\mathrm{Ker}\ C} \overset{*}{=} \mathcal{X})$, if there exists a proper algebraic variety $\boldsymbol{V} \subset \mathbf{R}^\nu$ such that, for any $(A, B, C) \in \boldsymbol{V}^c$, we have $\mathcal{V}^*_{b,\mathrm{Ker}\ C} = \mathcal{X}$ $(\mathcal{R}^*_{b,\mathrm{Ker}\ C} = \mathcal{X})$.

(ii) The infimal $\mathcal{L}_p$-almost conditionally invariant (complementary observability) subspace containing Im B is said to be generically equal to the zero space $\{0\}$, which is denoted by $\mathcal{S}^*_{b,\mathrm{Im}\ B} \overset{*}{=} \{0\}$ $(\mathcal{N}^*_{b,\mathrm{Im}\ B} \overset{*}{=} \{0\})$, if there exists a proper algebraic variety $\boldsymbol{V} \subset \mathbf{R}^\nu$ such that, for any $(A, B, C) \in \boldsymbol{V}^c$, we have $\mathcal{S}^*_{b,\mathrm{Im}\ B} = \{0\}$ $(\mathcal{N}^*_{b,\mathrm{Im}\ B} = \{0\})$.

Let us recall the notion of generic rank $\tilde{\rho}(\bullet)$ (Section 1.4). For a structured matrix $\tilde{M}$ with the associated parameter space $\mathbf{R}^\nu$, $\tilde{\rho}(\tilde{M}) = r$ if there exists a proper algebraic variety $\boldsymbol{V} \subset \mathbf{R}^\nu$ such that $\rho(M) = r$ for any $M \in \boldsymbol{V}^c$. When we consider a structured triple $(\tilde{A}, \tilde{B}, \tilde{C})$ with a parameter space $\mathbf{R}^\nu$, and state that

$$\tilde{\rho}\left(\begin{bmatrix} sI - \tilde{A} & \tilde{B} \\ \tilde{C} & 0 \end{bmatrix}\right) = r, \qquad \text{for almost all } s \in \mathbf{C}, \tag{5.55}$$

we mean that there exists a proper algebraic variety $\boldsymbol{V} \subset \mathbf{R}^\nu$ such that, for any $(A, B, C) \in \boldsymbol{V}^c$, we have

$$\rho\left(\begin{bmatrix} sI - A & B \\ C & 0 \end{bmatrix}\right) = r, \qquad \text{for almost all } s \in \mathbf{C}. \tag{5.56}$$

We also say that a triple $(\tilde{A}, \tilde{B}, \tilde{C})$, or a system $\tilde{\mathbf{S}}$, is standard if $\tilde{B}$ has generic full column rank and $\tilde{C}$ has generic full row rank.

From Theorem 5.6 and Definition 5.8, the following result is automatic:

5.9. THEOREM. Let a structured system $\tilde{\mathbf{S}}$ be standard. Then:

(i) $\mathcal{V}^*_{b,\mathrm{Ker}\ C} \overset{*}{=} \mathcal{X}$ if and only if

$$\tilde{\rho}\left(\begin{bmatrix} sI - \tilde{A} & \tilde{B} \\ \tilde{C} & 0 \end{bmatrix}\right) = n + \ell, \qquad \text{for almost all } s \in \mathbf{C}; \tag{5.57}$$

(ii) $\mathcal{R}^*_{b,\mathrm{Ker}\ C} \overset{*}{=} \mathcal{X}$ if and only if

$$\tilde{\rho}\left(\begin{bmatrix} sI - \tilde{A} & \tilde{B} \\ \tilde{C} & 0 \end{bmatrix}\right) = n + \ell, \qquad \text{for all } s \in \mathbf{C}; \tag{5.58}$$

(*iii*) $\mathcal{S}^*_{b,\mathrm{Im}\,B} \stackrel{*}{=} \{0\}$ if and only if

$$\tilde{\rho}\left(\begin{bmatrix} sI-\tilde{A} & \tilde{B} \\ \tilde{C} & 0 \end{bmatrix}\right) = n+m, \qquad \text{for almost all } s \in \mathbf{C}; \tag{5.59}$$

(*iv*) $\mathcal{N}^*_{b,\mathrm{Im}\,B} \stackrel{*}{=} \{0\}$ if and only if

$$\tilde{\rho}\left(\begin{bmatrix} sI-\tilde{A} & \tilde{B} \\ \tilde{C} & 0 \end{bmatrix}\right) = n+m, \qquad \text{for all } s \in \mathbf{C}. \tag{5.60}$$

Proof. Obvious from Theorem 5.6, Definition 5.8, and the definition of generic rank. Q.E.D.

The fact that the above generic rank conditions are difficult to test motivates a derivation of their graph-theoretic interpretation. This we consider next.

5.3. Graph-Theoretic Characterization

We first introduce some basic notation concerning the system graph. With a structured system $\tilde{\mathbf{S}}$, we associate a directed graph (digraph) $\mathbf{D} = (\mathtt{V}, \mathtt{E})$, where $\mathtt{V} = \mathtt{U} \times \mathtt{X} \times \mathtt{Y}$ is a set of vertices and $\mathtt{U} = \{u_1, u_2, \ldots, u_m\}$, $\mathtt{X} = \{x_1, x_2, \ldots, x_n\}$, and $\mathtt{Y} = \{y_1, y_2, \ldots, y_\ell\}$ are the input, state, and output vertices. $\mathtt{E} = \{(u_j, x_i)\colon \tilde{b}_{ij} \not\equiv 0\} \cup \{(x_j, x_i)\colon \tilde{a}_{ij} \not\equiv 0\} \cup \{(x_j, y_i)\colon \tilde{c}_{ij} \not\equiv 0\}$ is a set of edges, where $\tilde{a}_{ij}$, $\tilde{b}_{ij}$, and $\tilde{c}_{ij}$ are the elements of the structured matrices $\tilde{A}$, $\tilde{B}$, and $\tilde{C}$. A set of vertices $\mathtt{V}_s$ and edges $\mathtt{E}_s$ of a subgraph $\mathbf{D}_s = (\mathtt{V}_s, \mathtt{E}_s)$ are denoted by $\mathtt{V}(\mathbf{D}_s)$ and $\mathtt{E}(\mathbf{D}_s)$, respectively. A path from the vertex v_1 to the vertex v_k is a subgraph $\mathbf{P}_{v_1}^{v_k}$ with a set of vertices $\mathtt{V}(\mathbf{P}_{v_1}^{v_k}) = \{v_1, v_2, \ldots, v_k\}$ and a set of edges $\mathtt{E}(\mathbf{P}_{v_1}^{v_k}) = \{(v_1, v_2), (v_2, v_3), \ldots, (v_{k-1}, v_k)\}$. When the initial vertex v_1 or the terminal vertex v_k, or both, are not essential, we use the notation $\mathbf{P}^{v_k}$, $\mathbf{P}_{v_1}$, or $\mathbf{P}$, depending on the context. The length of a path $\mathbf{P}$ is the number of elements in $\mathtt{E}(\mathbf{P})$. A subgraph is called a cycle $\mathbf{C}$ if $\mathtt{V}(\mathbf{C}) = \{v_1, v_2, \ldots, v_k\}$ and $\mathtt{E}(\mathbf{C}) = \{(v_1, v_2), (v_2, v_3), \ldots, (v_{k-1}, v_k), (v_k, v_1)\}$. Subgraphs $\mathbf{D}_1, \mathbf{D}_2, \ldots, \mathbf{D}_r$ are mutually disjoint if $\mathtt{V}(\mathbf{D}_i) \cap \mathtt{V}(\mathbf{D}_j) = \emptyset$ for $i \neq j$.

Of special interest in our consideration is the notion of a matching in a digraph $\mathbf{D} = (\mathtt{V}, \mathtt{E})$ (*e.g.*, Chen, 1976). A subgraph is called a matching from a set of vertices $\mathtt{V}_{I_k} = \{v_{i_1}, v_{i_2}, \ldots, v_{i_k}\}$ to a set of vertices $\mathtt{V}_{J_k} = \{v_{j_1}, v_{j_2}, \ldots, v_{j_k}\}$, which we denote by $\mathbf{M}(\mathtt{V}_{I_k}, \mathtt{V}_{J_k})$, when it consists of k

mutually disjoint paths from $\mathtt{V}_{I_k}$ to $\mathtt{V}_{J_k}$. The length of a matching is the sum of the lengths of the corresponding paths, and we have the shortest matching $\mathbf{M}^*(\mathtt{V}_{I_k}, \mathtt{V}_{J_k})$ when we chose the shortest paths. Note that for given sets of $\mathtt{V}_{I_k}$ and $\mathtt{V}_{J_k}$, the shortest matching and, thus, any matching need not be unique.

The following result provides a graph-theoretic characterization of the parts (*i*) and (*iii*) of Theorem 5.6:

5.10. THEOREM. Let a structured system $\tilde{\mathbf{S}}$ be standard. Then:

(*i*) $\mathcal{V}^*_{b,\mathrm{Ker}\ C} \overset{*}{=} \mathcal{X}$ if and only if there exists a matching $\mathbf{M}(\mathtt{U}_{I_\ell}, \mathtt{Y})$ in $\mathbf{D}$;

(*ii*) $\mathcal{S}^*_{b,\mathrm{Ker}\ C} \overset{*}{=} \{0\}$ if and only if there exists a matching $\mathbf{M}(\mathtt{U}, \mathtt{Y}_{J_m})$ in $\mathbf{D}$.

Proof. (*i*) *Sufficiency*: Without loss of generality, we can assume that there exists a matching $\mathbf{M}(\{u_1, u_2, \ldots, u_\ell\}, \mathtt{Y})$ in $\mathbf{D}$. Let k be the total length of the shortest matching $\mathbf{M}^*(\{u_1, u_2, \ldots, u_\ell\}, \mathtt{Y})$, and denote a set of all the shortest matchings from $\{u_1, u_2, \ldots, u_\ell\}$ to $\mathtt{Y}$ by $\tilde{\mathbf{M}}^*$. Then it is straightforward to verify (Chen, 1976) that

$$\det \begin{bmatrix} sI_n - \tilde{A} & \tilde{B}_I \\ \tilde{C} & 0 \end{bmatrix} = \alpha_0(p)s^{n-k} + \alpha_1(p)s^{n-k-1} + \cdots + \alpha_{n-k}(p),$$
$$\alpha_0(p) = \sum_{\mathbf{D}_s \in \tilde{\mathbf{M}}^*} \pm f(\mathbf{D}_s), \tag{5.61}$$

where $\tilde{B}_I$ is a submatrix of $\tilde{B}$, which is composed of the first ℓ columns of $\tilde{B}$, p denotes the ν dimensional vector in the associated parameter space, and $f(\bullet)$ denotes the product of all nonfixed entries of $\tilde{A}$, $\tilde{B}$, $\tilde{C}$ corresponding to all edges contained in the indicated subgraph. Notice that all the $f(\mathbf{D}_s)$'s in (5.61) are monomials of p, and it holds that $f(\mathbf{D}_{s_1}) \not\equiv f(\mathbf{D}_{s_2})$ for $\mathbf{D}_{s_1}, \mathbf{D}_{s_2} \in \tilde{\mathbf{M}}^*$, $\mathbf{D}_{s_1} \neq \mathbf{D}_{s_2}$. This implies that $\alpha_0(p)$ in (5.61) is not identically zero as a polynomial of p because $\tilde{\mathbf{M}}^*$ is not empty by assumption. Therefore, a set $\boldsymbol{V} = \{p \in \mathbf{R}^\nu\colon\ \alpha_0(p) = 0\}$ is a proper variety in the parameter space, and it follows from (5.61) that for any $(A, B, C) \in \boldsymbol{V}^c$, $\det \begin{bmatrix} sI_n - A & B_I \\ C & 0 \end{bmatrix}$ is not identically zero as a polynomial of s, *i.e.*, for any $(A, B, C) \in \boldsymbol{V}^c$, the rank of $\begin{bmatrix} sI - A & B \\ C & 0 \end{bmatrix}$ is equal to $n + \ell$ for almost all $s \in \mathbf{C}$. From this and Theorem 5.6 (*i*), we get $\mathcal{V}^*_{b,\mathrm{Ker}\ C} \overset{*}{=} \mathcal{X}$.

Necessity: From Theorem 5.6 (*i*), $\mathcal{V}^*_{b,\mathrm{Ker}\ C} \overset{*}{=} \mathcal{X}$ implies that there exists at least one $(A, B, C) \in \mathbf{R}^\nu$ such that $\begin{bmatrix} sI_n - A & B \\ C & 0 \end{bmatrix}$ has full row rank for almost all $s \in \mathbf{C}$. This fact implies that $\begin{bmatrix} sI_n - \tilde{A} & \tilde{B} \\ \tilde{C} & 0 \end{bmatrix}$ has at least one minor with order $n + \ell$ that is not identically zero as a polynomial of s and p. Therefore, without loss of generality, assume that the minor that consists of the first n_1 columns of $\begin{bmatrix} sI_n - \tilde{A} \\ \tilde{C} \end{bmatrix}$ and the first n_2 columns of $\begin{bmatrix} \tilde{B} \\ 0 \end{bmatrix}$ $(n_1 + n_2 = n + \ell,\ n_1 \leq n,\ n_2 \geq \ell)$ is not identically zero. In other words, $\det \tilde{\Gamma} \not\equiv 0$, where $\tilde{\Gamma}$ is an $(n + m) \times (n + m)$ matrix defined by

$$\tilde{\Gamma} = \left[\begin{array}{cc|cc} \multicolumn{2}{c|}{sI_n - \tilde{A}} & \multicolumn{2}{c}{\tilde{B}} \\ \hline \multicolumn{2}{c|}{\tilde{C}} & \multicolumn{2}{c}{0} \\ \hline 0 & \tilde{I}_{n-n_1} & 0 & 0 \\ 0 & 0 & 0 & \tilde{I}_{m-n_2} \end{array}\right] . \tag{5.62}$$

Observe (Chen, 1976) that $\det \tilde{\Gamma} \not\equiv 0$ if and only if the Coates graph $\mathbf{G}_c(\tilde{\Gamma})$, which is associated with the matrix $\tilde{\Gamma}$, has a spanning subgraph $\mathbf{H} = \mathbf{C}_1 \cup \ldots \cup \mathbf{C}_r$, where the $\mathbf{C}_i$'s are mutually disjoint cycles. Notice that $\mathbf{G}_c(\tilde{\Gamma})$ is identical to a digraph that is obtained from $\mathbf{D}$ as follows: Add $(m - \ell)$ nodes $y_{\ell+1}, \ldots, y_m$ and edges (x_i, x_i) for $i = 1, \ldots, n$, $(x_{n_1+j}, y_{\ell+j})$ for $j = \ell \ldots, m - \ell$, (u_{n_2+k}, y_{n_2+k}) for $k = 1, \ldots, m - n_2$, to $\mathbf{D}$, and then identify u_i with y_i for $i = 1, 2, \ldots, m$. Therefore, it is easy to verify that the existence of the subgraph $\mathbf{H}$ in $\mathbf{G}_c(\tilde{\Gamma})$ implies the existence of a subgraph $\mathbf{D}_s = \mathbf{P}^{y_1}_{u_{i_1}} \cup \mathbf{P}^{y_2}_{u_{i_2}} \cdots \cup \mathbf{P}^{y_\ell}_{u_{i_\ell}}$ in $\mathbf{D}$, where the $\mathbf{P}^{y_j}_{u_{i_j}}$'s are mutually disjoint. Notice that the subgraph $\mathbf{D}_s$ is a matching $\mathbf{M}(\mathtt{U}_{I_\ell}, \mathtt{Y})$.

(*ii*) We omit the proof because it is similar to that of (*i*). Q.E.D.

5.11. REMARK. The statement (*i*) implies that the number of inputs should be greater than or equal to the number of outputs of $\mathbf{S}$. For (*ii*) to take place, the number of inputs should be less than or equal to the number of outputs.

Before proving the main result of this section, we briefly review some results (Matsumoto and Ikeda, 1983; Aoki *et al.* 1983) on structured controllability for a linear system in descriptor form

$$\mathbf{\Sigma}\colon\ K\dot{\varphi}(t) = L\varphi(t) + M\omega(t) \quad , \tag{5.63}$$

where $\varphi(t) \in \mathbf{R}^r$ and $\omega(t) \in \mathbf{R}^p$ are the descriptor state and input of $\boldsymbol{\Sigma}$, and K, L, and M are numerical matrices of appropriate dimensions.

Associated with the system $\boldsymbol{\Sigma}$ and the triple (K, L, M), consider a structured system $\tilde{\boldsymbol{\Sigma}}$ described by a triple of structural matrices $(\tilde{K}, \tilde{L}, \tilde{M})$. Notice that the generic rank of $\tilde{K}$ is not necessarily equal to r, but it is assumed that $s\tilde{K} - \tilde{L}$ is a regular pencil generically. Let q be the degree of $\det(s\tilde{K} - \tilde{L})$ with respect to s. Then, it is easy to verify that there exist two permutation matrices Q_1 and Q_2 such that

$$\tilde{K}': = Q_1\tilde{K}Q_2 = \begin{bmatrix} \tilde{I}_q & 0 \\ 0 & 0 \end{bmatrix} + \begin{bmatrix} \tilde{K}'_{11} & \tilde{K}'_{12} \\ \tilde{K}'_{21} & \tilde{K}'_{22} \end{bmatrix},$$

$$\tilde{L}': = Q_1\tilde{L}Q_2 = \begin{bmatrix} 0 & 0 \\ 0 & \tilde{I}_{r-q} \end{bmatrix} + \begin{bmatrix} \tilde{L}'_{11} & \tilde{L}'_{12} \\ \tilde{L}'_{21} & \tilde{L}'_{22} \end{bmatrix}, \tag{5.64}$$

$$\tilde{M}': = Q_1\tilde{M} = \begin{bmatrix} \tilde{M}'_1 \\ \tilde{M}'_2 \end{bmatrix},$$

where $\tilde{I}_q$ is a $q \times q$ structured diagonal matrix. With the system $\tilde{\boldsymbol{\Sigma}}$, we associate a digraph $\mathbf{D}_\Sigma = (\mathsf{V}_\Sigma, \mathsf{E}_\Sigma)$, where the set of vertices V_Σ is defined as $\mathsf{V}_\Sigma = \Phi \cup \Omega$, with $\Phi = \{\varphi_1, \varphi_2, \ldots, \varphi_r\}$ and $\Omega = \{\omega_1, \omega_2, \ldots, \omega_p\}$. The set of edges E_Σ is defined as $\mathsf{E}_\Sigma = \{(\omega_j, \varphi_i)\colon\ \tilde{m}'_{ij} \not\equiv 0\} \,\&\, \{(\varphi_j, \varphi_i)\colon\ \tilde{k}'_{ij} \not\equiv 0$ or $\tilde{\ell}'_{ij} \not\equiv 0\}$, where $\tilde{k}'_{ij}$, $\tilde{\ell}'_{ij}$, and $\tilde{m}'_{ij}$ indicate the (i, j) entries of $\tilde{K}'$, $\tilde{L}'$, and $\tilde{M}'$, respectively. With these preliminaries, the following result on structural controllability for linear systems in descriptor form has been obtained in (Matsumoto and Ikeda, 1983):

5.12. LEMMA. The following two conditions are equivalent:

(*i*) $\tilde{\rho}([s\tilde{K} - \tilde{L} \quad \tilde{M}]) = r, \qquad$ for all $s \in \mathbf{C}$.

(*ii*) The following two conditions hold:

(*ii*$_1$) $\tilde{\rho}([\tilde{L} \quad \tilde{M}]) = r$;

(*ii*$_2$) the nodes $\varphi_1, \ldots, \varphi_q$ among $\Phi = \{\varphi_1, \ldots, \varphi_q, \ldots, \varphi_r\}$ are reachable from the Ω-nodes in $\mathbf{D}_\Sigma$.

Next, we prove two more lemmas which we need for our main result (Theorem 5.15).

5.13. LEMMA.

$$\tilde{\rho}\left(\begin{bmatrix} sI_n - \tilde{A} & \tilde{B} \\ \tilde{C} & 0 \end{bmatrix}\right) = n + \ell, \qquad \text{for all } s \in \mathbf{C}, \tag{5.65}$$

if and only if

$$\tilde{\rho}\left(\begin{bmatrix} s\tilde{I}_n - \tilde{A} & \tilde{B} \\ \tilde{C} & 0 \end{bmatrix}\right) = n + \ell, \qquad \text{for all } s \in \mathbf{C}. \tag{5.66}$$

Proof. Sufficiency: trivial.

Necessity: (5.66) implies that there exist a quadruple (E_0, A_0, B_0, C_0) and a positive number ϵ such that, for any $(E, A, B, C) \in \mathcal{S} := \{(E, A, B, C) \in \mathbf{R}^{\nu+n}\colon \|E - E_0\| + \|A - A_0\| + \|B - B_0\| + \|C - C_0\| < \epsilon\}$,

$$\rho\left(\begin{bmatrix} sE - A & B \\ C & 0 \end{bmatrix}\right) = n + \ell, \qquad \text{for all } s \in \mathbf{C}, \tag{5.67}$$

where $\mathbf{R}^\nu$ is the associated parameter space of $(\tilde{A}, \tilde{B}, \tilde{C})$ and E is diagonal. Therefore, we can always find a quadruple $(E_1, A_1, B_1, C_1) \in \mathcal{S}$ with E_1 being nonsingular.

This implies that

$$\rho\left(\begin{bmatrix} sI_n - E_1^{-1}A_1 & E_1^{-1}B_1 \\ C_1 & 0 \end{bmatrix}\right) = n + \ell, \qquad \text{for all } s \in \mathbf{C}. \tag{5.68}$$

Notice that $(E_1^{-1}A_1, E_1^{-1}B_1, C_1) \in \mathbf{R}^\nu$ because E_1^{-1} is diagonal. Thus, (5.65) follows. Q.E.D.

To prove the following lemma, we need to use a subgraph $\mathbf{P}_u = \mathbf{P}_{u_{j_1}} \cup \mathbf{P}_{u_{j_2}} \cup \cdots \cup \mathbf{P}_{u_{j_p}}$ that is a union of mutually disjoint paths from vertices in $\mathtt{U}$ to arbitrary vertices of $\mathtt{X}$. Similarly, $\mathbf{P}_y = \mathbf{P}^{y_{i_1}} \cup \mathbf{P}^{y_{i_2}} \cup \cdots \cup \mathbf{P}^{y_{i_z}}$ is defined with respect to $\mathtt{X}$ and $\mathtt{Y}$. Finally, we need a subgraph $\mathbf{C}_x = \mathbf{C}_1 \cup \mathbf{C}_2 \cup \cdots \cup \mathbf{C}_r$ that is a union of mutually disjoint cycles of $\mathbf{D}$.

5.14. LEMMA.

$$\tilde{\rho}\begin{bmatrix} \tilde{A} & \tilde{B} \\ \tilde{C} & 0 \end{bmatrix} = n + \ell \tag{5.69}$$

if and only if there exists a subgraph $\mathbf{D}_s = \mathbf{P}_u \cup \mathbf{C}_x \cup \mathbf{M}(\mathtt{U}_{I_\ell}, \mathtt{Y})$ in $\mathbf{D}$ such that $\mathtt{V}_s \supset \mathtt{X}$ and $\mathbf{P}_u$, $\mathbf{C}_x$, and $\mathbf{M}(\mathtt{U}_{I_\ell}, \mathtt{Y})$ are mutually disjoint.

Proof. Necessity: By the same argument which we used in the proof of the necessity part of Theorem 5.10 (*i*), (5.69) implies that graph $\mathbf{G}_c(\tilde{\Delta})$ has a spanning subgraph $\mathbf{H} = \mathbf{C}_1 \cup \cdots \cup \mathbf{C}_r$, where the $\mathbf{C}_i$'s are mutually disjoint cycles, and $\tilde{\Delta}$ is an $(n+m) \times (n+m)$ matrix defined by

$$\tilde{\Delta} = \left[\begin{array}{cc|cc} \multicolumn{2}{c|}{\tilde{A}} & \multicolumn{2}{c}{\tilde{B}} \\ \hline \multicolumn{2}{c|}{\tilde{C}} & \multicolumn{2}{c}{0} \\ \hline 0 & \tilde{I}_{n-n_1} & 0 & 0 \\ 0 & 0 & 0 & \tilde{I}_{m-n_2} \end{array}\right] . \tag{5.70}$$

Notice that $\mathbf{G}_c(\tilde{\Delta})$ is identical to a digraph that is obtained from $\mathbf{D}$ as follows: Add $(m-\ell)$ nodes $y_{\ell+1}, \ldots, y_m$ and edges $(x_{n_1+j}, y_{\ell+j})$ for $j = \ell, \ldots, m-\ell$, (u_{n_2+k}, y_{n+k}) for $k = \ell, \ldots, m-n_2$, and identify u_i with y_i for $i = 1, 2, \ldots, m$. Therefore, it is easy to see that the existence of the subgraph $\mathbf{H}$ in $\mathbf{G}_c(\tilde{\Delta})$ implies the condition of the lemma.

Sufficiency: Assume that the condition of the lemma is satisfied. Without loss of generality, we can assume that $\mathbf{D}$ has a subgraph $\mathbf{D}_s = \mathbf{P}_u \cup \mathbf{C}_x \cup \mathbf{M}(\{u_1, \ldots, u_\ell\}, \mathtt{Y})$, where $\mathbf{P}_u = \mathbf{P}_{u_{\ell+1}} \cdots \mathbf{P}_{u_{\ell+p}}$, $\mathbf{C}_x = \mathbf{C}_1 \cup \cdots \cup \mathbf{C}_r$, and $\mathbf{M}(\{u_1, \ldots, u_\ell\}, \mathtt{Y}) = \mathbf{P}_{u_1}^{y_1} \cdots \mathbf{P}_{u_\ell}^{y_\ell}$. Furthermore, assume that the length of $\mathbf{P}_{u_{\ell+j}}$ is $\ell_{\ell+j}$ for $j = 1, \ldots, p$, the length of $\mathbf{C}_k$ is ℓ_k^c for $k = 1, \ldots, r$, and the length of $\mathbf{P}_{u_i}^{y_i}$ is ℓ_i for $i = 1, \ldots, \ell$. Then, let A, B, and C be matrices obtained from $\tilde{A}$, $\tilde{B}$, and $\tilde{C}$ by substituting 1's for all the nonfixed entries corresponding to the edges contained in $\mathbf{D}_s$, and substituting 0's for the other nonfixed entries. It is easy to verify that there exists a permutation matrix Q such that

$$\begin{bmatrix} Q^T & 0 \\ 0 & I_\ell \end{bmatrix} \begin{bmatrix} A & B \\ C & 0 \end{bmatrix} \begin{bmatrix} Q & 0 \\ 0 & I_m \end{bmatrix} = \left[\begin{array}{ccc|ccc} A_M & 0 & 0 & B_M & 0 & 0 \\ 0 & A_P & 0 & 0 & B_P & 0 \\ 0 & 0 & A_C & 0 & 0 & 0 \\ \hline C_M & 0 & 0 & 0 & 0 & 0 \end{array}\right], \tag{5.71}$$

where

$$A_M = \begin{bmatrix} N_{\ell_1} & & \bigcirc \\ & \ddots & \\ \bigcirc & & N_{\ell_\ell} \end{bmatrix}, \qquad b_M = \begin{bmatrix} e_{\ell_1}^c & & \bigcirc \\ & \ddots & \\ \bigcirc & & e_{\ell_\ell}^c \end{bmatrix},$$

$$C_M = \begin{bmatrix} e^r_{\ell_1} & & \bigcirc \\ & \ddots & \\ \bigcirc & & e^r_{\ell_\ell} \end{bmatrix}, \tag{5.72}$$

$$A_P = \begin{bmatrix} N_{\ell_{\ell+1}} & & \bigcirc \\ & \ddots & \\ \bigcirc & & N_{\ell_{\ell+p}} \end{bmatrix}, \qquad B_P = \begin{bmatrix} e^c_{\ell_{\ell+1}} & & \bigcirc \\ & \ddots & \\ \bigcirc & & e^c_{\ell_{\ell+p}} \end{bmatrix},$$

$$A_C = \begin{bmatrix} \bar{N}_{\ell_1} & & \bigcirc \\ & \ddots & \\ \bigcirc & & \bar{N}_{\ell_r} \end{bmatrix},$$

and N_i, e^c_i, e^r_i, and $\bar{N}_i$ are $i \times i$, $i \times 1$, $1 \times i$, and $i \times i$ matrices defined as

$$N_i = \begin{bmatrix} 0 & & & & \\ 1 & 0 & & \bigcirc & \\ & 1 & \ddots & & \\ & \bigcirc & \ddots & \ddots & \\ & & & 1 & 0 \end{bmatrix}, \qquad e^c_i = \begin{bmatrix} 1 \\ 0 \\ \vdots \\ 0 \end{bmatrix},$$

$$e^r_i = [0 \cdots\cdots 0 \quad 1], \tag{5.73}$$

$$\bar{N}_i = \begin{bmatrix} 0 & & & & 1 \\ 1 & 0 & & \bigcirc & \\ & 1 & \ddots & & \\ & \bigcirc & \ddots & \ddots & \\ & & & 1 & 0 \end{bmatrix}.$$

This implies that we have found a triple $(A, B, C) \in \mathbf{R}^\nu$ such that $\rho\left(\begin{bmatrix} A & B \\ C & 0 \end{bmatrix}\right) = n + \ell$. Thus (5.69) is satisfied. Q.E.D.

Finally, we are ready to present the main result of this section, obtained by Hayakawa and Šiljak (1988):

5.15. THEOREM. Let a structured system $\tilde{\mathbf{S}}$ be standard. Then,

(*i*) $\mathcal{R}^*_{b,\text{Ker } C} \overset{*}{=} \mathcal{X}$ if and only if the following two conditions hold:

(i_1) There exists a subgraph $\mathbf{D}_s = \mathbf{P}_u \cup \mathbf{C}_x \cup \mathbf{M}(\mathtt{U}_{I_\ell}, \mathtt{Y})$ in $\mathbf{D}$ such that $\mathtt{V}_s \supset \mathtt{X}$ and $\mathbf{P}_u$, $\mathbf{C}_x$, and $\mathbf{M}(\mathtt{U}_{I_\ell}, \mathtt{Y})$ are mutually disjoint; and

(i_2) For any vertex $x \in \mathtt{X} - \mathtt{V}(\mathbf{M}^*)$, there exists a matching $\mathbf{M}(\mathtt{U}_{I_\ell} \cup \{u\}, \mathtt{Y} \cup \{x\})$ in $\mathbf{D}$ for some $u \in \mathtt{U}$, and $\mathbf{M}^* = \mathbf{M}^*(\mathtt{U}_{I_\ell}, \mathtt{Y})$.

(*ii*) $\mathcal{N}^*_{b,\text{Im } B} \overset{*}{=} \{0\}$ if and only if the following two conditions hold:

(ii_1) There exists a subgraph $\mathbf{D}_s = \mathbf{P}_y \cup \mathbf{C}_x \cup \mathbf{M}(\mathtt{U}, \mathtt{Y}_{J_m})$ in $\mathbf{D}$ such that $\mathtt{V}_s \supset \mathtt{X}$, and $\mathbf{P}_y$, $\mathbf{C}_x$, and $\mathbf{M}(\mathtt{U}, \mathtt{Y}_{J_m})$ are mutually disjoint; and

(ii_2) For any vertex $x \in \mathtt{X} - \mathtt{V}(\mathbf{M}^*)$, there exists a matching $\mathbf{M}(\mathtt{U} \cup \{x\}, \mathtt{Y}_{J_m} \cup \{y\})$ for some $y \in \mathtt{Y}$, and $\mathbf{M}^* = \mathbf{M}^*(\mathtt{U}, \mathtt{Y}_{J_m})$.

Proof. Only part (*i*) will be proved because part (*ii*) can be derived by duality.

Necessity: Assume that $\mathcal{R}^*_{b,\text{Ker } C} \overset{*}{=} \mathcal{X}$. Then, it follows from Theorem 5.9 (*ii*) and Lemma 5.13 that a structured system $\tilde{\mathbf{\Sigma}}$, *i.e.*, a triple $(\tilde{K}, \tilde{L}, \tilde{M})$, satisfies condition (*i*) of Lemma 5.12, where

$$\tilde{K} = \begin{bmatrix} \tilde{I}_n & 0 \\ 0 & 0 \end{bmatrix}, \qquad \tilde{L} = \begin{bmatrix} \tilde{A} & \tilde{B}_I \\ \tilde{C} & 0 \end{bmatrix}, \qquad \tilde{M} = \begin{bmatrix} \tilde{B}_{II} \\ 0 \end{bmatrix}, \tag{5.74}$$

where $\tilde{B} = [\tilde{B}_I \; \tilde{B}_{II}]$. Therefore, condition (*ii*) of Lemma 5.12 follows.

By Lemma 5.14 it is easy to see that condition (ii_1) of Lemma 5.12 implies condition (i_1) of Theorem 5.15.

Next, we will prove that condition (ii_2) of Lemma 5.12 implies condition (i_2) of the theorem under condition (ii_1) of Lemma 5.12. To do this, we need to investigate a relationship between the two digraphs $\mathbf{D}_\Sigma$ and $\mathbf{D}$. Recall that condition (ii_1) of Lemma 5.12 implies condition (i_1) of the theorem. Therefore, without loss of generality, we can assume that there exists a matching $\mathbf{M}(\{u_1, u_2, \ldots, u_\ell\}, \mathtt{Y})$ in $\mathbf{D}$, and a shortest matching is $\mathbf{M}^* = \mathbf{P}^{y_1}_{u_1} \cup \cdots \cup \mathbf{P}^{y_\ell}_{u_\ell}$, where the length of $\mathbf{P}^{y_i}_{u_i}$ is ℓ_i for $i = 1, \ldots, \ell$. Let $q = n - \sum_{i=1}^{\ell} \ell_i$. Notice that q is the degree of det $[s\tilde{K} - \tilde{L}]$ with respect to s. Furthermore, without loss of generality, it can be assumed that $\tilde{A}$, $\tilde{B}$, and $\tilde{C}$ have the following structures:

$$\tilde{A} = \left[\begin{array}{c:ccc} \tilde{A}_{00} & \tilde{A}_{01} & \dots & \tilde{A}_{0\ell} \\ \hdashline \tilde{A}_{10} & \tilde{A}_{11} & \dots & \tilde{A}_{1\ell} \\ \vdots & \vdots & & \vdots \\ \tilde{A}_{\ell 0} & \tilde{A}_{\ell 1} & \dots & \tilde{A}_{\ell\ell} \end{array}\right], \qquad B = \left[\begin{array}{ccc:c} & B_0 & & \\ \hdashline \tilde{b}_{11} & \dots & \tilde{b}_{1\ell} & \\ & & & \tilde{B}_{II} \\ \vdots & & \vdots & \\ \tilde{b}_{\ell 1} & \dots & \tilde{b}_{\ell\ell} & \end{array}\right], \tag{5.75}$$

$$\tilde{C} = \left[\begin{array}{c:ccc} & \tilde{c}_{11} & \dots & \tilde{c}_{1\ell} \\ \tilde{C}_0 & \vdots & & \vdots \\ & \tilde{c}_{\ell 1} & \dots & \tilde{c}_{\ell\ell} \end{array}\right],$$

where $\tilde{A}_{ii}$, $\tilde{b}_{ii}$, and $\tilde{c}_{ii}$ are $\ell_i \times \ell_i$, $\ell_i \times 1$, $1 \times \ell_i$ structured matrices for $i = 1, \dots, \ell$ defined as

$$\tilde{A}_{ii} = \begin{bmatrix} \times & & & & \\ & \times & & & \times \\ \oplus & & \ddots & & \\ & \oplus & & \ddots & \\ \bigcirc & & \ddots & & \\ & & & \oplus & \times \end{bmatrix}, \qquad \tilde{b}_{ii} = \begin{bmatrix} \oplus \\ 0 \\ \vdots \\ 0 \end{bmatrix}, \tag{5.76}$$

$$\tilde{c}_{ii} = [0 \quad \dots \quad 0 \quad \oplus],$$

where the $\oplus$'s represent the nonfixed entries, the 0's represent the fixed (zero) entries, and the $\times$'s represent either the nonfixed or the fixed entries. Notice that the 0's in $\tilde{A}_{ii}$, $\tilde{b}_{ii}$, and $\tilde{c}_{ii}$ are due to the fact that $\mathbf{M}^*$ is the shortest matching, and $\tilde{A}_{00}$ is a $q \times q$ structured matrix. From (5.74) and (5.75), we can choose Q_1 and Q_2 in (5.64) as follows:

$$
Q_1 = \left[\begin{array}{c|c|c|c|c}
I_q & \bigcirc & \cdots & \bigcirc & \bigcirc \\ \hline
\bigcirc & I_{\ell_1} & \cdots & \bigcirc & \bigcirc \\ \hline
0\ldots0 & 0\ldots0 & \cdots & 0\ldots0 & 1\,0\ldots0 \\ \hline
\cdots & \cdots & \cdots & \cdots & \cdots \\ \hline
\bigcirc & \bigcirc & \cdots & I_{\ell_\ell} & \bigcirc \\ \hline
0\ldots0 & 0\ldots0 & \cdots & 0\ldots0 & 0\ldots0\,1
\end{array}\right],
$$

$$
Q_2 = \left[\begin{array}{c|c|c|c|c|c}
I_q & \begin{matrix}0\\ \vdots\\ 0\end{matrix} & \bigcirc & \cdots & \begin{matrix}0\\ \vdots\\ 0\end{matrix} & \bigcirc \\ \hline
\bigcirc & \begin{matrix}0\\ \vdots\\ 0\end{matrix} & I_{\ell_1} & \cdots & \begin{matrix}0\\ \vdots\\ 0\end{matrix} & \bigcirc \\ \hline
\cdots & \cdots & \cdots & \cdots & \cdots & \cdots \\ \hline
\bigcirc & \begin{matrix}0\\ \vdots\\ 0\end{matrix} & \bigcirc & \cdots & \begin{matrix}0\\ \vdots\\ 0\end{matrix} & I_{\ell_\ell} \\ \hline
\bigcirc & \begin{matrix}1\\ 0\\ \vdots\\ 0\end{matrix} & \bigcirc & \cdots & \begin{matrix}0\\ 0\\ \vdots\\ 1\end{matrix} & \bigcirc
\end{array}\right]. \tag{5.77}
$$

Therefore, by introducing the following notation about X-nodes of **D** and X-nodes of $\mathbf{D}_\Sigma$:

$$x_j^{(i)} := \begin{cases} x_j & \text{for } \ i = 0\,, \qquad 1 \le j \le q, \\ x_{q+\ell_1+\cdots+\ell_{i-1}+j} & \text{for } \ 1 \le i \le \ell\,, \quad 1 \le j \le \ell_i, \end{cases}$$

$$\varphi_k^{(i)} := \begin{cases} \varphi_k & \text{for } \ i = 0\,, \qquad 1 \le k \le q, \\ \varphi_{q+\ell_1+\cdots+\ell_{i-1}+(i-1)+k} & \text{for } \ 1 \le i \le \ell\,, \quad 1 \le k \le \ell_i + 1, \end{cases}$$

(5.78)

we get one-to-one correspondence between $\tilde{\mathrm{E}} := \mathrm{E} \cup \{(x_i,\, x_i) | i = 1, \ldots, n\}$ and E_Σ as follows:

$$\begin{array}{lcll} \mathrm{E} & \longrightarrow & \mathrm{E}_\Sigma & \\ (x_j^{(i)},\, x_k^{(h)}) & \longmapsto & \begin{cases} (\varphi_j^{(0)}\,, \quad \varphi_k^{(h)}) \\ (\varphi_{j+1}^{(i)}\,, \quad \varphi_k^{(h)}) \end{cases} & \begin{array}{l} \text{for } \ i = 0, \\ \text{for } \ 1 \le i \le \ell, \end{array} \\ (u_i,\, x_k^{(h)}) & \longmapsto & \begin{cases} (\varphi_1^{(i)}\,, \quad \varphi_k^{(h)}) \\ (\omega_{i-\ell}\,, \quad \varphi_k^{(h)}) \end{cases} & \begin{array}{l} \text{for } \ 1 \le i \le \ell, \\ \text{for } \ \ell+1 \le i \le m, \end{array} \\ (x_j^{(i)},\ y_k) & \longmapsto & \begin{cases} (\varphi_j^{(0)}\,, \quad \varphi_{\ell_k+1}^{(k)}) \\ (\varphi_{j+1}^{(i)}\,, \quad \varphi_{\ell_k+1}^{(k)}) \end{cases} & \begin{array}{l} \text{for } \ i = 0, \\ \text{for } \ 1 \le i \le \ell. \end{array} \end{array}$$

(5.79)

From the above relation between **D** and $\mathbf{D}_\Sigma$, it is easy to verify that condition (ii_2) of Lemma 5.12 implies condition (i_2) of the theorem.

Sufficiency: Omitted because we can use the proof of necessity part in the reverse order. Q.E.D.

5.16. REMARK. Condition (i) of the above theorem has two parts that correspond to the generic rank and input reachability properties of structural controllability (Section 1.4). Similarly, condition (ii) resembles the

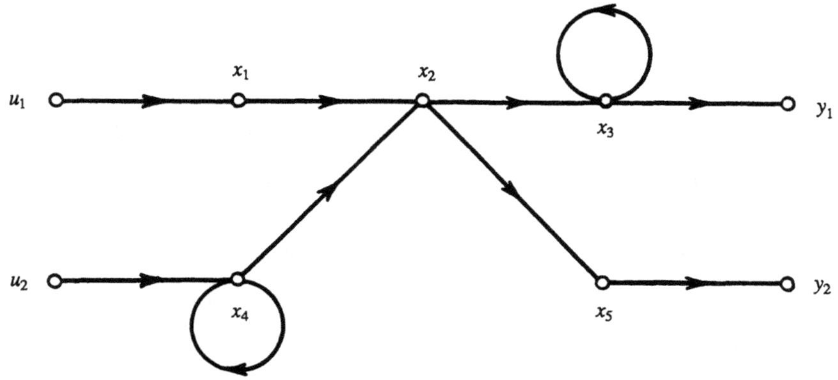

Fig. 5.1. Digraph **D** for Example 5.17.

structural observability conditions. It is not surprising, however, that Theorem 5.15 is more complex because it deals with both inputs and outputs simultaneously.

We illustrate Theorems 5.10 and 5.15 by several examples.

5.17. EXAMPLE. Consider a structured system $\tilde{\mathbf{S}}$ with the matrices

$$\tilde{A} = \begin{bmatrix} 0 & 0 & 0 & 0 & 0 \\ * & 0 & 0 & * & 0 \\ 0 & * & * & 0 & 0 \\ 0 & 0 & 0 & * & 0 \\ 0 & * & 0 & 0 & 0 \end{bmatrix}, \qquad \tilde{B} = \begin{bmatrix} * & 0 \\ 0 & 0 \\ 0 & 0 \\ 0 & * \\ 0 & 0 \end{bmatrix}, \qquad \tilde{C} = \begin{bmatrix} 0 & 0 & * & 0 & 0 \\ 0 & 0 & 0 & 0 & * \end{bmatrix}, \tag{5.80}$$

where as usual $*$ represent an independent real number, and the 0's are "hard zeros." The corresponding digraph **D** is shown in Figure 5.1. It is easy to see that there exists no matching from $\mathtt{U} = \{u_1, u_2\}$ to $\mathtt{Y} = \{y_1, y_2\}$ and, therefore, we conclude from Theorem 5.10 that $\mathcal{V}^*_{b,\mathrm{Ker}\ C} \stackrel{*}{=} \mathcal{X}$ does not hold. An implication of this fact is that the system $\tilde{\mathbf{S}}$ has no inverse no matter what the values of the free parameters are.

5.18. EXAMPLE. Let a system $\tilde{\mathbf{S}}$ be specified by

$$\tilde{A} = \begin{bmatrix} 0 & 0 & 0 & 0 & 0 & 0 \\ * & 0 & 0 & 0 & 0 & 0 \\ 0 & 0 & 0 & 0 & 0 & 0 \\ 0 & 0 & * & 0 & 0 & * \\ 0 & * & 0 & * & 0 & 0 \\ 0 & 0 & 0 & 0 & 0 & * \end{bmatrix}, \qquad \tilde{B} = \begin{bmatrix} * & 0 & 0 \\ 0 & 0 & 0 \\ 0 & * & 0 \\ 0 & 0 & 0 \\ 0 & 0 & 0 \\ 0 & * & * \end{bmatrix},$$

$$\tilde{C} = \begin{bmatrix} 0 & * & 0 & 0 & * & 0 \\ * & 0 & 0 & 0 & * & 0 \end{bmatrix}. \tag{5.81}$$

From digraph $\mathbf{D}$ of Figure 5.2, we see that there is a matching $\mathbf{M}(\{u_1, u_2\}, \{y_1, y_2\}) = \mathbf{P}_{u_1}^{y_1} \cup \mathbf{P}_{u_2}^{y_2}$, with $\mathbf{P}_{u_1}^{y_1} = (u_1, x_1, x_2, y_1)$ and $\mathbf{P}_{u_2}^{y_2} = (u_2, x_3, x_4, x_5, y_2)$. By Theorem 5.10, $\mathcal{V}_{b,\mathrm{Ker}\ C}^* \stackrel{*}{=} \mathcal{X}$. However, when we consider $\mathcal{R}_{b,\mathrm{Ker}\ C}^*$, we find that condition (i) of Theorem 5.15 is not satisfied. Although (i_1) is satisfied by $\mathbf{M}(\{u_1, u_2\}, \{y_1, y_2\})$ and a self-loop at x_6, condition (i_2) fails to hold. In fact, $\mathbf{M}^*(\{u_1, u_2\}, \{y_1, y_2\}) = \mathbf{P}_{u_1}^{y_2} \cup \mathbf{P}_{u_2}^{y_2}$ with $\mathbf{P}_{u_1}^{y_2} = (u_1, x_1, y_2)$ and $\mathbf{P}_{u_2}^{y_1} = (u_2, x_3, x_4, x_5, y_1)$, and $\mathtt{X} - \mathtt{V}(\mathbf{M}^*) = \{x_2, x_6\}$, so that we have $\mathbf{M}(\{u_1, u_2, u_3\}, \{y_1, y_2, x_6\})$ for x_6, but $\mathbf{M}(\{u_1, u_2, u_3\}, \{y_1, y_2, x_2\})$ for x_2 does not exist.

If we modify the digraph $\mathbf{D}$ of Figure 5.2 by moving the input u_3 from state x_6 to x_2, as shown in Figure 5.3, we can establish the condition $\mathcal{R}_{b,\mathrm{Ker}\ C}^* \stackrel{*}{=} \mathcal{X}$. The new matrix $\tilde{B}$, which is given as

$$\tilde{B} = \begin{bmatrix} * & 0 & 0 \\ 0 & 0 & * \\ 0 & * & 0 \\ 0 & 0 & 0 \\ 0 & 0 & 0 \\ 0 & * & 0 \end{bmatrix}, \tag{5.82}$$

leaves the set of vertices $\mathtt{X} - \mathtt{V}(\mathbf{M}^*) = \{x_2, x_6\}$ unchanged, but the two matchings mentioned above are now present. The reason is that, by moving u_3 to x_2, we create the path $\mathbf{P}_{u_3}^{x_2} = (u_3, x_2)$, which provides two required matchings in the modified graph $\mathbf{D}$ of Figure 5.3.

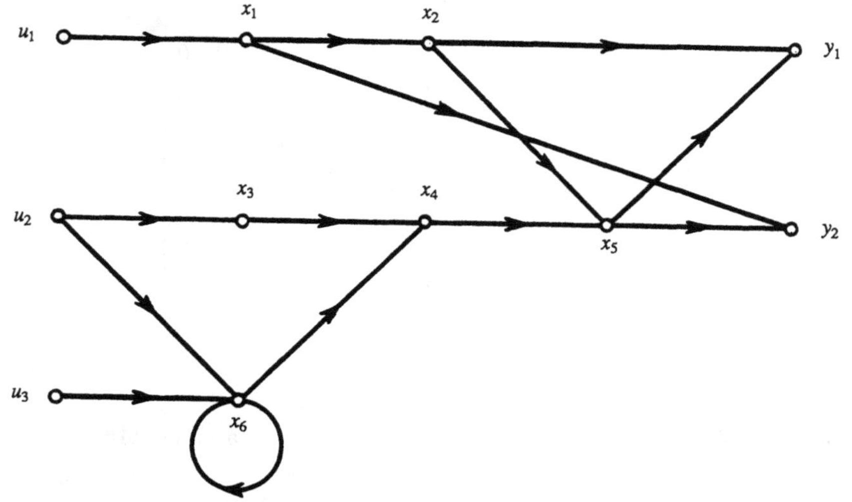

Fig. 5.2. Digraph **D** for Example 5.18.

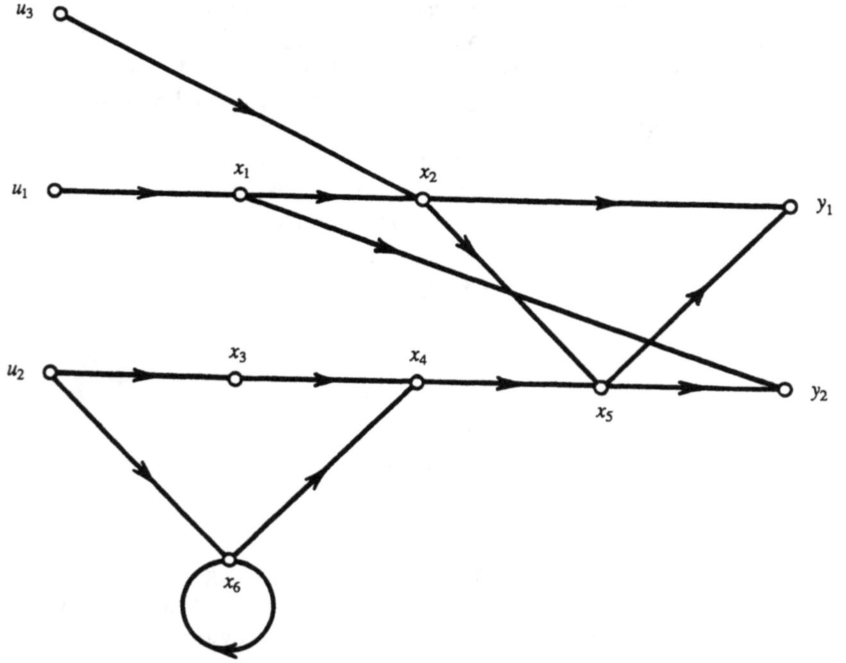

Fig. 5.3. Modified digraph **D** for Example 5.18.

5.19. REMARK. Special properties of almost invariant spaces are established by the existence of shortest matchings in a digraph **D**. This is computationally attractive because there are efficient algorithms for maximum flow in transportation networks that apply directly to this task (see, for example, Swamy and Thulasiraman, 1981). In fact, first add the source vertex s and the sink vertex t to **D**, then add edges from s to every node in U and edges from every vertex in Y to t. Finally, assign capacity 1 to every edge of the newly formed digraph, and apply any variant of the well-known Ford–Fulkerson's algorithm to get the desired result. Another way to test a directed graph for the existence of appropriate matchings is the generic rank test described in the Appendix.

5.4. Decentral Stabilizability

We now apply the obtained graph-theoretic conditions to determine when structured interconnected systems can be stabilized by decentralized controllers. For simplicity in presentation, we consider an input and output decentralized system composed of only two subsystems. An extension of the results obtained to the general case is obvious, but may involve some combinatorial representations (Young, 1985).

Let us consider an interconnected system

$$
\begin{aligned}
\mathbf{S}\colon\ \dot{x}_1 &= A_1x_1 + B_1u_1 + A_{12}x_2,\\
\dot{x}_2 &= A_2x_2 + B_2u_2 + A_{21}x_1,\\
y_1 &= C_1x_1,\\
y_2 &= C_2x_2,
\end{aligned}
\tag{5.83}
$$

where $x_i(t) \in \mathcal{X}_i := \mathbf{R}^{n_i}$, $u_i(t) \in \mathcal{U}_i := \mathbf{R}^{m_i}$, and $y_i(t) \in \mathcal{Y}_i := \mathbf{R}^{\ell_i}$ are the state, input, and output of the ith subsystem

$$
\begin{aligned}
\mathbf{S}_i\colon\ \dot{x}_i &= A_ix_i + B_iu_i,\\
y_i &= C_ix_i, \qquad i = 1,\, 2.
\end{aligned}
\tag{5.84}
$$

By using local controllers with dynamic output feedback

$$
\begin{aligned}
\mathbf{C}_i\colon\ \dot{z}_i &= F_iz_i + G_iy_i,\\
u_i &= H_iz_i + K_iy_i, \qquad i = 1,\, 2,
\end{aligned}
\tag{5.85}
$$

we obtain the overall closed-loop system

$$\hat{\mathbf{S}}\colon \quad \begin{bmatrix} \dot{x}_1 \\ \dot{z}_1 \\ \dot{x}_2 \\ \dot{z}_2 \end{bmatrix} = \left[\begin{array}{cc|cc} A_1 + B_1K_1C_1 & B_1H_1 & A_{12} & 0 \\ G_1C_1 & F_1 & 0 & 0 \\ \hline A_{21} & 0 & A_2 + B_2K_2C_2 & B_2H_2 \\ 0 & 0 & G_2C_2 & F_2 \end{array}\right] \begin{bmatrix} x_1 \\ z_1 \\ x_2 \\ z_2 \end{bmatrix}, \tag{5.86}$$

which is an interconnection of two closed-loop subsystems

$$\hat{\mathbf{S}}_i\colon \quad \begin{bmatrix} \dot{x}_i \\ \dot{z}_i \end{bmatrix} = \begin{bmatrix} A_i + B_iK_iC_i & B_iH_i \\ G_iC_i & F_i \end{bmatrix} \begin{bmatrix} x_i \\ z_i \end{bmatrix}, \qquad i = 1,\, 2. \tag{5.87}$$

In the context of connective stability (Section 2.1), we state the following:

5.20. DEFINITION. An interconnected system $\mathbf{S}$ is said to be decentrally connectively stabilizable if there exist controllers $\mathbf{C}_1$ and $\mathbf{C}_2$ such that the closed-loop system $\hat{\mathbf{S}}$ and subsystems $\hat{\mathbf{S}}_1$ and $\hat{\mathbf{S}}_2$ are *simultaneously* asymptotically stable.

The following lemma is a direct consequence of the result obtained by Willems and Ikeda (1984). In the lemma, we compute almost invariant subspaces relative to subsystems $\mathbf{S}_i = (A_i,\, B_i,\, C_i)$. Then, by $\mathcal{R}^*_{b,\mathrm{Ker}H}$ and $\mathcal{N}_{b,\mathrm{Im}G}$ w.r.t. $\mathbf{S}_i$, we denote the supremal $\mathcal{L}_p$-almost controllability subspace contained in Ker H computed relative to $(A_i,\, B_i)$, and the infimal $\mathcal{L}_o$-almost complementary observability subspace containing Im G computed relative to $(A_i,\, C_i)$, respectively. We also recall that a system $\mathbf{S}$ is complete if it is controllable and observable, and prove the following:

5.21. LEMMA. An interconnected system $\mathbf{S}$ is decentrally connectively stabilizable if one of the following conditions hold:

(*i*) $\mathbf{S}_1$ is complete and

$$\mathcal{R}^*_{b,\mathrm{Ker}\ A_{12}} = \mathcal{X}_2, \qquad \mathcal{N}^*_{b,\mathrm{Im}\ A_{21}} = \{0\}, \qquad \text{w. r. t. } \mathbf{S}_2; \tag{5.88}$$

(*ii*) $\mathbf{S}_2$ is complete and

$$\mathcal{R}^*_{b,\mathrm{Ker}\ A_{21}} = \mathcal{X}_1, \qquad \mathcal{N}^*_{b,\mathrm{Im}\ A_{12}} = \{0\}, \qquad \text{w. r. t. } \mathbf{S}_1. \tag{5.89}$$

Proof. Recall (Willems and Ikeda, 1984) that an interconnected system $\mathbf{S}$ is decentrally connectively stabilizable if $\mathbf{S}_1$ is stabilizable and detectable, $\mathbf{S}_2$ is stabilizable, and

$$\operatorname{Im} A_{21} \subset \mathcal{V}^+_{b,\operatorname{Ker}\, A_{12}}, \qquad \mathcal{S}^+_{b,\operatorname{Im}\, A_{21}} = \{0\}, \qquad \text{w. r. t.} \quad \mathbf{S}_2, \tag{5.90}$$

where $\mathcal{V}^+_{b,\operatorname{Ker}\, A_{12}}$ is the supremal asymptotically stable invariant subspace contained in Ker A_{12}. Therefore, it is obvious that the system $\mathbf{S}$ is decentrally connectively stabilizable if condition (i) of the lemma holds. Condition (ii) is also derived from Willems and Ikeda (1984). Q.E.D.

With a system $\mathbf{S}$ of (5.83), we associate a structured system $\tilde{\mathbf{S}}$ defined by the corresponding structured matrices. We assume that $\tilde{B}_1$ and $\tilde{B}_2$ have generic full column rank, and $\tilde{C}_1$ and $\tilde{C}_2$ have generic full row rank, and state the following:

5.22. DEFINITION. A structured interconnected system $\tilde{\mathbf{S}}$ is said to be decentrally connectively stabilizable if there exists a proper variety $\boldsymbol{V} \subset \mathbf{R}^{\nu}$ such that, for any $(A_i, B_i, C_i, A_{ij};\ i, j = 1, 2) \in \boldsymbol{V}^c$, the system $\mathbf{S}$ is decentrally connectively stabilizable.

Before we state the main result of this section, we need to clarify some facts regarding the interconnection matrices A_{ij}. Note that, in Lemma 5.21, the matrices A_{ij} do not necessarily have full rank. Thus, the structured matrices $\tilde{A}_{ij}$ need not have generic full rank either. In order to provide a structured version of Lemma 5.21, we have to modify the matrices $\tilde{A}_{ij}$ to obtain corresponding structured matrices having generic full rank. Let $\tilde{M}$ be a $p \times q$ matrix such that $\tilde{\rho}(\tilde{M}) = g$. Then, $\tilde{M}^r$ is an $g \times q$ submatrix of $\tilde{M}$ with $\tilde{\rho}(\tilde{M}^r) = g$, which can be obtained by removing $p - g$ rows of $\tilde{M}$. Similarly, we use $\tilde{M}^c$ of dimension $p \times g$ and $\tilde{\rho}(\tilde{M}^c) = g$ columns of $\tilde{M}$. For example, if

$$\tilde{M} = \begin{bmatrix} 0 & * & 0 & 0 \\ 0 & * & 0 & * \\ 0 & 0 & 0 & * \end{bmatrix}, \tag{5.91}$$

then any of the submatrices

$$\begin{bmatrix} 0 & * & 0 & 0 \\ 0 & * & 0 & * \end{bmatrix}, \quad \begin{bmatrix} 0 & * & 0 & 0 \\ 0 & 0 & 0 & * \end{bmatrix}, \quad \begin{bmatrix} 0 & * & 0 & * \\ 0 & 0 & 0 & * \end{bmatrix} \tag{5.92}$$

is a candidate for $\tilde{M}^r$. However, $\tilde{M}^c$ is the unique submatrix composed of the second and fourth column of $\tilde{M}$. We also use $\mathbf{D}(\tilde{A}, \tilde{B}, \tilde{C})$ to identify the digraph $\mathbf{D}$ of a structured system $\tilde{\mathbf{S}}$ having the triple $(\tilde{A}, \tilde{B}, \tilde{C})$.

The main result of Hayakawa and Šiljak (1988) is:

5.23. THEOREM. A structured interconnected system $\tilde{\mathbf{S}}$ is decentrally connectively stabilizable if one of the following conditions hold:

(*i*) $\tilde{\mathbf{S}}_1$ is structurally complete, and the digraphs $\mathbf{D}(\tilde{A}_2, \tilde{B}_2, \tilde{A}_{12}^r)$ and $\mathbf{D}(\tilde{A}_2, \tilde{A}_{21}^c, \tilde{C}_2)$ satisfy conditions (*i*) and (*ii*) of Theorem 5.15, respectively;

(*ii*) $\tilde{\mathbf{S}}_2$ is structurally complete, and the digraphs $\mathbf{D}(\tilde{A}_1, \tilde{B}_1, \tilde{A}_{21}^r)$ and $\mathbf{D}(\tilde{A}_1, \tilde{A}_{12}^c, \tilde{C}_1)$ satisfy conditions (*i*) and (*ii*) of Theorem 5.15, respectively.

Proof. From Lemma 5.21 and Definition 5.22, it is easy to see that the structured interconnected system $\mathbf{S}$ is decentrally connectively stabilizable if one of the following two conditions holds:

(*i*) $\mathbf{S}_1$ is structurally complete and

$$\mathcal{R}^*_{b,\mathrm{Ker}\ A^r_{12}} \overset{*}{=} \mathcal{X}_2, \qquad \mathcal{N}^*_{b,\mathrm{Im}\ A^c_{21}} \overset{*}{=} \{0\}, \qquad \text{w. r. t.} \quad \mathbf{S}_2; \tag{5.93}$$

(*ii*) $\mathbf{S}_2$ is structurally complete and

$$\mathcal{R}^*_{b,\mathrm{Ker}\ A^r_{21}} \overset{*}{=} \mathcal{X}_1, \qquad \mathcal{N}^*_{b,\mathrm{Im}\ A^c_{12}} \overset{*}{=} \{0\}, \qquad \text{w. r. t.} \quad \mathbf{S}_1. \tag{5.94}$$

Notice that (5.93) and (5.94) are characterized in terms of associated system graphs in Theorem 5.15. Therefore, the assertion is trivial. Q.E.D.

5.24. EXAMPLE. To illustrate the application of Theorem 5.23, let us consider a system $\tilde{\mathbf{S}}$ described by the following triple $(\tilde{A}, \tilde{B}, \tilde{C})$:

$$\tilde{A} = \left[\begin{array}{c|c} \tilde{A}_1 & \tilde{A}_{12} \\ \hline \tilde{A}_{21} & \tilde{A}_2 \end{array}\right] = \left[\begin{array}{ccc|ccc} 0 & 0 & 0 & 0 & 0 & 0 \\ * & 0 & * & * & 0 & 0 \\ 0 & 0 & 0 & * & 0 & 0 \\ \hline 0 & 0 & 0 & 0 & 0 & * \\ * & 0 & * & * & 0 & 0 \\ 0 & 0 & 0 & 0 & * & 0 \end{array}\right],$$

$$\tilde{B} = \left[\begin{array}{c|c} \tilde{B}_1 & 0 \\ \hline 0 & \tilde{B}_2 \end{array}\right] = \left[\begin{array}{cc|c} * & 0 & 0 \\ 0 & 0 & 0 \\ 0 & * & 0 \\ \hline 0 & 0 & 0 \\ 0 & 0 & 0 \\ 0 & 0 & * \end{array}\right], \tag{5.95}$$

$$\tilde{C} = \left[\begin{array}{c|c} \tilde{C}_1 & 0 \\ \hline 0 & \tilde{C}_2 \end{array}\right] = \left[\begin{array}{ccc|ccc} 0 & * & 0 & 0 & 0 & 0 \\ 0 & 0 & * & 0 & 0 & 0 \\ \hline 0 & 0 & 0 & 0 & 0 & * \end{array}\right].$$

The corresponding digraph $\mathbf{D}(\tilde{A}, \tilde{B}, \tilde{C})$ is given in Figure 5.4.

It is easy to see that

$$\tilde{A}_{21}^r = [* \quad 0 \quad *], \qquad \tilde{A}_{12}^c = \begin{bmatrix} 0 \\ * \\ * \end{bmatrix}. \tag{5.96}$$

Therefore, we get the digraphs $\mathbf{D}(\tilde{A}_1, \tilde{B}_1, \tilde{A}_{21}^r)$ and $\mathbf{D}(\tilde{A}_1, \tilde{A}_{12}^c, \tilde{C}_1)$ as shown in Figure 5.5. Since $\tilde{\mathbf{S}}_2$ is structurally complete, from Figure 5.5 and condition (ii) of Theorem 5.23, we conclude that the system $\tilde{\mathbf{S}}$ specified by (5.95) is decentrally connectively stabilizable. We note that the digraph $\mathbf{D}(\tilde{A}, \tilde{B}, \tilde{C})$ of Figure 5.4 does not satisfy condition (i) of Theorem 5.23.

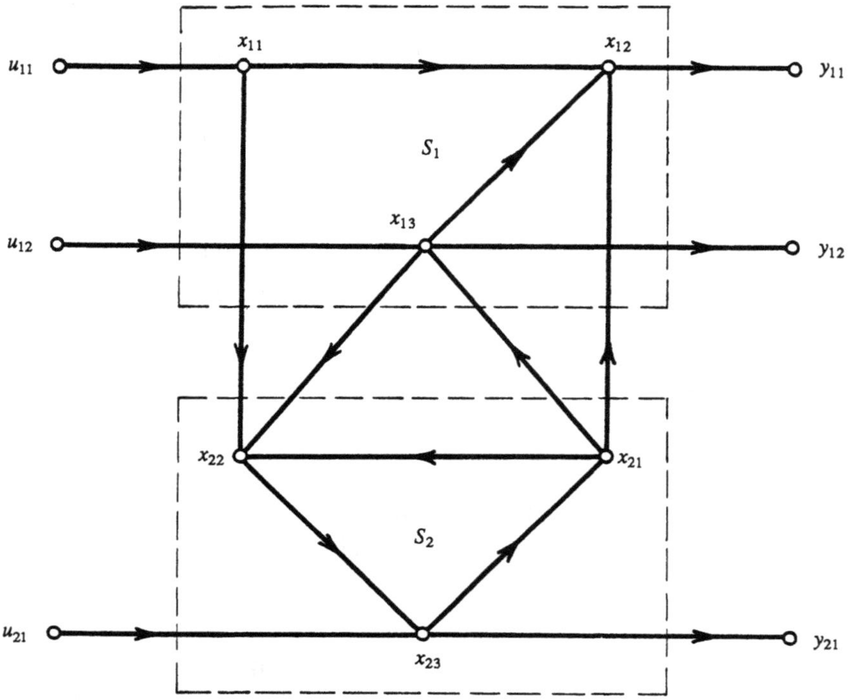

Fig. 5.4. Digraph $\mathbf{D}(\tilde{A}, \tilde{B}, \tilde{C})$ for Example 5.24.

5.5. Adaptive Control

We have shown that an effective way to handle uncertainty in complex systems is to use decentralized feedback. So far, however, feedback gains have been constant even though systems have been nonlinear and time-varying. Gain settings have been usually high to meet the worst case constraints. A considerable improvement in the gain allocation as well as performance of the closed-loop system can be achieved using adaptive decentralized feedback.

Let us consider again the system $\mathbf{S}$ composed of N subsystems

$$\begin{aligned} \mathbf{S}_i\colon\ \dot{x}_i &= A_i x_i + b_i u_i + P_i v_i, \\ y_i &= c_i^T x_i, \\ w_i &= Q_i x_i, \qquad i \in \boldsymbol{N}, \end{aligned} \tag{5.1}$$

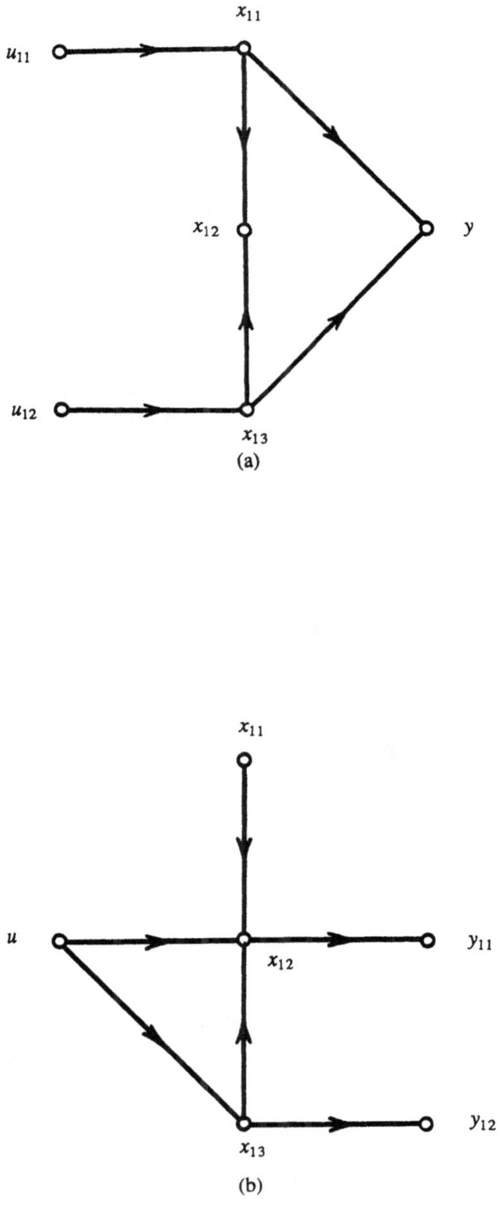

Fig. 5.5. Digraph for Example 5.24. (*a*) $(\tilde{A}_1, \tilde{B}_1, \tilde{A}_{21}^r)$; (*b*) $\mathbf{D}(\tilde{A}_1, \tilde{A}_{12}^c, \tilde{C}_1)$.

where the interaction function f: $\mathbf{R}\times\mathbf{R}^{\ell} \to \mathbf{R}^{m_i}$ is defined now as a function of time as well:

$$v_i = f_i(t, w), \qquad i \in \boldsymbol{N}. \tag{5.97}$$

The overall system $\mathbf{S}$ can be written in a compact form

$$\begin{aligned} \mathbf{S}\colon\ \dot{x} &= Ax + Bu + Pv, \\ y &= Cx, \\ w &= Qx, \end{aligned} \tag{5.98}$$

where $x(t) \in \mathbf{R}^n$, $u(t) \in \mathbf{R}^N$, and $y(t) \in \mathbf{R}^N$ are the state $x = (x_1^T, x_2^T, \dots, x_N^T)^T$, input $u = (u_1^T, u_2^T, \dots, u_N^T)^T$, and output $y = (y_1^T, y_2^T, \dots, y_N^T)^T$ of $\mathbf{S}$ at time $t \in \mathbf{R}$. The constant block diagonal matrices

$$\begin{aligned} A &= \mathrm{diag}\{A_1, A_2, \dots, A_N\}, \\ B &= \mathrm{diag}\{b_1, b_2, \dots, b_N\}, \\ C &= \mathrm{diag}\{c_1^T, c_2^T, \dots, c_N^T\}, \\ P &= \mathrm{diag}\{P_1, P_2, \dots, P_N\}, \\ Q &= \mathrm{diag}\{Q_1, Q_2, \dots, Q_N\} \end{aligned} \tag{5.99}$$

have appropriate dimensions. The vectors $v(t) \in \mathbf{R}^m$ and $w(t) \in \mathbf{R}^{\ell}$, which are defined as $v = (v_1^T, v_2^T, \dots, v_N^T)^T$ and $w = (w_1^T, w_2, \dots, w_N^T)^T$, are interconnected as

$$v = f(t, w). \tag{5.100}$$

The function f: $\mathbf{R} \times \mathbf{R}^{\ell} \to \mathbf{R}^m$ is continuous and bounded in both arguments, and sufficiently smooth so that the solution $x(t; t_0, x_0)$ of $\mathbf{S}$ is unique for all initial conditions $(t_0, x_0) \in \mathbf{R} \times \mathbf{R}^n$ and all piecewise continuous inputs $u(\bullet)$. In particular, we assume that there exist nonnegative, but unknown, numbers ξ_{ij} such that

$$\| f_i(t, w)\| \le \sum_{j=1}^{N} \xi_{ij} \| w_j \| \qquad \forall (t, w) \in \mathbf{R} \times \mathbf{R}^{\ell}, \qquad i \in \boldsymbol{N}. \tag{5.101}$$

We assume that the dynamics of the subsystems $\mathbf{S}_i$ are also unknown. That is, the matrices A_i and the vectors b_i and c_i are not specified, save for the fact that the pairs $(A_i,\, b_i)$ are controllable and the pairs $(A_i,\, c_i)$ are observable. This assumption guarantees a choice of the coordinate system such that the matrices of the decoupled systems have the companion form

$$A_i = \begin{bmatrix} 0 & 1 & 0 & \dots & 0 \\ 0 & 0 & 1 & \dots & 0 \\ \dots & \dots & \dots & \dots & \\ -a_1^i & -a_2^i & -a_3^i & \dots & -a_{n_i}^i \end{bmatrix}, \qquad b_i = \begin{bmatrix} 0 \\ \vdots \\ 0 \\ b^i \end{bmatrix}, \tag{5.102}$$

$$c_i^T = [\, c_1^i \quad c_2^i \quad \dots \quad c_{n_i}^i \,].$$

The uncertainty about subsystems $\mathbf{S}_i$, which is reduced to the unspecified coefficients of A_i, b_i, and c_i, compounds the essential uncertainty about the overall system $\mathbf{S}$ caused by our ignorance of the interconnections.

A measure of performance for $\mathbf{S}$ is provided by a set of local reference models

$$\begin{aligned} \mathbf{M}_i\colon \ \dot{x}_{mi} &= A_{mi} x_{mi} + b_{mi} r_i, \\ y_{mi} &= c_{mi}^T x_{mi}, \qquad i \in \boldsymbol{N}, \end{aligned} \tag{5.103}$$

which have the same dimensions as the corresponding subsystems $\mathbf{S}_i$, as well as stability characteristics we would like the subsystems to have. By stacking up the individual models $\mathbf{M}_i$, we obtain a reference model $\mathbf{M}$ for the overall system $\mathbf{S}$:

$$\begin{aligned} \mathbf{M}\colon \ \dot{x}_m &= A_m x_m + b_m r, \\ y_m &= c_m^T x_m, \end{aligned} \tag{5.104}$$

where $r(t) \in \mathbf{R}^N$ is the external (reference) input of $\mathbf{M}$. As in (5.99), we have

$$\begin{aligned} A_m &= \operatorname{diag}\{A_{m1},\, A_{m2},\, \dots,\, A_{mN}\}, \\ B_m &= \operatorname{diag}\{b_{m1},\, b_{m2},\, \dots,\, b_{mN}\}, \\ C_m &= \operatorname{diag}\{c_{m1}^T,\, c_{m2}^T,\, \dots,\, c_{mN}^T\}. \end{aligned} \tag{5.105}$$

Local coordinates can be chosen so that the triples $(A_{mi},\, b_{mi},\, c_{mi}^T)$ are in the companion (controller) canonical form as in (5.102). Obviously, in these coordinate frames we can choose vectors $k_i^* \in \mathbf{R}^{n_i}$, $k_{0i}^* \in \mathbf{R}$ such that

$$A_{mi} = A_i + b_i k_i^{*T}, \qquad b_{mi} = b_i k_{0i}^*. \tag{5.106}$$

We do not know these model-matching parameters simply because we do not know the subsystems dynamics. All we assume is that we know the sign of b^i and, without loss of generality, can set k_{0i}^* for all $i \in \boldsymbol{N}$. This is all well, because the presence of interconnections makes the ideal settings usually higher than those of isolated subsystems. This fact suggests that a decoupled set of local models may not be the best model for the overall system. Interconnected models were used in an early paper by Yoshikawa and Ikeda (1983), and may yet prove to result in a better performance of an adaptive system when nominal interconnections are different from zero and known *a priori*.

In this section, we start with the *state regulation problem* of a class of decentralized stabilizable systems, whose coupling parameters are within the range of the control inputs. This means that each matrix P_i can be factored as

$$P_i = b_i p_i^T, \qquad i \in \boldsymbol{N}, \tag{5.107}$$

for some constant vector $p_i \in \mathbf{R}^{m_i}$. This is the structural condition (5.4), which is essential for decentralized adaptive stabilizability when the size of interconnections is unknown.

To drive the state of $\mathbf{S}$ to zero we choose local state control laws

$$u_i = \theta_i^T x_i, \qquad i \in \boldsymbol{N}, \tag{5.108}$$

where $\theta_i(t) \in \mathbf{R}^{n_i}$ is the time-varying adaptation gain vector at time $t > t_0$. To arrive at a suitable adaptation law for $\theta_i(t)$, we choose the A_{mi}'s as stable matrices, so that for any symmetric positive definite matrix G_i there exists the unique symmetric positive definite matrix H_i as the solution of the Liapunov's matrix equation

$$A_{mi}^T H_i + H_i A_{mi} = G_i. \tag{5.109}$$

Then, we define $\bar{k}_i \in \mathbf{R}^{n_i}$ as

$$\bar{k}_i = H_i b_{mi}, \tag{5.110}$$

and propose to use the local adaptation laws

$$\dot{\theta}_i = -\Gamma_i \left(\bar{k}_i^T x_i\right) x_i, \tag{5.111}$$

where Γ_i is an $n_i \times n_i$ constant symmetric positive definite matrix and $\theta_i(t_0)$ is finite.

Finally, the closed-loop system is described as

$$\begin{aligned}\hat{\mathbf{S}}\colon\ \dot{x}_i &= (A_{mi} + b_i\phi_i^T)x_i + b_i p_i^T f_i,\\ \dot{\phi}_i &= -\Gamma_i(\bar{k}_i^T x_i)x_i,\end{aligned} \tag{5.112}$$

where $\phi_i(t) \in \mathbf{R}^{n_i}$ is the ith parameter adaptation error at time $t \in \mathbf{R}$, which is defined as

$$\phi_i = \theta_i - \theta_i^*, \tag{5.113}$$

and $\theta_i^* = k_i^*$ with k_i^* chosen as in (5.106). Relying on the decentralized stabilizability ensured by condition (5.107), which was established in Section 5.1, we define $\phi = (\phi_1^T, \phi_2^T, \ldots, \phi_N^T)^T$, denote the state of $\hat{\mathbf{S}}$ as $(x, \phi) \in \mathbf{R}^n \times \mathbf{R}^n$, and prove the following:

5.25. THEOREM. The solutions $(x, \phi)(t; t_0, x_0, \phi_0)$ of the system $\hat{\mathbf{S}}$ are globally bounded, and $x(t; t_0, x_0, \phi_0) \to 0$ as $t \to \infty$.

Proof. We define a function $\nu\colon \mathbf{R}^n \times \mathbf{R}^n \to \mathbf{R}_+$ as

$$\nu(x, \phi) = \sum_{i=1}^{N} [k_{0i}^* x_i^T H_i x_i + (\phi_i + \rho\bar{k}_i)^T \Gamma_i^{-1}(\phi_i + \rho\bar{k}_i)], \tag{5.114}$$

where the number $\rho > 0$ is not yet specified. The total time derivative $\dot{\nu}(x, \phi)$ of the function $\nu(x, \phi)$, which is computed with respect to (5.112), is obtained as

$$\dot{\nu}(x, \phi) = -x^T G x - 2\rho x^T \bar{K}^T \bar{K} x + 2x^T \bar{K}^T \tilde{P} f, \tag{5.115}$$

where

$$\begin{aligned} G &= \mathrm{diag}\{k_{01}^* G_1, k_{02}^* G_2, \ldots, k_{0N}^* G_N\},\\ \tilde{P} &= \mathrm{diag}\{p_1^T, p_2^T, \ldots, p_N^T\},\\ \bar{K} &= \mathrm{diag}\{\bar{k}_1^T, \bar{k}_2^T, \ldots, \bar{k}_N^T\}.\end{aligned} \tag{5.116}$$

Completing the square in (5.115), we get

$$\begin{aligned}\dot{\nu}(x, \phi) = &-x^T G x - \rho x^T \bar{K}^T \bar{K} x - \rho(\bar{K}x - \rho^{-1}\tilde{P}f)^T(\bar{K}x - \rho^{-1}\tilde{P}f)\\ &+ \rho^{-1} f^T \tilde{P}^T \tilde{P} f.\end{aligned} \tag{5.117}$$

By using the boundedness assumption (5.101) concerning $f(t, w)$, we get from (5.117) the inequality

$$\dot{\nu}(x, \phi) \leq -\left[\lambda_m(G) - \rho^{-1}\xi^2\|\tilde{P}\|^2\,\|Q\|^2\right]\,\|x\|^2, \qquad \forall (t, x, \phi) \in \mathbf{R}\times\mathbf{R}^n\times\mathbf{R}^n, \tag{5.118}$$

where $\lambda_m(G)$ is, as usual, the minimum eigenvalue of G, and $\xi = N\max_{i,j\in N}\,\xi_{ij}$. By selecting a sufficiently large number ρ^* so that

$$\rho^* > \lambda_m^{-1}(G)\xi^2\|\tilde{P}\|^2\,\|Q\|^2, \tag{5.119}$$

we get $\dot{\nu}(x, \phi) \leq 0$ for all $(t, x, \phi) \in \mathbf{R}\times\mathbf{R}^n\times\mathbf{R}^n$. Using standard arguments from the Liapunov's theory (*e.g.*, Rouche *et al.* 1977), we conclude that the solutions $(x, \phi)(t;\, t_0, x_0, \phi_0)$ are bounded.

From the boundedness of solutions and definition (5.114) of $\nu(x, \phi)$, we conclude that $\dot{\nu}(x, \phi)$ is uniformly continuous on $\mathbf{R}\times\mathbf{R}^n\times\mathbf{R}^n$. Furthermore, the function $\nu(t) = \nu[x(t), \phi(t)]$ is decreasing and is bounded from below. Hence,

$$\lim_{t\to\infty} \nu(t) = \inf_t \nu(t) = \nu_f \geq 0. \tag{5.120}$$

Denoting $\dot{\nu}(t) = \dot{\nu}[x(t), \phi(t)]$, we have

$$\lim_{t\to\infty} \int_{t_0}^{t} \dot{\nu}(\tau)\,d\tau = \nu_f - \nu_0 < \infty, \tag{5.121}$$

where $\nu_0 = \nu(x_0)$. Then, from (5.114), (5.118), and uniform continuity of $\dot{\nu}(t)$, we have $\lim_{t\to\infty}\dot{\nu}(t) = 0$, and $\lim_{t\to\infty} x(t;\, t_0, x_0, \phi_0) = 0$ for all $(t_0, x_0, \phi_0) \in \mathbf{R}\times\mathbf{R}^n\times\mathbf{R}^n$. Q.E.D.

5.26. REMARK. First, note that we did not use the M-matrix conditions of Theorem 2.4, because the structural condition (5.107) is all we need for stability. Second, it is not necessary to know or choose the ρ^* of (5.119). The local gains rise to whatever level is necessary to compensate for the effects of fluctuating size of interconnections. These are pleasing facts when the size and characterization of interconnections are neither known nor can be predicted during the operation of the overall system.

5.27. REMARK. The boundedness part of Theorem 5.25 provides for realizability of the proposed adaptive scheme.

In the *state tracking problem*, it is required that the state $x(t)$ of the plant $\mathbf{S}$ follows the state of the reference model $\mathbf{M}$ despite the fluctuating interaction levels among the subsystems $\mathbf{S}_i$. Let the tracking error be defined as

$$e(t) = x(t) - x_m(t), \tag{5.122}$$

where $e = (e_1^T, e_2^T, \ldots, e_N^T)^T$ and $e_i(t) \in \mathbf{R}^{n_i}$ is the tracking error of the subsystem $\mathbf{S}_i$ at $t \in \mathbf{R}$. To drive the error toward zero, we use the local control laws

$$u_i = \theta_i^T v_i, \tag{5.123}$$

where

$$\theta_i = (\hat{k}_i^T, \hat{k}_{0i})^T, \tag{5.124}$$

and

$$v_i = (e_i^T, r_i^T)^T. \tag{5.125}$$

The gains $\hat{k}_i$ and $\hat{k}_{0i}$ represent the estimates of k_i^* and k_{0i}^*.

5.28. REMARK. In this scheme, it is crucial to utilize, in the state regressor v_i, the tracking error e_i as proposed in (Gavel and Šiljak, 1989), rather than the state x_i, as it was used in centralized (Narendra and Valavani, 1978) and decentralized (Ioannou, 1986) adaptive control.

The resulting closed-loop system is

$$\hat{\mathbf{S}}\colon\ \dot{x}_i = A_{mi}x_i + b_{mi}r_i + b_i(\phi_i^T v_i - k_i^{*T} x_{mi} + p_i^T f_i), \qquad i \in \boldsymbol{N}, \tag{5.126}$$

where we have defined $\phi_i(t)$ as in (5.113), and

$$\theta_i^* = (k_i^{*T}, k_{0i}^*)^T. \tag{5.127}$$

Subtracting (5.104) from (5.126), we can write the error system as

$$\begin{aligned} \hat{\mathbf{S}}_e\colon\ \dot{e}_i &= A_{mi}e_i + b_i(\phi_i^T v_i - k_i^{*T} x_{mi} + p_i^T f_i), \\ \dot{\phi}_i &= -\sigma\Gamma_i\phi_i - \Gamma_i(\bar{k}_i^T e_i)v_i - \sigma\Gamma_i\theta_i^*, \qquad i \in \boldsymbol{N}, \end{aligned} \tag{5.128}$$

where $\sigma > 0$. The adaptation law was chosen as

$$\dot{\theta}_i = -\Gamma_i(\bar{k}_i^T e_i)v_i - \sigma\Gamma_i\theta_i, \qquad i \in \boldsymbol{N}, \tag{5.129}$$

which incorporates the "σ-modification," as originally suggested by Ioannou and Kokotović (1984). By $(e, \phi) \in \mathbf{R}^n \times \mathbf{R}^n$, we denote the state of $\hat{\mathbf{S}}_e$ and prove a result of Gavel and Šiljak (1989).

5.29. THEOREM. The solutions $(e, \phi)(t; t_0, e_0, \phi_0)$ of the system $\hat{\mathbf{S}}_e$ are globally ultimately bounded.

Proof. We use the same type of function ν: $\mathbf{R}^n \times \mathbf{R}^n \to \mathbf{R}_+$ as in (5.114):

$$\nu(e, \phi) = \sum_{i=1}^{N} \left[k_{0i}^* e_i^T H_i e_i + (\phi_i + \rho\bar{\theta}_i)^T \Gamma_i^{-1}(\phi_i + \rho\bar{\theta}_i)\right], \tag{5.130}$$

where $\bar{\theta}_i = (\bar{k}_i^T, 0) \in \mathbf{R}^{n_i+1}$, and ρ is a positive number. Using (5.128), we compute

$$\begin{aligned}\dot{\nu}(e, \phi) = \sum_{i=1}^{N} \Bigg[& - k_{0i}^* e_i^T G_i e_i - 2e_i^T \bar{k}_i k_i^{*T} x_{mi} \\ & + 2\sum_{j=1}^{N} e_i^T \bar{k}_i p_i^T f_i - 2\rho e_i^T \bar{k}_i \bar{k}_i^T e_i - 2\sigma(\phi_i + \rho\bar{\theta}_i)^T \theta_i \Bigg],\end{aligned} \tag{5.131}$$

where we have used the fact that $\bar{\theta}_i^T \nu_i = \bar{k}_i^T e_i$. Utilizing the block notation (5.116), (5.131) can be written as

$$\begin{aligned}\dot{\nu}(e, \phi) = & - e^T G e - 2e^T \bar{K}^T K^* x_m + 2e^T \bar{K}^T \tilde{P} f - 2\rho e^T \bar{K}^T \bar{K} e \\ & - 2\sigma(\phi + \rho\bar{\theta})^T (\phi + \rho\bar{\theta}) - 2\sigma(\phi + \rho\bar{\theta})^T (\theta^* - \rho\bar{\theta}),\end{aligned} \tag{5.132}$$

where $K^* = \text{diag}\{k_1^{*T}, k_2^{*T}, \ldots, k_N^{*T}\}$. Since $f(t, w) = f[t, Q(e + x_m)]$ is bounded as in (5.101), we can use

$$\| f_i(t, w)\| \leq \sum_{j=1}^{N} \xi_{ij} \| Q_j\| \left(\| e_i\| + \| x_{mi}\|\right) \tag{5.133}$$

in (5.132) to get the inequality

$$\begin{aligned}\dot{\nu}(e,\,\phi) \le &- e^T G e - 2\rho\|\bar{K}e\|^2 + 2\|\bar{K}e\|\;\|K^{*T}x_m\| \\ &+ 2\xi\|\bar{K}e\|\;\|\tilde{P}\|\;\|Q\|\;(\|e\| + \|x_m\|) - 2\sigma\|\phi + \rho\bar{\theta}\|^2 \\ &+ 2\sigma\|\phi + \rho\bar{\theta}\|\;\|\theta^* - \rho\bar{\theta}\|, \qquad \forall(t,\,e,\,\phi) \in \mathbf{R}\times\mathbf{R}^n\times\mathbf{R}^n.\end{aligned} \tag{5.134}$$

After completing the squares involving $\bar{K}e$ and $\phi + \rho\bar{\theta}$, and dropping negative terms, we obtain

$$\begin{aligned}\dot{\nu}(e,\,\phi) \le &- \left[\lambda_m(G) - \rho^{-1}\xi^2\|\tilde{P}\|^2\,\|Q\|^2\right]\,\|e\|^2 - \sigma\|\phi + \rho\bar{\theta}\|^2 \\ &+ \rho^{-1}\chi\left(\|K^*\| + \xi\|\tilde{P}\|\;\|Q\|\right)^2 + \sigma\|\theta^* - \rho\bar{\theta}\|^2 \\ &\qquad \forall(t,\,e,\,\phi) \in \mathbf{R}\times\mathbf{R}^n\times\mathbf{R}^n,\end{aligned} \tag{5.135}$$

where $\chi = \sup_t \|x_m(t)\|^2$.

Equation (5.135) can be written compactly as

$$\dot{\nu}(e,\,\phi) \le -\zeta\|e\|^2 - \sigma\|\phi + \rho\bar{\theta}\|^2 + \eta, \qquad \forall(t,\,e,\,\phi) \in \mathbf{R}\times\mathbf{R}^n\times\mathbf{R}^n, \tag{5.136}$$

where the constants ζ and η are defined by

$$\begin{aligned}\zeta &= \lambda_m(G) - \rho^{-1}\xi^2\|\tilde{P}\|^2\,\|Q\|^2, \\ \eta &= \rho^{-1}\chi\left(\|K^*\| + \xi\|\tilde{P}\|\;\|Q\|\right)^2 + \sigma\|\theta^* - \rho\bar{\theta}\|^2.\end{aligned} \tag{5.137}$$

Selecting ρ^* large enough so that $\zeta > 0$, we see that (5.136) implies

$$\dot{\nu}(e,\,\phi) \le \mu\nu(e,\,\phi) + \eta, \qquad \forall(t,\,e,\,\phi) \in \mathbf{R}\times\mathbf{R}^n\times\mathbf{R}^n, \tag{5.138}$$

where the positive number μ is given by

$$\mu \le \min\{\lambda_M^{-1}(H)\zeta,\; \lambda_m(\Gamma)\sigma\}, \tag{5.139}$$

and $\Gamma = \text{diag}\{\Gamma_i, \Gamma_2, \ldots, \Gamma_N\}$. From (5.138), it is clear that $\nu(e, \phi)$ decreases monotonically along any solution of $\hat{\mathbf{S}}_e$ until the solution reaches the compact set

$$\Omega_f = \{(e, \phi) \in \mathbf{R}^n \times \mathbf{R}^n \colon \nu(e, \phi) \leq \nu_f\}, \tag{5.140}$$

where

$$\nu_f = \mu^{-1}\eta. \tag{5.141}$$

Therefore, the solutions $(e, \phi)(t; t_0, e_0, \phi_0)$ of $\hat{\mathbf{S}}_e$ are globally ultimately bounded with respect to the bound ν_f. Q.E.D.

5.30. REMARK. Within Ω, we find that

$$\|e\| \leq \left[\lambda_m^{-1}(H)\nu_f\right]^{1/2}. \tag{5.142}$$

Now, if we choose $\sigma \propto \rho^{-3}$ and $\Gamma \propto \rho^3$, and let $\rho \to +\infty$, we find that $\zeta \to \lambda_m(G)$, $\eta \sim \rho^{-1}$, and, therefore, $\nu_f \sim \rho^{-1}$. This implies that the upper bound on the steady-state tracking error $e(t)$ can be made as small as desired by decreasing σ and increasing Γ sufficiently. We should note, however, that within Ω,

$$\|\phi + \rho\bar{\theta}| \leq [\lambda_M(\Gamma)\nu_f]^{1/2}, \tag{5.143}$$

and the upper bound on $\|\phi + \rho\bar{\theta}\|$ increases proportionally to the increase of ρ. Thus, making σ small and Γ large will allow adaptation gains to become high, resulting in a trade-off between small tracking errors and large gains in the proposed high-gain decentralized adaptive scheme.

We also note that again, as in Remark 5.26, it is not necessary for the designer to know or choose ρ^*. Local gains adjust automatically to counter the destabilizing effect of interconnections. The designer need only to tune the size of the residual set by adjusting Γ_i and σ as explained in Remark 5.30. Of course, these benefits would not be guaranteed without the structural restrictions (5.107) placed on the interconnections.

5.31. EXAMPLE. To illustrate the decentralized adaptive control, we consider again the two inverted penduli coupled by a spring (Figure 1.16). Choosing $g/\ell = 1$, $1/m\ell^2 = 1$, and $k/m = 2$, we get the system matrices

$$A_i = \begin{bmatrix} 0 & 1 \\ 1 & 0 \end{bmatrix}, \quad b_i = \begin{bmatrix} 0 \\ 1 \end{bmatrix}, \quad P_i = \begin{bmatrix} 0 \\ 1 \end{bmatrix}, \quad Q_i = [1 \quad 0], \qquad i = 1, 2, \tag{5.144}$$

or, in the block notation (5.98),

$$A = \left[\begin{array}{cc|cc} 0 & 1 & & \\ 1 & 0 & \multicolumn{2}{c}{\bigcirc} \\ \hline \multicolumn{2}{c|}{\bigcirc} & 0 & 1 \\ & & 1 & 0 \end{array}\right], \quad B = P = \left[\begin{array}{c|c} 0 & \\ & 0 \\ 1 & \\ \hline & 0 \\ 0 & \\ & 1 \end{array}\right], \quad Q = \left[\begin{array}{cc|cc} 1 & 0 & \multicolumn{2}{c}{\bigcirc} \\ \hline \multicolumn{2}{c|}{\bigcirc} & 1 & 0 \end{array}\right]. \tag{5.145}$$

As interconnection function, we choose

$$f(t,\, w) = 2\alpha(t) \begin{bmatrix} -1 & 1 \\ 1 & -1 \end{bmatrix} w, \tag{5.146}$$

where $\gamma(t) = a^2(t)/\ell^2$ and $w = (w_1,\, w_2)^T$. The function $a(t)$ is uncertain, and all we know is that $a(t)/\ell \in [0,\, 1]$.

Since $P_i = b_i$, $i = 1,\, 2$, the structural condition (5.107) for stabilizability of **S** by fixed decentralized feedback is satisfied. To build adaptive controllers, we choose the reference models (5.103) having the matrices

$$A_{mi} = \begin{bmatrix} 0 & 1 \\ -1 & -2 \end{bmatrix}, \quad b_{mi} = \begin{bmatrix} 0 \\ 1 \end{bmatrix}, \quad i = 1,\, 2, \tag{5.147}$$

and $\theta_i^* = (-2,\, -2,\, 1)^T$, $i = 1,\, 2$. Next, we choose G_i and compute H_i and $\bar{k}_i$ from (5.109) and (5.110) as

$$G_i = \begin{bmatrix} 1 & 0 \\ 0 & 1 \end{bmatrix}, \quad H_i = \begin{bmatrix} 1.5 & 0.5 \\ 0.5 & 0.5 \end{bmatrix}, \quad \bar{k}_i = \begin{bmatrix} 0.5 \\ 0.5 \end{bmatrix}, \quad i = 1,\, 2. \tag{5.148}$$

For the adaptation law (5.111), we select $\Gamma = I_3$, where I_3 is the 3×3 identity matrix, and set $\sigma = 0.01$.

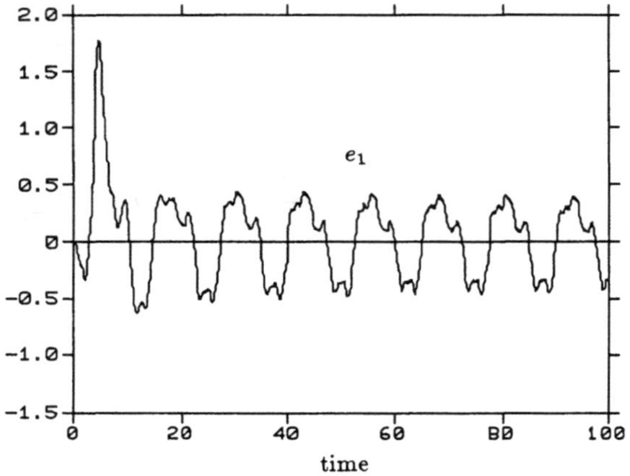

Fig. 5.6. Tracking error with the high-gain algorithm.

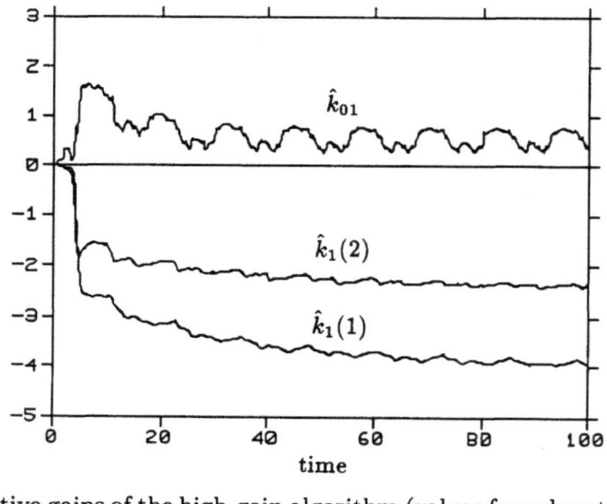

Fig. 5.7. Adaptive gains of the high-gain algorithm (values for subsystem 1 are shown).

With the chosen parameter settings, the results of a simulation are shown in Figures 5.6 and 5.7. The reference signals were

$$r_1(t) = \sin 20t + \sin 5t + \sin t,$$
$$r_2(t) = \sin 10t + \sin 2t + \sin 0.5t, \tag{5.149}$$

and $a(t)/\ell = 1$ for all time $t \geq t_0 = 0$. From the plots, we see that the presence of interconnections prevents the tracking errors from converging to zero and the feedback gains from converging to the model matching values. Instead, the gains tend toward a near steady-state level, $\hat{k}_1 = (-4.9, -3.2)^T$, which is higher than the local model-matching gains of $k_1^* = (-2, -2)^T$, which is indicative of the high-gain nature of the controller.

To contrast the result of Theorem 5.29 with the earlier approach (Ioannou, 1983) based upon the M-matrix conditions, we compute the 2×2 aggregate matrix $W = (w_{ij})$ as in Section 2.2:

$$w_{ij} = \begin{cases} -\lambda_m(G_i) + 2\lambda_M(H_i)\|P_i\|\xi_{ii}, & i = j, \\ 2\lambda_M(H_i)\|P_i\|\xi_{ij}, & i \neq j. \end{cases} \tag{5.150}$$

From the M-matrix conditions (2.25) applied on W, we derive the inequality

$$\frac{\lambda_m(G_i)}{\lambda_M(H_i)} > 4\|P_i\|\xi_{ij}, \qquad i = 1, 2. \tag{5.151}$$

Using the definition of ξ_{ij} in (5.101), we have $\xi_{ij} = 2$ for $i, j = 1, 2$. Then, from (5.151) we get

$$\frac{a(t)}{\ell} < \frac{1}{2}\left[\frac{m\lambda_m(G_i)}{k\|P_i\|\lambda_M(H_i)}\right]^{1/2} = 0.2706, \tag{5.152}$$

which means that by the M-matrix approach we guarantee stability only if the spring remains connected to the lower 27% of the length of the penduli. In our example, we allow $a(t)/\ell \equiv 1$ (the spring is moved all the way up to the bobs) making the M-matrix test inconclusive with regard to stability.

A simulation of the system using a controller design based on the M-matrix test was also performed in order to see the effect of regressor vectors $v_i = (x_i^T, r_i^T)^T$ that are different from those of (5.125), $v_i = (e_i^T, r_i^T)^T$, used in our example. The results are shown in Figures 5.8 and 5.9. The system appears stable, however, tracking residuals are considerably larger

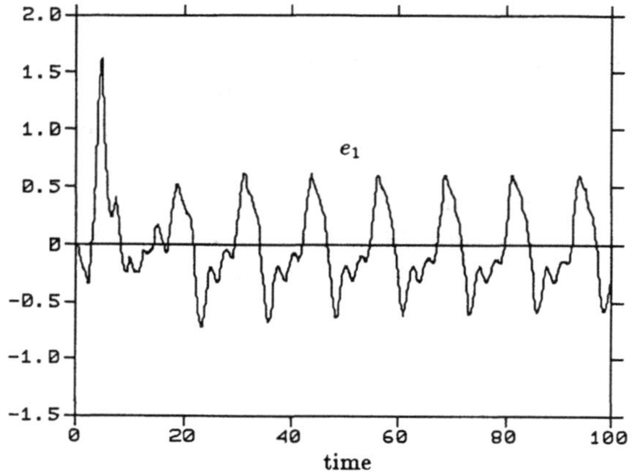

Fig. 5.8. Tracking error with the standard algorithm.

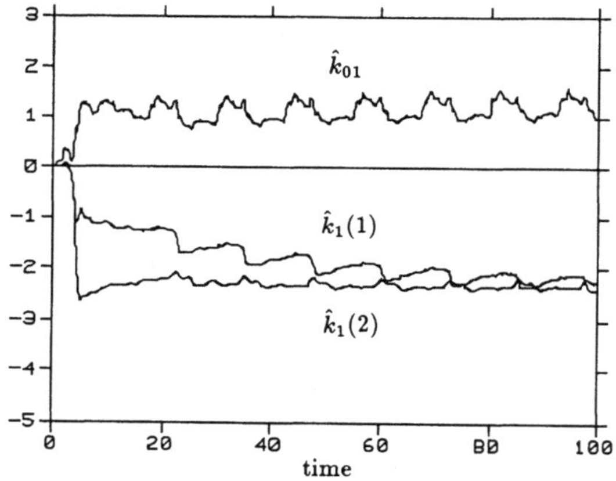

Fig. 5.9. Adaptive gains with the standard algorithm.

than before, indicating the fact that the high-gain approach may lead to a better-performing adaptive control scheme.

In another simulation run, we show the effect of a time-varying interconnection on adaptation gains and tracking residuals. We inject the jump

$$\frac{a(t)}{\ell} = \begin{cases} 0\,, & 0 \leq t < 50, \\ 1\,, & 50 \leq t \leq 100, \end{cases} \tag{5.153}$$

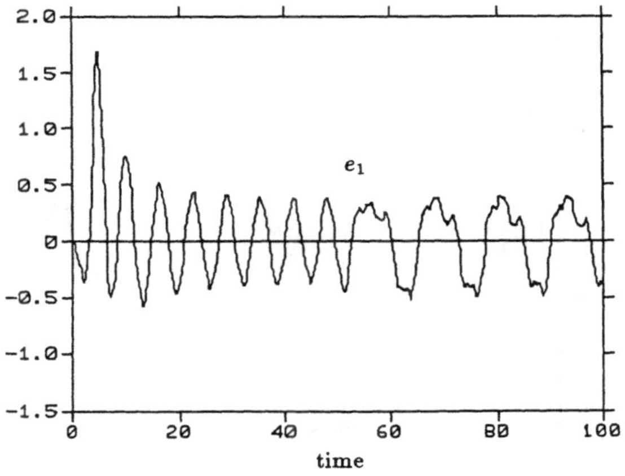

Fig. 5.10. Tracking error for the high-gain adaptive controller. Subsystems are isolated until $t = 50$, then interconnected.

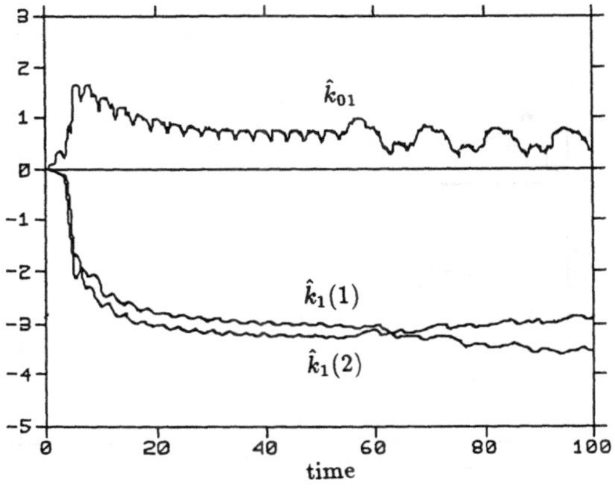

Fig. 5.11. Adaptive gains of the high-gain controller in response to increased interconnection strength at $t = 50$.

that is, the penduli are disconnected for the first half, and maximally interconnected for the second half of the experiment. Figures 5.10 and 5.11 show the results for the high-gain adaptive controllers, and Figures 5.12 and 5.13 display the results obtained by the standard adaptive scheme. The high-gain controllers maintain a tracking residual of about 0.5 throughout the run. As expected, the standard scheme behaves well when the subsystems are isolated. However, the residuals become large when the subsystems get

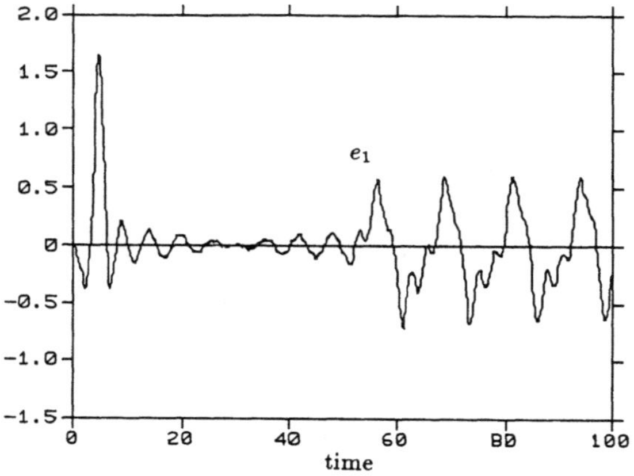

Fig. 5.12. Tracking error with the standard algorithm. Subsystems are isolated until $t = 50$, then interconnected.

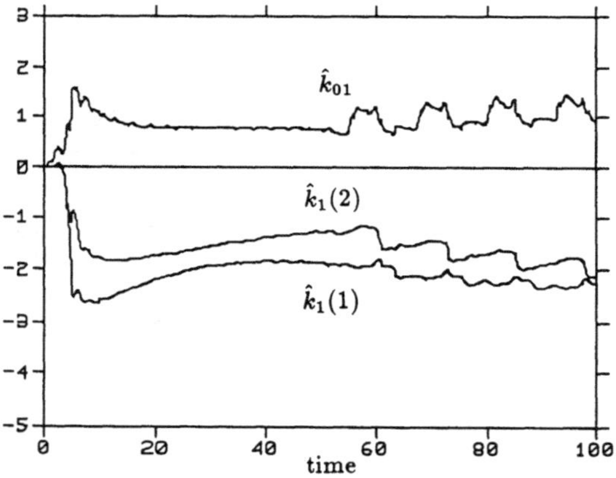

Fig. 5.13. Adaptive gains of the standard controller in response to increased interconnection strength at $t = 50$.

coupled. The high-gain adaptive controllers seem robust to time-varying interconnections, with local gains adjusting their values to maintain a consistent tracking residual.

Finally, we compare the proposed decentralized scheme to the standard centralized adaptive algorithm (Narendra and Valavani, 1978). For this purpose, we use the compact notation of the form (5.98) to describe the closed-loop system as

$$\hat{\mathbf{S}}: \ \dot{x} = Ax + Bx, \qquad u = Kx + K_0 r, \tag{5.154}$$

where

$$A = \begin{bmatrix} 0 & 1 & 0 & 0 \\ -1 & 0 & 2 & 0 \\ 0 & 0 & 0 & 1 \\ 2 & 0 & -1 & 0 \end{bmatrix}, \quad B = \begin{bmatrix} 0 & 0 \\ 1 & 0 \\ 0 & 0 \\ 0 & 1 \end{bmatrix}, \tag{5.155}$$

and K and K_0 are the gain matrices of appropriate dimensions. The model-matching gains are

$$K^* = \begin{bmatrix} 0 & -2 & -2 & 0 \\ -2 & 0 & 0 & -2 \end{bmatrix}, \quad K_0^* = \begin{bmatrix} 1 & 0 \\ 0 & 1 \end{bmatrix}. \tag{5.156}$$

Defining $\theta = (K, K_0)$ and $\nu = (x^T, r^T)^T$, the adaptation law is

$$\dot{\theta} = -\sigma\theta - \gamma \bar{K} e \nu, \tag{5.157}$$

where $\bar{K} = \text{diag}\{\bar{k}_1^T, \bar{k}_2^T\}$ as defined by (5.110), $e = (e_1^T, e_2^T)^T$, and σ and γ are positive numbers. For simulation purposes, we choose $\gamma = 1$, which corresponds to $\Gamma = I_3$ in our earlier runs. We set $\sigma = 0$ because there are no external disturbances or unmodeled interconnections, so that perfect model-matching is possible. Our reference signal is persistently exciting, so that $\phi(t) = [\theta(t) - \theta^*] \to 0$ as $t \to \infty$. Results are plotted in Figures 5.14 and 5.15 for $0 \leq t \leq 100$. The convergence rate appears to be roughly the same for the centralized and decentralized cases. Residual tracking errors are smaller (eventually zero) in the centralized case, which is to be expected since exact model-matching will occur. The centralized control law, however, requires twelve adaptation gains, compared to six in the decentralized case. In general, for the system composed of N interconnected subsystems, centralized controllers require $N(n + N)$ adaptation gains, where n is the total number of states. Decentralized controllers require only $n+N$ adaptation gains, and N sets of gain adaptation equations can be run in parallel. We are led to conclude that a decentralized adaptive controller has a far simpler algorithm than a centralized one, at the price of a relatively small decrease in performance.

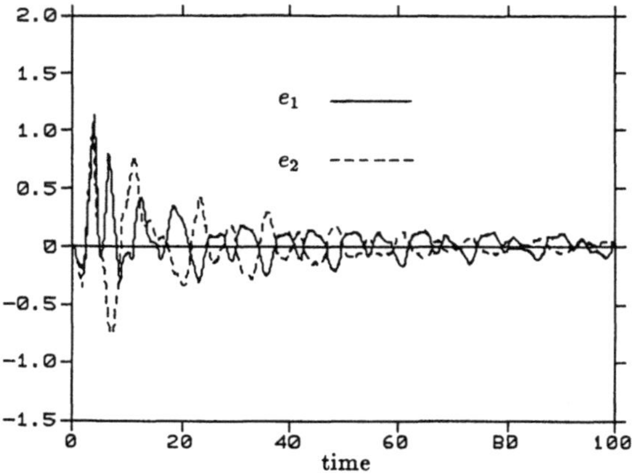

Fig. 5.14. Tracking error with centralized adaptive control.

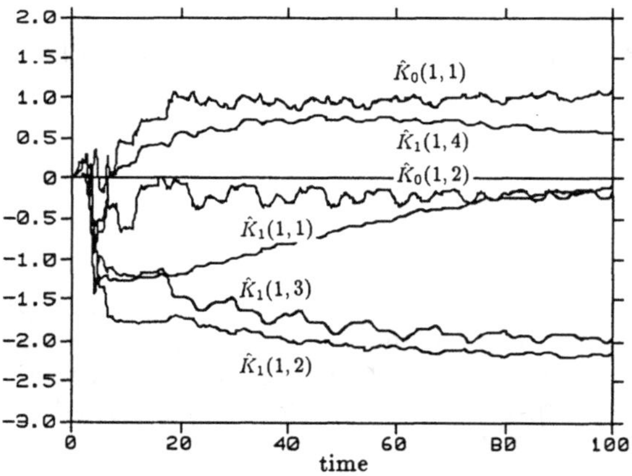

Fig. 5.15. Adaptive gains of the centralized adaptive controller.

5.6. Known Subsystems

In modeling of complex systems, it is standard (and natural!) to do it piece-by-piece leaving an essential uncertainty in the interactions among the system components or subsystems. Biased by this fact, one may look at the ignorance of subsystem dynamics, which was assumed in the preceding section, as an artifice to make the centralized algorithms work in the new decentralized setting rather than a genuine uncertainty about the subsystems. It is far from obvious, however, to see how one should take advantage of the knowledge about subsystem dynamics and come up with better adaptive schemes.

In this section, we outline an adaptive algorithm of (Gavel and Šiljak, 1989) that utilizes the knowledge of the subsystems to broaden the approach of the previous section and include Multi-Input–Multi-Output (MIMO) subsystems. Furthermore, the new algorithm is simpler to implement, requiring only one adaptation parameter per subsystem.

Consider the collection of MIMO subsystems

$$\begin{aligned} \mathbf{S}_i\colon\ \dot{x}_i &= A_i x_i + B_i u_i + P_i v_i, \\ y_i &= C_i x_i, \\ w_i &= Q_i x_i, \qquad\qquad i \in \boldsymbol{N}, \end{aligned} \tag{5.158}$$

where $x_i(t) \in \mathbf{R}^{n_i}$, $u_i(t) \in \mathbf{R}^{p_i}$, and $y_i(t) \in \mathbf{R}^{q_i}$ are the state, input, and output of the subsystem $\mathbf{S}_i$, and $v_i(t) \in \mathbf{R}^{m_i}$, and $w_i(t) \in \mathbf{R}^{\ell_i}$ are the interconnection inputs and outputs of $\mathbf{S}_i$ to and from other subsystems $\mathbf{S}_j$, $j \in \boldsymbol{N}$, at time $t \in \mathbf{R}$. We assume that the pair (A_i, B_i) is controllable, and the pair (A_i, C_i) is observable. The interconnection inputs and outputs v_i and w_i are related as before by interconnection functions defined in (5.97) and (5.101), where the numbers ξ_{ij} are fixed but unknown. Moreover, we assume that the interconnection matrices P_i are factorable as

$$P_i = B_i P_i', \tag{5.159}$$

for some constant (but not necessarily known) matrices P_i'. Our crucial assumption in this section is that the matrices A_i, B_i, and C_i are known.

Reference models are specified by first selecting local state feedback gain matrices $K_i \in \mathbf{R}^{p_i \times n_i}$ that would stabilize the local isolated subsystems. That is,

$$A_{mi} = A_i - B_i K_i, \qquad i \in \boldsymbol{N}, \tag{5.160}$$

is a stable matrix. The reference models are then described by

$$\begin{aligned} \mathbf{M}_i\colon\ \dot{x}_{mi} &= A_{mi} x_{mi} + B_i r_i, \\ y_{mi} &= C_i x_{mi}, \qquad i \in \boldsymbol{N}, \end{aligned} \tag{5.161}$$

where the r_i: $\mathbf{R} \to \mathbf{R}^{p_i}$ are bounded, piecewise continuous vector functions of time.

We now proceed with the design of an adaptive control law. The first step is to choose any symmetric positive definite $n_i \times n_i$ matrix G_i, and then find the symmetric positive definite solution H_i to the matrix Liapunov equation

$$A_{mi}^T H_i + H_i A_{mi} = -G_i. \tag{5.162}$$

A solution is guaranteed to exist since A_{mi} is stable. Now, choose a symmetric positive definite $p_i \times p_i$ matrix R_i, and define $\bar{K}_i$ as

$$\bar{K}_i = R_i^{-1} B_i^T H_i. \tag{5.163}$$

Using K_i of (5.160) and this $\bar{K}_i$, we formulate a local feedback control law

$$u_i(t) = -K_i x_i(t) - \alpha_i(t)\bar{K}_i e_i(t) + r_i(t), \tag{5.164}$$

where $\alpha_i(t)$ is a *scalar* adaptation gain at time $t > t_0$ obeying the adaptation law

$$\dot{\alpha}(t) = \gamma_i e_i^T(t)\bar{K}_i^T R_i \bar{K}_i e_i(t) - \sigma_i \alpha_i(t), \qquad \alpha_i(t_0) > 0, \tag{5.165}$$

γ_i and σ_i are positive constants, and $e_i(t) = x_i(t) - x_{mi}(t)$.

To analyze the stability of the above control scheme, we form an error differential equation by subtracting (5.161) from (5.158). Together with (5.164), this describes the motion of the entire system:

$$\begin{aligned} \hat{\mathbf{S}}_e\colon\ \dot{e}_i &= (A_{mi} - \alpha B_i \bar{K}_i)e_i + B_i P_i' v_i, \\ \dot{\alpha}_i &= \gamma_i e_i^T \bar{K}_i^T R_i \bar{K}_i e_i - \sigma_i \alpha_i, \qquad i \in \mathbf{N}. \end{aligned} \tag{5.166}$$

Now define $e = (e_1^T, e_2^T, \ldots, e_N^T)^T$, $\alpha = (\alpha_1, \alpha_2, \ldots, \alpha_N)^T$, and denote the state of $\hat{\mathbf{S}}_e$ as $(e, \alpha) \in \mathbf{R}^n \times \mathbf{R}^N$.

5.32. THEOREM. The solutions $(e, \alpha)\ (t; t_0, e_0, \alpha_0)$ of the system $\hat{\mathbf{S}}_e$ are globally ultimately bounded.

Proof. Using the Liapunov function

$$\nu(e, \alpha) = \sum_{i=1}^N e_i^T H_i e_i + \gamma_i^{-1}(\alpha_i - \alpha^*)^2, \tag{5.167}$$

we establish the theorem in pretty much the same way as Theorem 5.25. Q.E.D.

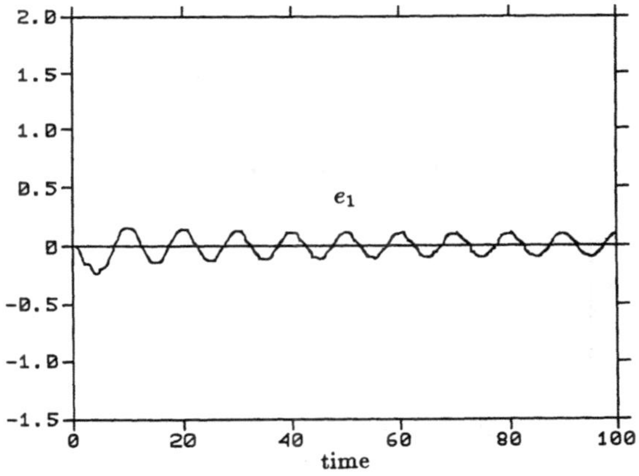

Fig. 5.16. Tracking error for the known subsystem scheme.

5.33. EXAMPLE. We continue Example 5.31, and compute the gain matrices

$$K_i = [2.0 \quad 2.0], \qquad \bar{K}_i = [0.5 \quad 0.5], \qquad i = 1,\, 2, \tag{5.168}$$

for use in the control law (5.164). The reference signal $r(t)$ is given by

$$\begin{aligned} r_1(t) &= \sin 0.2\pi t + \sin \pi t + \sin 2\pi t, \\ r_2(t) &= 0, \qquad\qquad t \geq t_0 = 0, \end{aligned} \tag{5.169}$$

that is, it is desired, starting at $t = 0$, to move the first pendulum sinusoidally, while holding the second pendulum unmoving in a vertical position. At $t = 0$, the adapted gain α_i is zero. We use $\gamma = 100$ and $\sigma = 0.01$ in the adaptation law (5.165).

The simulation results are shown in Figures 5.16 and 5.17. The pendulum positions closely follow the reference model, with accuracy improving over time. The gains $\alpha_i(t)$ adjust generally upward during a transient period, and then oscillate around a fixed positive value in steady-state.

For comparison, we simulated the same system in closed-loop, but without adaptation. Setting $\alpha_i(t) = 0$, $i = \{1, 2\}$, $t \geq 0$, we get the results shown in Figure 5.18. Notice that, without adaptation, a large amount of

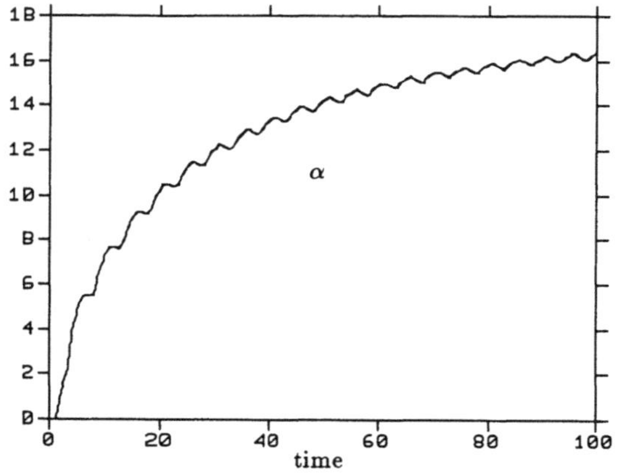

Fig. 5.17. Adaptation gain α for the known subsystem scheme.

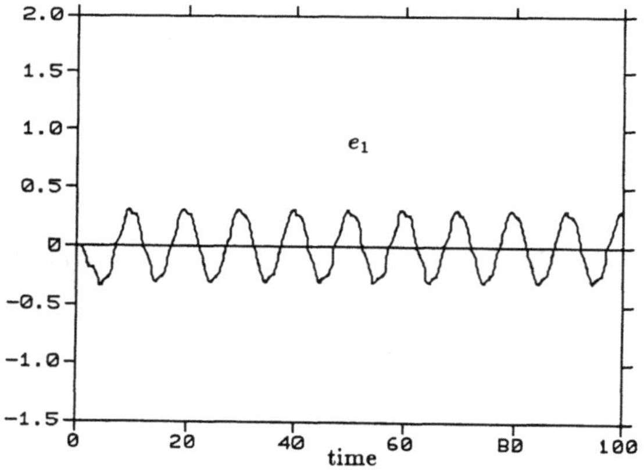

Fig. 5.18. Tracking error without adaptation ($\gamma = 0$).

signal intended to drive the first pendulum couples into the second pendulum through the connecting spring, whereas with adaptation, local gains increase in order to reduce the coupling effect.

This adaptive control law is the simplest of the ones described so far, since only two parameters are adapted in the overall system, as compared to six for the decentralized adaptive controller described in the preceding section, and twelve for the centralized adaptive controller. For a system with N subsystems, there are N adaptation gains, and the N adaptation gain equations can be run in parallel.

5.7. Adaptive Output Feedback

We now remove the assumption that all states of each subsystem are available for control and consider local output feedback. A centralized adaptive controller using output feedback was proposed by Monopoli (1973) and proven stable by Narendra *et al.* (1980). A decentralized version of this controller, which was introduced by Ioannou (1983), is suitable for our use.

Each subsystem controller consists of a precompensator $\mathbf{C}_i^p$ and a feedback compensator $\mathbf{C}_i^f$ leading to the open-loop subsystem description

$$\begin{aligned} \mathbf{S}_i\colon\ \dot{x}_i &= A_i x_i + b_i u_i + P_i v_i, \\ y_i &= c_i^T x_i, \\ w_i &= Q_i x_i, \\ \mathbf{C}_i^p\colon\ \dot{z}_{pi} &= F_i z_{pi} + g_i u_i, \\ \mathbf{C}_i^f\colon\ \dot{z}_{fi} &= F_i z_{fi} + g_i y_i, \qquad i \in \boldsymbol{N}, \end{aligned} \tag{5.170}$$

where $x_i(t) \in \mathbf{R}^{n_i}$, $z_{pi}(t) \in \mathbf{R}^{n_i-1}$, and $z_{fi}(t) \in \mathbf{R}^{n_i-1}$ are the states of $\mathbf{S}_i$, $\mathbf{C}_i^p$ and $\mathbf{C}_i^f$ at time $t \in \mathbf{R}$, $u_i(t) \in \mathbf{R}^{p_i}$ and $y_i(t) \in \mathbf{R}^{q_i}$ are the input and output of the subsystem $\mathbf{S}_i$, and $v_i(t) \in \mathbf{R}^{m_i}$ and $w_i(t) \in \mathbf{R}^{\ell_i}$ are the interconnection inputs and outputs of $\mathbf{S}_i$ to and from other subsystems $\mathbf{S}_j$, $j \in \boldsymbol{N}$, at time $t \in \mathbf{R}$. We assume that the pair $(A_i,\, b_i)$ is controllable and the pair $(A_i,\, c_i^T)$ is observable. The assumptions about interconnection functions remain unchanged.

To use the stabilizability conditions of Section 5.1, we assume that there exists a subset $\boldsymbol{I} \subset \boldsymbol{N}$ such that

$$\begin{aligned} P_i &= b_i p_i^T, \qquad i \in \boldsymbol{I}, \\ Q_i &= q_i c_i^T, \qquad i \in \boldsymbol{N} - \boldsymbol{I}, \end{aligned} \tag{5.4}$$

for some constant vectors $p_i \in \mathbf{R}^{m_i}$, $q_i \in \mathbf{R}^{\ell_i}$. For ease of notation, we number the subsystems so that $\boldsymbol{I} = \{1,\, 2,\, \ldots,\, I\}$ and $\boldsymbol{N} - \boldsymbol{I} = \{I+1,\, I+2,\, \ldots,\, N\}$.

An adaptive decentralized control law for the defined class of interconnected systems is given by

$$u_i = \theta_i^T v_i, \tag{5.171}$$

where $v_i = (\tilde{y}_i,\, z_{pi}^T,\, z_{fi}^T,\, r_i)^T$ is a vector of available signals and $\theta_i = (\hat{d}_{fi},\, \hat{c}_{pi}^T,\, \hat{c}_{fi}^T,\, \hat{k}_{0i})^T$ is the adaptation gains vector. In $v_i(t)$, $r_i(t)$ is the external reference input, which is a uniformly bounded and piecewise continuous function of time, and $\tilde{y}_i = y_i - y_{mi}$ is the model output following error. Stable reference models $\mathbf{M}_i$ are specified as before:

$$\mathbf{M}_i\colon\ \dot{x}_{mi} = A_{mi}x_{mi} + b_{mi}r_i, \qquad y_{mi} = c_{mi}^T x_{mi}, \qquad i \in \boldsymbol{N}. \tag{5.103}$$

Now, define the transfer functions

$$\mathbf{S}_i\colon\ \frac{y_i(s)}{u_i(s)} = \varphi_i(s) = \kappa_i \frac{\beta_i(s)}{\alpha_i(s)},$$

$$\hat{\mathbf{S}}_i\colon\ \frac{y_i(s)}{r_i(s)} = \hat{\varphi}_i(s,\, \theta) = \hat{\kappa}_i(\theta)\frac{\hat{\beta}_i(s,\, \theta)}{\hat{\alpha}_i(s,\, \theta)}, \tag{5.172}$$

$$\mathbf{M}_i\colon\ \frac{y_{mi}(s)}{r_i(s)} = \varphi_{mi}(s) = \kappa_{mi}\frac{\beta_{mi}(s)}{\alpha_{mi}(s)}, \qquad i \in \boldsymbol{N},$$

which describe the open-loop plant, closed-loop plant, and the reference model, respectively. The linear time-invariant system transfer function $\hat{\varphi}_i(s,\, \theta)$ is defined for fixed values of $\theta \in \mathbf{R}^{2n_i+2}$. In order for a decentralized adaptive control design to be feasible, the following conditions must be met:

(*i*) *The plant is minimum phase*; that is, the monic polynomial $\beta_i(s)$ is Hurwitz (all zeros have negative real parts).

(*ii*) *The plant has relative degree one*; that is, the degree of $\beta_i(s)$ is $n_i - 1$, where n_i is the degree of $\alpha_i(s)$.

(*iii*) *The sign of* κ_i *is known.* Without loss of generality, we assume that $\kappa_i > 0$.

In addition to the above conditions on the open-loop plant, we must choose reference models that are *strictly positive real*, which means that $\alpha_{mi}(s)$ and $\beta_{mi}(s)$ are Hurwitz, and Re $\{\varphi_{mi}(j\omega)\} > 0$ for all $\omega \in [0,\, \infty)$ (Šiljak, 1971).

Under these conditions, it has been established (Narendra and Valavani, 1978) that there exists a unique $\theta_i^* = (d_{fi}^*,\, c_{pi}^*,\, c_{fi}^*,\, k_{0i}^*)^T$ such that the closed-loop isolated subsystem $\hat{\mathbf{S}}_i$ has the same input/output behavior as the reference model $\mathbf{M}_i$; that is, $\hat{\varphi}_i(s,\, \theta_i^*) = \varphi_{mi}(s)$, $i \in \boldsymbol{N}$.

In fact, however, we do not know θ_i^*, so we use its estimate $\theta_i(t)$ in the control law (5.171). The proposed decentralized adaptation law for $\theta_i(t)$ is

$$\dot{\theta}_i = -\Gamma_i(\sigma\theta_i + v_i\tilde{y}_i), \qquad i \in \boldsymbol{N}, \tag{5.173}$$

where Γ_i is a constant symmetric positive definite matrix of dimension $(3n_i - 2) \times (3n_i - 2)$, and σ is a constant positive scalar.

With $\theta_i(t)$ different from θ_i^*, the closed-loop isolated subsystem acts like the reference model, but with the added disturbance input $\phi_i^T(t)v_i(t)$, where $\phi_i(t) = \theta_i(t) - \theta_i^*$. The closed-loop interconnected system is thus described by

$$\begin{aligned} \hat{\mathbf{S}}_i\colon\ \dot{\hat{x}}_i &= \hat{A}_i(\theta_i^*)\hat{x}_i + \hat{b}_i\phi_i^T v_i + \hat{b}_i r_i + \hat{P}_i v_i - \hat{b}_{pi} d_{fi}^* y_{mi}, \\ y_i &= \hat{c}_i^T \hat{x}_i, \\ w_i &= \hat{Q}_i \hat{x}_i, \qquad i \in \boldsymbol{N}, \end{aligned} \tag{5.174}$$

where $\hat{x}_i = (x_i^T, z_{pi}^T, z_{fi}^T)^T$, and

$$\begin{aligned} \hat{A}_i(\theta_i^*) &= \begin{bmatrix} A_i + b_i d_i^* c_i^T & b_i c_{pi}^{*T} & b_i c_{fi}^{*T} \\ g_i d_i^* c_i^T & F_i + g_i c_{pi}^{*T} & g_i c_{fi}^{*T} \\ g_i c_i^T & 0 & F_i \end{bmatrix}, \\ \hat{b}_i &= (b_i^T, g_i^T, 0)^T, \\ \hat{c}_i^T &= (c_i^T, 0, 0), \\ \hat{P}_i &= (P_i^T, 0, 0)^T, \\ \hat{Q}_i &= (Q_i, 0, 0). \end{aligned} \tag{5.175}$$

The triple $\{\hat{A}_i(\theta_i^*), \hat{b}_i, \hat{c}_i^T\}$ is a nonminimal realization of the reference model:

$$\begin{aligned} \hat{\mathbf{M}}_i\colon\ \dot{\hat{x}}_{mi} &= \hat{A}_i(\theta_i^*)\hat{x}_{mi} + \hat{b}_i r_i, \\ y_{mi} &= \hat{c}_{mi}^T \hat{x}_{mi}, \qquad i \in \boldsymbol{N}. \end{aligned} \tag{1.176}$$

The reference model is strictly positive real, so that, from the Kalman–Yakubovich Lemma (*e.g.*, Lefschetz, 1965; Šiljak, 1969), it follows that

$\{\hat{A}_i(\theta_i^*),\ b_i,\ \hat{c}_i^T\}$ satisfies the equations

$$\begin{aligned} \hat{A}_i^T(\theta_i^*)\hat{H}_i + \hat{H}_i\hat{A}_i(\theta_i^*) &= \hat{\ell}_i\hat{\ell}_i^T - \epsilon_i\hat{L}_i, \\ \hat{H}_i\hat{b}_i &= \hat{c}_i, \qquad i \in \boldsymbol{N}, \end{aligned} \tag{5.177}$$

for some constant, symmetric positive definite matrices $\hat{H}_i$ and $\hat{L}_i$, some constant vector $\ell_i \in \mathbf{R}^{3n_i-2}$, and sufficiently small number $\epsilon_i > 0$.

An equation for the tracking error $\hat{e}_i(t) = \hat{x}_i(t) - \hat{x}_{mi}(t)$ can be derived by subtracting (5.176) from (5.174):

$$\begin{aligned} \hat{\mathbf{S}}_{ei}\colon\ \dot{\hat{e}}_i &= \hat{A}_i(\theta_i^*)\hat{e}_i + \hat{b}_i\phi_i^T v_i + \hat{P}_i v_i - \hat{b}_{pi}d_{fi}^* y_{mi}, \\ \tilde{y}_i &= \hat{c}_i^T\hat{e}_i, \\ w_i &= \hat{Q}_i(\hat{e}_i + \hat{x}_{mi}), \qquad i \in \boldsymbol{N}, \end{aligned} \tag{5.178}$$

The decentralized stabilizability conditions (5.4) are stated with respect to the open-loop system $\mathbf{S}$. The closed-loop system $\hat{\mathbf{S}}$ with the local dynamic compensators is a structurally different large-scale system; thus, we must restate the stabilizability conditions in terms of the new structure. The condition

$$\hat{Q}_i = q_i\hat{c}_i^T, \qquad i \in \boldsymbol{N} - \boldsymbol{I}, \tag{5.179}$$

follows immediately; however, it is *not* true that $\hat{P}_i = \hat{b}_i p_i^T$, $i \in \boldsymbol{I}$. This is because interconnection disturbances, which enter at the input to the plant, are not available signals for input to the precompensator as well (see Figure 5.19a). We can, however, *reflect* the disturbances to the input of the closed-loop system as shown in Figure 5.19b.

From (5.169), we calculate the transfer function for the *open-loop* precompensator as

$$\mathbf{C}_i^p\colon\ \varphi_{pi}(s) = c_{pi}^{*T}(sI - F_i)^{-1}g_i, \qquad i \in \boldsymbol{I}. \tag{5.180}$$

When the loop is closed, the precompensator acts as a prefilter for the feedback, reference, and adaptation error signals. The prefilter has the transfer function

$$\hat{\mathbf{C}}_i^p\colon\ \hat{\varphi}_{pi}(s) = [1 - \varphi_{pi}(s)]^{-1} = c_{pi}^{*T}(sI - F_i - g_i c_{pi}^{*T})^{-1}g_i, \qquad i \in \boldsymbol{I}. \tag{5.181}$$

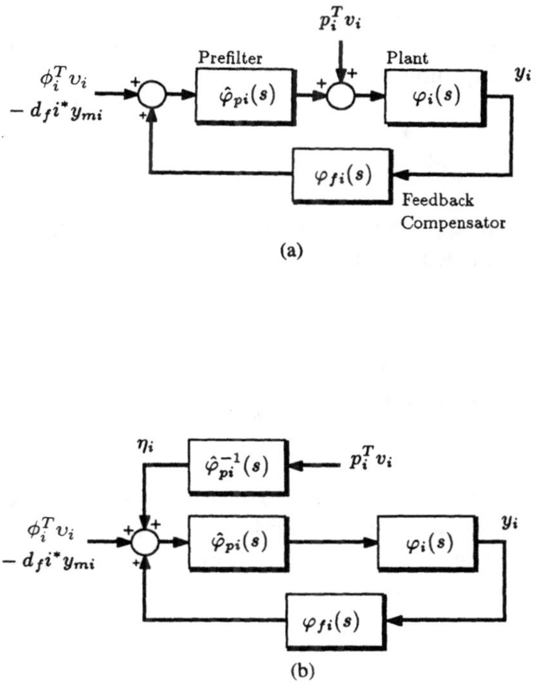

Fig. 5.19. Reflecting the interconnection disturbance to the input of the closed loop subsystem.

In (Narendra and Valavani, 1978), it is shown that

$$\hat{\varphi}_{pi}(s) = \frac{\beta_{mi}(s)}{\beta_i(s)}, \qquad i \in \boldsymbol{I}; \tag{5.182}$$

therefore, since both $\beta_i(s)$ and $\beta_{mi}(s)$ are assumed Hurwitz, we have that $\hat{\varphi}_{pi}(s)$ and $\hat{\varphi}_{pi}^{-1}(s)$ are stable transfer functions.

For each subsystem $\mathbf{S}_{ei}$, $i \in \boldsymbol{I}$, we reflect the interconnection signal $p_i^T v_i$ to the input of the closed-loop subsystem by disconnecting it from the input of the plant, filtering it through the *inverse* of the prefilter $\hat{\mathbf{C}}_i^p$, then feeding the resulting filtered signal to the input of the closed-loop subsystem, as shown in Figure 5.19b. In doing this, we have introduced $\boldsymbol{I}$ new subsystems into the analysis whose transfer functions are $\hat{\varphi}_{pi}^{-1}(s)$:

$$\begin{aligned} \mathbf{F}_i\colon\ \dot{z}_i &= F_i z_i + g_i p_i^T v_i, \qquad z_i(t_0) = 0, \\ \eta_i &= -c_{pi}^{*T} z_i + p_i^T v_i, \qquad\qquad i \in \boldsymbol{I}, \end{aligned} \tag{5.183}$$

where $z_i \in \mathbf{R}^{n_i-1}$ is the state, $p_i^T v_i$ is the input, and $\eta_i \in \mathbf{R}$ is the output of $\mathbf{F}_i$. Since F_i is a stable matrix by design, the subsystems $\mathbf{F}_i$ have, associated with them, the Liapunov equations

$$F_i^T H_i' + H_i' F_i = -G_i', \qquad i \in \boldsymbol{I}, \tag{5.184}$$

where both H_i' and G_i' are positive definite $(n_i - 1) \times (n_i - 1)$ matrices.

With this modification, the closed-loop error systems $\hat{\mathbf{S}}_{ei}$, $i \in \boldsymbol{I}$, are now described by

$$\hat{\mathbf{S}}_{ei}\colon\ \dot{\hat{e}}_i = \hat{A}_i(\theta^*)\hat{e}_i + \hat{b}_i \phi_i^T v_i + \hat{b}_i \eta_i - \hat{b}_{pi} d_{fi}^* y_{mi}, \qquad i \in \boldsymbol{I}. \tag{5.185}$$

We are now in a position to fully take advantage of the interconnection structure, as originally stated by conditions (5.4).

In summary, the differential equations for the large-scale, interconnected, adaptively controlled system are:

$$\begin{aligned}
\hat{\mathbf{S}}_e\colon\ \dot{\hat{e}}_i &= \hat{A}_i \hat{e}_i + \hat{b}_i[\phi_i^T v_i + \eta_i - d_{fi}^* y_{mi}], && i \in \boldsymbol{I},\\
\dot{z}_i &= F_i z_i + g_i p_i^T v_i, && i \in \boldsymbol{I},\\
\dot{\hat{e}}_i &= \hat{A}_i \hat{e}_i + \hat{b}_i[\phi_i^T v_i - d_{fi}^* y_{mi}] + \hat{P}_i v_i, && i \in \boldsymbol{N} - \boldsymbol{I},\\
\dot{\phi}_i &= -\Gamma_i \sigma(\phi_i + \theta_i^*) - \Gamma_i(\hat{c}_i^T \hat{e}_i) v_i, && i \in \boldsymbol{N},
\end{aligned} \tag{5.186}$$

with interconnections given by

$$\begin{aligned}
\eta_i &= c_{pi}^{*T} z_{pi} + p_i^T v_i, && i \in \boldsymbol{I},\\
w_i &= \hat{Q}_i(\hat{e}_i + \hat{x}_{mi}), && i \in \boldsymbol{I},\\
w_i &= \hat{q}_i \hat{c}_i^T(\hat{e}_i + \hat{x}_{mi}), && i \in \boldsymbol{N} - \boldsymbol{I},\\
v_i &= f_i(t, w), && i \in \boldsymbol{N}.
\end{aligned} \tag{5.187}$$

We denote the state of the overall system $\hat{\mathbf{S}}_e$ as $(\hat{e},\, z,\, \phi) \in \Sigma$, where

$$\begin{aligned}
\Sigma &= \mathbf{R}^{(3n-2N)} \times \mathbf{R}^k \times \mathbf{R}^{2n}, \\
k &= \sum_{i=1}^{I} n_i, \\
\hat{e} &= (\hat{e}_1^T,\, \hat{e}_2^T,\, \ldots,\, \hat{e}_N^T)^T v, \\
\phi &= (\phi_1^T,\, \phi_2^T,\, \ldots,\, \phi_N^T)^T, \\
z &= (z_1^T,\, z_2^T,\, \ldots,\, z_I^T)^T,
\end{aligned} \tag{5.188}$$

and state the following result obtained by Gavel and Šiljak (1989):

5.34. THEOREM. The solutions $(\hat{e},\, z,\, \phi)\, (t;\, t_0,\, \hat{e}_0,\, z_0,\, \phi_0)$ of the system $\mathbf{S}_e$ are globally ultimately bounded.

Proof. Define the function ν: $\Sigma \to \mathbf{R}_+$ as

$$\nu(\hat{e},\, z,\, \phi) = \sum_{i=1}^{N} \delta_i \left[k_{0i}^* \hat{e}_i^T \hat{H}_i \hat{e}_i + (\phi_i + \rho\bar{\theta}_i)^T \Gamma_i^{-1} (\phi_i + \rho\bar{\theta}_i) \right] + \sum_{i=1}^{I} \delta_i' z_i^T H_i' z_i, \tag{5.189}$$

where the $\bar{\theta}_i = (1,\, 0,\, \ldots,\, 0)^T$ are vectors having the same dimension as θ_i for $i \in \boldsymbol{N}$, and ρ, δ_i, and δ_i' are as yet unspecified positive constants. Taking the time derivative of $\nu(\hat{e},\, z,\, \phi)$ with respect to (5.186), the theorem is established by a lengthy proof. See (Gavel and Šiljak, 1989) for details. Q.E.D.

5.35. REMARK. Again, if we choose $\sigma \propto \rho^{-3}$ and $\Gamma \propto \rho^3$, and increase ρ, we find that $\nu_f \sim \rho^{-1}$ as before. However, note that this time, *a priori* knowledge of the interconnection bounds is not sufficient to calculate ν_f, and hence the size of the residual set. This is because the constants μ and μ' depend on $\hat{H}$, the solution to (5.177), and, in order to solve (5.177), we would have to know θ^*. This is counter to our assumption that the subsystem dynamics are unknown.

5.36. EXAMPLE. We continue with the two penduli example and assume that the only measurable variables are

$$y_i = [\,1.0 \quad 0.1\,] x_i, \qquad i = 1,\, 2. \tag{5.190}$$

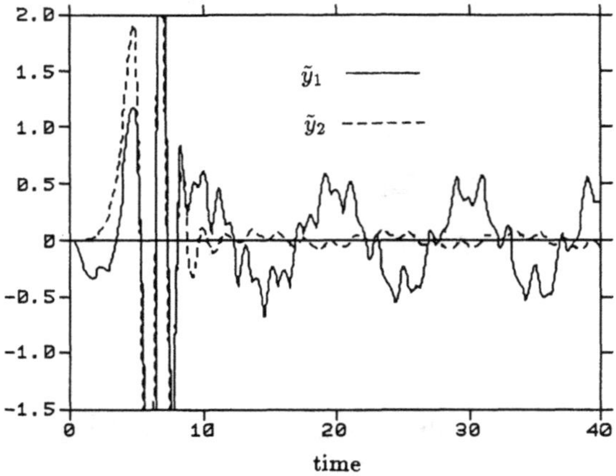

Fig. 5.20. Tracking errors using the output feedback adaptive controller.

We want the penduli to track the motion of the reference model specified by

$$A_{mi} = \begin{bmatrix} 0 & 1 \\ -1 & -2 \end{bmatrix}, \qquad b_{mi} = \begin{bmatrix} 0 \\ 1 \end{bmatrix}, \qquad c_{mi}^T = [\,0.5 \quad 0.5\,], \qquad i = 1,\, 2. \tag{5.191}$$

The pre- and post-compensators are given by

$$\begin{aligned} \mathbf{C}_i^p\colon\ \dot{z}_{pi} &= -z_{pi} + u_i, \\ \mathbf{C}_i^f\colon\ \dot{z}_{fi} &= -z_{fi} + y_i, \qquad i = 1,\, 2, \end{aligned} \tag{5.192}$$

and we apply the reference signal described by (5.169). The results are shown in Figures 5.20 and 5.21. Notice that initially, the pendulums begin to fall, but as the parameters adjust, the tracking error approaches the low magnitudes that were observed with the state feedback algorithm.

5.8. Notes and References

If states of interconnected plants are not available for control and, at the same time, the communication costs of state reconstruction are prohibitive, then the only alternative is to build decentralized controllers using

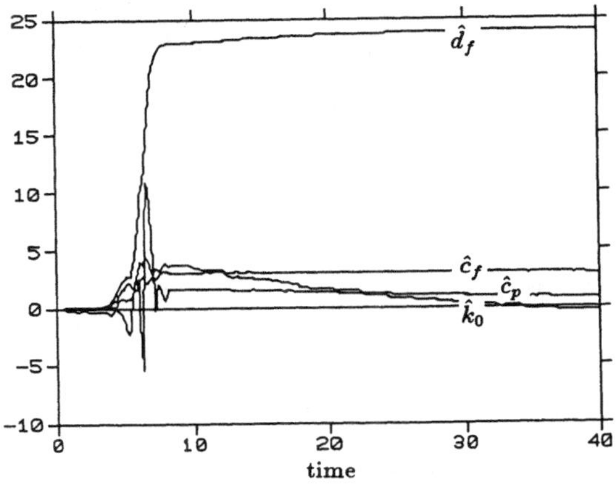

Fig. 5.21. Adapted gains for the output feedback controller.

local output feedback. A suitable control strategy is to associate dynamic compensators with subsystems and boost the gains, so that the interconnection effect is neutralized. Of course, this high-gain strategy works only if the interconnections satisfy certain structural conditions (Hüseyin *et al.* 1982: Section 5.1). A variety of these conditions (sometimes called "matching conditions," *e.g.*, Leitmann, 1979; Šiljak, 1989) were used in Chapter 2 to build decentralized state controllers for interconnected plants. The conditions work equally well for both static and dynamic output feedback (Saberi and Khalil, 1985; see also the survey of high-gain feedback results by Saberi and Sannuti, 1988b).

It was made clear in (Šiljak, 1978) that we cannot rely on interconnections to produce a stabilization effect when plants are inherently unstable, because a loss of interconnections may cause instability: Robust decentralized control is achieved by considering interconnections as a disturbance of subsystems. Both exact (Wonham, 1979) and almost (Willems, 1981, 1982) disturbance decoupling methods can be used in this context (Willems and Ikeda, 1984; Linneman, 1985; Young, 1985; Saberi and Sannuti, 1988a). The stabilizability conditions, however, are parameter dependent, because the relevant subspaces vary as functions of system parameters. A way to overcome this dependency is to select the cases in which the relevant subspaces are invariant under parameter perturbations. This led to the investigation of the almost invariant subspaces of structured systems (Hayakawa and Šiljak, 1988), as presented in Sections 5.2–4.

High-gain feedback may become excessive, causing severe noise problems. Yet, lower gain values may be just as satisfactory for other system performance characteristics. This fact makes the decentral-adaptive control an attractive design concept. The research in this area (Hmamed and Radouane, 1983; Ioannou, 1983, 1986; Ioannou and Kokotović, 1983; Ioannou and Reed, 1988) has been directed to applications of model reference adaptive controllers (Parks, 1966; Monopoli, 1973; Narendra and Valavani, 1978; Narendra *et al.* 1980; Yoshikawa and Ikeda, 1983) to the control of unknown subsystems as if they were decoupled from each other. Then (as in Chapter 2), the standard M-matrix conditions have been used to establish the bounds on interconnections for stability of the overall system.

There are at least three reasons why the M-matrix approach to decentralized adaptation may be unsuitable for application. First, it may be difficult, if not impossible, to predict and limit the size of the coupling among the subsystems within the bounds established by M-matrix conditions. Second, even if we know the bounds on interconnection strengths, it is not clear how the local reference models should be chosen for the overall system to satisfy the M-matrix conditions. Third, in the case of tracking problems, we would like to have the capability of controlling the size of the residual set. For these reasons, a high-gain adaptive scheme was proposed (Gavel and Šiljak, 1989: Sections 5.5–7), which was based on structural conditions (*i.e.*, matching conditions) for decentral stabilizability. Under these conditions, the feedback gains are adaptively adjusted to the levels necessary to neutralize the interconnections and, at the same time, drive the unknown subsystems toward the performance of the locally chosen reference models, as close as desired. It is important to note that the resulting adaptive schemes retain the connective stability property and robustness to nonlinearities that is enjoyed by the fixed-gain decentralized controllers of Section 5.1. Furthermore, the adaptive schemes are likely to use lower gains than those set by conservative fixed-gain design procedures. This fact was confirmed when adaptive decentralized control was used in robotics (Gavel and Hsia, 1988; Gavel, 1988; Mills and Goldenberg, 1988).

FREQUENCY DOMAIN METHODS

A popular alternative to time domain design of output feedback is to use the Nyquist and inverse Nyquist array methods of frequency domain variety proposed originally by Rosenbrock (1974). One of the reasons for the development of these methods is the fact that *the controllers are designed decentrally*, resulting in procedures of outstanding conceptual and numerical simplicity. Only a glimpse of a vast number of results in this area is offered here, relying on a paper by Ohta *et al.* (1986), where a short

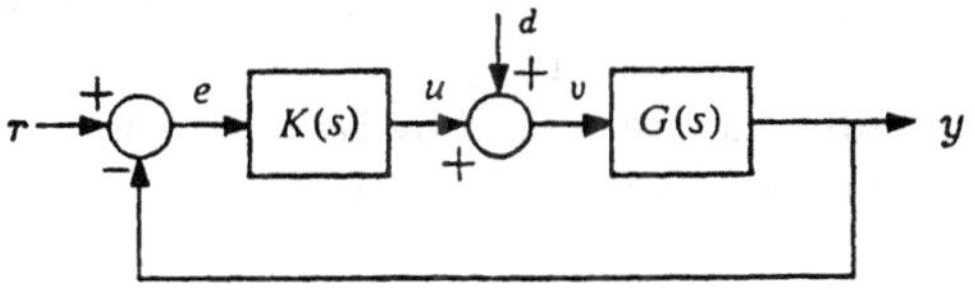

Fig. 5.22. Feedback system.

up-to-date survey of the design procedures is provided (see also Bennett and Baras, 1985).

Let us consider a feedback system **S** of Figure 5.22, which consists of

$$\begin{aligned} &\text{plant} \quad \mathbf{P}\colon \; y = G(s)v, \\ &\text{controller} \quad \mathbf{C}\colon \; u = K(s)e, \\ &\text{interconnection} \quad \mathbf{I}\colon \; v = u + d, \qquad e = r - y, \end{aligned} \tag{5.193}$$

where the matrices $G(s)$, $K(s) \in \text{Mat}\, \mathcal{R}(s)$ have dimensions $m \times \ell$ and $\ell \times m$, respectively, and belong to the set of real rational matrices. The system equation of **S** is given by

$$\mathbf{S}\colon \begin{bmatrix} I_m & G(s) \\ -K(s) & I_\ell \end{bmatrix} \begin{bmatrix} e \\ v \end{bmatrix} = \begin{bmatrix} r \\ d \end{bmatrix}. \tag{5.194}$$

Suppose that the feedback system **S** is well posed, that is, $\det[I_m + G(s)K(s)] \not\equiv 0$. Let $H(G, K)(s)$ be the transfer function matrix of **S** relating (r, d) to (e, v):

$$H(G, K)(s) = \begin{bmatrix} I_m & G(s) \\ -K(s) & I_\ell \end{bmatrix}^{-1}. \tag{5.195}$$

We use $\mathcal{S}(s)$ to denote the subset of $\mathcal{R}(s)$ consisting of proper rational functions whose poles are all in $\mathbf{C}_-$, and say that **S** is stable if $H(G, K)(s) \in \text{Mat}\, \mathcal{S}(s)$, that is, **C** stabilizes **P**.

The system is decomposed as

$$\left.\begin{aligned} &\mathbf{P}\colon \; y_i = G_{ii}(s)v_i + \sum_{j \neq i}^{N} G_{ij}(s)v_j, \\ &\mathbf{C}\colon \; u_i = K_{ii}(s)e_i + \sum_{j \neq i}^{N} K_{ij}(s)e_j, \\ &\mathbf{I}\colon \; v_i = u_i + d_i, \qquad e_i = r_i - y_i, \end{aligned}\right\} \quad i \in \boldsymbol{N}, \tag{5.196}$$

where $G_{ij}(s)$ and $K_{ij}(s)$ are submatrices of $G(s)$ and $K(s)$ having dimensions $m_i \times \ell_j$ and $\ell_i \times m_j$, respectively.

The following result was established by Bennett and Baras (1980), which is a block-matrix analog of (Rosenbrock, 1974):

5.37. THEOREM. Let $R(s) = R_D(s) + R_C(s)$, where

$$\begin{aligned} R_D(s) &= \text{diag}\{R_1(s),\, R_2(s),\, \dots,\, R_N(s)\} \\ &= I_m + \text{diag}\{G_{11}(s)K_{11}(s),\, G_{22}(s)K_{22}(s),\, \dots,\, G_{NN}(s)K_{NN}(s)\}, \\ R_C(s) &= [R_{ij}(s)] = G(s)K(s) - G_D(s)K_D(s), \\ G_D(s) &= \text{diag}\{G_{11}(s),\, G_{22}(s),\, \dots\, G_{NN}(s)\}, \\ K_D(s) &= \text{diag}\{K_{11}(s),\, K_{22}(s),\, \dots,\, K_{NN}(s)\}. \end{aligned} \tag{5.197}$$

Assume the following:

(*i*) $\{G_D(s),\, K_D(s)\}$ and $\{G(s),\, K(s)\}$ have the same number of poles in $\mathbf{C}_+$;

(*ii*) $H(G_D,\, K_D)(s) \in \text{Mat}\, \mathcal{S}(s)$;

(*iii*) $\det[R_D(s) + \mu R_C(s)] \neq 0 \quad \forall s \in \mathcal{D}, \quad \forall \mu \in [0,\, 1]$, where $\mathcal{D}$ is the standard Nyquist $\mathcal{D}$-contour.

Then, **S** is stable, that is, $H(G,\, K)(s) \in \text{Mat}\, \mathcal{S}(s)$.

The central question is: Under what conditions does (*iii*) hold? Using the concept of block diagonal dominance introduced by Feingold and Varga (1962) and Fiedler (1961), Bennett and Baras (1980), Limebeer (1982, 1983), Limebeer and Hung (1983), and Hung and Limebeer (1984) derived sufficient conditions for (*iii*). Ohta *et al.* (1986) generalized these results by applying the *quasi block diagonal dominance* (Ohta and Šiljak, 1985) of Definition 7.14:

5.38. THEOREM. If the Perron–Frobenius root is such that

$$\lambda_{PF}\{B(s)\} < 1, \qquad \forall s \in \mathcal{D}, \tag{5.198}$$

where the matrix $B = (b_{ij})$ is defined as

$$b_{ij}(s) = \| R_{ij}(s) R_j^{-1}(s) \|, \tag{5.199}$$

then **S** is stable.

Although the condition (5.198) is the least restrictive, it is not clear how $K(s)$ affects stability of $\mathbf{S}$ even if $K(s)$ were a block diagonal matrix. A suitable set of conditions is provided by Ohta *et al.* (1986), under which a matrix $K(s)$, when restricted to a diagonal matrix $K_D(s) = \text{diag}\{K_{11}(s), K_{22}(s), \ldots, K_{NN}(s)\}$, can be chosen "piece-by-piece" to stabilize $\mathbf{S}$, involving only the independent choice of each block $K_{ii}(s)$, one at a time. The procedure of Ohta *et al.* (1986) was formulated by generalizing certain results of Araki and Nwokah (1975), and using either the characteristic loci method (MacFarlane and Belletrutti, 1973) or the Nyquist array technique (Rosenbrock, 1974; Araki *et al.* 1981).

It is more or less clear that the design procedure of Ohta *et al.* (1985) is basically a *loop-shaping* by quasi block diagonal dominance. This fact opens up a possibility of using H^∞ theory in this context. Wu and Mansour (1988, 1989) observed this fact and proposed an interesting parametrization of the decentralized problem, which leads to a numerically effective design method.

A new emerging frequency-domain approach to decentralized control design, which has been advanced recently by Tan and Ikeda (1987), is based upon the stable factorization method (Vidyasagar, 1985). The decentralized control problem is essentially a simultaneous stabilization problem of finding a set of controllers that stabilize all individual subsystems and the overall system at the same time. Since the stable factorization method provides a suitable parametrization of all stabilizing controllers, Tan and Ikeda (1987) showed how a decentralized controller set can be obtained sequentially using this method. With some care (but with increased conservativeness), their procedure can assure connective stability of the overall system (Ikeda and Tan, 1989; Tan and Ikeda, 1990).

Bibliography

Aoki, T., S. Hosoe, and Y. Hayakawa (1983). Structural controllability for linear systems in descriptor form (in Japanese). *Transactions SICE*, 19, 628–635.

Araki, M., and O. I. Nwokah (1975). Bounds for closed-loop transfer functions of multivariable systems. *IEEE Transactions*, AC-20, 666–670.

Araki, M., K. Yamamoto, and B. Kondo (1981). GG-pseudo-band method for the design of multivariable control systems. *Proceedings of the 8th IFAC World Congress*, 3, 137–142.

Bennett, W. H., and J. S. Baras (1980). Block diagonal dominance and design of decentralized compensators. *Proceedings of the IFAC Symposium on Large Scale Systems: Theory and Applications*, Toulouse, France, 93–102.

Bennett, W. H., and J. S. Baras (1985). Decomposition and decentralized control system design: A review of frequency domain methods. *Proceedings of the 24th Conference on Decision and Control*, Fort Lauderdale, Florida, 1828–1835.

Chen, K. W. (1976). *Applied Graph Theory*. North-Holland, New York.

Cho, Y. J., and Z. Bien (1989). Reliable control via an additive redundant controller. *International Journal of Control*, 50, 385–398.

Feingold, D. G., and R. C. Varga (1962). Block diagonally dominant matrices and generalizations of the Gerschgorin circle theorem. *Pacific Journal of Mathematics*, 12, 1241–1250.

Fiedler, M. (1961). Some estimates of spectra of matrices. *Proceedings of the Symposium on Numerical Treatment of Ordinary Differential Integral, and Integro-Differential Equations*, Birkhauser-Verlag, Basel, Switzerland, 33–36.

Gavel, D. T. (1988). *Decentralized Adaptive Control of Large Scale Systems, with Application to Robotics*. Ph.D. Thesis, Report No. UCRL-53866, Lawrence Livermore National Laboratory, Livermore, California.

Gavel, D. T., and T. C. Hsia (1988). Decentralized adaptive control of robot manipulators. *Proceedings of the IEEE International Conference on Robotics and Automation*, Philadelphia, Pennsylvania, 1230–1235.

Gavel, D. T., and D. D. Šiljak (1989). Decentralized adaptive control: Structural conditions for stability. *IEEE Transactions*, AC-34, 413–426.

Hayakawa, Y., and D. D. Šiljak (1988). On almost invariant subspaces of structured systems and decentralized control. *IEEE Transactions*, AC-33, 931–939.

Hmamed, A., and L. Radouane (1983). Decentralized nonlinear adaptive feedback stabilization of large-scale interconnected systems. *IEE Proceedings*, 130-D, 57–62.

Hung, Y. S., and D. J. N. Limebeer (1984). Robust stability of additively perturbed interconnected systems. *IEEE Transactions*, AC-29, 1069–1075.

Hüseyin, Ö., M. E. Sezer, and D. D. Šiljak (1982). Robust decentralized control using output feedback. *IEE Proceedings*, 129-D, 310–314.

Ikeda, M., and X. L. Tan (1989). Connective stabilization of large-scale systems: A stable factorization approach. *Mathematical Theory of Networks and Systems* (to appear).

Inan, K. (1978). *Asymptotic Root Loci in Linear Multivariable Control Systems*. Dissertation, Middle East Technical University, Ankara, Turkey.

Ioannou, P. A. (1983). Design of decentralized adaptive controllers. *Proceedings of the IEEE Conference on Decision and Control*, San Antonio, Texas, 205–210.

Ioannou, P. A. (1986). Decentralized adaptive control of interconnected systems. *IEEE Transactions*, AC-31, 291–298.

Ioannou, P. A., and P. V. Kokotović (1983). *Adaptive Systems with Reduced Models*. Springer, New York.

Ioannou, P. A., and P. V. Kokotović (1984). Robust redesign of adaptive control. *IEEE Transactions*, AC-29, 202–211.

Ioannou, P. A., and J. S. Reed (1988). Discrete-time decentralized adaptive control. *Automatica*, 24, 419–421.

Lefschetz, S. (1965). *Stability of Nonlinear Control Systems*. Academic Press, New York.

Leitmann, G. (1979). Guaranteed asymptotic stability for some linear systems with bounded uncertainties. *Transactions of ASME, Journal of Dynamic Systems, Measurement, and Control*, 101, 212–216.

Limebeer, D. J. N. (1982). The application of generalized diagonal dominance to linear system stability theory. *International Journal of Control*, 36, 185–212.

Limebeer, D. J. N. (1983). An inclusion region for the eigenvalues of partitioned matrices. *International Journal of Control*, 37, 429–436.

Limebeer, D. J. N., and Y. S. Hung (1983). Robust stability of interconnected systems. *IEEE Transactions*, CAS-30, 397–403.

Linneman, A (1985). Decentralized stabilization by high gain feedback. *Real Time Control of Large Scale Systems*, G. Schmidt, M. G. Singh, and A. Titli (eds.). Springer, New York, 316–325.

MacFarlane, A. G. J., and J. J. Belletrutti (1973). The characteristic locus design method. *Automatica*, 9, 575–588.

Matsumoto, T., and M. Ikeda (1983). Structural controllability based on intermediate standard form (in Japanese). *Transactions SICE*, 19, 601–606.

Mills, J. K., and A. A. Goldenberg (1988). Global connective stability of a class of robotic manipulators. *Automatica*, 24, 835–839.

Monopoli, R. V. (1973). The Kalman–Yakubovich lemma in adaptive control system design. *IEEE Transactions*, AC-18, 527–529.

Morse, A. S. (1973). Structural invariants of linear multivariable systems. *SIAM Journal of Control*, 11, 446–465.

Narendra, K. S., and J. H. Taylor (1973). *Frequency Domain Criteria for Absolute Stability*. Academic Press, New York.

Narendra, K. S., and L. S. Valavani (1978). Stable adaptive controller design—direct control. *IEEE Transactions*, AC-23, 570–583.

Narendra, K. S., Y. H. Lin, and L. S. Valavani (1980). Stable adaptive controller design; Part II: Proofs of stability. *IEEE Transactions*, AC-25, 440–448.

Ohta, Y., and D. D. Šiljak (1985). Overlapping diagonal dominance and existence of Lyapunov functions. *Journal of Mathematical Analysis and Applications*, 112, 396–410.

Ohta, Y., D. D. Šiljak, and T. Matsumoto (1986). Decentralized control using quasi-block diagonal dominance of transfer function matrices. *IEEE Transactions*, AC-31, 420–430.

Parks, P. C. (1966). Liapunov redesign of model reference adaptive control systems. *IEEE Transactions*, AC-11, 362–367.

Radenković, M. S., (1989). Decentralized stochastic adaptive control of interconnected systems. *International Journal of Adaptive Control and Signal Processing*, 3 (to appear).

Radenković, M. S., and D. D. Šiljak (1989). Robust self-tuning control. *Technical Report* EECS-881327304, School of Engineering, Santa Clara University, Santa Clara, California.

Rosenbrock, H. H. (1974). *Computer-Aided Control System Design*. Academic Press, New York.

Rouche, N., P. Habets, and M. Lalloy (1977). *Stability Theory by Liapunov's Direct Method*. Springer, New York.

Saberi, A., and H. Khalil (1985). Decentralized stabilization of interconnected systems using output feedback. *International Journal of Control*, 41, 1461–1475.

Saberi, A., and P. Sannuti (1988a). Global stabilization with almost disturbance decoupling of a class of uncertain non-linear systems. *International Journal of Control*, 47, 717–727.

Saberi, A., and P. Sannuti (1988b). Time-scale structure assignment in linear multivariable systems by high-gain feedback. *Proceedings of the 27th Conference on Decision and Control*, Austin, Texas, 423–430.

Šiljak, D. D. (1969). *Nonlinear Systems: The Parameter Analysis and Design*. Wiley, New York.

Šiljak, D. D. (1971). New algebraic criteria for positive realness. *Journal of the Franklin Institute*, 291, 109–120.

Šiljak, D. D. (1978). *Large-Scale Dynamic Systems: Stability and Structure*. North-Holland, New York.

Šiljak, D. D. (1989). Parameter space methods for robust control design: A guided tour. *IEEE Transactions*, AC-34, 674–688.

Swamy, M. N. S., and K. Thulasiraman (1981). *Graphs, Networks, and Algorithms*. Wiley, New York.

Tan, X. L., and M. Ikeda (1987). Decentralized stabilization of large-scale interconnected systems: A stable factorization approach. *Proceedings of the 26th Conference on Decision and Control*, Los Angeles, California, 2295–2300.

Tan, X. L., and M. Ikeda (1990). Decentralized stabilizations for expanding construction of large-scale systems. *IEEE Transactions*, AC–35, 644–651.

Vidyasagar, M. (1985). *Control System Synthesis: A Factorization Approach*. MIT Press, Cambridge, Massachusetts.

Willems, J. C. (1981). Almost invariant subspaces: An approach to high-gain feedback design; Part I: Almost controlled invariant subspaces. *IEEE Transactions*, AC-26, 235–252.

Willems, J. C. (1982). Almost invariant subspaces: An approach to high-gain feedback design; Part II: Almost conditionally invariant subspaces. *IEEE Transactions*, AC-27, 1071–1085.

Willems, J. C., and M. Ikeda (1984). Decentralized stabilization of large-scale interconnected systems. *Proceedings of the 6th International Conference on Analysis and Optimizations of Systems*, Nice, France, 236–244.

Wonham, W. M. (1979). *Linear Multivariable Control: A Geometric Approach*. Springer, New York.

Wu, Q. H., and M. Mansour (1988). An application of H^∞ theory in decentralized control. *Proceedings of the 27th Conference on Decision and Control*, Austin, Texas, 1335–1340.

Wu, Q. H., and M. Mansour (1989). Decentralized robust control using H^∞-optimization technique. *Information and Decision Technologies*, 15, 59–76.

Yoshikawa, T., and M. Ikeda (1983). Decentralized adaptive control of large-scale systems: A parameter adaptation method. *Electronics and Communications in Japan*, 66-A, 25–32.

Young, K. D. (1985). A decentralized control synthesis method for large-scale interconnected systems. *Proceedings of the American Control Conference*, Boston, Massachusetts, 574–579.

Zames, G. (1966). On the input–output stability of time-varying nonlinear feedback systems; Part I: Conditions derived using concepts of loop gain, conicity, and positivity. *IEEE Transactions*, AC-11, 228–238.

Chapter 6 Hierarchical LBT Decompositions

So far, we have assumed that a system can be decomposed into subsystems on the basis of physical insight, intuition, or experience. During the modeling process, subsystems can be identified as natural parts of the system, and the choice where to tear a system is either obvious or can be made after a few simple experiments.

When a system is complex having many states, inputs, and outputs, it may be difficult, or impossible, to obtain partitions by physical reasoning. A more appealing alternative is to develop systematic methods, which can be used to decompose a system by extracting information from the structure of the state equations themselves. Decompositions of this type can take advantage of the sparseness of the system matrices, and identify hidden structures that are conducive to conceptual and numerical simplifications of control and estimation design. In this way, decompositions of dynamic systems are embedded in the broad and powerful mathematical framework of sparse matrix research.

In this chapter, we shall describe and use a graph-theoretic algorithm for partitioning a linear dynamic system into a Lower Block Triangular (LBT) form. The resulting system is obtained as a hierarchy of input reachable subsystems, and stabilization of the overall system can be accomplished by stabilizing decentrally a number of low-order subsystems. By involving the outputs in the tearing process, we can produce LBT partitions having subsystems with independent inputs and outputs. The subsystems are input and output reachable, and we will show how local observers can be built for a piece-by-piece stabilization using output feedback.

We shall demonstrate the utility of LBT decompositions by a sequential LQG optimization of hierarchically structured systems. The plant is assumed to be a linear discrete-time stochastic system, which is decomposed into an LBT form. With subsystems, we associate quadratic costs, and minimize each sequentially, one at a time, assuming that feedback loops are closed at higher levels of the hierarchy. A methodology is described for deriving optimal control algorithms based on a construction of state models for each stage of the optimization process. This approach allows for the standard LQG techniques to be used sequentially from top to bottom of the hierarchy.

Finally, we discuss information structures for distributed control of LBT forms, and expose the intricate relationship between the physical constraints of the system and the communication capabilities of the computer network. Special attention is paid to the sequential information structure where fast top-down communication makes data available from upper to lower levels of the hierarchy. The structure is contrasted to the standard global as well as the totally decentralized structure, which appear at the two opposite extremes of communication requirements.

6.1. Input Decompositions

We begin a presentation of the LBT design scheme by a description of the graph-theoretic procedure for decomposition of linear systems into LBT structures (Sezer and Šiljak, 1981). The resulting system, being a hierarchical interconnection of input reachable subsystems, can be stabilized by decentralized local feedback applied to each individual subsystem. Before stabilization is attempted, the subsystems matrices should be tested for full generic rank to make sure that the subsystems are structurally controllable (Section 1.4). The fact that input reachability and full generic rank of subsystems matrices imply not only structural controllability of the subsystems but the overall LBT system as well makes the LBT decompositions appealing in stabilization of large-scale systems. When a system is in an LBT form, even ordinary controllability is implied by subsystems controllability; this makes it easy to remove any doubt about controllability of the overall system without resorting to numerically sensitive controllability criteria (Nour–Eldin, 1987) applied to the full-scale matrices of the original system.

Let us consider a linear time-invariant system

$$\mathbf{S}\colon\ \dot{x} = Ax + Bu, \tag{6.1}$$

where, as usual, $x(t) \in \mathbf{R}^n$ and $u(t) \in \mathbf{R}^m$ are the state and input of $\mathbf{S}$ at time $t \in \mathbf{R}$. With $\mathbf{S}$, we associate a directed graph $\mathbf{D} = (\mathtt{U} \cup \mathtt{X}, \mathtt{E})$. The adjacency and input reachability matrices of $\mathbf{D}$ have the form

$$E = \begin{bmatrix} \bar{A} & \bar{B} \\ 0 & 0 \end{bmatrix}, \qquad R = \begin{bmatrix} F & G \\ 0 & 0 \end{bmatrix}. \tag{6.2}$$

We consider the problem of decomposing the system into an LBT form

$$\mathbf{S}\colon\ \dot{x}_k = \sum_{j=1}^{k} A_{kj}x_j + \sum_{j=1}^{k} B_{kj}u_j, \qquad k \in \boldsymbol{N}, \tag{6.3}$$

which is a hierarchical ordering of the subsystems

$$\mathbf{S}_k\colon\ \dot{x}_k = A_{kk}x_k + B_{kk}u_k, \qquad k \in \boldsymbol{N}, \tag{6.4}$$

where $x_k(t) \in \mathbf{R}^{n_k}$ and $u_k(t) \in \mathbf{R}^{m_k}$ are the state and input of $\mathbf{S}_k$ at time $t \in \mathbf{R}$, and $n = \sum_{k=1}^{N} n_k$, $m = \sum_{k=1}^{N} m_k$, $\boldsymbol{N} = \{1, 2, \dots, N\}$. The decomposition (6.3) is obtained by permuting the states of $\mathbf{S}$ until the matrices $\bar{A}$ and $\bar{B}$ are brought into compatible LBT forms. Then, the interconnection matrix E has the obvious LBT representation

$$E = \left[\begin{array}{cccc|cccc} \bar{A}_{11} & & & \bigcirc & \bar{B}_{12} & & & \bigcirc \\ \bar{A}_{21} & \bar{A}_{22} & & & \bar{B}_{21} & \bar{B}_{22} & & \\ \cdots & \cdots & \cdots & \cdots & \cdots & \cdots & \cdots & \cdots \\ \bar{A}_{N1} & \bar{A}_{N2} & \dots & \bar{A}_{NN} & \bar{B}_{N1} & \bar{B}_{N2} & \dots & \bar{B}_{NN} \\ \hline & \bigcirc & & & & \bigcirc & & \end{array}\right]. \tag{6.5}$$

In terms of the digraph $\mathbf{D} = (\mathtt{U} \cup \mathtt{X},\ \mathtt{E})$, the LBT form of E corresponds to an input acyclic partition $\mathbf{P}_I$ of $\mathbf{D}$ into subgraphs $\mathbf{D}_k = (\mathtt{U}_k \cup \mathtt{X}_k,\ \mathtt{E}_k)$ associated with the block diagonal pairs $(\bar{A}_{kk},\ \bar{B}_{kk})$ of the subsystem $\mathbf{S}_k$.

6.1. DEFINITION. A partition $\mathbf{P}_I$ of a digraph $\mathbf{D} = (\mathtt{U} \cup \mathtt{X},\ \mathtt{E})$ is said to be an input acyclic partition of $\mathbf{D}$ if:

(*i*) The interconnection matrix E has LBT form.

(*ii*) Each subgraph $\mathbf{D}_k = (\mathtt{U}_k \cup \mathtt{X}_k,\ \mathtt{E}_k)$ is input reachable.

If, in addition, none of the subgraphs $\mathbf{D}_k$ can be further partitioned into acyclic subgraphs, the partition $\mathbf{P}_I$ is said to be irreducible.

To obtain irreducible partitions $\mathbf{P}_I$, we assume that the digraph $\mathbf{D}$ is input reachable. Our decomposition scheme, however, can easily be modified to remove the input unreachable part from $\mathbf{D}$ before the beginning of decomposition process. To avoid trivial cases, we assume that each input vertex of **D** is adjacent to at least one state vertex. Under these assumptions, there always exists an irreducible partition $\mathbf{P}_I$, although that might be the trivial partition with $N = 1$. Obviously, we will be interested in nontrivial irreducible partitions of **D**, for the existence of which a necessary condition is that $n > 1$ and $m > 1$. Finally, we note that neither the irreducible partitions nor the number of subgraphs in such partitions need be unique, as we demonstrate by examples in Section 6.3.

The first result is concerned with the existence of acyclic partitions that are not necessarily irreducible:

6.2. LEMMA. Given a digraph $\mathbf{D} = (\mathtt{U} \cup \mathtt{X}, \mathtt{E})$ with $n > 1$ and $m > 1$. Then, **D** has a nontrivial partition $\mathbf{P}_I$ if and only if the submatrix G of the reachability matrix R contains at least one zero element.

Proof. Necessity follows by direct computation of R (see Section 1.2). To prove sufficiency, we assume that G contains a zero in the (i, j)th position and partition the set $\mathtt{U}$ into two disjoint subsets

$$\mathtt{U}_1 = \mathtt{U} - \{u_j\}, \qquad \mathtt{U}_2 = \{u_j\}. \tag{6.6}$$

Let $\mathtt{X}_2$ be the largest subset of $\mathtt{X}$ that is reachable from u_j, and let $\mathtt{X}_1 = \mathtt{X} - \mathtt{X}_2$. These subsets have the following properties:

(*i*) $\mathtt{U}_2 \neq \emptyset$ by construction and $\mathtt{U}_1 \neq \emptyset$ by the assumption that $m > 1$.

(*ii*) $\mathtt{X}_1 \neq \emptyset$ for $x_i \in \mathtt{X}_1$, and $\mathtt{X}_2 \neq \emptyset$ by the assumption that each input vertex is adjacent to at least one state vertex.

(*iii*) By construction, no $x_\ell \in \mathtt{X}_1$ is adjacent from u_j.

(*iv*) No $x_\ell \in \mathtt{X}_1$ is adjacent from some $x_k \in \mathtt{X}_2$, for otherwise x_ℓ would be reachable from u_j, contradicting the construction of $\mathtt{X}_2$.

The properties (*i*)–(*iv*) imply that the chosen subsets of $\mathtt{U}$ and $\mathtt{X}$ induce a decomposition (6.5) of E with $N = 2$. Furthermore, both subgraphs $\mathbf{D}_1 = (\mathtt{U}_1 \cup \mathtt{X}_1, \mathtt{E}_1)$ and $\mathbf{D}_2 = (\mathtt{U}_2 \cup \mathtt{X}_2, \mathtt{E}_2)$ are input reachable; $\mathbf{D}_1$ because of the properties (*iii*) and (*iv*) and $\mathbf{D}_2$ by construction. Therefore, $\mathbf{D}_1$ and $\mathbf{D}_2$ constitute a partition $\mathbf{P}_I$ of the digraph **D**. Q.E.D.

The construction of $\mathbf{P}_I$ described in the proof of Lemma 6.2 is the core of the following algorithm proposed by Sezer and Šiljak (1981) as Input Decomposition (IDEC).

6.3. ALGORITHM.

IDEC

Assume $n > 1$, $m > 1$, and calculate input reachability matrix G.
IF $G(i, j) \neq 0, \forall i, j$ **RETURN**
Define the index sets
$\boldsymbol{I} = \{i_1, i_2, \dots, i_k\} \subset \{1, 2, \dots, n\}$
$\boldsymbol{J} = \{j_1, j_2, \dots, j_\ell\} \subset \{1, 2, \dots, m\}$
such that
(1) rows $i_1, i_2, \dots, i_k$ of G are the same and contain 1 in columns $j_1, j_2, \dots, j_\ell$ of R and 0 in other columns;
(2) when rows $i_1, i_2, \dots, i_k$ and columns $j_1, j_2, \dots, j_\ell$ of G are deleted, the resulting submatrix contains no zero rows.
Define the state and input partitions
$\mathtt{X}_1 = \{x_i, i \in \boldsymbol{I}\}$, $\mathtt{X}_2 = \mathtt{X} - \mathtt{X}_1$
$\mathtt{U}_1 = \{u_j, j \in \boldsymbol{J}\}$, $\mathtt{U}_2 = \mathtt{U} - \mathtt{U}_1$
OUTPUT irreducible partition $\{\mathtt{U}_1, \mathtt{X}_1\}$

END IDEC

It is clear now that the assumptions on n, m, and G guarantee the existence of the index sets $\boldsymbol{I}$ and $\boldsymbol{J}$. Letting

$$\begin{aligned} \mathtt{U}_1 &= \{u_{j_1}, u_{j_2}, \dots, u_{j_\ell}\}, \qquad & \mathtt{U}_2 &= \mathtt{U} - \mathtt{U}_1, \\ \mathtt{X}_1 &= \{x_{i_1}, x_{i_2}, \dots, x_{i_k}\}, \qquad & \mathtt{X}_2 &= \mathtt{X} - \mathtt{X}_1, \end{aligned} \tag{6.7}$$

we prove the following result:

6.4. THEOREM. The subgraphs $\mathbf{D}_k = (\mathtt{U}_k \cup \mathtt{X}_k, \mathtt{E})$, $k = 1, 2$, where $\mathtt{U}_k$ and $\mathtt{X}_k$ are defined by (6.7), constitute a partition $\mathbf{P}_I$ of $\mathbf{D} = (\mathtt{U} \cup \mathtt{X}, \mathtt{E})$ such that $\mathbf{D}_1$ is irreducible.

Proof. The proof that $\mathbf{D}_k = (\mathtt{U}_k \cup \mathtt{X}_k, \mathtt{E})$, $k = 1, 2$, is an acyclic input reachable partition is similar to the proof of the sufficiency part of Lemma 6.2. Since the submatrix of G consisting of rows $i_1, i_2, \dots, i_k$ and columns $j_1, j_2, \dots, j_\ell$ is the input reachability matrix of $\mathbf{D}_1$ and contains no 0's, Lemma 6.2 implies that $\mathbf{D}_1$ is irreducible. Q.E.D.

The main feature of the decomposition scheme leading to Theorem 6.4 is that it can be used recursively to remove an irreducible subgraph from $\mathbf{D}$ at each step, the existence of which is guaranteed by Theorem 6.4. This recursive process produces eventually an irreducible partition $\mathbf{P}_I$. Furthermore, if at each step we consider all possible choices of subgraphs to be removed, we can generate all irreducible partitions.

A considerable reduction in the computation which is involved in Algorithm 6.3 can be achieved if we obtain the condensation $\mathbf{D}^*$ of $\mathbf{D}$ first, and apply the scheme to $\mathbf{D}^*$. For this purpose, we partition the set $\mathtt{X}$ of $\mathbf{D}$ as

$$\mathtt{X} = \bigcup_{i=1}^{p} \mathtt{X}_i^\bullet, \tag{6.8}$$

so that each $\mathtt{X}_i^\bullet$ is a vertex set of a strong component of the subgraph $\mathbf{D}_x = (\mathtt{X}, \mathtt{E})$ which was termed the state truncation of $\mathbf{D}$ (Section 1.3). This decomposition of $\mathbf{D}_x$ into strong components induces a partition of the input nodes:

$$\mathtt{U} = \bigcup_{i=1}^{q} \mathtt{U}_i^\bullet. \tag{6.9}$$

Each $\mathtt{U}_i^\bullet$ contains all the input vertices that are adjacent exactly to those state vertices that occur in the same strong component of $\mathbf{D}_x$. By $\mathbf{D}^* = (\mathtt{U}^* \cup \mathtt{X}^*, \mathtt{E}^*)$, we denote the condensation of $\mathbf{D}$ with respect to the partitions of the sets $\mathtt{X}$ and $\mathtt{U}$ in (6.8) and (6.9), and prove the following:

6.5. THEOREM. To any irreducible partition $\mathbf{P}_I$ of $\mathbf{D}^*$ into N subgraphs, there corresponds a unique partition $\mathbf{P}_I$ of $\mathbf{D}$ into the same number of subgraphs and vice versa.

Proof. Consider an irreducible $\mathbf{P}_I^*$ of $\mathbf{D}^* = (\mathtt{U}^* \cup \mathtt{X}^*, \mathtt{E}^*)$ into subgraphs $\mathbf{D}_k^* = (\mathtt{U}_k^* \cup \mathtt{X}_k^*, \mathtt{E}_k^*)$, $k \in \boldsymbol{N}$, according to which $\mathbf{D}^*$ has the interconnection matrix

$$E^* = \left[\begin{array}{cccc|cccc} \bar{A}_{11}^* & & & \bigcirc & \bar{B}_{11}^* & & & \bigcirc \\ \bar{A}_{21}^* & \bar{A}_{22}^* & & & \bar{B}_{21}^* & \bar{B}_{22}^* & & \\ \cdots & \cdots & \cdots & \cdots & \cdots & \cdots & \cdots & \cdots \\ \bar{A}_{N1}^* & \bar{A}_{N2}^* & \cdots & \bar{A}_{NN}^* & \bar{B}_{N1}^* & \bar{B}_{N2}^* & \cdots & \bar{B}_{NN}^* \\ \hline & \bigcirc & & & & \bigcirc & & \end{array}\right]. \tag{6.10}$$

We define the subsets $\mathtt{U}_k$, $\mathtt{X}_k$, $k \in \boldsymbol{N}$, of $\mathtt{U}$ and $\mathtt{X}$ as

$$\begin{aligned} \mathtt{U}_k &= \{u_i \in \mathtt{U}\colon\ u_i \in \mathtt{U}_j^\bullet \text{ for some } u_j^* \in \mathtt{U}_k^*\}, \\ \mathtt{X}_k &= \{x_i \in \mathtt{X}\colon\ x_i \in \mathtt{X}_j^\bullet \text{ for some } x_j^* \in \mathtt{X}_k^*\}, \end{aligned} \tag{6.11}$$

where $\mathtt{U}_j^\bullet$ and $\mathtt{X}_j^\bullet$ are defined in (6.8) and (6.9). It is easy to see that $\mathtt{U}_k$ and $\mathtt{X}_k$ are uniquely defined subsets satisfying

$$
\begin{aligned}
&\mathtt{U}_i \cap \mathtt{U}_j = \emptyset, \qquad i \neq j, \qquad \bigcup_{k=1}^{N} \mathtt{U}_k = \mathtt{U}, \\
&\mathtt{X}_i \cap \mathtt{X}_j = \emptyset, \qquad i \neq j, \qquad \bigcup_{k=1}^{N} \mathtt{X}_k = \mathtt{X},
\end{aligned}
\tag{6.12}
$$

so that they induce a partition $\mathbf{P}_I$ of $\mathbf{D} = (\mathtt{U} \cup \mathtt{X},\, \mathtt{E})$ into $\mathbf{D}_k = (\mathtt{U}_k \cup \mathtt{X}_k,\, \mathtt{E}_k)$, $k \in \boldsymbol{N}$. The subgraphs $\mathbf{D}_k$ have the following properties:

(*i*) No $x_i \in \mathtt{X}_k$ is adjacent to some $x_j \in \mathtt{X}_\ell$ for $k > \ell$, for, otherwise, some $x_r^* \in \mathtt{X}_k^*$, for which $x_i \in \mathtt{X}_r^\bullet$, would be adjacent to some $x_s^* \in \mathtt{X}_\ell^*$, for which $x_j \in \mathtt{X}_s^\bullet$, contradicting the structure of E^* in (6.10). Similarly, no $u_i \in \mathtt{U}_k$ is adjacent to some $u_j \in \mathtt{U}_\ell$ for $k > \ell$.

(*ii*) Each subgraph $\mathbf{D}_k^*$ of $\mathbf{D}^*$ is the condensation of the corresponding subgraph $\mathbf{D}_k$ of $\mathbf{D}$, where condensation is performed in exactly the same way as $\mathbf{D}$ is condensed to $\mathbf{D}^*$. Since the $\mathbf{D}_k^*$'s are input reachable by the construction of $\mathbf{P}_I^*$, it can be shown (Theorem 1.19) that the $\mathbf{D}_k$'s are also input reachable.

(*iii*) By the construction of $\mathbf{P}_I^*$, the $\mathbf{D}_k^*$'s are irreducible. Thus, by Lemma 6.2, each $u_i^* \in \mathtt{U}_k^*$ reaches all $x_j^* \in \mathtt{X}_k$ which, in turn, implies that each $u_r \in \mathtt{U}_k$ reaches all $x_s \in \mathtt{X}_k$. Using Lemma 6.2 once more, we show that the $\mathbf{D}_k$'s are irreducible.

Properties (*i*)–(*iii*) imply that the partition induced by the partitions of the sets $\mathtt{U}$ and $\mathtt{X}$ in (6.8) and (6.9) is an irreducible partition $\mathbf{P}_I$.

To prove the converse result, consider any irreducible partition $\mathbf{P}_I$ of $\mathbf{D} = (\mathtt{U} \cup \mathtt{X},\, \mathtt{E})$ into $\mathbf{D}_k = (\mathtt{U}_k \cup \mathtt{X}_k,\, \mathtt{E}_k)$, $k \in \boldsymbol{N}$, according to which the interconnection matrix of $\mathbf{D}$ has the LBT form (6.5). We note that any two state vertices that belong to the same strong component of $\mathbf{D}_x$ should occur in the same subgraph $\mathbf{D}_k$, as well as any two input vertices that are adjacent to exactly the same state vertices. Thus, the definition

$$
\begin{aligned}
\mathtt{U}_k^* &= \{u_i^* \in \mathtt{U}^*\colon\ \mathtt{U}_i^\bullet \subseteq \mathtt{U}_k\}, \qquad k \in \boldsymbol{N}, \\
\mathtt{X}_k^* &= \{x_i^* \in \mathtt{X}^*\colon\ \mathtt{X}_i^\bullet \subseteq \mathtt{X}_k\}, \qquad k \in \boldsymbol{N},
\end{aligned}
\tag{6.13}
$$

of the subsets $\mathtt{U}_k^*$ and $\mathtt{X}_k^*$ of $\mathtt{U}^*$ and $\mathtt{X}^*$ is consistent, and these subsets induce a unique partition $\mathbf{P}^*$ of $\mathbf{D}^* = (\mathtt{U}^* \cup \mathtt{X},\, \mathtt{E}^*)$ into $\mathbf{D}_k^* = (\mathtt{U}_k^* \cup \mathtt{X}_k^*,\, \mathtt{E}_k^*)$, $k \in \boldsymbol{N}$. Moreover, it can be shown that properties (*i*)–(*iii*) described above still hold when the roles of $\mathtt{X}_k$, $\mathtt{U}_k$, $\mathbf{D}_k$ and $\mathtt{X}_k^*$, $\mathtt{U}_k^*$, $\mathbf{D}_k^*$ are interchanged. Therefore, $\mathbf{P}^*$ is an irreducible partition $\mathbf{P}_I^*$ of $\mathbf{D}^*$. Q.E.D.

Simply stated, Theorem 6.5 says that the decomposition problems of $\mathbf{D}$ and $\mathbf{D}^*$ are interchangeable. However, $\mathbf{D}^*$ is usually much smaller than $\mathbf{D}$, and decomposition of $\mathbf{D}^*$ instead of $\mathbf{D}$ is an equally smaller computational effort. Furthermore, identifying strong components of $\mathbf{D}$ makes the stabilization scheme applicable to the subsystems $\mathbf{S}_k$ directly as explained in the next section.

Once a system is obtained in the LBT form with input reachable subsystems, we should proceed by testing the generic rank of the subsystems to make sure that none of them are of Form II (Definition 1.27). If one or more subsystems have a deficient generic rank, that is, the corresponding subgraphs have a dilation (Remark 1.31), we have to modify the resulting partition to eliminate the dilation from the subgraphs. For this purpose, we denote by $\mathbf{D}^* = (\mathtt{V}^*,\ \mathtt{E}^*)$ the condensation of the system digraph $\mathbf{D}$ with respect to some partition $\mathbf{P}_I$, where $\mathtt{V}_k = \mathtt{U}_k \cup \mathtt{X}_k$. Then, we assign levels to the vertices of $\mathbf{D}^*$ in such a way that any vertex that is not reachable from some other vertex has the assigned level 0, and any vertex v_i^* has the assigned level ℓ if the length of the longest path from some vertex at level 0 to v_i^* is ℓ. Since $\mathbf{D}^*$ is an acyclic digraph by the construction of $\mathbf{P}_I$, the level assignment procedure is consistent in that every vertex v_i^* has a unique assigned level (see Theorem 1.18). Corresponding to the levels of vertices of $\mathbf{D}^*$, each subgraph $\mathbf{D}_k$ of the partition $\mathbf{P}_I$ has a unique assigned level, which is the same as the level of v_k^* in $\mathtt{V}^*$. In Figure 6.1, the level assignment is illustrated by eight-vertex condensation.

Testing of the subgraphs $\mathbf{D}_k$ for dilation starts from the lowest level. If a subgraph $\mathbf{D}_i$ is found to have a dilation, that is, the corresponding structured matrix $[\tilde{A}_i\ \ \tilde{B}_i]$ has Form II, then $\mathbf{D}_i$ can be combined with another subgraph $\mathbf{D}_j$ at a lower level (if v_i^* is adjacent from v_j^*) to obtain a composite subgraph

$$\mathbf{D}_{ij} = (\mathtt{U}_i \cup \mathtt{U}_j \cup \mathtt{X}_i \cup \mathtt{X}_j,\ \mathtt{E}_i \cup \mathtt{E}_j \cup \mathtt{E}_{ij}) \tag{6.14}$$

of $\mathbf{D}$, where $\mathtt{E}_{ij}$ is the set of edges of $\mathbf{D}$ that connect vertices of $\mathbf{D}_j$ to vertices of $\mathbf{D}_i$. Since dilations occur due to the lack of edges in $\mathbf{D}_i$, additional edges $\mathtt{E}_{ij}$ may be sufficient to prevent the dilations of $\mathbf{D}_i$ from appearing in $\mathbf{D}_{ij}$. In combining the subgraphs $\mathbf{D}_i$ and $\mathbf{D}_j$, we should not destroy the acyclic property of $\mathbf{D}^*$. Therefore, we should merge together with $\mathbf{D}_i$ and $\mathbf{D}_j$ all other subgraphs $\mathbf{D}_k$ such that v_k^* appears on any path from v_j^* to v_i^*. For example, referring to Figure 6.1, if we combine $\mathbf{D}_7$ with $\mathbf{D}_3$, then we should include $\mathbf{D}_5$ as well. Once $\mathbf{D}_3$, $\mathbf{D}_5$, and $\mathbf{D}_7$ are combined to form $\mathbf{D}_{357}$, we

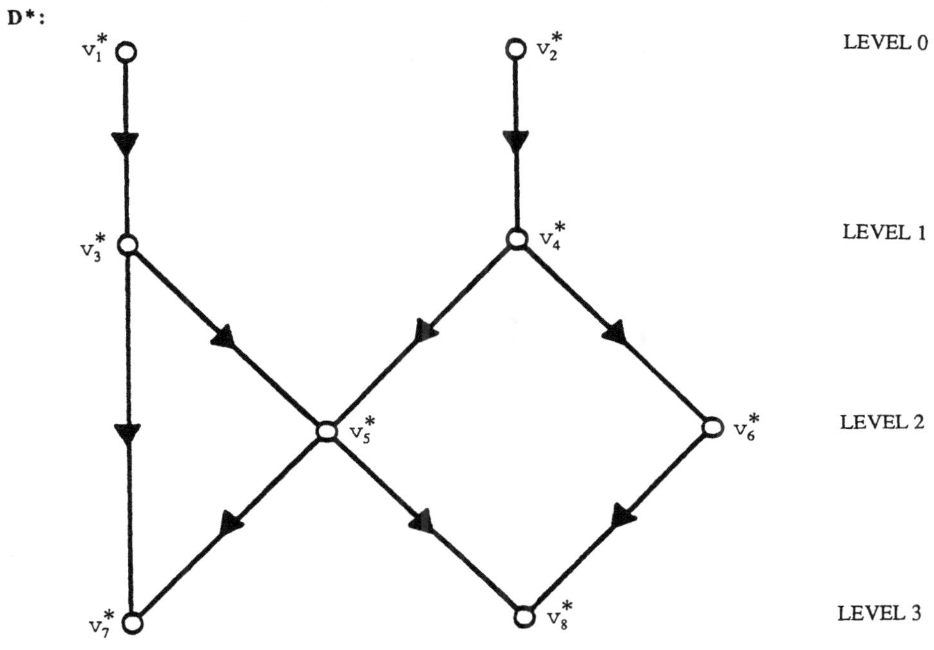

Fig. 6.1. Level assignment.

obtain another acyclic partition $\hat{\mathbf{P}}_I$ with respect to which the condensation $\hat{\mathbf{D}}^*$ of $\mathbf{D}$ is shown in Figure 6.2 together with its level assignment.

We note that if the dilation in a subgraph cannot be avoided by merging it with the adjacent subgraphs, then more complex combinations can be tried. For example, if the subgraph $\hat{\mathbf{D}}_8$ corresponding to v_8^* in Figure 6.2 contains a dilation that also appears in $\hat{\mathbf{D}}_{68}$, then we can try the combination of $\hat{\mathbf{D}}_{68}$ with $\hat{\mathbf{D}}_4$. Provided $\mathbf{D}$ contains no dilation, this process leads to an acyclic partition $\mathbf{P}_I$ of $\mathbf{D}$, the subgraph of which has no dilation. The corresponding system $\mathbf{S}$ is in an LBT form with structurally controllable subsystems, and we may proceed to stabilize the system by stabilizing each subsystem separately.

Before closing this section, we should mention an alternative approach to graph-theoretic decomposition of systems with inputs, which has been proposed by Vidyasagar (1980). After the system digraph is partitioned into strong components with respect to states, the inputs are identified as attached to the components and included in the subsystems. This secondary role of inputs may cause a problem, because after decomposition some subsystems may have no inputs of their own or no inputs at all. In

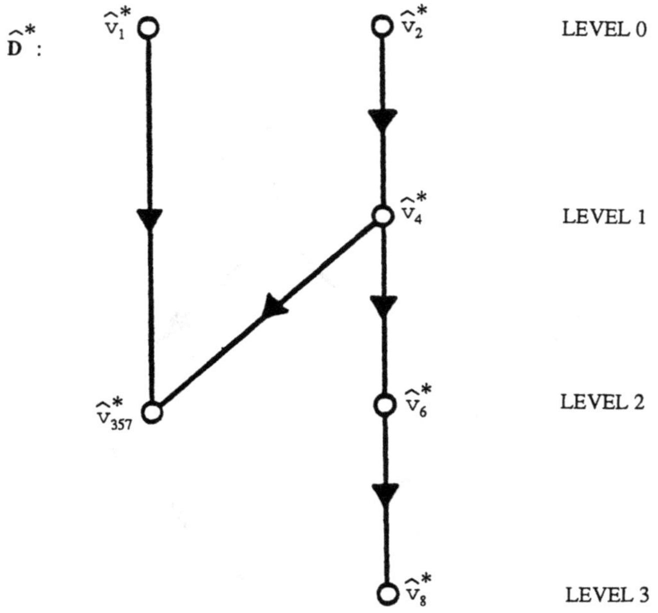

Fig. 6.2. Level assignment.

the decomposition scheme outlined in this section, the partitioning of the system digraph is performed directly in terms of inputs *via* input reachable components, which ensures that each subsystem has its own input, this being the crucial requirement in the decomposition approach to stabilization.

6.2. Stabilization

When a system **S** is in an LBT form, stability of each subsystem implies stability of the overall system. For this reason, we describe only a subsystem stabilization by considering a typical subsystem

$$\mathbf{S}\colon\ \dot{x} = Ax + Bu, \tag{6.15}$$

where the subscripts are dropped for simplicity. Controllability of the subsystem **S** implies that there exists a feedback law

$$u = -Kx, \tag{6.16}$$

which places the eigenvalues of **S** at prescribed locations. We recall that in the process of obtaining an acyclic decomposition of the overall system, the strong components of each subsystem have been identified, and the subsystem matrix A in (6.15) is also in an LBT form. Using this fact, we can give an efficient procedure to obtain the feedback matrix K in (6.16).

Let us first assume that the subsystem **S** in (6.15) has two strong components, so that the A matrix is of the form

$$A = \begin{bmatrix} A_{11} & 0 \\ A_{21} & A_{22} \end{bmatrix}. \tag{6.17}$$

The matrices B and K can be partitioned accordingly as

$$B = \begin{bmatrix} B_1 \\ B_2 \end{bmatrix}, \qquad K = [\, K_1 \quad K_2 \,]. \tag{6.18}$$

From (6.17), it follows that controllability of the pair $(A,\, B)$ implies controllability of the pair $(A_{11},\, B_1)$. Therefore, there exists a matrix F_1 such that

$$\Lambda(A_{11} - B_1 F_1) = \mathcal{L}_1, \tag{6.19}$$

where $\Lambda(A_{11} - B_1 F_1)$ denotes the set of eigenvalues of $A_{11} - B_1 F_1$, and $\mathcal{L}_1$ is a symmetric set of prescribed eigenvalues for $A_{11} - B_1 F_1$ such that

$$\Lambda(A_{22}) \cap \mathcal{L}_1 = \emptyset. \tag{6.20}$$

We let $\hat{A}_{11} = A_{11} - B_1 F_1$, and consider the equation

$$A_{22} T_1 - T_1 \hat{A}_{11} = -A_{21} + B_2 F_1. \tag{6.21}$$

Equations (6.19) and (6.20) guarantee that (6.21) has a unique solution for T_1 (Gantmacher, 1959). We now have the following:

6.6. LEMMA. The pair $(A_{22},\, B_2 - T_1 B_1)$ is controllable.

Proof. Controllability of the pair (A, B) implies (Rosenbrock, 1970)

$$\begin{aligned} n_1 + n_2 &= \operatorname{rank} \begin{bmatrix} \lambda I - A_{11} & 0 & B_1 \\ -A_{21} & \lambda I - A_{22} & B_2 \end{bmatrix} \\ &= \operatorname{rank} \left(\begin{bmatrix} I & 0 \\ -T_1 & I \end{bmatrix} \begin{bmatrix} \lambda I - A_{11} & 0 & B_1 \\ -A_{21} & \lambda I - A_{22} & B_2 \end{bmatrix} \begin{bmatrix} I & 0 & 0 \\ T_1 & I & 0 \\ F_1 & 0 & I \end{bmatrix} \right) \\ &= \operatorname{rank} \begin{bmatrix} \lambda I - \tilde{A}_{11} & 0 & B_1 \\ 0 & \lambda I - A_{22} & B_2 - T_1 B_1 \end{bmatrix}, \qquad \forall \lambda \in \mathbf{C}, \end{aligned} \tag{6.22}$$

where n_1 and n_2 are the dimensions of the square matrices A_{11} and A_{22}, and the last equality follows from (6.21). Thus, we have

$$\operatorname{rank}\,[\lambda I - A_{22},\ B_2 - T_1 B_1] = n_2, \qquad \forall \lambda \in \mathbf{C}. \tag{6.23}$$

Q.E.D.

Lemma 6.6 implies that for any symmetric set $\mathcal{L}_2$ of prescribed eigenvalues, there exists K_2 such that

$$\Lambda[A_{22} - (B_2 - T_1 B_1) K_2] = \mathcal{L}_2. \tag{6.24}$$

With K_2 chosen as above, we set

$$K_1 = F_1 - K_2 T_1, \tag{6.25}$$

and prove the following:

6.7. THEOREM. $\Lambda(A - BK) = \mathcal{L}_1 \cup \mathcal{L}_2$.

Proof. The proof is immediate upon noting that for

$$T = \begin{bmatrix} I & 0 \\ T_1 & I \end{bmatrix}, \qquad T^{-1} = \begin{bmatrix} I & 0 \\ -T_1 & I \end{bmatrix}, \tag{6.26}$$

we have, from (6.21) and (6.25),

$$T^{-1}(A - BK) = \begin{bmatrix} \hat{A}_{11} & -B_1 K_2 \\ 0 & \hat{A}_{22} \end{bmatrix}, \tag{6.27}$$

where $\hat{A}_{22} = A_{22} - (B_2 - T_1 B_1) K_2$. Q.E.D.

The argument leading to Theorem 6.7 provides a recursive procedure for assigning the eigenvalues of a subsystem having two strong components. Since this procedure requires only transformation of $(A_{11},\ B_1)$ and $(A_{22},\ B_2 - T_1B_1)$ into suitable forms to satisfy (6.19) and (6.24), and the solution of (6.21), it is more efficient than any single step procedure that requires a transformation of the whole subsystem.

To complete our discussion, we consider the general case, where the matrices A and B of (6.15) have the form

$$A = \begin{bmatrix} A_{11} & 0 & \dots & 0 \\ A_{21} & A_{22} & & \bigcirc \\ \dots & \dots & & \\ A_{k1} & A_{k2} & \dots & A_{kk} \end{bmatrix} = \begin{bmatrix} A_{11} & 0 \\ A'_{21} & A'_{22} \end{bmatrix}, \tag{6.28}$$

$$B = \begin{bmatrix} B_1 \\ B_2 \\ \vdots \\ B_k \end{bmatrix} = \begin{bmatrix} B_1 \\ B'_2 \end{bmatrix}. \tag{6.29}$$

Partitioning the feedback matrix accordingly as

$$K = [\,K_1 \quad K_2 \quad \dots \quad K_k\,] = [\,K_1 \quad K'_2\,], \tag{6.30}$$

and noting that the A matrix in (6.28) has the form (6.17), we observe that the eigenvalue assignment problem can be solved by following the steps:

(1) Choose F_1 so that $\Lambda(A_{11} - B_1F_1) = \mathcal{L}_1$, where $\mathcal{L}_1$ is a symmetric set of eigenvalues such that $\Lambda(A'_{22}) \cap \mathcal{L}'_1 = \emptyset$.

(2) Solve

$$A'_{22}T_1 - T_1\hat{A}_{11} = -A'_{21} + B'_2F_1, \tag{6.31}$$

for T_1, where $\hat{A}_{11} = A_{11} - B_1F_1$.

(3) Choose K'_2 such that $\Lambda[A'_{22} - (B'_2 - T_1B_1)K'_2] = \mathcal{L}'_2$, where $\mathcal{L}'_2$ is a symmetric set of eigenvalues.

(4) Compute $K_1 = F_1 - K'_2T_1$. Then, $\Lambda(A - BK) = \mathcal{L}_1 \cup \mathcal{L}'_2$.

Only steps (2) and (3) may involve operations on relatively large matrices. However, writing (6.31) explicitly as

$$\begin{bmatrix} A_{22} & & & \bigcirc \\ A_{32} & A_{33} & & \\ \cdots & \cdots & \cdots & \cdots \\ A_{k2} & A_{k3} & \dots & A_{kk} \end{bmatrix} \begin{bmatrix} T_{12} \\ T_{13} \\ \vdots \\ T_{1k} \end{bmatrix} - \begin{bmatrix} T_{12} \\ T_{13} \\ \vdots \\ T_{1k} \end{bmatrix} \hat{A}_{11} = - \begin{bmatrix} A_{21} - B_2 F_1 \\ A_{31} - B_3 F_1 \\ \vdots \\ A_{k1} - B_k F_1 \end{bmatrix}, \tag{6.32}$$

we observe that the blocks T_{12}, T_{13}, ..., T_{1k} of T_1 can be obtained by solving recursively the following equations:

$$\begin{aligned} & A_{22}T_{12} - T_{12}\hat{A}_{11} = -A_{21} + B_2F_1, \\ & A_{ii}T_{1i} - T_{1i}\hat{A}_{11} = -A_{i1} + B_iF_1 - \sum_{j=2}^{i-1} A_{ij}T_{1j}, \qquad i = 3,\, 4, \dots,\, k, \end{aligned} \tag{6.33}$$

Furthermore, the problem of computing K_2' in step (3) is similar to the original problem of computing K, where A, B, and K, are replaced by lower dimensional matrices A_{22}', $B_2' - T_1B_1$, and K_2', respectively. Thus, the blocks K_1, K_2, ..., K_k of the gain matrix K can also be obtained recursively by solving a set of equations as in (6.33), resulting in a considerable saving of the computational effort.

6.8. EXAMPLE. To illustrate the stabilization procedure let us consider the system **S** described by the equation

$$\begin{bmatrix} \dot{x}_1 \\ \dot{x}_2 \\ \dot{x}_3 \\ \dot{x}_4 \\ \dot{x}_5 \\ \dot{x}_6 \\ \dot{x}_7 \\ \dot{x}_8 \end{bmatrix} = \begin{bmatrix} 0 & 0 & 0 & 0 & 0 & 0 & 0 & 0 \\ 0 & a_{22} & a_{23} & 0 & a_{25} & a_{26} & a_{27} & a_{28} \\ 0 & 0 & a_{33} & 0 & 0 & 0 & 0 & a_{38} \\ 0 & 0 & 0 & a_{44} & 0 & 0 & a_{47} & 0 \\ 0 & 0 & 0 & 0 & 0 & 0 & 0 & 0 \\ a_{61} & a_{62} & a_{63} & a_{64} & a_{65} & a_{66} & 0 & 0 \\ 0 & 0 & 0 & 0 & 0 & 0 & a_{77} & 0 \\ 0 & 0 & a_{83} & 0 & a_{85} & 0 & 0 & 0 \end{bmatrix} \begin{bmatrix} x_1 \\ x_2 \\ x_3 \\ x_4 \\ x_5 \\ x_6 \\ x_7 \\ x_8 \end{bmatrix}$$

$$
+\begin{bmatrix} 0 & b_{12} & b_{13} \\ 0 & b_{22} & 0 \\ 0 & 0 & 0 \\ 0 & b_{42} & 0 \\ 0 & b_{52} & b_{53} \\ b_{61} & 0 & 0 \\ b_{71} & 0 & 0 \\ 0 & 0 & 0 \end{bmatrix} \begin{bmatrix} u_1 \\ u_2 \\ u_3 \end{bmatrix}, \tag{6.34}
$$

where a_{ij}'s and b_{ij}'s denote arbitrary nonzero elements.

Applying the input decomposition scheme, we obtain two irreducible acyclic partitions:

$$
\mathbf{P}_I^1\colon \begin{array}{ll} \mathtt{X}_1^1 = \{x_7\}, & \mathtt{U}_1^1 = \{u_1\}, \\ \mathtt{X}_2^1 = \{x_4\}, & \mathtt{U}_2^1 = \{u_2\}, \\ \mathtt{X}_3^1 = \{x_1,\, x_2,\, x_6,\, x_3,\, x_8,\, x_5\}, & \mathtt{U}_3^1 = \{u_3\}; \end{array} \tag{6.35}
$$

$$
\mathbf{P}_I^2\colon \begin{array}{ll} \mathtt{X}_1^2 = \{x_1,\, x_5,\, x_3,\, x_8\}, & \mathtt{U}_1^2 = \{u_2,\, u_3\}, \\ \mathtt{X}_2^2 = \{x_7,\, x_4,\, x_2,\, x_6\}, & \mathtt{U}_2^2 = \{u_1\}. \end{array} \tag{6.36}
$$

The partition $\mathbf{P}_I^1$ of $\mathbf{D}$ is shown in Figure 6.3. It can easily be verified that the subgraph $\mathbf{D}_3^1$ contains a dilation, that is, the structured matrix $[\tilde{A}_{33}\ \ \tilde{B}_3]$ does not have the full generic rank. Dilation takes place whenever a subgraph $\mathbf{D}_k = (\mathtt{U}_k \cup \mathtt{X}_k,\, \mathtt{E}_k)$ contains a subset of vertices $\tilde{\mathtt{X}}_i \subseteq \mathtt{X}_k$ such that the number of distinct vertices of $\mathtt{U}_k \cup \mathtt{X}_k$ from which a vertex of $\tilde{\mathtt{X}}_i$ is reachable is less than the number of vertices of $\tilde{\mathtt{X}}_i$. From Figure 6.3, we conclude by inspection that $\tilde{\mathtt{X}}_i = \{x_1,\, x_5\}$ is such a subset. This graphical characterization of dilation is equivalent to the generic rank characterization: $\mathbf{D}_k$ contains a dilation if and only if the $n_k \times (n_k + m_k)$ structured matrix $[\tilde{A}_{kk}\ \ \tilde{B}_{kk}]$ contains a zero submatrix of order $r \times (n_k + m_k + 1 - r)$ for some r such that $1 \le r \le n_k$ (see Lemma 1.54). To eliminate the dilation in $\mathbf{D}_3^1$, we can combine the subgraphs $\mathbf{D}_3^1$ and $\mathbf{D}_2^1$ into a single digraph $\mathbf{D}_{23}^1$. In this way, we obtain a partition of $\mathbf{D}$ into two structurally controllable components.

The other irreducible partition $\mathbf{P}_I^2$ of $\mathbf{D}$ given in (6.36) has two subgraphs $\mathbf{D}_1^2$ and $\mathbf{D}_2^2$ that are structurally controllable. Permuting the states

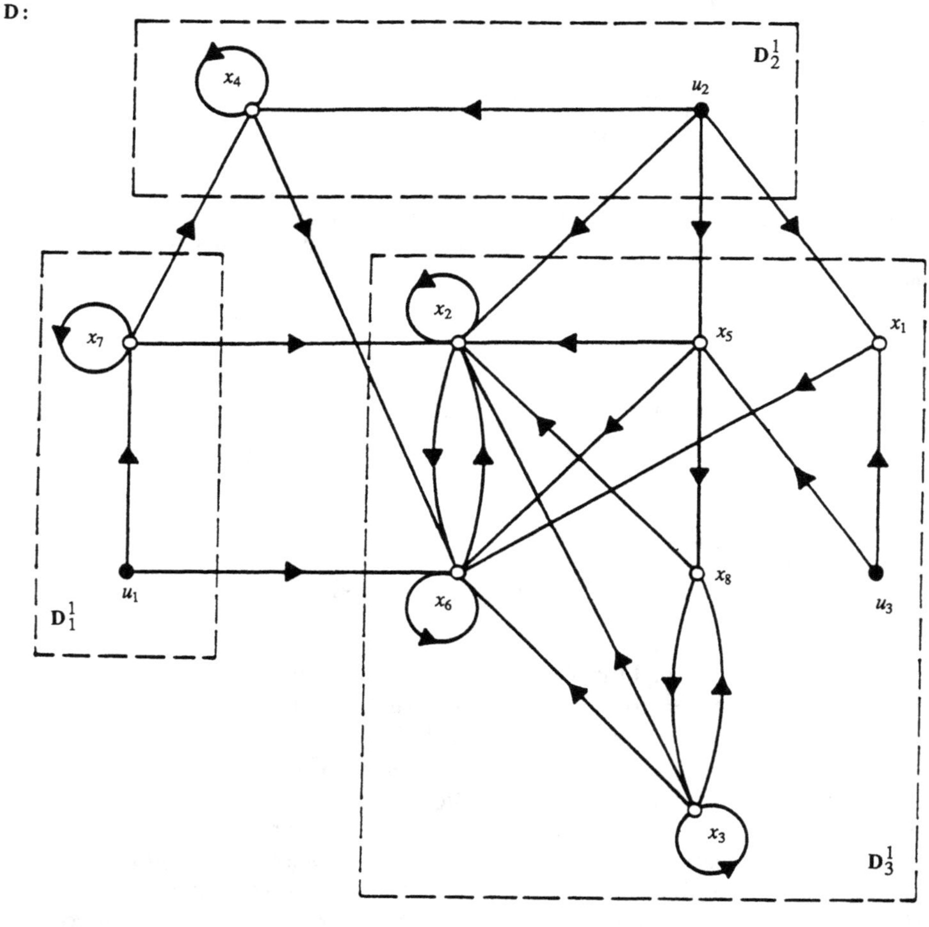

Fig. 6.3. Partition $\mathbf{P}_I^1$.

and inputs of $\mathbf{S}$ according to $\mathbf{P}_I^2$, we obtain the LBT representation of $\mathbf{S}$ as

$$
\begin{bmatrix} \dot{x}_1 \\ \dot{x}_5 \\ \dot{x}_3 \\ \dot{x}_8 \\ \dot{x}_7 \\ \dot{x}_4 \\ \dot{x}_2 \\ \dot{x}_6 \end{bmatrix}
=
\left[\begin{array}{cccc:c:c:cc}
0 & 0 & 0 & 0 & & & & \\
0 & 0 & 0 & 0 & & \bigcirc & & \\
0 & 0 & a_{33} & a_{38} & & & & \\
0 & a_{85} & a_{83} & 0 & & & & \\
\hdashline
0 & 0 & 0 & 0 & a_{77} & 0 & 0 & 0 \\
\hdashline
0 & 0 & 0 & 0 & a_{47} & a_{44} & 0 & 0 \\
\hdashline
0 & a_{25} & a_{23} & a_{28} & a_{27} & 0 & a_{22} & a_{26} \\
a_{61} & a_{65} & a_{63} & 0 & 0 & a_{64} & a_{62} & a_{66}
\end{array}\right]
\begin{bmatrix} x_1 \\ x_5 \\ x_3 \\ x_8 \\ x_7 \\ x_4 \\ x_2 \\ x_6 \end{bmatrix}
$$

$$
+\left[\begin{array}{cc:c}
b_{12} & b_{13} & \\
b_{52} & b_{53} & \bigcirc \\
0 & 0 & \\
0 & 0 & \\
\hdashline
0 & 0 & b_{71} \\
\hdashline
b_{42} & 0 & 0 \\
\hdashline
b_{22} & 0 & 0 \\
0 & 0 & b_{61}
\end{array}\right]
\begin{bmatrix} u_2 \\ u_3 \\ u_1 \end{bmatrix}. \tag{6.37}
$$

To illustrate the eigenvalue assignments for subsystems, let us consider the subsystem $\mathbf{S}_2$ in (6.37) and assume the following numerical values for a_{ij}'s and b_{ij}'s:

$$
A = \left[\begin{array}{c:ccc}
1 & 0 & 0 & 0 \\
\hdashline
-3 & -4 & 0 & 0 \\
1 & 0 & -1 & -2 \\
0 & 5 & 3 & -1
\end{array}\right]
= \left[\begin{array}{c:c}
A_{11} & 0 \\
\hdashline
A_{21}' & A_{22}'
\end{array}\right],
$$

$$
B = \left[\begin{array}{c}
1 \\
\hdashline
0 \\
0 \\
-1
\end{array}\right]
= \left[\begin{array}{c}
B_1 \\
\hdashline
B_2'
\end{array}\right]. \tag{6.38}
$$

Let the set of required eigenvalues for the closed-loop subsystem be

$$
\mathcal{L} = \{-1\} \cup \{-1,\ -1 + j1,\ -1 - j1\}, \tag{6.39}
$$

and let us partition the feedback matrix to be computed as

$$K = [K_1 \;\; K_2']. \tag{6.40}$$

Following the steps outlined above, we first choose a matrix F_1 such that

$$\Lambda(A_{11} - B_1 F_1) = \mathcal{L}_1 = \{-1\}, \tag{6.41}$$

which can easily be obtained as

$$F_1 = 2. \tag{6.42}$$

With this F_1, Equation (6.31) becomes

$$\begin{bmatrix} -4 & 0 & 0 \\ 0 & -1 & -2 \\ 5 & 3 & -1 \end{bmatrix} T_1 - T_1(-1) = -\begin{bmatrix} -3 \\ 1 \\ 0 \end{bmatrix} - \begin{bmatrix} 0 \\ 0 \\ -1 \end{bmatrix} 2, \tag{6.43}$$

the solution of which can be computed as

$$T_1 = [-1 \;\; 1 \;\; 1/2]^T. \tag{6.44}$$

At this stage, the problem reduces to computing K_2' such that

$$\Lambda[A_{22}' - (B_2' - T_1 B_1)K_2'] = \mathcal{L}_2', \tag{6.45}$$

where

$$A_{22}' = \left[\begin{array}{c|cc} -4 & 0 & 0 \\ \hline 0 & -1 & -2 \\ 5 & 3 & -1 \end{array}\right], \qquad B_2' - T_1 B_1 = \left[\begin{array}{c} 1 \\ \hline -1 \\ -3/2 \end{array}\right], \tag{6.46}$$

and

$$\mathcal{L}_2' = \{-1\} \cup \{-1 + j1, \; -1 - j1\}. \tag{6.47}$$

Repeating the entire procedure for A_{22}', $B_2' - T_1 B_1$, K_2', and $\mathcal{L}_2'$ instead of A, B, K, and $\mathcal{L}$, we can compute

$$K_2' = [0 \;\; 0 \;\; 2]. \tag{6.48}$$

Finally, using (6.25), we obtain K_1 in (6.40):

$$K_1 = F_1 - K_2' T_1 = 1. \tag{6.49}$$

The required feedback matrix is

$$K = [1 \;\; 0 \;\; 0 \;\; 2]. \tag{6.50}$$

From this example, we conclude that LBT decompositions can be useful in a number of ways. First, the stabilization problem of the overall system is reduced to stabilization of a number of smaller problems associated with the subsystems. Second, the special structure of the subsystems allows for a "piece-by-piece" computation of the feedback gains, providing an additional possibility for dimensionality reduction of the stabilization problem. Furthermore, once the block triangular structure is determined, the interconnections, which are outside the subsystems, can be identified as interconnections that have no effect on stability of the overall closed-loop systems, either in the size or sign of their elements. It should also be mentioned that the proposed decomposition scheme, being independent of the nature of the equations (nonlinear, discrete, *etc.*), can be used for stabilization of wider classes of systems than presently considered, provided the obtained subsystems can be stabilized by known methods.

6.3. Input–Output Decompositions

The output decompositions are obvious duals of input decompositions, which can be used in simplifying the *state determination* problem of large sparse systems. If we want to solve simultaneously the state determination and control problems in a subsystem-by-subsystem fashion, we need the input–output decomposition scheme introduced by Pichai *et al.* (1983), which produces hierarchically ordered subsystems with independent inputs and outputs. If such a decomposition is attempted by the input and output decompositions separately, the result may be unsatisfactory because the input and output partitions would not be compatible. The subsystems would share the inputs and outputs or even the states. The purpose of this section is to describe the input–output reachable acyclic decompositions, which can be used for an efficient and numerically attractive design of decentralized estimators and controllers for large sparse systems.

Let us consider a linear dynamic system

$$\mathbf{S}\colon \begin{aligned} z &= Ax + Bu, \\ y &= Cx, \end{aligned} \tag{6.51}$$

where, as usual, $x(t) \in \mathbf{R}^n$ is the state, $u(t) \in \mathbf{R}^m$ is the input, and $y(t) \in \mathbf{R}^\ell$ is the output of $\mathbf{S}$ at time $t \in \mathbf{T}$, and A, B, and C are constant matrices of appropriate dimensions. In (6.51), we use $z(t) \in \mathbf{R}^n$ to represent the change of the state $x(t)$ in time. When $z(t) \equiv dx(t)/dt$, $\mathbf{T}$ is the nonnegative real line $\mathbf{R}_+$, and $\mathbf{S}$ is a continuous-time dynamic system. In case $z(t) \equiv x(t+1)$, $\mathbf{T}$ is a set of nonnegative integers $\{0, 1, 2, \ldots, \}$, and $\mathbf{S}$ is a discrete-time system.

We consider the problem of decomposing the system $\mathbf{S}$ into the Input–Output LBT form

$$\mathbf{S}\colon \begin{aligned} z_i &= \sum_{j=1}^{i} A_{ij}x_j + \sum_{j=1}^{i} B_{ij}u_j, \\ y_i &= \sum_{j=1}^{i} C_{ij}x_j, \qquad i \in \boldsymbol{N}, \end{aligned} \tag{6.52}$$

which is a hierarchical interconnection of N subsystems

$$\mathbf{S}_i\colon \begin{aligned} z_i &= A_{ii}x_i + B_{ii}u_i, \\ y_i &= C_{ii}x_i, \qquad i \in \boldsymbol{N}, \end{aligned} \tag{6.53}$$

where $x_i(t) \in \mathbf{R}^{n_i}$, $u_i(t) \in \mathbf{R}^{m_i}$, and $y_i(t) \in \mathbf{R}^{\ell_i}$ are the state, input, and output of the subsystem $\mathbf{S}_i$ at time $t \in \mathbf{T}$, and $\sum_{i=1}^N n_i = n$, $\sum_{i=1}^N m_i = m$, and $\sum_{i=1}^N \ell_i = \ell$.

With the dynamic system $\mathbf{S}$ of (6.51), we associate a digraph $\mathbf{D} = (\mathtt{U} \cup \mathtt{X} \cup \mathtt{Y}, \mathtt{E})$ with the interconnection matrix

$$E = \begin{bmatrix} \bar{A} & \bar{B} & 0 \\ 0 & 0 & 0 \\ \bar{C} & 0 & 0 \end{bmatrix}, \tag{6.54}$$

and the reachability matrix

$$R = \begin{bmatrix} F & G & 0 \\ 0 & 0 & 0 \\ H & \theta & 0 \end{bmatrix}, \tag{6.55}$$

which can be computed by Algorithm 1.10.

The decomposition (6.52) is achieved by a permutation of the variables of $\mathbf{S}$, which brings the matrices A, B, C of (6.51) into an LBT form. Accordingly, the interconnection matrix E of (6.54) becomes

$$E = \left[\begin{array}{cccc|cccc|c}
\bar{A}_{11} & & & \bigcirc & \bar{B}_{11} & & & \bigcirc & \\
\bar{A}_{21} & \bar{A}_{22} & & & \bar{B}_{21} & \bar{B}_{22} & & & \bigcirc \\
\cdots & \cdots & \cdots & \cdots & \cdots & \cdots & \cdots & \cdots & \\
\bar{A}_{N1} & \bar{A}_{N2} & \cdots & \bar{A}_{NN} & \bar{B}_{N1} & \bar{B}_{N2} & \cdots & \bar{B}_{NN} & \\
\hline
 & \bigcirc & & & & & \bigcirc & & \bigcirc \\
\hline
\bar{C}_{11} & & & \bigcirc & & & & & \\
\bar{C}_{21} & \bar{C}_{22} & & & & & \bigcirc & & \bigcirc \\
\cdots & \cdots & \cdots & \cdots & & & & & \\
\bar{C}_{N1} & \bar{C}_{N2} & \cdots & \bar{C}_{NN} & & & & &
\end{array}\right]. \tag{6.56}$$

In terms of the associated digraph $\mathbf{D} = (\mathtt{U} \cup \mathtt{X} \cup \mathtt{Y}, \mathtt{E})$, the above decomposition of $\mathbf{S}$ corresponds to a partitioning of $\mathbf{D}$ into subgraphs $\mathbf{D}_i = (\mathtt{U}_i \cup \mathtt{X}_i \cup \mathtt{Y}_i, \mathtt{E}_i)$ that identify the triplets $(\bar{A}_{ii}, \bar{B}_{ii}, \bar{C}_{ii})$ of (6.53) and, thus, the subsystems $\mathbf{S}_i$.

For purposes of control and estimation of $\mathbf{S}$, we decompose the system $\mathbf{S}$ into a hierarchically ordered input–output reachable subsystem. The input–output reachability (see Remark 1.12) is the essential for controllability–observability of $\mathbf{S}$, and it is the first part of structural controllability and observability of $\mathbf{S}$. The other part is the full generic rank of the matrices $[\tilde{A}\ \ \tilde{B}]$ and $[\tilde{A}^T\ \ \tilde{C}^T]^T$, which should be tested on the subsystems levels after the LBT form of $\mathbf{S}$ is obtained for the reasons explained in the context of input LBT decompositions in Section 6.1. Thus, we start with:

6.9. DEFINITION. A partition $\mathbf{P}_{\mathrm{IO}}$ of a digraph $\mathbf{D} = (\mathtt{U} \cup \mathtt{X} \cup \mathtt{Y}, \mathtt{E})$ is said to be an input–output acyclic partition if:

(*i*) The interconnection matrix E is in an input–output LBT form.

(*ii*) Each subgraph $\mathbf{D}_i$ is input–output reachable.

An obvious necessary condition for existence of IO reachable partitions is that the digraph $\mathbf{D}$ itself be IO reachable. If this is not the case, the decomposition procedure can easily be modified to remove the unreachable parts of $\mathbf{D}$ before the decomposition is started. To avoid trivial cases, we assume that each input vertex of $\mathbf{D}$ is adjacent to at least one state vertex, and each output vertex is adjacent from at least one state vertex. These assumptions imply that **D** itself is a trivial partition $\mathbf{P}_{\mathrm{IO}}$ with $N = 1$. To consider nontrivial partitions of **D**, we assume that $n > 1$, $m > 1$, and $\ell > 1$.

Let us assume that $\mathbf{D} = (\mathtt{U} \cup \mathtt{X} \cup \mathtt{Y},\ \mathtt{E})$ has an acyclic partition into two subgraphs $\mathbf{D}_i = (\mathtt{U}_i \cup \mathtt{X}_i \cup \mathtt{Y}_i,\ \mathtt{E}_i)$, $i = 1, 2$. Then, the reachability matrix **R** of **D** can be obtained from the interconnection matrix E of (6.56), with $N = 2$, by direct computation (Algorithm 1.10) as

$$R = \left[\begin{array}{cc|cc|cc} F_{11} & 0 & G_{11} & 0 & 0 & 0 \\ F_{21} & F_{22} & G_{21} & G_{22} & 0 & 0 \\ \hline 0 & 0 & 0 & 0 & 0 & 0 \\ 0 & 0 & 0 & 0 & 0 & 0 \\ \hline H_{11} & 0 & \theta_{11} & 0 & 0 & 0 \\ H_{12} & H_{22} & \theta_{21} & \theta_{22} & 0 & 0 \end{array}\right], \qquad (6.57)$$

where, by assumption, G_{11} and G_{22} have no zero rows, H_{11} and H_{22} have no zero columns, and θ_{11} and θ_{22} each have at least one nonzero element. From (6.57), we observe that a necessary condition for the existence of an acyclic IO reachable partition is that the input–output reachability submatrix θ contains at least one zero element. Such a simple condition on G was necessary and sufficient for the existence of acyclic input reachable partitions of **D**, which were described in the preceding section. For IO reachability, this is not even close to sufficiency, and we proceed to consider a Boolean OR operation on the rows and columns of R corresponding to $\mathtt{U}_1$, $\mathtt{U}_2$, $\mathtt{Y}_1$, and $\mathtt{Y}_2$. This yields a reduced reachability matrix

$$\begin{array}{c} \qquad\quad \mathtt{X}_1 \quad\ \mathtt{X}_2 \quad\ \hat{u}_1 \quad\ \hat{u}_2 \quad\ \hat{y}_1 \quad\ \hat{y}_2 \\ \hat{R} = \left[\begin{array}{cc|cc|cc} * & 0 & e_1 & 0 & 0 & 0 \\ * & * & * & e_2 & 0 & 0 \\ \hline 0 & 0 & 0 & 0 & 0 & 0 \\ 0 & 0 & 0 & 0 & 0 & 0 \\ \hline e_1^T & 0 & 1 & 0 & 0 & 0 \\ * & e_2^T & * & 1 & 0 & 0 \end{array}\right] \begin{array}{c} \mathtt{X}_1 \\ \mathtt{X}_2 \\ \hat{u}_1 \\ \hat{u}_2 \\ \hat{y}_1 \\ \hat{y}_2 \end{array} = \begin{bmatrix} F & \hat{G} & 0 \\ 0 & 0 & 0 \\ \hat{H} & \hat{\theta} & 0 \end{bmatrix}, \end{array} \qquad (6.58)$$

where $e_i = (1, 1, \ldots, 1)^T$, $i = 1, 2$, and $*$ denotes submatrices of elements that are of no concern at this point. Finally, taking the Boolean AND of $\hat{G}$

and $\hat{H}^T$ in (6.58), that is, $\hat{S} = \hat{G} \wedge \hat{H}^T$, we can get the matrix

$$\hat{S} = \begin{array}{c} (\hat{u}_1 \wedge \hat{y}_1) \quad (\hat{u}_2 \wedge \hat{y}_2) \\ \begin{bmatrix} e_1 & 0 \\ 0 & e_2 \end{bmatrix} \end{array}, \tag{6.59}$$

and state the following:

6.10. THEOREM. A given digraph $\mathbf{D} = (\mathtt{U} \cup \mathtt{X} \cup \mathtt{Y}, \mathtt{E})$ has a $\mathbf{P}_{\text{IO}}$ partition if and only if U and Y can be partitioned into $\mathtt{U}_1$, $\mathtt{U}_2$, and $\mathtt{Y}_1$, $\mathtt{Y}_2$ such that $\hat{\theta}$ has the form in (6.58), and the matrix $\hat{S}$ contains exactly a single 1 in each row.

Proof. Necessity is obvious from the construction of the matrices $\hat{R}$ and $\hat{S}$. On the other hand, if the matrix $\hat{S} = (\hat{s}_{ij})$ has exactly a single 1 in each row, defining $\mathtt{X}_j = \{x_i \in \mathtt{X}\colon\ \hat{s}_{ij} = 1,\ i \in \boldsymbol{N}\}$, $j = 1, 2$, we obtain a partition of X into $\mathtt{X}_1$ and $\mathtt{X}_2$. A permutation of X according to this partition puts $\hat{S}$ into the form of (6.59), which implies that $\mathtt{X}_j$ is reachable from $\mathtt{U}_j$ and reaches $\mathtt{Y}_j$. Furthermore, the form of $\hat{R}$ in (6.58) implies that neither $\mathtt{U}_2$ reaches $\mathtt{X}_1$ nor $\mathtt{X}_2$ reaches $\mathtt{X}_1$ or $\mathtt{Y}_1$. Thus, we obtain a $\mathbf{P}_{\text{IO}}$ partition of $\mathbf{D}$ into $\mathbf{D}_j = \{\mathtt{U}_j \cup \mathtt{X}_j \cup \mathtt{Y}_j, \mathtt{E}_j\}$, $j = 1, 2$. This completes the sufficiency and the proof. Q.E.D.

To get the maximum benefit from the decomposition we would like to obtain irreducible $\mathbf{P}_{\text{IO}}$ partitions of $\mathbf{D}$, in which no subgraph $\mathbf{D}_i$ can be further decomposed into acyclic IO reachable subgraphs. For this purpose, we need the following definition:

6.11. DEFINITION. We say that in the $\ell \times m$ matrix $\theta = (\theta_{ir})$ of (6.55), the pth row dominates the qth row if $\theta_{pr} = 1$ whenever $\theta_{qr} = 1$, and for at least one r, $\theta_{pr} = 1$, but $\theta_{qr} = 0$, $r \in \boldsymbol{M} = \{1, 2, \ldots, m\}$.

The following result is automatic:

6.12. THEOREM. Under the conditions of Theorem 6.10, the subgraph $\mathbf{D}_1$ of the $\mathbf{P}_{\text{IO}}$ partition induced by the matrix $\hat{S}$ is irreducible if and only if no row of the matrix θ corresponding to $\mathtt{Y}_1$ dominates any other row corresponding to $\mathtt{Y}_2$.

In other words, if we start the partitioning of the digraph $\mathbf{D}$ from those rows of θ that are not dominated by any other row of θ, we guarantee irreducibility of the first digraph $\mathbf{D}_1$ of the resulting partition $\mathbf{P}_{\mathrm{IO}}$, provided such a partition exists. Once such a partition is obtained, the algorithm can be applied successively to the remaining subgraph $\mathbf{D}_2$ to generate the irreducible partitions of the entire digraph $\mathbf{D}$.

The following algorithm of Pichai *et al.* (1983) can be used to generate all irreducible acyclic partitions of a given digraph $\mathbf{D}$:

6.13. ALGORITHM.

```
PROCEDURE DECOMP
        Calculate reachability matrix R
        Form IO reachability matrix θ
        IF  θ(p, q) ≠ 0 ∀p, q, THEN RETURN
            ELSE k = 0,  j = 0
1       IF  j = ℓ  THEN RETURN
            ELSE i = j + 1
2       IF  θ(i, *) labeled THEN j = j + 1 GO TO 1
            ELSE continue
3       IF  θ(i, *) dominates (+) labeled θ(p, *) ∀p
            THEN label (+)θ(i, *),  i = j + 1,  GO TO 2
        IF  θ(i, *) dominates unlabeled θ(p, *) ∀p
            THEN  i = p  GO TO 3
            ELSE continue
        Define the index sets
            I = {p = 1, 2, ..., ℓ:  θ(i, *) dominates θ(p, *)}
            J = {j = 1, 2, ..., m:  θ(i, j) = 1}
        Define the input and output partitions
            U1 = {uj:  j ∈ J},  U2 = U − U1
            Y1 = {yj:  j ∈ I},  Y2 = Y − Y1
        Form reduced reachability matrices R, G̃, H̃
        Compute S̃ = G̃ ∧ H̃^T
        IF  S̃ has one and only one nonzero in each row
            THEN  Define state partition
                  X1 = {xp:  S̃(*, 1) has nonzero in pth row}
                  X2 = X − X1
                  k = k + 1,  OUTPUT  {U1 ∪ X1 ∪ Y1}
                  label (+) θ(i, *) GO TO 1
            ELSE  label (−) θ(i, *) GO TO 1
END DECOMP
```

Now, several remarks are in order.

6.14. REMARK (Existence and Uniqueness). From the algorithm we see that if the IO reachability matrix θ contains no zero elements, then **D** has no acyclic partitions. On the other hand, if **D** has such partitions, they are not necessarily unique. For example, the digraph of Figure 6.4 has two distinct decompositions, one having two and the other three components. In Algorithm 6.13, k denotes the number of distinct acyclic partitions of **D**.

6.15. REMARK (Condensations). As in the input reachable decompositions of Section 6.1, the computations involved in Algorithm 6.13 can be considerably reduced by considering a condensation of **D** instead of **D** itself. For this purpose, we partition the vertex sets of **D** as

$$\mathtt{X} = \bigcup_{i=1}^{p} \mathtt{X}_i^\bullet, \qquad \mathtt{U} = \bigcup_{i=1}^{q} \mathtt{U}_i^\bullet, \qquad \mathtt{Y} = \bigcup_{i=1}^{r} \mathtt{Y}_i^\bullet, \tag{6.60}$$

where $\mathtt{X}_i^\bullet$ are the vertex sets of the strong components of the state truncation subgraph $\mathbf{D}_x = (\mathtt{X}, \mathtt{E}_x)$ defined in Section 1.3, and each $\mathtt{U}_i^\bullet(\mathtt{Y}_i^\bullet)$ contains all the input (output) vertices that are adjacent to (from) those state vertices that occur in the same strong component of $\mathbf{D}_x$, but no others. If we denote the condensations of **D** with respect to the above partition by $\mathbf{D}^* = (\mathtt{U}^* \cup \mathtt{X}^* \cup \mathtt{Y}^*, \mathtt{E}^*)$, then it is not difficult to show that, to any irreducible acyclic partition of **D** into N subgraphs, there corresponds a unique irreducible acyclic partition of $\mathbf{D}^*$ into the same number of subgraphs, and conversely. Intuitively, this follows from the fact that any two state vertices of **D** that belong to the same strong component of the truncation $\mathbf{D}_x$ should occur in the same subgraph of any irreducible acyclic decomposition of **D**, as well as any two input (output) vertices that are adjacent exactly to (from) the same strong component of $\mathbf{D}_x$. In case of input decompositions, this result has been established by Theorem 6.5.

The fact outlined above simply states that the decomposition problems of **D** and $\mathbf{D}^*$ are interchangeable. However, since $\mathbf{D}^*$ is considerably smaller than **D**, irreducible acyclic partitions of **D** can be generated much more easily by first obtaining $\mathbf{D}^*$, and then applying the decomposition scheme to $\mathbf{D}^*$. An additional advantage of using such an approach is that the strong components of each subgraph $\mathbf{D}_k^*$ of the decoupled subsystems would also be identified as a by-product of the decomposition scheme, allowing for a further simplification in the subsequent control design (see Section 6.2).

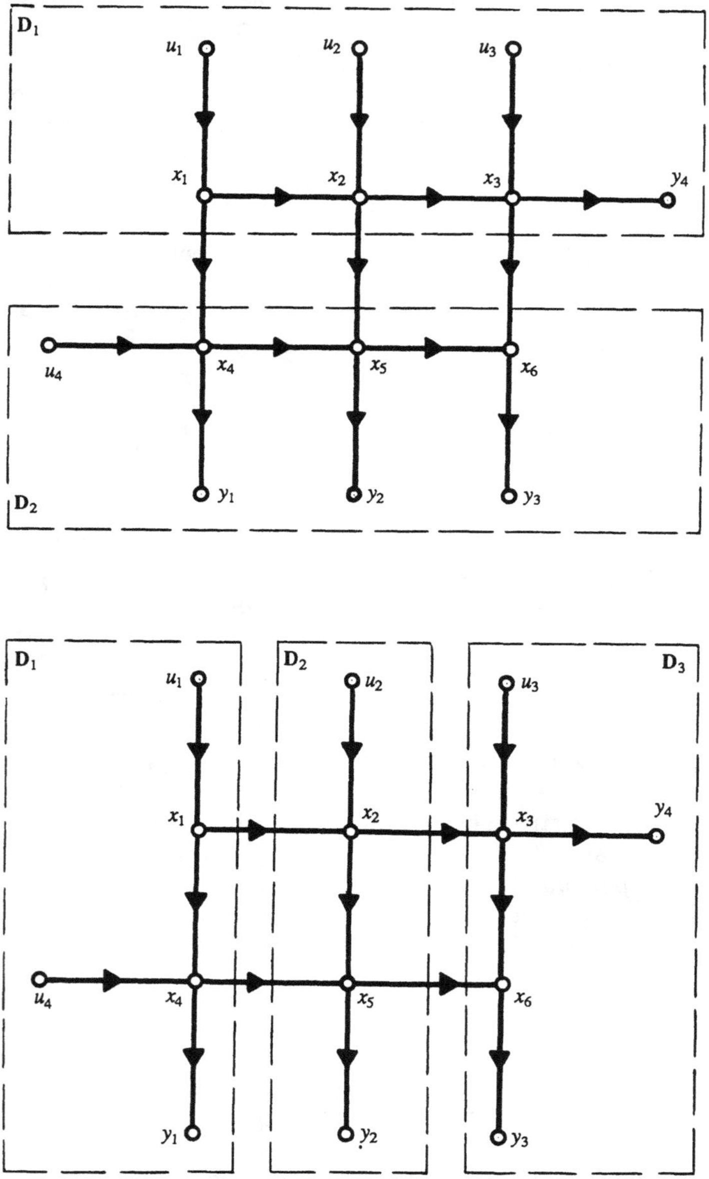

Fig. 6.4. Two distinct IO partitions.

6.16. REMARK (Input and Output Reachable Partitions). In general, decompositions of a digraph into IO reachable subgraphs are more restrictive than decompositions into only input (output) reachable subgraphs without considerations of outputs (inputs). For example, the digraph of Figure 6.5a has no IO reachable partitions, but it does have an input reachable partition as shown in Figure 6.5b, as well as an output reachable partition as given in Figure 6.5c. This observation suggests that each subgraph $\mathbf{D}_k$ of an IO reachable partition can be further decomposed into input (output) reachable subgraphs. This way, we obtain a multilevel partition of $\mathbf{D}$ as illustrated by the corresponding interconnection matrix of Figure 6.6.

6.17. REMARK (IO Decomposition is NP-complete). It was shown by Tarjan (1984) that our IO decomposition scheme is NP-complete. In the studies of algorithm design and computational complexity, the NP-complete problems are of central interest, and Tarjan's result calls for an explanation of what it means in the context of estimation and control designs. We start by recognizing the fact that in combinatorial problems one prefers polynomial to exponential time algorithms, because the latter type have a running time which grows exponentially with the length of the input and can rapidly put a problem beyond available computational means. For this reason, one is interested in finding a polynomial algorithm for a given problem, or showing that no such algorithm is possible and the problem is "intractable." Due to a lack of such strong results, one looks for a class of problems that are equally difficult and can serve as a reference in evaluating the complexity of a given problem. Such class is the set of NP-complete problems (Garey and Johnson, 1979), which are polynomially transferable into each other, and which are *probably* intractable. Although the NP-complete problems have been difficult to solve, we stress the fact that it is not known that there are no polynomial algorithms for their solution. Furthermore, even if they are intractable, the meaning of "intractability" should not be taken literally, because the 2^n algorithm is faster than n^5 algorithm for $n \leq 20$. Since our IO decomposition involves a condensation digraph $\mathbf{D}^* = (\mathtt{U}^* \cup \mathtt{X}^* \cup \mathtt{Y}^*, \mathtt{E}^*)$, if we assume, for example, that each super-state $x^* \in \mathtt{X}^*$ represents on the average ten states of the strong component of the truncation $\mathbf{D}_x = (\mathtt{X}, \mathtt{E}_x)$, then it follows that we can consider a system $\mathbf{S}$ having two hundred states without any concern about intractability. We can safely say that the dynamic part of the control problems in large-scale systems would saturate numerically well before the IO decomposition algorithm would experience some difficulties, if any at all (Pichai *et al.* 1984). Finally, we should note that the IO decomposition is NP-complete because it involves matching of three sets: input, state, and

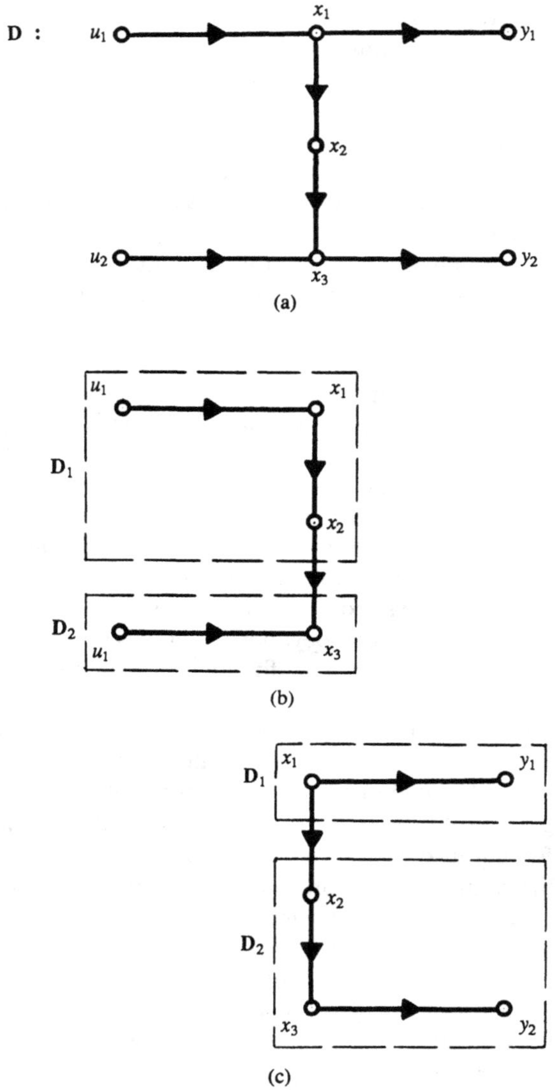

Fig. 6.5. Digraph without IO reachable partition.

output vertices of $\mathbf{D}^*$. On the other hand, the input decomposition and output decompositions are "two dimensional" problems, and can be solved in polynomial time. For this reason, one can use two-stage decompositions for building estimators and controllers. First, an output decomposition is

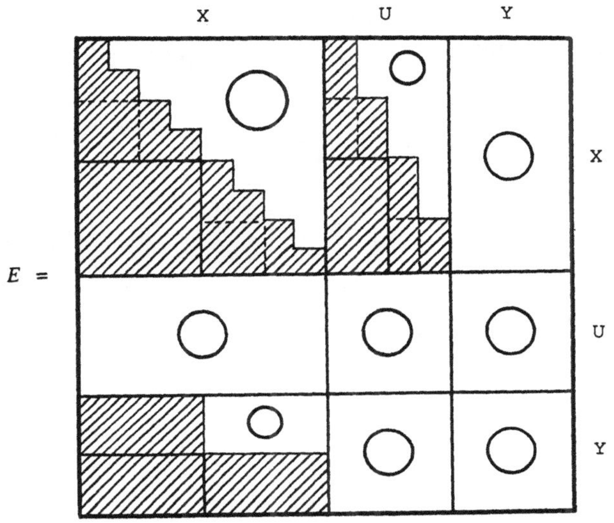

Fig. 6.6. Interconnection matrix.

used to build an estimator, and then the variables are permuted into an input decomposition to design a controller. This two-stage design is possible because the separation principle for the estimator–controller combination is valid. How this can be done even when there is no IO decompositions at all, is shown in Example 6.25.

6.18. REMARK (Structural Controllability and Observability). We recall that the objective of the IO decomposition is to produce an acyclic partition of **S** into input and output reachable subsystems. After such a partition is obtained, the subsystem can be checked for full generic rank to ensure that all the subsystems are structurally controllable and observable (Section 1.4). Of course, the subsystems can be checked for standard controllability and observability, which suffice to establish these properties for the overall system **S**, because it has the lower block triangular form. This should normally be an easy task since the subsystems are of lower dimensions. If we discover a lack of full generic rank in any of the subsystems, the corresponding subsystems can be fused into larger subsystems in order to satisfy the full rank condition. This procedure has been explained in Example 6.8 which dealt with input reachable partitions.

6.19. EXAMPLE. To illustrate the application of the IO decompositions, we consider a dynamic system described by the equations

$$
\mathbf{S}\colon\ \dot{x}(t) = \begin{bmatrix}
 & & & & a_{15} & & & & & & & \\
 & & & & & & & & & & & \\
 & a_{32} & & & & & & & & a_{3,10} & & \\
 & & & & & & & a_{48} & a_{49} & & & \\
 & & & & & & & & a_{59} & & & \\
 & & & & & & & & a_{69} & & & \\
 & a_{72} & & & & a_{76} & & & & & & \\
 & & & & & & & & & & & \\
 & & & & a_{95} & & & & & & & \\
 & & a_{10,3} & & & & a_{10,7} & & & & & \\
a_{11,1} & & & & & & & & & & & \\
 & & & & & & & a_{12,8} & & & &
\end{bmatrix} x(t)
+ \begin{bmatrix}
 & & & & b_{15} \\
 & & & b_{24} & \\
b_{31} & & & & \\
 & & & & \\
 & & & & \\
 & b_{62} & & & b_{65} \\
 & & & & \\
 & & b_{83} & & \\
 & & & & b_{95} \\
 & & & & \\
 & & & & \\
b_{12,1} & & & &
\end{bmatrix} u(t),
$$

$$
y(t) = \begin{bmatrix}
 & & & & & & & & & & & c_{1,12} \\
 & & & c_{24} & & & & & & & & \\
 & & & & & & & & & & c_{3,11} & \\
 & c_{42} & & & & & c_{47} & & & & & \\
 & & & & & & & & & c_{5,10} & &
\end{bmatrix} x(t), \tag{6.61}
$$

which are *sparse*; that is, there are considerably less nonzero than zero elements in the system matrices.

The first step is to compute the reachability matrix R of (6.55), and get the submatrix θ as

$$\theta = \begin{bmatrix} 1 & 0 & 1 & 0 & 0 \\ 0 & 0 & 1 & 0 & 1 \\ 0 & 0 & 0 & 0 & 1 \\ 0 & 1 & 0 & 1 & 1 \\ 1 & 1 & 0 & 1 & 1 \end{bmatrix}. \tag{6.62}$$

Following Algorithm 6.13, we set $i = 1$. Since the first row of θ does not dominate any other row, we define the sets $\boldsymbol{I} = \{1\}$, $\boldsymbol{J} = \{1, 3\}$. The matrix $\hat{S}$ can be obtained as

$$\tilde{S} = \begin{bmatrix} 0 & 0 & 0 & 0 & 0 & 0 & 0 & 1 & 0 & 0 & 0 & 1 \\ 1 & 1 & 1 & 1 & 1 & 1 & 1 & 0 & 1 & 1 & 1 & 0 \end{bmatrix}^T, \tag{6.63}$$

which indicates that the partitions

$$\begin{aligned} \mathtt{U}_1 &= \{u_1, u_3\}, & \mathtt{U}_2 &= \{u_2, u_4, u_5\}, \\ \mathtt{X}_1 &= \{x_8, x_{12}\}, & \mathtt{X}_2 &= \{x_1, x_2, x_3, x_4, x_5, x_6, x_7, x_9, x_{10}, x_{11}\}, \\ \mathtt{Y}_1 &= \{y_1\}, & \mathtt{Y}_2 &= \{y_2, y_3, y_4, y_5\} \end{aligned} \tag{6.64}$$

yield an acyclic IO reachable decomposition of $\mathbf{D}$ into two components $\mathbf{D}_i^1 = \{\mathtt{U}_i \cup \mathtt{X}_i \cup \mathtt{Y}_i, \mathtt{E}_i\}$, $i = 1, 2$, where $\mathbf{D}_1^1$ is irreducible.

Having achieved a decomposition, we label the first row of θ by (+), and check the second row. Since the second row dominates the unlabeled third row, which does not dominate any other row, we set $i = 3$, and define $\boldsymbol{I} = \{3\}$, $\boldsymbol{J} = \{5\}$. This gives

$$\hat{S} = \begin{bmatrix} 1 & 0 & 0 & 0 & 1 & 0 & 0 & 0 & 1 & 0 & 1 & 0 \\ 0 & 1 & 1 & 1 & 0 & 1 & 1 & 1 & 0 & 1 & 0 & 1 \end{bmatrix}^T, \tag{6.65}$$

indicating that the partitions

$$\begin{aligned} \mathtt{U}_1 &= \{u_5\}, & \mathtt{U}_2 &= \{u_1, u_2, u_3, u_4\}, \\ \mathtt{X}_1 &= \{x_1, x_5, x_9, x_{11}\}, & \mathtt{X}_2 &= \{x_2, x_3, x_4, x_6, x_7, x_8, x_{10}, x_{12}\}, \\ \mathtt{Y}_1 &= \{y_3\}, & \mathtt{Y}_2 &= \{y_1, y_2, y_4, y_5\} \end{aligned} \tag{6.66}$$

provide another acyclic decomposition of $\mathbf{D}$ into $\mathbf{D}_1^2$ and $\mathbf{D}_2^2$, where $\mathbf{D}_1^2$ is irreducible. We label the third row $(+)$.

Now, we go back to the second row of θ. Since it dominates the $(+)$ labeled third row, we label it $(+)$ also. The fourth and fifth rows of θ are also labeled $(+)$, since they both dominate the third row, and Algorithm 6.13 stops. This first round of the algorithm produces two distinct acyclic IO reachable decompositions, $\mathbf{D} = \mathbf{D}_1^1 \cup \mathbf{D}_2^1$ and $\mathbf{D} = \mathbf{D}_1^2 \cup \mathbf{D}_2^2$, where $\mathbf{D}_1^1$ and $\mathbf{D}_1^2$ are irreducible. In the second round, we apply the algorithm to $\mathbf{D}_2^1$ and $\mathbf{D}_2^2$, separately, to decompose them further into acyclic IO reachable components. The process is continued until all components are irreducible. Table 6.1 summarizes all possible $\mathbf{P}_{\mathrm{IO}}$ partitions for the system $\mathbf{S}$ of (6.61).

In Table 6.1, we observe that only the partition $\mathbf{P}_{\mathrm{IO}}^2$ has no dilation in any of its components, and they are all structurally controllable (SC) and observable (SO), which is indicated by the symbol $\surd$, while the symbol X indicates the lack of these properties. The partition $\mathbf{P}_{\mathrm{IO}}^2$ is shown in Figure 6.7. Although structural controllability and observability can be established by merging components in the other partitions (*e.g.*, by merging $\mathbf{D}_2^5$, $\mathbf{D}_3^5$, and $\mathbf{D}_4^5$ in $\mathbf{P}_{\mathrm{IO}}^5$), this produces coarser decompositions, resulting in a reduction of computational simplification offered by acyclic IO partitions.

Let us show how we can use the acyclic decompositions for a sequential design of an observer for the system $\mathbf{S}$ of (6.61). Choosing some arbitrary numbers for the nonzero elements of the system matrices in (6.61), the partition $\mathbf{P}_{\mathrm{IO}}^2$ of Table 6.1 after permutation yields the system $\mathbf{S}$ as

$$
\mathbf{S}\colon\ \dot{x}(t) = \left[\begin{array}{cccc|cc|c|ccccc}
0 & -1 & 0 & 0 & & & & & & & & \\
1 & 0 & 0 & 0 & & & & & & \bigcirc & & \\
3 & 0 & 0 & 0 & & & & & & & & \\
0 & 0 & 1 & 0 & & & & & & & & \\
\hline
0 & 0 & 0 & 0 & 0 & 0 & & & & & & \\
0 & 4 & 0 & 0 & -2 & 0 & & & & & & \\
\hline
 & \bigcirc & & & \bigcirc & & 0 & & & & & \\
\hline
0 & 0 & 0 & 0 & & & & 0 & 0 & 0 & 0 & 0 \\
0 & 3 & 0 & 0 & & & & 0 & 0 & 0 & 0 & 0 \\
0 & 0 & 0 & 0 & \bigcirc & & 0 & 1 & 1 & 0 & 0 & 0 \\
0 & 0 & 0 & 0 & & & & 4 & 0 & 0 & 0 & 2 \\
0 & 0 & 0 & 0 & & & & 0 & 0 & 3 & -1 & 0
\end{array}\right] x(t)
$$

$$+ \begin{bmatrix} 0 & & & & \\ 1 & & & \bigcirc & \\ 2 & & & & \\ 0 & & & & \\ \hline 0 & -1 & & & \\ & 0 & & & \\ \hline 0 & 0 & 1 & & \\ \hline 0 & & & 1 & 0 \\ 2 & & & 0 & -1 \\ 0 & 0 & 0 & 0 & 0 \\ 0 & & & 0 & 0 \\ 0 & & & 0 & 0 \end{bmatrix} u(t),$$

$$y(t) = \begin{bmatrix} 0 & 0 & 0 & 2 & & & & & & \\ & & & & 0 & 1 & & & \bigcirc & \\ & & & & & & 1 & & & \\ & & \bigcirc & & & & & 0 & 0 & 0 & 0 & 1 \\ & & & & & & & 2 & 0 & 1 & 0 & 0 \end{bmatrix} x(t),$$

$$\tag{6.67}$$

where

$$x = (x_5,\ x_9,\ x_1,\ x_{11};\ x_8,\ x_4;\ x_{12};\ x_2,\ x_6,\ x_7,\ x_3,\ x_{10})^T,$$

$$u = (u_5;\ u_3;\ u_1;\ u_4,\ u_2)^T, \tag{6.68}$$

$$y = (y_3;\ y_2;\ y_1;\ y_5,\ y_4)^T.$$

To explain how the Luenberger observer can be built sequentially for the system **S** of (6.67), let us consider a system

$$\textbf{S}:\ \dot{x}_i = A_{ii}x_i + \sum_{j=1}^{i-1} A_{ij}x_j + \sum_{j=1}^{i} B_{ij}u_j,$$

$$\tag{6.69}$$

$$y_i = C_{ii}x_i + \sum_{j=1}^{i-1} C_{ij}x_j, \qquad i \in \textbf{N},$$

Table 6.1. Distinct partitions.

$\mathbf{P}^k_{IO}$	$\mathbf{D}^k_1$	$\mathtt{U}_i$	$\mathtt{X}_i$	$\mathtt{Y}_i$	SC	SO
$\mathbf{P}^1_{IO}$	$\mathbf{D}^1_1$	$\{u_1, u_3\}$	$\{x_8, x_{12}\}$	$\{y_1\}$	√	√
	$\mathbf{D}^1_2$	$\{u_5\}$	$\{x_1, x_4, x_5, x_9, x_{11}\}$	$\{y_2, y_3\}$	X	√
	$\mathbf{D}^1_3$	$\{u_2, u_4\}$	$\{x_2, x_3, x_6, x_7, x_{10}\}$	$\{y_4, y_5\}$	√	√
$\mathbf{P}^2_{IO}$	$\mathbf{D}^2_1$	$\{u_5\}$	$\{x_1, x_5, x_9, x_{11}\}$	$\{y_3\}$	√	√
	$\mathbf{D}^2_2$	$\{u_3\}$	$\{x_4, x_8\}$	$\{y_2\}$	√	√
	$\mathbf{D}^2_3$	$\{u_1\}$	$\{x_{12}\}$	$\{y_1\}$	√	√
	$\mathbf{D}^2_4$	$\{u_2, u_4\}$	$\{x_2, x_3, x_6, x_7, x_{10}\}$	$\{y_4, y_5\}$	√	√
$\mathbf{P}^3_{IO}$	$\mathbf{D}^3_1$	$\{u_5\}$	$\{x_1, x_5, x_9, x_{11}\}$	$\{y_3\}$	√	√
	$\mathbf{D}^3_2$	$\{u_3\}$	$\{x_4, x_8\}$	$\{y_2\}$	√	√
	$\mathbf{D}^3_3$	$\{u_2, u_4\}$	$\{x_2, x_6, x_7\}$	$\{y_4\}$	√	X
	$\mathbf{D}^3_4$	$\{u_1\}$	$\{x_3, x_{10}, x_{12}\}$	$\{y_1, y_5\}$	√	√
$\mathbf{P}^4_{IO}$	$\mathbf{D}^4_1$	$\{u_5\}$	$\{x_1, x_5, x_9, x_{11}\}$	$\{y_3\}$	√	√
	$\mathbf{D}^4_2$	$\{u_2, u_4\}$	$\{x_2, x_6, x_7\}$	$\{y_4\}$	√	X
	$\mathbf{D}^4_3$	$\{u_3\}$	$\{x_4, x_8\}$	$\{y_2\}$	√	√
	$\mathbf{D}^4_4$	$\{u_1\}$	$\{x_3, x_{10}, x_{12}\}$	$\{y_1, y_5\}$	√	√
$\mathbf{P}^5_{IO}$	$\mathbf{D}^5_1$	$\{u_5\}$	$\{x_1, x_5, x_9, x_{11}\}$	$\{y_3\}$	√	√
	$\mathbf{D}^5_2$	$\{u_2, u_4\}$	$\{x_2, x_6, x_7\}$	$\{y_4\}$	√	X
	$\mathbf{D}^5_3$	$\{u_1\}$	$\{x_3, x_{10}\}$	$\{y_5\}$	√	√
	$\mathbf{D}^5_4$	$\{u_3\}$	$\{x_4, x_8, x_{12}\}$	$\{y_{1,2}\}$	X	√

which has LBT form, and assume that the pairs (A_{ii}, C_{ii}) are observable. Treating all the summation terms in (6.69) as inputs, a reduced order observer can be designed sequentially as

$$\hat{\mathbf{S}}\colon\ \dot{\tilde{x}}_i = \tilde{A}_{ii}\tilde{x}_i + \sum_{j=1}^{i-1} \tilde{M}_{ij}\hat{x}_j + \sum_{j=1}^{i} \tilde{B}_{ij}u_j + \tilde{L}_{ii}\left(y_i - \sum_{j=1}^{i-1} C_{ij}\hat{x}_j\right),$$

$$\hat{x}_i = \hat{M}_{ii}\tilde{x}_i + \hat{L}_{ii}\left(y_i - \sum_{j=1}^{i-1} C_{ij}x_j\right), \qquad i \in \boldsymbol{N}, \tag{6.70}$$

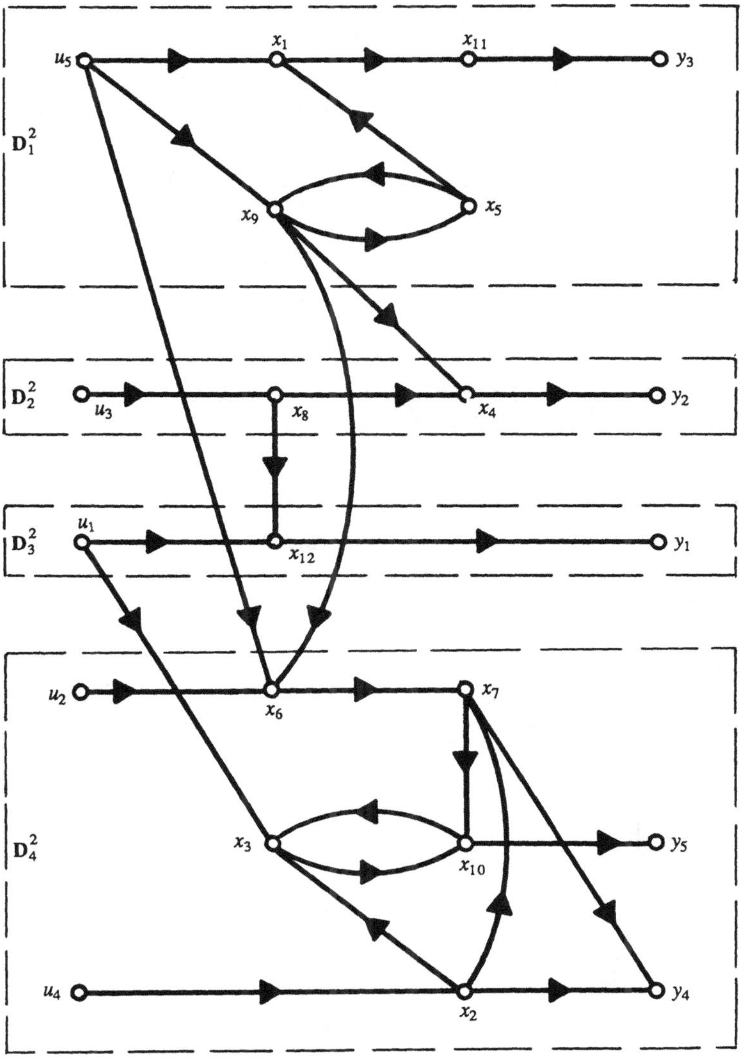

Fig. 6.7. Irreducible partition $\mathbf{P}_{\mathrm{IO}}^2$.

where $\tilde{x}_i(t) \in \mathbf{R}^{n_i-\ell_i}$ is the state of the ith component of the observer, and $\hat{x}_i(t)$ is used to estimate $x_i(t)$. The matrices appearing in (6.70) are computed in a standard way (*e.g.*, O'Reilly, 1983). The matrices $\tilde{T}_{ii}$ and $\tilde{L}_{ii}$ are chosen such that $[\tilde{T}_{ii}^T \;\; C_{ii}^T]$ is nonsingular, and the equation

$$\tilde{A}_{ii}\tilde{T}_{ii} - \tilde{T}_{ii}A_{ii} + \tilde{L}_{ii}C_{ii} = 0, \qquad i \in \boldsymbol{N}, \tag{6.71}$$

is satisfied for some matrix $\tilde{A}_{ii}$ with specified eigenvalues. Other matrices in (6.71) are then computed as

$$\tilde{A}_{ij} = \tilde{T}_{ii} A_{ij}, \qquad \tilde{B}_{ij} = \tilde{T}_{ii} B_{ij}, \qquad [\tilde{T}_{ii}^T \;\; C_{ii}^T]^{-1} = [\hat{M}_{ii} \;\; \hat{L}_{ii}]^T. \tag{6.72}$$

With this choice of the matrices and defining $e_i(t) = \tilde{x}_i - \tilde{T}_{ii} x_i$, it takes a little matrix algebra to show that

$$\dot{e} = \tilde{A}_{ii} e_i + \sum_{j=1}^{i-1} \tilde{A}_{ij} e_j, \tag{6.73}$$

and

$$\hat{x}_i - x_i = \hat{M}_{ii} e_i - \hat{L}_{ii} \sum_{j=1}^{i-1} C_{ij} (\hat{x}_j - x_j), \tag{6.74}$$

where the matrices $\tilde{A}_{ij}$ are defined in terms of $\tilde{L}_{ii}$, $\hat{M}_{jj}$, C_{pq}, $j < p \leq i$, $j \leq q < i$, and $\hat{L}_{rr}$, $j < r < i$. From (6.73) and (6.74) it is clear that $\hat{x}_i(t) \to x_i(t)$ as $t \to +\infty$. In this way, an observer design is achieved by designing small individual observers for the acyclic components of the system. Equation (6.70) indicates that the observer has the same hierarchical (LBT) structure as the system itself.

Now, we go back to our specific system **S** of (6.67), and assume that it is required that the overall observer have all its eigenvalues at -1. Applying the above procedure, we obtain the following subsystem observers:

$$O_1\colon \; \dot{\tilde{x}}_1 = \begin{bmatrix} 0 & -1 & -2/3 \\ 1 & 0 & -2/3 \\ -3 & 0 & -3 \end{bmatrix} \tilde{x}_1 + \begin{bmatrix} 0 \\ 2 \\ 4 \end{bmatrix} u_5 + \begin{bmatrix} -8/3 \\ -4/3 \\ -11 \end{bmatrix} y_3,$$

$$\begin{bmatrix} \hat{x}_5 \\ \hat{x}_9 \\ \hat{x}_1 \\ \hat{x}_{11} \end{bmatrix} = \begin{bmatrix} 1/2 & 0 & 0 \\ 0 & 1/2 & 0 \\ 0 & 0 & 1/2 \\ 0 & 0 & 0 \end{bmatrix} \tilde{x}_1 + \begin{bmatrix} 1/3 \\ 1/3 \\ 3/2 \\ 1/2 \end{bmatrix} y_3;$$

$$O_2\colon \; \dot{\tilde{x}}_2 = -\tilde{x}_2 + 2x_9 - u_3 + \frac{1}{2} y_2,$$

$$\begin{bmatrix} \hat{x}_8 \\ \hat{x}_4 \end{bmatrix} = \begin{bmatrix} 1 \\ 0 \end{bmatrix} \tilde{x}_2 + \begin{bmatrix} -2 \\ 1 \end{bmatrix} y_2;$$

$$O_3\colon \; \hat{x}_{12} = y_1;$$

$$O_4:\ \dot{\tilde{x}}_4 = \begin{bmatrix} -3/2 & 0 & -1/4 \\ -1 & -1 & 0 \\ 1 & 0 & -1/2 \end{bmatrix}\tilde{x}_4 + \begin{bmatrix}0\\3\\0\end{bmatrix} x_9 + \begin{bmatrix}0\\2\\0\end{bmatrix}u_5 + \begin{bmatrix}0\\0\\-2\end{bmatrix}u_1$$

$$+ \begin{bmatrix}1&0\\-2&-1\\0&0\end{bmatrix}\begin{bmatrix}u_4\\u_2\end{bmatrix} + \begin{bmatrix}1/2&3/4\\1/4&-1\\2&3/2\end{bmatrix}\begin{bmatrix}y_5\\y_4\end{bmatrix},$$

$$\begin{bmatrix}\hat{x}_2\\ \hat{x}_6\\ \hat{x}_7\\ \hat{x}_3\\ \hat{x}_{10}\end{bmatrix} = \begin{bmatrix}1&0&0\\0&1&0\\-2&0&0\\0&0&1\\0&0&0\end{bmatrix}\tilde{x}_4 + \begin{bmatrix}-1/4&0\\0&1\\1/2&1\\1/2&1\\-1/2&0\\1&0\end{bmatrix}\begin{bmatrix}y_5\\y_4\end{bmatrix}. \tag{6.75}$$

With a little manipulation, an overall representation of the dynamic part of the observer can be obtained as

$$O:\ \begin{bmatrix}\dot{\tilde{x}}_1\\ \dot{\tilde{x}}_2\\ \dot{\tilde{x}}_4\end{bmatrix} = \left[\begin{array}{ccc|c|ccc} 0 & -1 & -2/3 & & & & \\ 1 & 0 & -2/3 & & & \bigcirc & \\ -3 & 0 & -3 & & & & \\ \hline 0 & 1 & 0 & -1 & & & \\ \hline 0 & 0 & 0 & & -3/2 & 0 & -1/4 \\ 0 & 3/2 & 0 & \bigcirc & -1 & -1 & 0 \\ 0 & 0 & 0 & & 1 & 0 & -1/2 \end{array}\right]\begin{bmatrix}\tilde{x}_1\\ \tilde{x}_2\\ \tilde{x}_4\end{bmatrix}$$

$$+ \left[\begin{array}{c|c|c|cc} 0 & & & & \\ 2 & & \bigcirc & & \\ 4 & & & & \\ \hline 0 & -1 & & & \\ \hline 0 & & 0 & 1 & 0 \\ 2 & \bigcirc & 0 & -2 & -1 \\ 0 & & -2 & 0 & 0 \end{array}\right] u(t)$$

$$+\left[\begin{array}{c:c:c:cc} -8/3 & & & & \\ -4/3 & & \bigcirc & & \\ -11 & & & & \\ \hdashline 2/3 & 1/2 & & & \\ \hdashline 0 & & & 1/2 & 3/4 \\ 1 & \bigcirc & \bigcirc & 1/4 & -1 \\ 0 & & & 2 & 3/2 \end{array}\right] y(t), \tag{6.76}$$

which has the same hierarchical (LBT) structure as **S** itself, save that the expression for $\dot{\hat{x}}_3$ is omitted as it is not necessary for the implementation of the observer O.

6.4. Sequential Optimization

Optimization of hierarchically structured control systems may proceed in several different ways, depending on the choice of information distribution among the subsystems. We shall consider the three basic information structure constraints: global, local, and sequential. Special attention will be given to the sequential information structure where fast top-down communication is available for fast data transmission from the upper to the lower levels of the hierarchy. The sequential processing and control design will be contrasted to the standard global and totally decentralized local structures in the context of system performance.

We shall consider a hierarchically structured plant represented by a discrete-time linear stochastic system. With each subsystem, we associate a quadratic cost, which we minimize sequentially one at a time assuming that feedback control loops are closed at higher levels of the hierarchy. Optimal control algorithms are formulated as a scheme (Stanković and Šiljak, 1990), which builds state models for each optimization stage allowing for the standard LQG method to be used sequentially. The scheme was introduced by Özgüner and Perkins (1978) for control design of LBT deterministic continuous-time systems with complete state information.

A discrete-time system **S** is hierarchically structured as

$$\mathbf{S}\colon\ \begin{aligned} x_1(t+1) &= A_{11}x_1(t) + B_{11}u_1(t) + w_1(t), \\ y_1(t) &= C_{11}x_1(t) + v_1(t), \end{aligned} \tag{6.77}$$

$$x_i(t+1) = A_{ii}x_i(t) + B_{ii}u_i(t) + \sum_{j=1}^{i-1} A_{ij}x_j(t) + \sum_{j=1}^{i-1} B_{ij}u_j(t) + w_i(t),$$

$$y_i(t) = C_{ii}x_i(t) + \sum_{j=1}^{i-1} C_{ij}x_j(t) + v_i(t), \qquad i = 2, 3, \ldots, N, \tag{6.78}$$

where $x_i(t) \in \mathbf{R}^{n_i}$, $u_i(t) \in \mathbf{R}^{m_i}$, and $y_i(t) \in \mathbf{R}^{\ell_i}$ are the state, input, and output of the subsystem $\mathbf{S}_i$ at time $t = 0, 1, 2, \ldots$, which is described as

$$\begin{aligned} \mathbf{S}_i\colon\ x_i(t+1) &= A_{ii}x_i(t) + B_{ii}u_i(t) + w_i(t), \\ y_i(t) &= C_{ii}x_i(t) + v_i(t), \qquad i \in \boldsymbol{N}. \end{aligned} \tag{6.79}$$

The input noise $w_i(t) \in \mathbf{R}^{n_i}$ and measurement noise $v_i(t) \in \mathbf{R}^{\ell_i}$ are independent white Gaussian processes with zero mean and covariances R_w^{ii} and R_v^{ii}, respectively. Matrices A_{ij}, B_{ij}, and C_{ij} are constant and of appropriate dimensions.

The system $\mathbf{S}$ can be written in a compact form:

$$\begin{aligned} \mathbf{S}\colon\ x(t+1) &= Ax(t) + Bu(t) + w(t), \\ y(t) &= Cx(t) + v(t), \end{aligned} \tag{6.80}$$

where $x(t) \in \mathbf{R}^{n}$, $u(t) \in \mathbf{R}^{m}$, and $y(t) \in \mathbf{R}^{\ell}$ are the state, input, and output of $\mathbf{S}$, $w(t) \in \mathbf{R}^{n}$ and $v(t) \in \mathbf{R}^{\ell}$ are the state and measurement noise, and the system matrices are

$$A = \begin{bmatrix} A_{11} & & & \bigcirc \\ A_{21} & A_{22} & & \\ \cdots & \cdots & \cdots & \cdots \\ A_{N1} & A_{N2} & \ldots & A_{NN} \end{bmatrix}, \qquad B = \begin{bmatrix} B_{11} & & & \bigcirc \\ B_{21} & B_{22} & & \\ \cdots & \cdots & \cdots & \cdots \\ B_{N1} & B_{N2} & \ldots & B_{NN} \end{bmatrix},$$

$$C = \begin{bmatrix} C_{11} & & & \bigcirc \\ C_{21} & C_{22} & & \\ \cdots & \cdots & \cdots & \cdots \\ C_{N1} & C_{N2} & \ldots & C_{NN} \end{bmatrix}. \tag{6.81}$$

With each subsystem $\mathbf{S}_i$, we associate the performance criterion

$$\mathcal{E}J_i = \mathcal{E}\left\{ \lim_{T\to\infty} \frac{1}{T} \sum_{t=0}^{T-1} x_i^T(t)Q_x^{ii}x_i(t) + u_i^T(t)Q_u^{ii}u_i(t) \right\}, \tag{6.82}$$

where the constant matrix Q_x^{ii} is nonnegative definite, and the constant matrix Q_u^{ii} is positive definite. Moreover, we assume that each pair $[A_{ii},\ (Q_x^{ii})^{1/2}]$ is detectable.

Optimization of the system **S** is performed sequentially, going from the top to the bottom of the hierarchy. Once the optimal control law for $\mathbf{S}_1$ is determined and implemented, the optimization of $\mathbf{S}_2$ is carried out. The optimization of $\mathbf{S}_3$ is performed after the optimal control law for $\mathbf{S}_2$ is implemented, and so on until the bottom of the hierarchy is reached. In this way, the optimization process takes the advantage of the lower block triangular structure of the system.

The optimal control law for subsystem $\mathbf{S}_i$ depends crucially upon the information structure constraints of the overall system. Let $\mathcal{Y}_i^{t-1} = \{y_i(0), y_i(1), \ldots, y_i(t-1)\}$ denote the output information of $\mathbf{S}_i$ at time t. Three basic structures are of interest: sequential, local, and global.

6.20. SEQUENTIAL STRUCTURE. Only the outputs $\mathcal{Y}_j^{t-1}$ from the same and higher hierarchical levels ($j \leq i$) are available at $\mathbf{S}_i$.

The first step in the sequential algorithm is to find the optimal control law $u_1^*(t)$ for $\mathbf{S}_1$ that minimizes $\mathcal{E}J_1$ under the assumption that we have $\mathcal{Y}_1^{t-1}$, that is, $u_1(t)$ is measurable with respect to the σ-algebra generated by $\mathcal{Y}_1^{t-1}$. Since $\mathbf{S}_1$ is decoupled from the rest of the system, we use the standard LQG solution

$$u_1^*(t) = -K_1\hat{x}_1(t|1), \tag{6.83}$$

where $\hat{x}_1(t|1) = \mathcal{E}\{x_1(t)|\mathcal{Y}_1^{t-1}\}$, and $K_1 = M_{11}^{-1}B_{11}^TP^1A_{11}$. P^1 is the solution of the Riccati equation

$$P^1 = A_{11}^TP^1A_{11} + Q_x^{11} - A_{11}^TP^1B_{11}M_{11}^{-1}B_{11}^TP^1A_{11}, \tag{6.84}$$

and $M_{11} = B_{11}^TP^1B_{11} + Q_u^{11}$. For control law $u_1^*(t)$, $\hat{x}_1(t|1)$ is generated by the Kalman filter

$$\hat{\mathbf{S}}_1\colon\ \hat{x}_1(t+1|1) = A_{11}\hat{x}_1(t|1) + B_{11}u_1^*(t) + L_1[y_1(t) - C_{11}x_1(t|1)], \tag{6.85}$$

where L_1 is the steady-state optimal gain. Under the standard stabilizability and detectability assumption, the closed-loop system driven by the optimal control law (6.83) is asymptotically stable (Åström, 1970; Goodwin and Sin, 1984).

As we announced above, we implement $u_1^*(t)$ and optimize $\mathbf{S}_2$ assuming the measurements $\{\mathcal{Y}_1^{t-1}, \mathcal{Y}_2^{t-1}\}$ are at our disposal, that is, the admissible controls $u_2(t)$ are measurable with respect to the σ-algebra generated by

$\{\mathcal{Y}_1^{t-1}, \mathcal{Y}_2^{t-1}\}$. At this second stage, we need a model $\mathbf{M}_2$ of $\mathbf{S}_2$ reflecting these assumptions. It is important to note that, after the implementation of $u_1^*(t)$ in $\mathbf{S}_1$, we have to consider the augmented state $X_2(t) = [\hat{x}_1^T(t|1), x_1^T(t), x_2^T(t)]^T$, and obtain

$$\mathbf{M}_2\colon \begin{bmatrix} \hat{x}_1(t+1|1) \\ x_1(t+1) \\ x_2(t+1) \end{bmatrix} = \begin{bmatrix} A_{11} - B_{11}K_1 - L_1C_{11} & L_1C_{11} & 0 \\ -B_{11}K_1 & A_{11} & 0 \\ -B_{21}K_1 & A_{21} & A_{22} \end{bmatrix} \begin{bmatrix} \hat{x}_1(t|1) \\ x_1(t) \\ x_2(t) \end{bmatrix}$$
$$+ \begin{bmatrix} 0 \\ 0 \\ B_{22} \end{bmatrix} u_2(t) + \begin{bmatrix} L_1 v_1(t) \\ w_1(t) \\ w_2(t) \end{bmatrix}. \tag{6.86}$$

In a compact form, $\mathbf{M}_2$ is

$$\mathbf{M}_2\colon\ X_2(t+1) = \bar{A}^2 X_2(t) + \bar{B}^2 u_2(t) + N_2(t), \tag{6.87}$$

where the matrices

$$\bar{A}^2 = \begin{bmatrix} \bar{A}_{11}^2 & 0 \\ \bar{A}_{21}^2 & \bar{A}_{22}^2 \end{bmatrix}, \quad \bar{B}^2 = \begin{bmatrix} 0 \\ B_{22} \end{bmatrix}, \tag{6.88}$$

and $N_2(t)$ are obvious from (6.87), with $\bar{A}_{22}^2 = A_{22}$. We also note that $N_2(t)$ is a white zero-mean noise independent of $X_2(t)$.

The output equation, which is compatible with the assumed information structure constraints, is

$$Y_2(t) = \bar{C}^2 X_2(t) + V_2(t), \tag{6.89}$$

where $Y_2(t) = [y_1^T(t), y_2^T(t)]^T$,

$$\bar{C}^2 = \begin{bmatrix} 0 & C_{11} & 0 \\ 0 & C_{21} & C_{22} \end{bmatrix}, \tag{6.90}$$

and $V_2(t) = [v_1^T(t), v_2^T(t)]^T$.

The criterion $\mathcal{E}J_2$ is expressed as

$$\mathcal{E}J_2 = \left\{ \lim_{T\to\infty} \sum_{t=0}^{T-1} X_2^T(t)\bar{Q}_x^2 X_2(t) + u_2^T(t) Q_u^{22} u_2(t) \right\}, \tag{6.91}$$

where

$$\bar{Q}_x^2 = \begin{bmatrix} 0 & 0 \\ 0 & Q_x^{22} \end{bmatrix}. \tag{6.92}$$

The problem of finding the optimal control law $u_2^*(t)$, as formulated above, is reduced to the standard LQG problem. The solution is

$$u_2^*(t) = -K_2\hat{X}_2(t|2), \tag{6.93}$$

where $\hat{X}_2(t|2) = \mathcal{E}\{X_2(t)|\mathcal{Y}_1^{t-1}, \mathcal{Y}_2^{t-1}\}$, and $K_2 = M_{22}^{-1}(\bar{B}^2)^T P^2 \bar{A}^2$. The matrix P^2 is the solution of the Riccati equation

$$P^2 = (\bar{A}^2)^T P^2 \bar{A}^2 + \bar{Q}_x^2 - (\bar{A}^2)^T P^2 \bar{B}^2 M_{22}^{-1} (\bar{B}^2)^T P^2 \bar{A}^2, \tag{6.94}$$

where $M_{22} = (\bar{B}^2)^T P^2 \bar{B}^2 + Q_u^{22}$.

Let us analyze the structure of the obtained optimal control law $u_2^*(t)$. For this purpose, we decompose P^2 as

$$P^2 = \begin{bmatrix} P_{11}^2 & P_{12}^2 \\ P_{21}^2 & P_{22}^2 \end{bmatrix}, \tag{6.95}$$

where the blocks are compatible with the blocks of $\bar{A}^2$ delineated in (6.86). Then,

$$u_2^*(t) = -M_{22}^{-1} B_{22}^T \left[\sum_{j=1}^{2} P_{2j}^2 \bar{A}_{j1}^2,\ P_{22}^2 A_{22} \right] \hat{X}_2(t|2). \tag{6.96}$$

Obviously, calculation of $u_2^*(t)$ requires the last row of P^2 in (6.95), which is obtained by computing

$$P_{22}^2 = A_{22}^T P_{22}^2 A_{22} + Q_x^{22} - A_{22}^T P_{22}^2 B_{22} M_{22}^{-1} B_{22}^T P_{22}^2 A_{22}, \tag{6.97}$$

and

$$P_{21}^2 = \tilde{A}_{22}^T P_{21}^2 \bar{A}_{11}^2 + \tilde{A}_{22}^T P_{22}^2 \bar{A}_{21}^2, \tag{6.98}$$

where $\tilde{A}_{22} = A_{22} - B_{22} M_{22}^{-1} B_{22}^T P_{22}^2 A_{22}^T$.

The Riccati Equation (6.97) corresponds to the local optimization problem of $\mathbf{S}_2$ when the interconnections are ignored. The solution P_{22}^2 exists under the local stabilizability assumption concerning $\mathbf{S}_2$. The Equation (6.98) has the unique solution in the case when

$$1 - \lambda_p(\tilde{A}_{22})\lambda_q(\bar{A}_{11}^2) \neq 0, \qquad p = 1, 2, \ldots, 2n_1; \qquad q = 1, 2, \ldots, n_2, \tag{6.99}$$

where $\lambda_p(\tilde{A}_{22})$ and $\lambda_q(\bar{A}_{11}^2)$ are the eigenvalues of the indicated matrices (Gantmacher, 1959). The matrix $\bar{A}_{11}^2$ is the closed-loop state transition matrix of $\mathbf{S}_1$ after implementation of $u_1^*(t)$ and is, therefore, a stable matrix (that is, all its eigenvalues are inside the unit circle). On the other hand, $\tilde{A}_{22}$ is the state transition matrix of $\mathbf{S}_2$ under the same control law $u_1^*(t)$ obtained as if the whole state $x_1(t)$ of $\mathbf{S}_1$ is known, and it is also stable. Stability of $\bar{A}_{11}^2$ and $\tilde{A}_{22}$ implies condition (6.99) which, in turn, implies the existence and uniqueness of the solution P_{21}^2 to (6.98). These facts regarding the algebraic Equations (6.97) and (6.98) follow closely those of Özgüner and Perkins (1978).

The estimate $\hat{X}_2(t|2)$ appearing in the control law $u_2^*(t)$ can be expressed as $\hat{X}_2(t|2) = [\mathcal{E}\{\hat{x}_1^T(t|1)|\mathcal{Y}_1^{t-1}, \mathcal{Y}_2^{t-1}\}, \mathcal{E}\{x_1^T(t)|\mathcal{Y}_1^{t-1}, \mathcal{Y}_2^{t-1}\}, \mathcal{E}\{x_2^T(t)|\mathcal{Y}_1^{t-1}, \mathcal{Y}_2^{t-1}\}]^T$. Having in mind that the σ-algebra generated by $\mathcal{Y}_1^{t-1}$ is a subset of the σ-algebra generated by $\{\mathcal{Y}_1^{t-1}, \mathcal{Y}_2^{t-1}\}$, one concludes that $\mathcal{E}\{\hat{x}_1(t|1)|\mathcal{Y}_1^{t-1}, \mathcal{Y}_2^{t-1}\} = \hat{x}_1(t|1)$. Computation of the remaining terms of $\hat{X}_2(t|2)$ can proceed by using the Kalman filter for $\xi_2(t) = [x_1^T(t), x_2^T(t)]^T$ given $\{\mathcal{Y}_1^{t-1}, \mathcal{Y}_2^{t-1}\}$, to get

$$\hat{\mathbf{S}}_2\text{:}\;\; \hat{\xi}_2(t+1|2) = A^2\hat{\xi}_2(t|2) + B^2U_2^*(t) + L_2[Y_2(t) - C^2\hat{\xi}(t|2)], \tag{6.100}$$

where $U_2^*(t) = [u_1^{*T}(t), u_2^{*T}(t)]^T$,

$$A^2 = \begin{bmatrix} A_{11} & 0 \\ A_{21} & A_{22} \end{bmatrix}, \qquad B^2 = \begin{bmatrix} B_{11} & 0 \\ B_{21} & B_{22} \end{bmatrix}, \qquad C^2 = \begin{bmatrix} C_{11} & 0 \\ C_{21} & C_{22} \end{bmatrix}, \tag{6.101}$$

and L_2 is the optimal steady-state Kalman gain, which produces a stable $\hat{\mathbf{S}}_2$ under the local stabilizability and detectability assumptions, due to the lower block triangular structure of the matrices (6.101).

From the above analysis of the structure of the optimal control law $u_2^*(t)$, it follows that it can be broken into two components. The first component is the locally optimal control law for $\mathbf{S}_2$ with respect to $\mathcal{E}J_2$ when the interconnection with $\mathbf{S}_1$ is ignored. The second component is the feedforward control from $\mathbf{S}_1$ to $\mathbf{S}_2$, which takes into account the interconnection. It is crucial to observe that the sequentially optimal control laws stabilize the interconnected pair $(\mathbf{S}_1, \mathbf{S}_2)$ because the pair has a hierarchical structure, each closed-loop subsystem is stable, and the feedforward gain is bounded.

The optimization of the rest of the subsystems $\mathbf{S}_3, \mathbf{S}_4, \ldots, \mathbf{S}_i, \ldots, \mathbf{S}_N$ can be performed in an analogous way. The state vector, which is relevant for the optimization of $\mathbf{S}_i$ when the optimal control for $\mathbf{S}_1, \mathbf{S}_2, \ldots, \mathbf{S}_{i-1}$ is implemented, is $X_i(t) = [\hat{x}_1^T(t|1), \hat{x}_1^T(t|2), \hat{x}_2^T(t|2), \ldots, \hat{x}_1^T(t|i-1), \ldots, \hat{x}_{i-1}^T(t|i-1), x_1^T(t), \ldots, x_{i-1}^T(t), x_i^T(t)]^T$, where $\hat{x}_p(t|q) = \mathcal{E}\{x_p(t)|\mathcal{Y}_1^{t-1}, \mathcal{Y}_2^{t-1}, \ldots, \mathcal{Y}_q^{t-1}\}$. Proceeding in the same manner as in the case of the subsystem $\mathbf{S}_2$, it is possible to construct a state model for $X_i(t)$ after putting together the models for $\xi_i(t)$ and the corresponding Kalman filters generating the optimal estimates in $X_i(t)$. The resulting equation has the following form:

$$X_i(t+1) = \begin{bmatrix} \bar{A}_{11}^i & 0 \\ \bar{A}_{21}^i & \bar{A}_{22}^i \end{bmatrix} X_i(t) + \begin{bmatrix} 0 \\ B_{ii} \end{bmatrix} u_i(t) + N_i(t), \tag{6.102}$$

where the matrices $\bar{A}_{11}^i$ and $\bar{A}_{21}^i$ have the dimensions $[in_1 + (i-1)n_2 + \cdots + 2n_{i-1}] \times [in_1 + (i-1)n_2 + \cdots + 2n_{i-1}]$ and $n_i \times [in_1 + (i-1)n_2 + \cdots + 2n_{i-1}]$, respectively, while $\bar{A}_{22}^i = A_{ii}$, and $N_i(t)$ represents a white noise sequence. The corresponding observation equation is

$$Y_i(t) = C^i \xi_i(t) + V_i(t), \tag{6.103}$$

which has the same form as (6.89).

After expressing $\mathcal{E}J_i$ in terms of $X_i(t)$ in the same way as $\mathcal{E}J_2$ has been expressed in terms of $X_2(t)$, it can be shown that the optimal control law on the ith level is given as

$$u_i^*(t) = -K_i \hat{X}_i(t|i), \tag{6.104}$$

where

$$K_i = M_{ii}^{-1} B_{ii}^T \left[\sum_{j=1}^{2} P_{2j}^i A_{j1}^i, \quad P_{22}^i A_{ii} \right], \tag{6.105}$$

and $\hat{X}_i(t|i) = \mathcal{E}\{X_i(t)|\mathcal{Y}_1^{t-1}, \mathcal{Y}_2^{t-1}, \ldots, \mathcal{Y}_i^{t-1}\}$. The matrices P_{21}^i and P_{22}^i are obtained from (6.97) and (6.98) after replacing A_{22} by A_{ii} and $\bar{A}_{21}^2$ by $\bar{A}_{21}^i$. Obviously, the control law has always the same structure: the local feedback and the feedforward part. The estimates $\hat{x}_i(t|1), \ldots, \hat{x}_1(t|i-1), \ldots, \hat{x}_{i-1}(t|i-1), \hat{x}_1(t|i), \ldots, \hat{x}_i(t|i)$, which are needed for the control law, are generated by a bank of Kalman filters.

The above optimization procedure can be concisely formulated as follows (Stanković and Šiljak, 1989):

6.21. PROPOSITION. A solution to the sequential optimization problem for the system $\mathbf{S}$, which is composed of hierarchically ordered subsystems $\mathbf{S}_i$, with performance criteria $\mathcal{E}J_i$ and the sequential information structure, exists, is unique, and the entire closed-loop system is asymptotically stable provided all pairs (A_{ii}, B_{ii}), $[A_{ii}, (R_w^{ii})^{1/2}]$ are stabilizable and all pairs (A_{ii}, C_{ii}), $[A_{ii}, (Q_x^{ii})^{1/2}]$ are detectable.

Let us consider the alternative information structure, which is very much in tune with our general approach to decentralized control of complex systems: the local information structure constraint. To our disappointment, this constraint is not providing us with the usual simplification that was suggested by Hodžić and Šiljak (1985).

6.22. LOCAL STRUCTURE. Only the local output $\mathcal{Y}_i^{t-1}$ is available at $\mathbf{S}_i$.

If only local output can be used, then we have to modify the sequential design. After the first step, is completed, which remains obviously unchanged, we consider again the state model $\mathbf{M}_2$, but the observation equation becomes

$$y_2(t) = [0 \quad C_{21} \quad C_{22}]\, X_2(t) + v_2(t). \tag{6.106}$$

The solution to the corresponding LQG problem is now given by

$$u_2^*(t) = -K_2 \mathcal{E}\{X_2(t) | \mathcal{Y}_2^{t-1}\}, \tag{6.107}$$

where the matrix K_2 is defined as in (6.93). The optimal state estimate, which is conditioned by the available measurements, is composed of $\mathcal{E}\{\hat{x}_1(t|1)|\mathcal{Y}_2^{t-1}\}$, $\mathcal{E}\{x_1(t)|\mathcal{Y}_2^{t-1}\}$, and $\mathcal{E}\{x_2(t)|\mathcal{Y}_2^{t-1}\}$. In general, the σ-algebra generated by $\mathcal{Y}_1^{t-1}$ is not contained in the σ-algebra generated by $\mathcal{Y}_2^{t-1}$. Therefore, the estimation algorithm cannot be decomposed in this case into two independent Kalman filters $\hat{\mathbf{S}}_1$ and $\hat{\mathbf{S}}_2$ producing $\hat{\xi}_1(t|1)$ and $\hat{\xi}_2(t|2)$, respectively [see (6.85) and (6.100)]. Rather, the estimates $\hat{X}_2^*(t|2) = \mathcal{E}\{X_2(t)|\mathcal{Y}_2^{t-1}\}$ have to be obtained directly on the basis of $\mathbf{M}_2$, that is,

$$\hat{X}_2^*(t+1|2) = \bar{A}^2 \hat{X}_2^*(t|2) + \bar{B}^2 u_2^*(t) + L_2^*\{y_2(t) - [0 \quad C_{21} \quad C_{22}]\, \hat{X}_2^*(t|2)\}, \tag{6.108}$$

where L_2^* represents the corresponding steady-state Kalman gain. This gain exists and the estimator is stable under appropriate stabilizability and detectability conditions for the model (6.86) with (6.106) (Goodwin and Sin, 1984).

The third and subsequent optimization steps are different from those of the sequential algorithm. The state model, which is relevant for the third step, defines the state evolution of the first three subsystems in which the optimal controls $u_1^*(t)$ and $u_2^*(t)$ are already implemented as (6.83) and (6.107). The augmented state is defined as $X_3^*(t) = [\hat{x}_1^T(t|1),\ \hat{X}_2^{*T}(t|2),\ x_1^T(t),\ x_2^T(t),\ x_3^T(t)]^T$, and the corresponding model $\mathbf{M}_3^*$ is obtained by joining (6.77) with both (6.85) and (6.108). The optimal control for $\mathbf{S}_3$ is now

$$u_3^*(t) = -K_3 \hat{X}_3^*(t|3), \tag{6.109}$$

where $\hat{X}_3^*(t|3) = \mathcal{E}\{X_3^*(t)|\mathcal{Y}_3^{t-1}\}$. Owing to the fortunate fact that the state model $\mathbf{M}_3^*$ preserves its lower block triangular structure, the matrix L_3^* can still be decomposed into two distinct blocks. The first block defines the local feedback when the interconnections are ignored, while the second block specifies the feedforward path from the estimate of $[\hat{x}_1^T(t|1),\ X_2^{*T}(t|2),\ x_1^T(t),\ x_2^T(t)]^T$, which is available on the basis of the local measurements. Although the gain matrix calculations are similar for all stages, the complexity of the estimation algorithm becomes prohibitive as the number of systems increases. For example, in $\mathbf{M}_3^*$, $\dim\{\hat{X}_3^*(t)\} = 3n_1 + 2n_2 + n_3$ and, in general, $\dim\{X_i^*(t)\} = in_1 + (i-1)n_2 + \cdots + n_i$. The cause for this dimensionality explosion is the lack of information about the outputs at higher hierarchical levels. In practice, it might be worthwhile to examine the relative importance of the information contained in the outputs at various levels, especially when interconnections are relatively weak among the individual levels of the hierarchy.

6.23. GLOBAL STRUCTURE. All outputs $\mathcal{Y}_j^{t-1}$, $j \in \boldsymbol{N}$, are available at each subsystem $\mathbf{S}_i$, $i \in \boldsymbol{N}$.

We now consider the case when the whole vector $\mathcal{Y}^{t-1} = \{\mathcal{Y}_1^{t-1}, \mathcal{Y}_2^{t-1}, \ldots, \mathcal{Y}_N^{t-1}\}$ is available at each subsystem $\mathbf{S}_i$, $i \in \boldsymbol{N}$. Both control and estimation algorithms of the sequential optimization have to be altered. At the first optimization stage, the optimal control is given as

$$u_1^*(t) = -K_1 \hat{x}_1(t|\mathcal{Y}), \tag{6.110}$$

where K_1 is defined as in (6.83), but $\hat{x}_1(t|\mathcal{Y})$ represents the optimal estimate of $x_1(t)$ given all the measurements, that is, $\hat{x}_1(t,\ \mathcal{Y}) = \mathcal{E}\{x_1(t)|\mathcal{Y}_1^{t-1},\ \mathcal{Y}_2^{t-1},\ \ldots,\ \mathcal{Y}_N^{t-1}\}$. Owing to the lower block triangular structure of the system, one can formulate the following expression representing one part of the recursion for the optimal estimate of the whole vector $x(t)$, given $\mathcal{Y}^{t-1}$:

$$\hat{x}_1(t+1|\mathcal{Y}^{t-1}) = (A_{11} - B_{11}L_1)\, \hat{x}_1(t|\mathcal{Y} + L_1^{\mathcal{Y}} \nu(t|\mathcal{Y}), \tag{6.111}$$

where $\nu(t|\mathcal{Y}) = y(t) - C\hat{x}(t|\mathcal{Y})$ is the innovation vector for the global Kalman filter, while $L_1^{\mathcal{Y}}$ represents the corresponding block of the steady-state gain $L_{\mathcal{Y}} = P_{\mathcal{Y}}C^T(CP_{\mathcal{Y}}C^T + R_v)^{-1}$, in which $P_{\mathcal{Y}}$ is the steady-state prediction error covariance matrix.

The second optimization stage is based on the construction of a state model for $\mathbf{S}_1$ and $\mathbf{S}_2$ when the control (6.110) is implemented. Combining (6.111) with the corresponding part of $\mathbf{S}$ in (6.80), one obtains

$$\begin{bmatrix} \hat{x}_1(t+1|\mathcal{Y}) \\ x_1(t+1) \\ x_2(t+1) \end{bmatrix} = \begin{bmatrix} A_{11} - B_{11}K_1 & 0 & 0 \\ -B_{11}K_1 & A_{11} & 0 \\ 0 & A_{21} & A_{22} \end{bmatrix} \begin{bmatrix} \hat{x}_1(t|\mathcal{Y}) \\ x_1(t) \\ x_2(t) \end{bmatrix} + \begin{bmatrix} L_{\mathcal{Y}}^1 \nu(t|\mathcal{Y}) \\ w_1(t) \\ w_2(t) \end{bmatrix}. \tag{6.112}$$

In (6.112), the sequence $\{\nu(t)|\mathcal{Y}\}$ is zero-mean and uncorrelated and, moreover, satisfies the condition $\mathcal{E}\{\nu(t|\mathcal{Y})|\mathcal{Y}_1^{t-1}, \mathcal{Y}_2^{t-1}, \ldots, \mathcal{Y}_N^{t-1}\} = 0$. These are the essential prerequisites for the application of LQG methodology, since they allow the construction of the mean-square optimal predictor of $X_2^{\mathcal{Y}}(t) = [\hat{x}_1^T(t|\mathcal{Y}),\, x_1^T(t),\, x_2^T(t)]^T$ on the basis of the system matrices appearing in (6.112) together with the *a priori* noise statistics (*e.g.*, Åström, 1970). The resulting optimal control is given by

$$u_2^*(t) = -K_2^{\mathcal{Y}}\hat{X}_2^{\mathcal{Y}}(t|\mathcal{Y}), \tag{6.113}$$

where $K_2^{\mathcal{Y}}$ is the gain matrix obtained by solving the Riccati equation corresponding to (6.112). The structure of the model allows again the decomposition of the optimal control law into two parts, the first of which represents the stable local feedback, and the second defines the feedforward path from the estimates of the states of the subsystems at higher hierarchical levels. Consequently, the local feedback results from the Riccati Equation (6.97), while the feedforward block is the solution to the Liapunov equation analogous to (6.98). The estimate $\hat{X}_2^{\mathcal{Y}}(t|\mathcal{Y})$ contains now $\mathcal{E}\{\hat{x}_1(t|\mathcal{Y})|\mathcal{Y}_1^{t-1}, \ldots, \mathcal{Y}_N^{t-1}\} = \hat{x}_1(t|\mathcal{Y})$ and, therefore, the whole vector can be constructed by picking up the corresponding components from the estimate generated by the global Kalman filter.

In general, at the ith optimization stage, a model structurally similar to (6.112) can be constructed. The state vector is composed of $\hat{x}_1(t|\mathcal{Y})$, $\ldots$, $\hat{x}_{i-1}(t|\mathcal{Y})$, $x_1(t)$, $\ldots$, $x_{i-1}(t)$, and $x_i(t)$, while the corresponding gain in the equation analogous to (6.111) becomes $L_{i-1}^{\mathcal{Y}}$, consisting of the first $i-1$ blocks of $L_{\mathcal{Y}}$ related to the first $i-1$ subsystems. The gain matrix $K_i^{\mathcal{Y}}$ is again decomposable; it can be found by solving the corresponding Riccati

$$
A_c = \begin{array}{c} \begin{array}{cccccccccccccc} x_1 & x_2 & x_3 & x_4 & x_5 & x_6 & x_7 & x_8 & x_9 & x_{10} & x_{11} & x_{12} & x_{13} & x_{14} \end{array} \\ \left[\begin{array}{cccccccccccccc}
0 & 0 & 0 & 0 & 0 & 0 & 0 & 0 & 0 & 0.0068 & 0 & 0.0021 & 0 & 0 \\
0 & -0.0333 & 0 & 0 & 0 & 0 & 0 & 0 & 0 & 0 & 0 & 0.0767 & 0 & 0 \\
0 & 0.00182 & -0.00182 & 0 & 0 & 0 & 0 & 0 & 0 & -0.00263 & 0 & 0 & 0 & 0 \\
0 & 0 & 0 & -0.00263 & 0 & 0 & 0 & 0 & 0 & 0 & 0 & 0 & 0 & 0 \\
0 & 0 & 0 & 0 & -0.02 & 0 & 0 & 0 & 0 & 0 & 0 & 0 & 0 & 0 \\
0 & 0 & 0 & 0 & -0.00769 & -0.00769 & 0 & 0 & 0 & -0.00923 & 0 & 0 & 0 & 0 \\
0 & 0 & 0 & 0 & 0 & 0 & -0.01945 & 0.01 & 0 & 0 & 0 & -0.000945 & 0 & 0.00378 \\
0 & 0 & 0 & 0 & 0 & 0 & -0.011 & 0 & 0 & 0 & 0 & 0.00556 & 0 & -0.00222 \\
0 & 0 & 0 & 0 & 0 & 0 & 0 & 0 & -0.4 & 0 & 0 & 0 & 0 & 0 \\
0 & 0 & -0.1 & 0.1 & 0 & 0 & 0 & 0 & 0.667 & -0.667 & 0 & 0 & 0 & 0 \\
0 & 0 & 0 & 0 & 0 & 0 & 0 & 0 & 0 & 0 & -1.0 & 0 & 0 & 0 \\
0 & 0 & 0 & 0 & 0 & 0 & 0 & 0 & 0 & 0 & 0.5 & -0.5 & 0 & 0 \\
0 & 0 & 0 & 0 & 0 & 0 & 0 & 0 & 0 & 0 & 0 & 0 & -0.667 & 0 \\
0 & 0 & -0.04 & 0.04 & 0 & 0 & 0 & 0 & 0 & 0 & 0 & 0 & 0.4 & -0.4
\end{array} \right] \end{array}
\begin{array}{c} x_1 \\ x_2 \\ x_3 \\ x_4 \\ x_5 \\ x_6 \\ x_7 \\ x_8 \\ x_9 \\ x_{10} \\ x_{11} \\ x_{12} \\ x_{13} \\ x_{14} \end{array}
\tag{6.115}
$$

and Liapunov equations. All the required state estimates in all the stages of the optimization procedure can be obtained from the global Kalman filter.

6.24. COMPLETE STATE INFORMATION. All states $x_1(t)$, $x_2(t)$, ..., $x_i(t)$ are known at each $\mathbf{S}_i$ for every $t = 0, 1, 2, \dots$.

In this case, which was assumed by Özgüner and Perkins (1978), calculations of the optimal controls are greatly simplified. For details, see (Stanković and Šiljak, 1989), where computational aspects of all types of the sequential optimization were discussed.

6.25. EXAMPLE. To illustrate the LBT/LQG sequential design, let us consider a 14th order continuous-time model of a ship boiler proposed by Chemouil and Wahdan (1980). The model has the form

$$\begin{aligned} \mathbf{S}_c\colon\ \dot{x}(t) &= A_c x(t) + B_c u(t) + \Gamma_c w(t), \\ y(t) &= C_c x(t) + v(t), \end{aligned} \tag{6.114}$$

where the matrices A_c, B_c, C_c, and Γ_c are given by Equations (6.115)–(6.118):

$$B_c^T = \begin{bmatrix} 0 & 0 & 0 & 0 & 0 & 0 & 0 & 0 & 1 & 0 & 0 & 0 & 0 & 0 \\ 0 & 0 & 0 & 0 & 0 & 0 & 0 & 0 & 0 & 0 & 1 & 0 & 0 & 0 \\ 0 & 0 & 0 & 0 & 0 & 0 & 0 & 0 & 0 & 0 & 0 & 0 & 1 & 0 \end{bmatrix} \begin{matrix} u_1 \\ u_2 \\ u_3 \end{matrix} \tag{6.116}$$

$$\Gamma_C^T = [-0.0044 \quad 0 \quad 0 \quad 0.004474 \quad 0.00708 \quad 0.000415 \quad 0 \quad 0 \quad 0 \quad 0 \quad 0 \quad 0 \quad 0 \quad 0], \tag{6.117}$$

$$C_c = \begin{bmatrix} 1 & 0 & 0 & 0 & 0 & 0 & 0 & 0 & 0 & 0 & 0 & 0 & 0 & 0 \\ 0 & 0 & 1 & -1 & 0 & 0 & 0 & 0 & 0 & 0 & 0 & 0 & 0 & 0 \\ 0 & 0 & 0 & 0 & 0 & 1 & 1 & 0 & 0 & 0 & 0 & 0 & 0 & 0 \\ 0 & 0 & 0 & 0 & 0 & 0 & 0 & 0 & 0 & 1 & 0 & 0 & 0 & 0 \\ 0 & 0 & 0 & 0 & 0 & 0 & 0 & 0 & 0 & 0 & 0 & 1 & 0 & 0 \\ 0 & 0 & 0 & 0 & 0 & 0 & 0 & 0 & 0 & 0 & 0 & 0 & 0 & 1 \end{bmatrix} \begin{matrix} y_1 \\ y_2 \\ y_3 \\ y_4 \\ y_5 \\ y_6 \end{matrix} \tag{6.118}$$

$$
\tilde{A}_c = \begin{array}{c} \\ \end{array}
\begin{array}{cccccccccccccc|c}
x_{11} & x_{12} & x_2 & x_3 & x_4 & x_9 & x_{10} & x_1 & x_5 & x_6 & x_{13} & x_{14} & x_7 & x_8 & \\
-1 & 0 & 0 & & & & & & & & & \bigcirc & & & x_{11} \\
0.5 & -0.5 & 0 & & & & & & & & & & & & x_{12} \\
0 & 0.0767 & -0.033 & & & & & & & & & & & & x_2 \\
0 & 0 & 0.00182 & -0.00182 & 0 & 0 & -0.00263 & 0 & 0 & 0 & & & & & x_3 \\
0 & 0 & 0 & 0 & -0.00263 & 0 & 0 & 0 & 0 & 0 & & & & & x_4 \\
0 & 0 & 0 & 0 & 0 & -0.4 & 0 & 0 & 0 & 0 & & & & & x_9 \\
0 & 0 & 0 & -0.1 & 0.1 & 0.667 & -0.667 & 0 & 0 & 0 & & & & & x_{10} \\
0 & 0.0021 & 0 & 0 & 0 & 0 & 0.0068 & 0 & 0 & 0 & & & & & x_1 \\
0 & 0 & 0 & 0 & 0 & 0 & 0 & 0 & -0.02 & 0 & & & & & x_5 \\
0 & 0 & 0 & 0 & 0 & 0 & -0.00923 & 0 & -0.00769 & 0.00769 & & & & & x_6 \\
0 & 0 & 0 & 0 & 0 & 0 & 0 & 0 & 0 & 0 & -0.667 & 0 & 0 & 0 & x_{13} \\
0 & 0 & 0 & -0.04 & 0.04 & 0 & 0 & 0 & 0 & 0 & 0.4 & -0.4 & 0 & 0 & x_{14} \\
0 & -0.000945 & 0 & 0 & 0 & 0 & 0 & 0 & 0 & 0 & 0 & 0.000378 & -0.01945 & 0.01 & x_7 \\
0 & 0.00556 & 0 & 0 & 0 & 0 & 0 & 0 & 0 & 0 & 0 & -0.00222 & -0.011 & 0 & x_8
\end{array}
\tag{6.119}
$$

To apply the proposed LQG methodology, we need to transform first the system $\mathbf{S}_c$ into an input–output LBT form, if such a form exists. For this purpose, we apply Algorithm 6.13 and conclude that such an LBT does not exist in this case, but both input and output LBT forms are present, although with different numbers of subsystems and incompatible input and output structures. For this reason, we build the controllers and estimators separately at cost of an inferior performance of the overall system.

For control design, the system $\mathbf{S}_c$ is decomposed into three hierarchically ordered subsystems with independent inputs resulting in $\tilde{A}_c$ and $\tilde{B}_c$ given in (6.119) and

$$\tilde{B}_c^T = \left[\begin{array}{c|c|c} 1\;\;0\;\;0 & \multicolumn{2}{c}{\bigcirc} \\ \cline{1-2} \bigcirc & 0\;\;0\;\;1\;\;0\;\;0\;\;0\;\;0 & \\ \cline{2-3} & & 1\;\;0\;\;0\;\;0 \end{array}\right] \begin{array}{l} u_2 \\ u_1 \\ u_3 \end{array}. \tag{6.120}$$

All three subsystems and, thus, the overall system, are stabilizable. After discretization with sampling period $T = 0.25$ sec, the matrices $\tilde{A}_c$ and $\tilde{B}_c$ become $\tilde{A}_d$ and $\tilde{B}_d$ of (6.121) and (6.122), retaining the LBT form:

$$
\tilde{A}_d = \left[\begin{array}{ccc|ccccccc|cccc}
7.7880 & 0 & 0 & & & & & \mathbf{0} & & & & & & \\
1.0370 & 8.8250 & 0 & & & & & & & & & & & \\
1.0560 & 1.7949 & 3.9171 & & & & & & & & & & & \\
\hdashline
1.6530 & 4.1737 & 4.5301 & 9.9955 & -7.7774 & -5.0177 & -6.0547 & 0 & 0 & 0 & & & & \\
0 & 0 & 0 & 0 & 9.9934 & 0 & 0 & 0 & 0 & 0 & & & & \\
0 & 0 & 0 & 0 & 0 & 9.0484 & 0 & 0 & 0 & 0 & & & & \\
-1.0183 & -3.3739 & -5.3682 & -2.3021 & 2.3019 & 1.4596 & 8.4642 & 0 & 0 & 0 & & & & \\
2.8995 & 4.9351 & -3.0861 & -2.0113 & 2.0112 & 1.2976 & 1.5658 & 1.0000 & 0 & 0 & & & & \\
0 & 0 & 0 & 0 & 0 & 0 & 0 & 0 & 9.9501 & 0 & & & & \\
4.7878 & 1.9764 & 4.1909 & 2.7319 & -2.7317 & -1.7624 & -2.1275 & 0 & -1.9195 & 1.0019 & & & & \\
\hdashline
0 & 0 & 0 & 0 & 0 & 0 & 0 & 0 & 0 & 0 & 8.4641 & 0 & 0 & 0 \\
-4.1278 & -1.3720 & -2.1945 & -9.5141 & 9.5131 & 1.6676 & 3.0091 & 0 & 0 & 0 & 8.7530 & 9.0484 & 0 & 0 \\
-1.2960 & -2.1987 & -6.9410 & -4.5407 & 4.5404 & 3.9968 & 9.6433 & 0 & 0 & 0 & 4.2968 & 8.9037 & 9.9515 & 2.4939 \\
7.6779 & 1.3069 & 4.0970 & 2.6848 & -2.6146 & -2.3589 & -5.6924 & 0 & 0 & 0 & -2.5408 & -5.2828 & -2.7433 & 1.0000
\end{array}\right]
\tag{6.121}
$$

$$\tilde{B}_d = \left[\begin{array}{c:c:c}
2.2120 & & \bigcirc \\
1.3807 & & \\
9.0837 & & \\
\hdashline
1.0529 & -4.2752 & \\
0 & 0 & \\
0 & 2.3791 & \\
-5.1855 & 1.9082 & \\
2.4923 & 1.1055 & \\
0 & 0 & \\
2.0209 & -1.5013 & \\
\hdashline
0 & 0 & 2.3027 \\
-2.0973 & 1.0616 & 1.1443 \\
-1.1160 & 2.0243 & 3.6671 \\
6.5995 & -1.1929 & -2.1647
\end{array}\right]. \tag{6.122}$$

The state noise covariance is $R_w = 0.025$.

The criteria for synthesizing the sequentially optimal control have the following matrices:

$$\begin{aligned}
Q_x^{11} &= \mathrm{diag}\{1,\ 10,\ 100\}, \\
Q_x^{22} &= \mathrm{diag}\{100,\ 1,\ 100,\ 100,\ 100,\ 1,\ 100\}, \\
Q_x^{33} &= \mathrm{diag}\{1,\ 10,\ 100,\ 100\}, \\
Q_u^{ii} &= 1, \qquad i = 1,\ 2,\ 3.
\end{aligned} \tag{6.123}$$

To control the system three control algorithms have been utilized: global, sequential, and decentralized. At first, it has been assumed that all states are available for control. The global algorithm is the standard LQG scheme applied to the pair $(\tilde{A}_d, \tilde{B}_d)$, the sequential algorithm is described above, and the local optimization algorithm of Section 4.5 is the LQG scheme for the pair $(\tilde{A}_D, \tilde{B}_D)$ that is obtained from $(\tilde{A}_d, \tilde{B}_d)$ by neglecting the off-diagonal blocks.

The figure of merit of the above algorithms is

$$J = \text{tr } (Q_x + K^T Q_u K) Px, \tag{6.124}$$

where $Q_x = \text{diag}\{Q_x^{11}, Q_x^{22}, Q_x^{33}\}$, $Q_u = \text{diag}\{Q_u^{11}, Q_u^{22}, Q_u^{33}\}$, while P_x is the steady-state value of $\mathcal{E}\{x(t)x^T(t)\}$. The results are presented in the first column of Table 6.2. We observe that there is negligible difference between the algorithms, which is the consequence of weak interconnections between the subsystems.

Table 6.2. State feedback.

ρ / Algorithm	1	10	100
Global	$7.5117\,10^{-4}$	$6.2924\,10^{-4}$	$6.3318\,10^{-4}$
Sequential	$8.2014\,10^{-4}$	$8.2205\,10^{-4}$	$1.0133\,10^{-3}$
Decentralized	$8.2019\,10^{-4}$	$8.2662\,10^{-4}$	$1.4705\,10^{-3}$

In order to compare the algorithms, we introduce a parameter ρ which multiplies the off-diagonal blocks of $\tilde{A}_d$ and $\tilde{B}_d$. Two values of ρ are used and the results are shown in Table 6.2. Superiority of the sequential algorithm over the decentralized one increases with the increase of ρ, as expected. At the same time, the inferiority of sequential and decentralized schemes with respect to the globally optimal one increases with ρ.

To design the output feedback, we use the output LBT form

$$\begin{aligned} \bar{\mathbf{S}}_d\colon\ x(t+1) &= \bar{A}_d x(t) + \bar{B}_d u(t) + \bar{\Gamma}_d w(t), \\ y(t) &= \bar{C}_d x(t) + v(t), \end{aligned} \tag{6.125}$$

where $\bar{A}_d$, $\bar{C}_d$, and $\bar{\Gamma}_d$ are given in (6.126)–(6.128):

$$\bar{A}_d = \left[\begin{array}{cc|ccccc|c|cc|cccc}
7.7880 & 0 & & & & & & & & & & & & O \\
1.0370 & 8.8250 & & & & & & & & & & & & \\
\hline
1.0570 & 1.7949 & 3.9171 & 0 & 0 & 0 & 0 & & & & & & & \\
1.6530 & 4.1737 & 4.5301 & 9.9955 & -7.7774 & -5.0177 & -6.0547 & & & & & & & \\
0 & 0 & 0 & 0 & 9.9934 & 0 & 0 & & & & & & & \\
0 & 0 & 0 & 0 & 0 & 9.0484 & 0 & & & & & & & \\
-1.0183 & -3.3739 & -5.3682 & -2.3021 & 2.3019 & 1.4596 & 8.4642 & & & & & & & \\
\hline
2.8995 & 4.9351 & -3.0861 & -2.0113 & 2.0112 & 1.2976 & 1.5658 & 1.0000 & & & & & & \\
\hline
0 & 0 & 0 & 0 & 0 & 0 & 0 & 0 & 8.4641 & 0 & & & & \\
-4.1278 & -1.3720 & -2.1945 & -9.5141 & 9.5131 & 1.6676 & 3.0091 & 0 & 8.7530 & 9.0484 & & & & \\
\hline
0 & 0 & 0 & 0 & 0 & 0 & 0 & 0 & 0 & 0 & 9.9501 & 0 & 0 & 0 \\
-4.7878 & 1.9764 & 4.1909 & 2.7319 & -2.7317 & -1.7624 & -2.1275 & 0 & 0 & 0 & -1.9195 & 1.0019 & 0 & 0 \\
-1.2960 & -2.1987 & -6.9410 & -4.5407 & 4.5404 & 3.9968 & 9.6433 & 0 & 4.2968 & 8.9307 & 0 & 0 & 9.9515 & 2.4939 \\
7.6779 & 1.3069 & 4.0970 & 2.6848 & -2.6146 & -2.3589 & -5.6924 & 0 & -2.5408 & -5.2828 & 0 & 0 & -2.7433 & 1.0000
\end{array}\right] \quad (6.126)$$

$$\bar{\Gamma}_d^T = [0. \quad 0. \quad 0. \quad -1.518 \quad 8.946 \quad 0. \quad 8.562 \quad -8.800 \quad 0. \quad 3.485 \quad 1.413 \quad 8.197 \quad 8.806 \quad -5.193]\,. \quad (6.127)$$

$$\bar{C}_d = \left[\begin{array}{cc|ccccc|cc|cc|ccc} 0 & 1 & 0 & 0 & 0 & 0 & 0 & 0 & 0 & 0 & 0 & 0 & 0 & 0 \\ \hline 0 & 0 & 0 & 1 & -1 & 0 & 0 & 0 & 0 & 0 & 0 & 0 & 0 & 0 \\ 0 & 0 & 0 & 0 & 0 & 0 & 1 & 0 & 0 & 0 & 0 & 0 & 0 & 0 \\ \hline 0 & 0 & 0 & 0 & 0 & 0 & 0 & 1 & 0 & 0 & 0 & 0 & 0 & 0 \\ \hline 0 & 0 & 0 & 0 & 0 & 0 & 0 & 0 & 0 & 1 & 0 & 1 & 0 & 0 \\ \hline 0 & 0 & 0 & 0 & 0 & 0 & 0 & 0 & 0 & 0 & 0 & 0 & 1 & 0 \end{array}\right]. \quad (6.128)$$

The matrix $\bar{B}_d$ is obtained from $\tilde{B}_d$ by permuting the corresponding rows and columns, while the output noise covariance is $R_v = \text{diag}\{1, 0.005, 0.005, 0.005, 7 \times 10^{-7}, 1\}$. The performance index for the output control is

$$J = \text{tr}\left\{\begin{bmatrix} Q_x & 0 \\ 0 & K^T Q_u K \end{bmatrix} P_x\right\}, \quad (6.129)$$

where P_x is the steady-state value of $\mathcal{E}\{X(t)X^T(t)\}$ where $X = (x^T, \hat{x}^T)^T$.

Table 6.3. Output feedback.

Estimation / Control	Optimal	Decentralized
Optimal	6.7536	6.7620
Sequential	6.9438	6.9527
Decentralized	7.3535	7.3631

Again, the globally optimal LQG algorithm is taken as reference. The completely decentralized scheme consists of a decentralized estimator based on (6.125) and a decentralized controller derived using $\tilde{A}_d$ and $\tilde{B}_d$. The decentralized estimator is made up of five decoupled Kalman filters for the five subsystems in (6.125); it provides optimal estimates for the system $\bar{\mathbf{S}}_d$ in which the off-diagonal blocks are removed. These estimates have to be permuted before they are used for control compatible with $(\tilde{A}_d, \tilde{B}_d)$.

The sequential LQG algorithm has been adjusted to the two-fold decomposition of the system. It consists of the sequential control algorithm in conjunction with the globally optimal Kalman filter or the above-described decentralized estimator. The indices computed using (6.129) are given the Table 6.3. Due to weak interconnections and large estimation errors, the numbers in Table 6.3 do not show a significant distinction between the performance of the algorithms. The gain matrices for the global, sequential, and decentralized control are given in (6.130)–(6.132), and the estimator gains for the global and decentralized cases are specified in (6.133) and (6.134). This completes the LQG design.

$$K_{\text{opt}}^T = \begin{bmatrix} 9.3443 & -2.1397 & -1.5269 \\ 1.9237 & -8.9103 & -6.1934 \\ 7.0316 & 7.2500 & 3.3662 \\ 3.3481 & -5.1174 & 4.0812 \\ -3.5635 & 6.3010 & 1.2643 \\ -8.2555 & 3.4370 & -1.2219 \\ -8.0873 & 1.3542 & -1.1842 \\ 3.5604 & -3.2474 & 5.8332 \\ -6.9589 & 3.8840 & -1.1832 \\ 2.4648 & -1.4616 & 4.1838 \\ -1.5263 & -3.2125 & 8.9167 \\ -6.4563 & -1.4102 & 1.4459 \\ -4.5291 & -2.8164 & 1.9071 \\ 2.2917 & 4.8847 & -5.6750 \end{bmatrix}, \tag{6.130}$$

$$K_{\text{seq}}^{T} = \left[\begin{array}{c|c|c}
9.3250 & -5.7876 & -2.8182 \\
1.9156 & -2.2631 & -1.0995 \\
7.1933 & 6.8756 & 3.1401 \\
\hline
 & -3.3880 & -1.1991 \\
 & 6.3080 & 1.1895 \\
 & 3.4370 & 1.8572 \\
 & 1.3550 & 1.9850 \\
 & -3.5635 & 3.7850 \\
 & 3.9494 & 3.8666 \\
 & -1.4876 & 1.4695 \\
\cline{2-3}
 & & 8.9180 \\
 & \bigcirc & 1.4466 \\
 & & 2.0133 \\
 & & -5.9116
\end{array}\right], \tag{6.131}$$

$$
K_{\text{dec}}^T = \left[\begin{array}{c|c|c}
9.3250 & & \\
1.9156 & & \bigcirc \\
7.1933 & & \\
\hline
 & -3.3880 & \\
 & 6.3080 & \\
 & 3.4370 & \\
 & 1.3550 & \\
 & -3.5635 & \\
 & 3.9494 & \\
 & -1.4846 & \\
\hline
 & & 8.9180 \\
 & \bigcirc & 1.4466 \\
 & & 2.0133 \\
 & & -5.9116
\end{array}\right], \tag{6.132}
$$

$$L_{\text{opt}} = \begin{bmatrix} 0 & 0 & 0 & 0 & 0 & 0 \\ 0 & 0 & 0 & 0 & 0 & 0 \\ 0 & 0 & 0 & 0 & 0 & 0 \\ -1.0357 & 8.7857 & -1.3212 & 1.2804 & -6.3059 & 1.5656 \\ -6.6091 & -6.1553 & 8.9004 & -6.2581 & 4.1425 & -5.8609 \\ 0 & 0 & 0 & 0 & 0 & 0 \\ -8.2594 & -9.0875 & 1.3367 & -9.3070 & 6.2484 & -9.1714 \\ -9.5655 & 6.3882 & -9.2558 & 7.1696 & -4.3135 & 6.3089 \\ 0 & 0 & 0 & 0 & 0 & 0 \\ -4.1278 & -5.9247 & 8.7617 & -6.0735 & 4.1083 & -6.1759 \\ 7.8161 & -5.9038 & 8.3258 & -5.8536 & 3.8137 & -2.6672 \\ 4.3725 & 1.2138 & -1.8531 & 1.2674 & 8.9070 & 3.8450 \\ 6.4483 & 6.1644 & -0.2863 & 1.0552 & -4.4498 & 1.6542 \\ -7.9434 & 5.0446 & -7.5885 & 5.9504 & -3.6231 & 1.0250 \end{bmatrix}, \tag{6.133}$$

$$
L_{\text{dec}} =
\begin{bmatrix}
0 & & & & & & \\
0 & & & & \bigcirc & & \\
& 0 & 0 & & & & \\
& 5.8266 & -8.7638 & & & & \\
& -1.4511 & 2.1467 & & & & \\
& 0 & 0 & & & & \\
& -2.2399 & 3.3385 & & & & \\
& & & 1.9658 & & & \\
& & & & 0 & & \\
& & & & 2.1652 & & \\
& & & & & & -9.6058 \\
& & \bigcirc & & & & 3.8451 \\
& & & & & & 7.9666 \\
& & & & & & 1.6725
\end{bmatrix}
. \qquad (6.134)
$$

6.5. Notes and References

Powerful sparsity-oriented methods have been developed over the years to solve large systems of linear *algebraic equations* with considerable reduction of information storage and retrieval, as well as computational simplifications which decrease liability to errors in the end results. Numerous practical applications have been successfully solved in this context (Tewarson, 1973; see the survey and the book by Duff, 1977, 1981). Motivated by this development, there has been a number of papers dealing with *sparsity in dynamic systems*. The initial results (Özgüner and Perkins, 1975; Callier *et al.* 1976) introduced decomposition schemes to obtain system models in LBT form so that stability and eigenvalues can be determined by considering only diagonal blocks. This approach has been extended to nonlinear systems as well (Michel *et al.* 1978; Vidyasagar, 1980; Michel and Miller, 1983).

The role of control in the context of LBT systems has been explored by several people (Kevorkian, 1975; Vidyasagar, 1980; Özgüner and Perkins, 1978). An algorithm was proposed by Sezer and Šiljak (1981: Section 6.1) that produces LBT systems with input reachable subsystems. Input reachability, being a minimal requirement for controllability, makes the LBT form conducive to stabilization by local decentralized control (Sezer and Šiljak, 1981: Section 6.2; Khurana *et al.* 1986).

If inputs and outputs are handled simultaneously during decomposition, the resulting LBT system has subsystems with independent inputs and outputs (Pichai *et al.* 1983: Section 6.3). This allows for a bank of Kalman filters to be built to estimate the states of subsystems (Hodžić and Šiljak, 1985), and solve a decentralized LQG problem for the overall system. In fact, the decentralized Kalman filtering scheme can be obtained by an iterative process that resembles those used to solve linear algebraic equations (*e.g.*, Hageman and Young, 1981). Jacobi and Gauss–Seidel estimators have been formulated in this context (Hodžić and Šiljak, 1986b), which can be implemented by multiprocessor schemes for parallel multirate processing (Hodžić and Šiljak, 1986a; Sezer and Šiljak, 1989).

The overall performance of the decentralized LQG scheme for LBT systems can be improved by a sequential design initiated by Özgüner and Perkins (1978) for deterministic systems. A stochastic version of the design was formulated by Stanković and Šiljak (1989), as described in Section 6.4, where the improvement was demonstrated by comparing performances of global, local, and sequential schemes

Bibliography

Åström, K. J. (1970). *Introduction to Stochastic Control.* Academic Press, New York.

Callier, F. M., W. S. Chan, and C. A. Desoer (1976). Input–output stability theory of interconnected systems using decomposition techniques. *IEEE Transactions*, AC-23, 714–729.

Candy, J. C. (1986). *Signal Processing: The Model-Based Approach.* McGraw-Hill, New York.

Chemouil, P., and A. M. Wahdan (1980). Output feedback control of systems with slow and fast modes. *Large Scale Systems*, 1, 257–264.

Duff, I. S. (1977). A survey of sparse matrix research. *Proceedings of IEEE*, 65, 500–535.

Duff, I. S., Ed. (1981). *Sparse Matrices and Their Uses.* Academic Press, New York.

Gantmacher, F. R. (1959). *The Theory of Matrices.* Chelsea, New York.

Gardner, W. T., and C. T. Leondes (1990). Gain transfer: an algorithm for decentralized hierarchical estimation. *International Journal of Control*, **52**, 279–292.

Garey, M. R., and D. S. Johnson (1979). *Computers and Intractability: A Guide to the Theory of NP-Completeness.* Freeman, San Francisco, California.

Goodwin, G. C., and K. S. Sin (1984). *Adaptive Filtering, Prediction and Control.* Prentice-Hall, Englewood Cliffs, New Jersey.

Hageman, L. A., and D. M. Young (1981). *Applied Iterative Methods.* Academic Press, New York.

Hodžić, M., and D. D. Šiljak (1985). Estimation and control of large sparse systems. *Automatica*, 21, 277–292.

Hodžić, M., and D. D. Šiljak (1986a). A parallel estimation algorithm. *Parallel Processing Techniques for Simulation*, M. G. Singh, A. Y. Allidina, and B. K. Daniels (eds.), Plenum Press, New York, 113–121.

Hodžić, M., and D. D. Šiljak (1986b). Iterative methods for parallel-multirate estimation. *Proceedings of the Fourth IFAC/IFOR Symposium on Large Scale Systems*, Zürich, Switzerland, 169–175.

Kevorkian, A. K. (1975). Structural aspects of large dynamic systems. *Preprints of the Sixth IFAC World Congress*, Paper No. 19-3, Boston, Massachusetts.

Khurana, H., S. I. Ahson, and S. S. Lamba (1986). Variable structure control design for large-scale systems. *IEEE Transactions*, SMC-16, 573–576.

Michel, A. N., and R. K. Miller (1983). On stability preserving mappings. *IEEE Transactions*, CAS-30, 671–679.

Michel, A. N., R. K. Miller, and W. Tang (1978). Lyapunov stability of interconnected systems: Decomposition into strongly connected subsystems. *IEEE Transactions*, CAS-25, 799–809.

Nour-Eldin, H. A. (1987). Linear multivariable systems controllability and observability: Numerical aspects. *Systems and Control Encyclopedia*, M. G. Singh (ed.), Pergamon Press, Oxford, UK, 2816–2827.

O'Reilly, J. (1983). *Observers for Linear Systems.* Academic Press, New York.

Özgüner, Ü., and W. R. Perkins (1975). Graph theory in the analysis of large scale composite systems. *Proceedings of the International Symposium on Circuits and Systems*, Boston, Massachusetts, 121–123.

Özgüner, Ü., and W. R. Perkins (1978). Optimal control of multilevel large-scale systems. *International Journal of Control*, 28, 967–980.

Pichai, V., M. E. Sezer, and D. D. Šiljak (1983). A graph-theoretic algorithm for hierarchical decomposition of dynamic systems with applications to estimation and control. *IEEE Transactions*, SMC-13, 197–207.

Pichai, V., M. E. Sezer, and D. D. Šiljak (1984). Reply to "Input–output decomposition of dynamic systems is NP-complete." *IEEE Transactions*, AC-29, 864.

Rosenbrock, H. H. (1970). *State Space and Multivariable Theory*, Nelson, London, UK.

Sezer, M. E., and D. D. Šiljak (1981). On structural decomposition and stabilization of large-scale control systems. *IEEE Transactions*, AC-26, 439–444.

Sezer, M. E. and D. D. Šiljak (1990). Decentralized multirate control. *IEEE Transactions*, AC-34 (to appear).

Stanković, S. S., and D. D. Šiljak (1989). Sequential LQG optimization of hierarchically structured systems. *Automatica*, 25, 545–559.

Tarjan, R. E. (1984). Input-output decomposition of dynamic systems is NP-complete. *IEEE Transactions*, AC-29, 863–864.

Tewarson, R. P. (1973). *Sparse Matrices.* Academic Press, New York.

Vidyasagar, M. (1980). Decomposition techniques for large-scale systems with nonadditive interactions: Stability and stabilizability. *IEEE Transactions*, AC-25, 773–779.

Chapter 7 | Nested Epsilon Decompositions

In the modeling of complex systems with a large number of variables, it has long been recognized that most of the variables are weakly coupled, if coupled at all. This fact has been observed in fields as diverse as economics and electric power systems, social sciences and computer systems, where it has been argued that most often the behavior of the overall system is dominated by strongly connected variables. A considerable conceptual insight and numerical simplification can be gained by grouping these variables into subsystems, solving the subsystems independently, and resolving subsequently the effects of interconnections to get the overall solution. The ultimate success of this recipe depends crucially upon our ability to identify groups of strongly coupled variables, cluster the variables to form subsystems, and represent the overall system as a weak interconnection of the resulting subsystems. It is absolutely essential that, during the decomposition–clustering process, there is a considerable freedom in choosing conveniently the *strength of coupling* among the subsystems, because it is this strength that controls the size and character of the subsystems and, ultimately, their dominance in the solution of the overall problem.

The objective of this chapter is to present a graph-theoretic decomposition of large systems into weakly coupled subsystems, which was proposed in the context of dynamic systems (Sezer and Šiljak, 1986). A full range of nested decompositions can be generated by controlling the threshold of the interconnection strength among the subsystems. The underlying idea is to associate a graph with a given system, disconnect the edges corresponding to interconnections with strength smaller than a prescribed threshold ϵ, and identify the disconnected subgraphs (components) of the resulting graph. The obtained components correspond to the subsystems with mutual coupling smaller than or equal to ϵ. Since components of a graph can be determined in linear time, the epsilon decomposition algorithm has an extraordinary simplicity requiring binary computations only.

Let us emphasize that the epsilon decomposition is developed for systems with inputs and outputs. We are interested in producing weakly coupled subsystems, which can be locally stabilized by state or output feedback. Weak coupling should ensure that stability of subsystems implies stability of the overall system. If this is not certain, we should apply the method of vector Liapunov functions to establish the implication. In other words, epsilon decompositions produce nested hierarchies which, unlike "vertical" LBT hierarchies of the preceding chapter, require testing for stability. For this purpose, we present a construction of hierarchical Liapunov functions (Ikeda and Šiljak, 1985), which is ideally suited for nested structures, and can be used to establish *nested connective stability.* It is interesting to note that this kind of "stratified" stability is promoted by Simon (1962) and Bronowski (1970) as an essential ingredient in the evolution of both the natural and man-made complex systems. Later, in Notes and References, we shall say more about epsilon decompositions and nested stability, emphasizing a variety of extensions and applications of the idea of weak coupling and nested connective stability of dynamic systems.

7.1. Epsilon Decomposability

There are three important aspects of epsilon decompositions. First, we can prescribe a coupling strength among the subsystems. Second, for different levels of coupling, decompositions are nested. Third, decompositions are identified by graphs. To make these aspects transparent and, at the same time, show the extraordinary simplicity of epsilon decomposition for partitioning of complex systems, we use the following:

7.1. EXAMPLE. Let us assume that we are given a matrix

$$
\begin{array}{c}
\begin{array}{cccccccc} 1 & 2 & 3 & 4 & 5 & 6 & 7 & 8 \end{array} \\
A = \left[\begin{array}{cccccccc}
0 & 0 & .0.01 & 0 & 2.15 & 0 & 0.04 & 0 \\
0.07 & 1.10 & 0.03 & 0 & -0.06 & -0.02 & 0 & 0.25 \\
0 & 0 & 1.70 & 0.40 & 0.01 & 0 & 0 & 0.01 \\
0.02 & -0.01 & 0 & 3.00 & 0 & 0.07 & 1.95 & 0 \\
1.75 & 0.01 & 0 & -0.01 & 0.45 & 0 & 0.01 & 0.06 \\
0 & 0.03 & -0.02 & 0.45 & 0 & 0.95 & 0 & 0 \\
0 & 0 & 0.35 & 0.08 & -0.01 & 0.02 & -0.85 & 0 \\
0.02 & -0.03 & 0 & 0 & 0 & 0.01 & 0 & 1.35
\end{array} \right]
\begin{array}{c} 1 \\ 2 \\ 3 \\ 4 \\ 5 \\ 6 \\ 7 \\ 8 \end{array}
\end{array}, \qquad (7.1)
$$

and we are asked to rearrange the rows and columns of A until we identify diagonal blocks with couplings no larger than a given threshold. Let us say, we choose a threshold $\epsilon_1 = 0.05$. Then, all we have to do is associate a digraph $\mathbf{D}$ with A and disconnect the edges of $\mathbf{D}$ with weights less than 0.05. With this action we obtain from $\mathbf{D}$ a digraph $\mathbf{D}^{0.05}$ in Figure 7.1a, where we identify two disconnected components. Permuting the rows and columns of A according to the two components, we get the matrix

$$\bar{A} = \begin{array}{c} \begin{array}{cccc|cccc} 1 & 2 & 5 & 8 & 3 & 4 & 6 & 7 \end{array} \\ \left[\begin{array}{cccc|cccc} 0 & 0 & 2.15 & 0 & 0.01 & 0 & 0 & 0.04 \\ 0.07 & 1.10 & -0.06 & 0.25 & 0.03 & 0 & -0.02 & 0 \\ 1.75 & 0.01 & 0.45 & 0.06 & 0 & -0.01 & 0 & 0.01 \\ 0.02 & -0.03 & 0 & 1.35 & 0 & 0 & 0.01 & 0 \\ \hline 0 & 0 & 0.01 & 0.01 & 1.70 & 0.40 & 0 & 0 \\ 0.02 & -0.01 & 0 & 0 & 0 & 3.00 & 0.07 & 1.95 \\ 0 & 0.03 & 0 & 0 & -0.02 & 0.45 & 0.95 & 0 \\ 0 & 0 & -0.01 & 0 & 0.35 & 0.08 & 0.02 & -0.85 \end{array}\right] \end{array} \begin{array}{c} \\ 1 \\ 2 \\ 5 \\ 8 \\ 3 \\ 4 \\ 6 \\ 7 \end{array}, \qquad (7.2)$$

with two diagonal blocks and off-diagonal blocks having no elements larger than 0.05 in absolute value.

Let us now increase the threshold to $\epsilon_2 = 0.1$. A pleasing fact is that now we need only consider the individual diagonal blocks and not the whole matrix A. We see that the first component of $\mathbf{D}^{0.05}$ breaks down further into two components of $\mathbf{D}^{0.1}$ shown in Figure 7.1b, while the second component of $\mathbf{D}^{0.05}$ remains unchanged. When we select $\epsilon_3 = 0.5$, we get $\mathbf{D}^{0.5}$ of Figure 7.1c and conclude that this decomposition stage effects the second and third component of $\mathbf{D}^{0.1}$. The final result is the matrix

$$\bar{\bar{A}} = \begin{array}{c} \begin{array}{cc:cc|cccc} 1 & 5 & 2 & 8 & 3 & 4 & 7 & 6 \end{array} \\ \left[\begin{array}{cc:cc|cccc} 0 & 2.15 & 0 & 0 & 0.01 & 0 & 0.04 & 0 \\ 1.75 & 0.45 & 0.01 & 0.06 & 0 & -0.01 & 0.01 & 0 \\ \hdashline 0.07 & -0.06 & 1.10 & 0.25 & 0.03 & 0 & 0 & -0.02 \\ 0.02 & 0 & -0.03 & 1.35 & 0 & 0 & 0 & 0.01 \\ \hline 0 & 0.01 & 0 & 0.01 & 1.70 & 0.40 & 0 & 0 \\ 0.02 & 0 & -0.01 & 0 & 0 & 3.00 & 1.95 & 0.07 \\ 0 & -0.01 & 0 & 0 & 0.35 & 0.08 & -0.85 & 0.02 \\ 0 & 0 & 0.03 & 0 & -0.02 & 0.45 & 0 & 0.95 \end{array}\right] \end{array} \begin{array}{c} \\ 1 \\ 5 \\ 2 \\ 8 \\ 3 \\ 4 \\ 7 \\ 6 \end{array} \qquad (7.3)$$

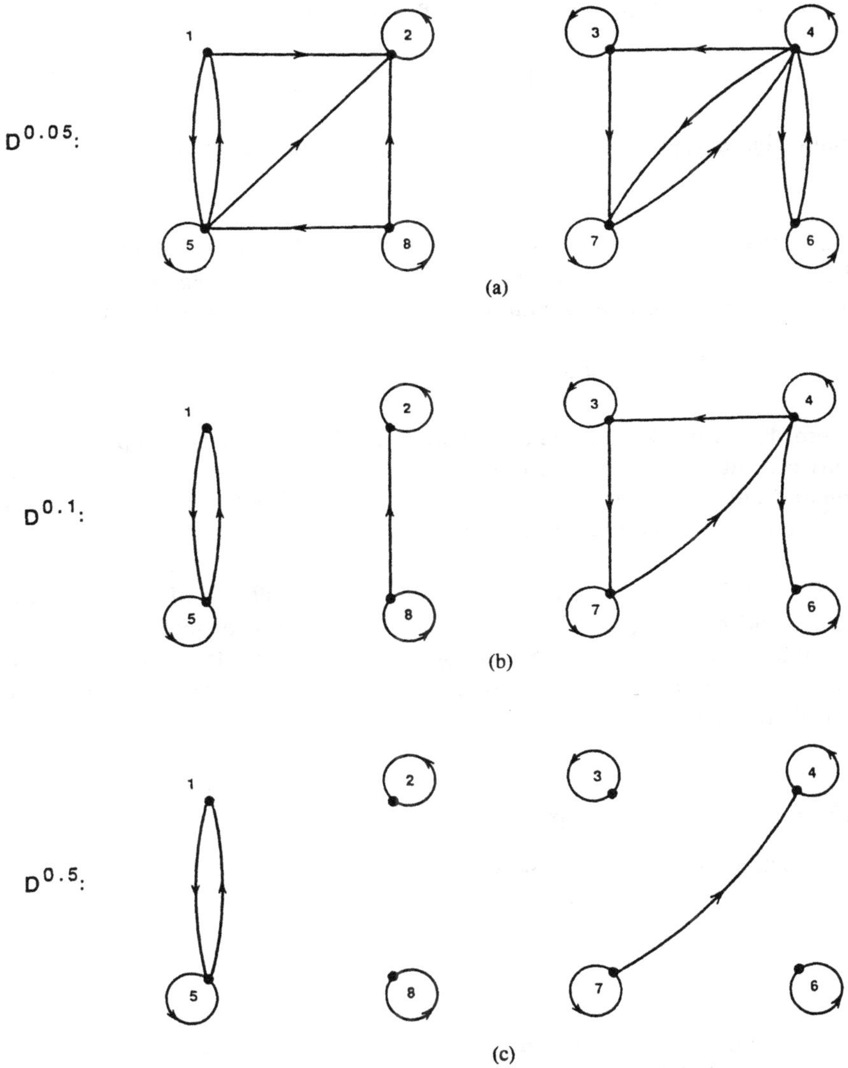

Fig. 7.1. Digraphs for Example 7.1.

It is important to note that at each stage of epsilon decomposition we need only to determine components of a digraph, which can be done in polynomial time (*e.g.*, Mehlhorn, 1984). Due to nestedness of decompositions, this process requires less computation at each stage.

A convenient representation of the matrix $\bar{\bar{A}}$ is

$$\bar{\bar{A}} = \bar{A}_0 + \epsilon_1 \bar{A}_1 + \epsilon_2 \bar{A}_2 + \epsilon_3 \bar{A}_3, \tag{7.4}$$

where the matrices $\bar{A}_0$, $\bar{A}_1$, $\bar{A}_2$, and $\bar{A}_3$ are obvious from (7.3).

7.2. DEFINITION. A matrix A has a K-fold epsilon decomposition if there are $K > 0$ positive numbers $\epsilon_1 > \epsilon_2 > \cdots > \epsilon_K$ such that, by permutations of rows and columns, the matrix A can be represented as

$$\bar{A} = \bar{A}_0 + \epsilon_1 \bar{A}_1 + \cdots + \epsilon_K \bar{A}_K, \tag{7.5}$$

where $\bar{A}_0$ is a block diagonal matrix and $\bar{A}_1$, $\bar{A}_2$, ..., $\bar{A}_K$ are all partitioned matrices with compatible blocks, such that each nonzero block appears in one and only one matrix $\bar{A}_k$, $k = 1, 2, \ldots, K$, and none of the elements of any $\bar{A}_k$ is larger than one in absolute value.

Our immediate interest is in epsilon decompositions of systems with inputs and outputs, which are conducive to the use of feedback control. We will return to decompositions of type (7.5) in Section 7.4, where we show stability of a dynamic system on each level of the threshold ϵ_k, that is, nested connective stability.

7.2. Decomposition Algorithm

Consider now a linear system

$$\mathbf{S}\colon\ \dot{x} = Ax + Bu,$$
$$y = Cx, \tag{7.6}$$

where $x(t) \in \mathbf{R}^n$, $u(t) \in \mathbf{R}^m$, and $y(t) \in \mathbf{R}^\ell$ are the state, input, and output of $\mathbf{S}$ at time $t \in \mathbf{R}$, and $A = (a_{ij})$, $B = (b_{ij})$, and $C = (c_{ij})$ are constant matrices of appropriate dimensions. Define a system matrix $M = (m_{ij})$ of $\mathbf{S}$ as the $s \times s$ matrix

$$M = \begin{bmatrix} A & B & 0 \\ 0 & 0 & 0 \\ C & 0 & 0 \end{bmatrix}, \tag{7.7}$$

where $s = n + m + \ell$. The matrix M becomes the interconnection matrix $E = (e_{ij})$ of Definition 1.1 when nonzero elements of M are replaced by ones.

We state the following:

7.3. PROBLEM. Given a system $\mathbf{S}$. Find a permutation matrix P such that the matrix $\tilde{M} = (\tilde{M}_{pq})$ defined by

$$\tilde{M} = P^T M P \tag{7.8}$$

is epsilon decomposable as

$$\tilde{M} = \tilde{M}_0 + \epsilon \tilde{M}_1, \tag{7.9}$$

where ϵ is a prescribed positive number.

If we assume that the matrix $\tilde{M}$ is an $N \times N$ block matrix with $N_p \times N_q$ blocks $\tilde{M}_{pq} = (\tilde{m}_{ij}^{pq})$, then

$$\begin{aligned} |\tilde{m}_{ij}^{pq}| \le \epsilon, \qquad p,\, q \in \boldsymbol{N}, \qquad p \neq q, \\ i \in \boldsymbol{N}_p, \qquad j \in \boldsymbol{N}_q, \end{aligned} \tag{7.10}$$

where $\boldsymbol{N} = \{1, 2, \ldots, N\}$, $\boldsymbol{N}_p = \{1, 2, \ldots, N_p\}$, $\boldsymbol{N}_q = \{1, 2, \ldots, N_q\}$. In other words, Problem 7.3 means that we should identify diagonal square blocks $\tilde{M}_{pp}$ of $\tilde{M}$ outside of which all elements are not larger in absolute value than a prescribed threshold ϵ.

To solve Problem 7.3 by a graph-theoretic algorithm, we associate with $\mathbf{S}$ a directed graph $\mathbf{D} = (\mathtt{V}, \mathtt{E})$ in the usual way (Definition 1.2), and recall that a subgraph $\mathbf{D}_i = (\mathtt{V}_i, \mathtt{E}_i)$ of $\mathbf{D}$ is a digraph such that $\mathtt{V}_i \subseteq \mathtt{V}$ and $\mathtt{E}_i \subseteq \mathtt{E}$. If $\mathtt{E}_i$ includes all the edges in $\mathtt{E}$ that connect the vertices in $\mathtt{V}_i$, then we say that $\mathbf{D}_i$ *is formed by* $\mathtt{V}_i$. If $\mathtt{V}_i$ are the equivalent classes defined by connectedness then the $\mathbf{D}_i$ formed by $\mathtt{V}_i$ are called the connected components of $\mathbf{D}$. A partition $\mathbf{P}$ of $\mathtt{V}$ is a collection $\{\mathtt{V}_1, \mathtt{V}_2, \ldots, \mathtt{V}_N\}$ of N disjoint subsets $\mathtt{V}_i$ such that $\cup_{i=1}^N \mathtt{V}_i = \mathtt{V}$. If $\mathbf{D}_i$ is a subgraph formed by $\mathtt{V}_i$, $i \in \boldsymbol{N}$, then the collection $\mathbf{P} = \{\mathbf{D}_1, \mathbf{D}_2, \ldots, \mathbf{D}_N\}$ is a partition of $\mathbf{D}$. A condensation of $\mathbf{D}$ with respect to a partition $\mathbf{P}$ is a digraph $\mathbf{D}^* = (\mathtt{V}^*, \mathtt{E}^*)$, where $\mathtt{V}^* = \{v_1^*, v_2^*, \ldots, v_N^*\}$ is a set of super-nodes with v_i^* representing the subset $\mathtt{V}_i$, and $(v_q^*, v_p^*) \in \mathtt{E}^*$ if and only if $q \neq p$ and at least one vertex in the subset $\mathtt{V}_q$ reaches a vertex in the subset $\mathtt{V}_p$.

If $\mathbf{P}$ is a decomposition of an undirected (directed) graph $\mathbf{D}$ into its connected components (strong components), then a suitable reordering of the vertices of $\mathbf{D}$ brings its adjacency matrix E into a block diagonal (lower

$\bar{M} =$

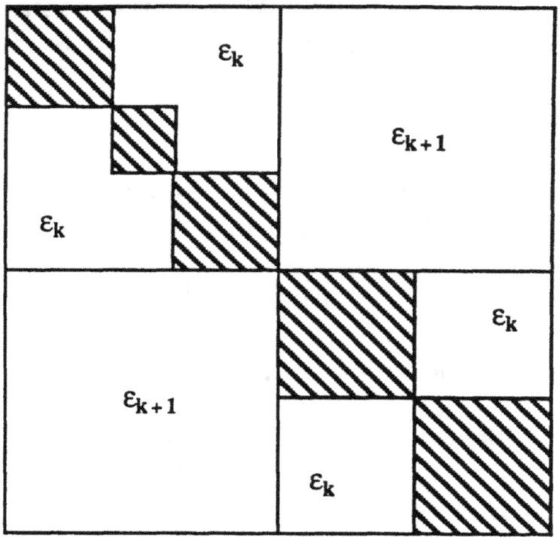

Fig. 7.2. Nestedness.

block triangular) form. That is, there exists a permutation matrix P defined by $\mathbf{P}$ such that $P^T E P$ is block diagonal (lower block triangular).

Now, going back to Problem 7.3, we see that finding a permutation matrix P for epsilon decomposition of M is equivalent to partitioning $\mathbf{D}$ into N subgraphs $\mathbf{D}_i$, $i \in \boldsymbol{N}$, such that the edges interconnecting the subgraphs correspond to the elements of M with magnitude no larger than ϵ. Such a partition $\mathbf{P}^\epsilon$ of $\mathbf{D}$ is called an *epsilon partition* of $\mathbf{D}$. It is easy to see (Example 7.1) that if the subgraphs $\mathbf{D}^\epsilon = (\mathtt{V}, \mathtt{E}^\epsilon)$ of $\mathbf{D}$ are defined by deleting all edges corresponding to values weaker than ϵ, then an epsilon partition is obtained simply by identifying the connected components of $\mathbf{D}^\epsilon$, for it is the same partition of $\mathtt{V}$ that defines the connected components of $\mathbf{D}^\epsilon$ and the ϵ-coupled subgraph of $\mathbf{D}$.

Observe that since $\mathbf{D}$ has s vertices, then there can be at most s distinct ϵ-decompositions of $\mathbf{D}$. Moreover, if two vertices v_i and v_j occur in the same connected component of $\mathbf{D}^\epsilon$ for some $\epsilon = \epsilon_k$, then they occur in the same connected component of all $\mathbf{D}^\epsilon$ for $\epsilon < \epsilon_k$. This implies that there exists a finite sequence $\max_{i \neq j} \{|m_{ij}|\} = \epsilon_1 > \epsilon_2 > \cdots > \epsilon_K = 0$, $K \leq s$, such that every $\epsilon \in [\epsilon_k, \epsilon_{k-1})$, $k \in \boldsymbol{K} = \{1, 2, \ldots, K\}$, yields the same partitions $\mathbf{P}^\epsilon_k$, $k \in \boldsymbol{K}$, and that these partitions are *nested*: $\mathbf{P}^\epsilon_1 \subset \mathbf{P}^\epsilon_2 \subset \cdots \subset \mathbf{P}^\epsilon_K$. The partition $\mathbf{P}^\epsilon_1$ is the finest ($N = s$) and $\mathbf{P}^\epsilon_K$ is the coarsest. Nesting of epsilon decompositions of matrix M is illustrated in Figure 7.2 for two values of ϵ, namely ϵ_k and ϵ_{k+1}. A numerical example is provided by the matrix $\bar{\bar{A}}$ of (7.3).

We already noted that one of the most appealing numerical features of epsilon decomposition is the fact that they are nested: $\mathbf{P}_{k+1}^{\epsilon}$ is obtained from the triple $(\mathbf{D}_k^*, M_k^*, \epsilon_{k+1})$ instead of $(\mathbf{D}, M, \epsilon_{k+1})$, where $\mathbf{D}_k^*$ is the condensation of $\mathbf{D}$ with respect to $\mathbf{P}_k^{\epsilon}$, and $M_k = (m_{pq}^k)$ is a condensation matrix of $M = (m_{ij})$ defined as

$$m_{pq}^k = \begin{cases} 0, & p = q, \\ \max\{|m_{ij}|\}, & p \neq q, \end{cases}$$
$$v_i \in \mathsf{V}_p^k,$$
$$v_j \in \mathsf{V}_q^k, \tag{7.11}$$

where V_p^k and V_q^k are the subsets of V in the partition $\mathbf{P}_k^{\epsilon}$. In this way, instead of dealing with digraphs of s vertices at each step, digraphs of sizes $s = N_1$, $N_2, \ldots, N_{K-1}$ are processed, where N_K is the number of subgraphs of $\mathbf{D}$ in $\mathbf{P}_k^{\epsilon}$, and $N_{K-1} < N_{K-2} < \cdots < N_1 = s$, which results in an enormous computational saving when large systems are considered.

On the basis of the above observations, an algorithm of Sezer and Šiljak (1986) is now presented which can generate all possible epsilon decompositions of a digraph associated with a matrix M.

7.4. ALGORITHM.

1. (*i*) Set $k = 1$, $N_k = s$.

 (*ii*) $\mathbf{P}_k^{\epsilon}$ is defined as

$$\mathbf{P}_k^{\epsilon}\colon\ \mathsf{V} = \bigcup_{p \in \boldsymbol{N}_k} \mathsf{V}_p^k,$$

$$\mathsf{V}_p^k = \{v_p\colon\ p \in \boldsymbol{N}_k\}.$$

 (*iii*) Define $M_k = (m_{pq}^k)$ as

$$m_{pq}^k = \begin{cases} 0, & p = q, \\ |m_{pq}|, & p \neq q, \end{cases}$$

 and compute

$$\epsilon_k = \max_{p,q}\{m_{pq}^k\}.$$

2. (*i*) Let $\mathbf{D}_k^* = (\mathsf{V}_k^*, \mathsf{E}_k^*)$ be the digraph associated with M_k, where $\mathsf{V}_k^* = \{v_{k1}^*, v_{k2}^*, \ldots, v_{kN_k}^*\}$.

 (*ii*) Print k; ϵ_k; N_k; V_p^k, $p \in \boldsymbol{N}$.

(*iii*) If $\epsilon_k = 0$ or $N_k = 1$, stop.

(*iv*) Let $\epsilon = \epsilon_k - \delta$, where $\delta > 0$ is an arbitrarily small number and obtain an epsilon partition of $\mathbf{D}_k^*$ (that is, identify connected components of $\mathbf{D}_k^*$):

$$\mathbf{P}_{k+1}\colon\ \mathsf{V}_k^* = \bigcup_{p\,\in\,\boldsymbol{N}_{k+1}} \mathsf{V}_{k+1,p}^*.$$

(*v*) $\mathbf{P}_{k+1}^{\epsilon}$ is defined as

$$\mathbf{P}_{k+1}^{\epsilon}\colon\ \mathsf{V} = \bigcup_{p\,\in\,\boldsymbol{N}_{k+1}} \mathsf{V}_p^{k+1}, \text{ where}$$

$$\mathsf{V}_p^{k+1} = \{v_i\colon\ v_i \in \mathsf{V}_j^k \ \text{ for some } v_{kj}^* \in \mathsf{V}_{k+1,p}^*\}.$$

(*vi*) Compute $M_{k+1} = (m_{pq}^{k+1})$ as

$$m_{pq}^{k+1} = \left\{\begin{array}{ll} 0, & p = q, \\ \max_{\substack{v_{ki}^* \,\in\, \mathsf{V}_{kp}^* \\ v_{kj}^* \,\in\, \mathsf{V}_{kq}^*}} \{m_{ij}^k\}, & p \neq q, \end{array}\right\}$$

$$\epsilon_{k+1} = \max_{p,q} \{m_{pq}^{k+1}\}.$$

(*vii*) Set $k = k + 1$ and go to 2(*ii*).

Note that in Algorithm 7.4, $\mathbf{D}_k^*$ is the condensation of $\mathbf{D}$ with respect to $\mathbf{P}_k^{\epsilon}$, which is the same as the condensation of $\mathbf{D}_{k-1}^*$ with respect to $\mathbf{P}_k^*$. Note also that in a computer implementation of the algorithm, there is no need to construct $\mathbf{D}_k^*$; $\mathbf{P}_{k+1}^*$ can be obtained from the matrix M_k directly (see Appendix).

7.3. Control Applications

We turn our attention to control design based on epsilon decompositions. Only state feedback is considered, but estimators or dynamic output feedback can be designed in pretty much the same way.

A linear system

$$\mathbf{S}\colon\ \dot{x} = Ax + Bu, \tag{7.12}$$

is decomposed by Algorithm 7.4 as

$$\bar{\mathbf{S}}^{\epsilon}\colon\ \dot{x}_p = A_{pp}x_p + B_{pp}u_p + \epsilon\left(\sum_{\substack{q=1\\ q\neq p}}^{N} A_{pq}x_q + B_{pq}u_q\right), \qquad p \in \boldsymbol{N}. \tag{7.13}$$

Control objectives may present some problems in using $\bar{\mathbf{S}}^{\epsilon}$. When the subgraph $\mathbf{D}^{\epsilon}$ of $\mathbf{D} = (\mathtt{V},\ \mathtt{E})$ is obtained, the partition

$$\mathbf{P}^{\epsilon}\colon\ \mathtt{V} = \bigcup_{p\,\in\,\boldsymbol{N}} \mathtt{V}_p, \qquad \mathtt{V}_p = \mathtt{X}_p \bigcup \mathtt{U}_p, \tag{7.14}$$

may have some of the sets $\mathtt{X}_p$ or $\mathtt{U}_p$ empty. To consider this type of situation, let us define index sets $\boldsymbol{N}_x$, $\boldsymbol{N}_u$, $\boldsymbol{N}_{xu} \subset \boldsymbol{N}$ as

$$\begin{aligned} \boldsymbol{N}_x &= \{p \in \boldsymbol{N}\colon\ \mathtt{X}_p \neq \emptyset\},\\ \boldsymbol{N}_u &= \{p \in \boldsymbol{N}\colon\ \mathtt{U}_p \neq \emptyset\},\\ \boldsymbol{N}_{xu} &= \boldsymbol{N}_x \cap \boldsymbol{N}_u. \end{aligned} \tag{7.15}$$

Since $\mathtt{V}_p \neq \emptyset$, obviously, $\boldsymbol{N}_x \cup \boldsymbol{N}_u = \boldsymbol{N}$. Five distinct cases may take place:

7.5. CASE $\boldsymbol{N}_{xu} = \boldsymbol{N}$. In this case, $\mathbf{P}^{\epsilon}$ is the required partition, as each subgraph $\mathbf{D}_p^{\epsilon} = (\mathtt{U}_p \cup \mathtt{X}_p,\ \mathtt{E}_p)$ has at least one state and one input vertex corresponding to a decoupled subsystem

$$\bar{\mathbf{S}}_p^{\epsilon}\colon\ \dot{x}_p = A_{pp}x_p + B_{pp}u_p. \tag{7.16}$$

At this point, each decoupled subsystem may be checked for controllability. If a subsystem fails the test, it can be combined with another subsystem to form a single subsystem that has the desired characteristic.

7.6. CASE $\boldsymbol{N}_{xu} = \boldsymbol{N}_u \subset \boldsymbol{N}$. Now, $\mathtt{U}_p = \emptyset$ for $p \in \boldsymbol{N} - \boldsymbol{N}_u$; that is, the corresponding subgraphs $\mathbf{D}_p^{e}$ contain no input vertices. The decomposition should be modified by combining each subgraph with one containing inputs. This yields a new partition

$$\mathbf{P}_1^{\epsilon}\colon\ \mathtt{V} = \bigcup_{p\,\in\,\boldsymbol{N}_u} \mathtt{V}_p^1, \qquad \mathtt{V}_p^1 = \mathtt{X}_p^1 \cup \mathtt{U}_p^1, \tag{7.17}$$

where $\mathtt{U}_p^1 = \mathtt{U}_p$, and $\mathtt{X}_p^1$ is the union of $\mathtt{X}_p$ with one or more $\mathtt{X}_q$, $q \in \boldsymbol{N} - \boldsymbol{N}_u$. Although the choice of $\mathtt{X}_p$ to be combined with an $\mathtt{X}_q$ is arbitrary, it may be desirable to interconnect an $\mathtt{X}_q$ with that $\mathtt{X}_p$ with which $\mathtt{X}_q$ interacts most strongly, or to satisfy other criteria (stabilizability, a balanced distribution of inputs among the subsystems, *etc.*).

7.7. CASE $\boldsymbol{N}_{xu} = \boldsymbol{N}_x \subset \boldsymbol{N}$. This is a dual of Case 7.6, and can be treated similarly.

7.8. CASE $\boldsymbol{N}_{xu} = \emptyset$. In this case, each subgraph contains either only state vertices or only input vertices, indicating a poor choice of ϵ. A smaller ϵ should be tried.

7.9. CASE $\boldsymbol{N}_{xu} \neq \emptyset, \boldsymbol{N}_x, \boldsymbol{N}_u$. Under this condition, each subgraph $\mathbf{D}_p^\epsilon$, $p \in \boldsymbol{N}_{xu}$, contains both input and state vertices, and other subgraphs have the property $\boldsymbol{N}_{xu} = \emptyset$ (Case 7.8). The partition is modified as

$$\mathbf{P}_2^\epsilon\text{:}\ \ \mathtt{V} = \bar{\mathtt{V}} \cup \left(\bigcup_{p \in \boldsymbol{N}_{xu}} \mathtt{V}_p \right), \qquad \bar{\mathtt{V}} = \bigcup_{q \in \boldsymbol{N} - \boldsymbol{N}_{xu}} \mathtt{V}_q, \tag{7.18}$$

that is, we form a subsystem for each $\mathbf{D}_p^\epsilon$, $p \in \boldsymbol{N}_{xu}$, and an additional one from the combination of the remaining ones.

Observe that, in general, Cases 7.6–7.9 may result in rather large subsystems even for relatively high values of ϵ. Unfortunately, they are more likely to occur than the Case 7.5, especially when the elements of the matrices A and B are of very different order. An improvement can be achieved by *scaling* the matrix B before the decomposition is attempted, which would not matter in a control design. This is equivalent to using two different values ϵ_A and ϵ_B for A and B. To implement this modification, a subgraph $\mathbf{D}^{\epsilon_A \epsilon_B} = (\mathtt{V}, \mathtt{E}^{\epsilon_A} \cup \mathtt{E}^{\epsilon_B})$ is obtained from $\mathbf{D} = (\mathtt{V}, \mathtt{E})$ by removing all the edges corresponding to those elements of A and B with $|a_{ij}| \leq \epsilon_A$, $i \neq j$, and $|b_{ij}| \leq \epsilon_B$. By fixing ϵ_A, a state part of $\mathbf{D}^{\epsilon_A \epsilon_B}$ is formed. Then, ϵ_B can be chosen to maximize the input-state set $\boldsymbol{N}_{xu}$ defined in (7.15). Although a better way is to scale each column of B separately, it may be a difficult task due to a wide choice of possible combinations of scale factors.

Once a system is in the form (7.13) with a relatively small value of ϵ, it is obvious that we can stabilize or optimize the system piece-by-piece using decentralized feedback. Methods presented in Chapters 2–5 are efficient when applied to weakly coupled systems. An illustration of the design process is given by the following:

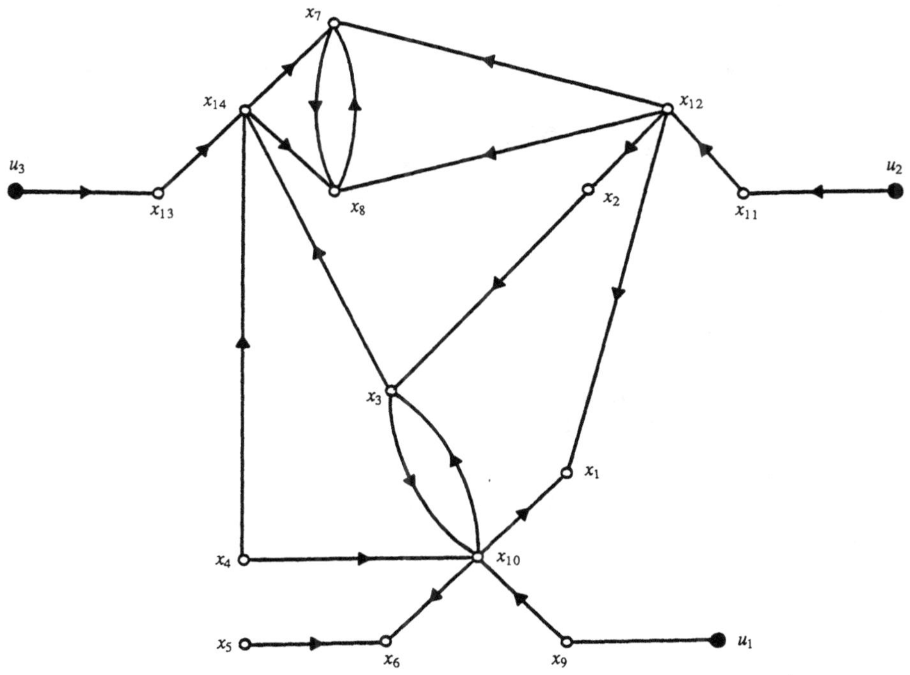

Fig. 7.3. System digraph.

7.10. EXAMPLE. We again consider the model of Example 6.25 with matrices A and B given in (6.115) and (6.116). The system digraph $\mathbf{D}$ is shown in Figure 7.3. From $\mathbf{D}$, it is obvious that the states x_4 and x_5 are not input reachable and can be ignored in further considerations.

Since each input of the system $\mathbf{S}$ is associated with only one state variable, as seen from Figure 7.3, there is no need to consider the inputs in the decomposition procedure. Applying Algorithm 7.4 to the state-truncated digraph $\mathbf{D}_x = (\mathtt{X}, \mathtt{E}_x)$, we get the nested epsilon decompositions listed in Table 7.1. In this table, only those subsets $\mathtt{X}_p^k$ that contain more than one state vertex are shown. It can be observed that, at every step up to $\epsilon_6 = 0.04$, either a new subset $\mathtt{X}_p^k$ with at least two vertices is formed, or a new vertex is added to one of such existing subsets. However, as ϵ gets smaller, partitioning of $\mathtt{X}$ becomes unbalanced in terms of the size of the subsets $\mathtt{X}_p^k$. Therefore, $\epsilon = 0.04$ is the best compromise between the number and size of the subsystems; this kind of compromise being a common issue in any decomposition method.

With ϵ chosen as $\epsilon_6 = 0.04$ the states x_1, x_6, x_7, and x_8 are decoupled

Table 7.1. Nested ϵ-decompositions.

k	ϵ_k	$\mathtt{X}_p^k$ (Partial listing)
1	0.667	$\{x_1\}, \{x_2\}, \dots, \{x_{14}\}$
2	0.5	$\{x_9, x_{10}\}$
3	0.4	$\{x_9, x_{10}\}, \{x_{11}, x_{12}\}$
4	0.1	$\{x_9, x_{10}\}, \{x_{11}, x_{12}\}, \{x_{13}, x_{14}\}$
5	0.767	$\{x_3, x_9, x_{10}\}, \{x_{11}, x_{12}\}, \{x_{13}, x_{14}\}$
6	0.04	$\{x_3, x_9, x_{10}\}, \{x_2, x_{11}, x_{12}\}, \{x_{13}, x_{14}\}$
7	0.011	$\{x_3, x_9, x_{10}, x_{13}, x_{14}\}, \{x_2, x_{11}, x_{12}\}$
8	0.00923	$\{x_3, x_9, x_{10}, x_{13}, x_{14}\}, \{x_2, x_{11}, x_{12}\}, \{x_7, x_8\}$
9	0.0068	$\{x_3, x_6, x_9, x_{10}, x_{13}, x_{14}\}, \{x_2, x_{11}, x_{12}\}, \{x_7, x_8\}$
10	0.00556	$\{x_1, x_3, x_6, x_9, x_{10}, x_{13}, x_{14}\}, \{x_2, x_{11}, x_{12}\}, \{x_7, x_8\}$
11	0.00222	$\{x_1, x_3, x_6, x_9, x_{10}, x_{13}, x_{14}\}, \{x_2, x_7, x_8, x_{11}, x_{12}\}$
12	0	$\{x_1, x_2, \dots, x_{14}\}$

from the inputs (Case 7.6). To make all components of this decomposition input reachable, each of these states is combined with one of $\mathtt{X}_p^6$ listed in Table 7.1. The modified decomposition consists of the components

$$\begin{aligned} \mathtt{V}_1^6 &= \{x_1, x_3, x_6, x_9, x_{10};\ u_1\}, \\ \mathtt{V}_2^6 &= \{x_2, x_{11}, x_{12};\ u_2\}, \\ \mathtt{V}_3^6 &= \{x_7, x_8, x_{13}, x_{14};\ u_3\}, \end{aligned} \tag{7.19}$$

where now the inputs are also included in $\mathtt{V}_p^6$ for completeness. The decomposition of $\mathbf{D}$ according to the partition in (7.19) is illustrated in Figure 7.4. The subsystems are connected by lines having weights no larger than 0.04, which was the goal of the decomposition. The decomposed system is

$$\begin{aligned} \bar{S}^{0.04}\colon\ \dot{x} = &\begin{bmatrix} A_{11} & & \bigcirc \\ & A_{22} & \\ \bigcirc & & A_{33} \end{bmatrix} x + \begin{bmatrix} B_{11} & & \bigcirc \\ & B_{22} & \\ \bigcirc & & B_{33} \end{bmatrix} u \\ &+ 0.04 \begin{bmatrix} 0 & A_{12} & 0 \\ 0 & 0 & 0 \\ A_{31} & A_{32} & 0 \end{bmatrix} x, \end{aligned} \tag{7.20}$$

where the state vectors of the three subsystems $\bar{S}_i^{0.04}$, $i = 1, 2, 3$, are listed in (7.19), and the submatrices A_{pq} and B_{pp} can be identified from the matrices $\bar{A}$ and $\bar{B}$ of (7.21). Only nonzero elements are shown save for a zero in the first row and column of A.

$$
\bar{A} = \left[\begin{array}{ccccc|cccc|cccc}
0 & & & & 0.0068 & & & & 0.0021 & & & & \\
 & -0.00182 & & & -0.00263 & 0.00182 & & & & & & & \\
 & & -0.00769 & & -0.00923 & & & & & & \bigcirc & & \\
 & & & -0.4 & & & & & & & & & \\
-0.1 & & & 0.667 & -0.667 & & & & & & & & \\
\hline
 & & & & & -0.0333 & & & 0.0767 & & & & \\
 & & \bigcirc & & & & -1 & & & & \bigcirc & & \\
 & & & & & & & 0.5 & -0.5 & & & & \\
\hline
 & & & & & & & & -0.000945 & -0.01945 & 0.01 & & 0.000378 \\
 & & & & & & & & 0.00556 & -0.011 & 0 & & 0.00222 \\
 & & & & & & & & & & & -0.667 & \\
-0.04 & & & & & & & & & & & 0.4 & -0.4
\end{array}\right]
\quad
\bar{B} = \left[\begin{array}{c|c|c}
 & & \\
 & & \\
 & & \bigcirc \\
0.4 & & \\
 & & \\
\hline
 & & \\
 & 0.9 & \\
 & & \\
\hline
 & & \\
\bigcirc & & \\
 & & 0.533 \\
 & &
\end{array}\right]
\tag{7.21}
$$

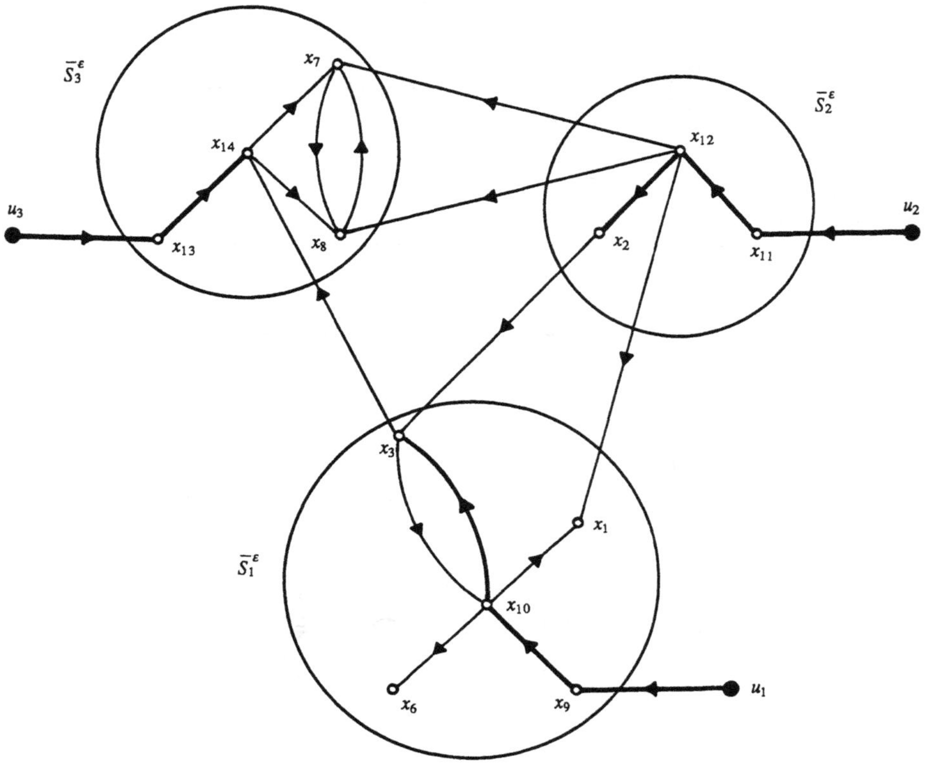

Fig. 7.4. Epsilon decomposition.

To design suboptimal decentralized controllers (Section 3.2), we choose $Q_D = I_{12}$, $R_D = I_3$, and compute the gain matrices K_1, K_2, and K_3 for $K_D = \text{diag}\{K_1, K_2, K_3\}$ as

$$\begin{aligned} K_1 &= [\,1.0000 \quad -0.3169 \quad -0.2246 \quad 0.5955 \quad 0.1636\,], \\ K_2 &= [\,0.4521 \quad 0.4858 \quad 0.2841\,], \\ K_3 &= [\,-0.0490 \quad 0.1480 \quad 0.4405 \quad 0.1975\,]. \end{aligned} \tag{7.22}$$

To determine stability of the closed-loop system

$$\hat{\mathbf{S}}^{0.04}\colon\ \dot{x} = (\bar{A} - \bar{B}K_D)x, \tag{7.23}$$

we use the vector Liapunov method (Section 2.2). Then, the test matrix is computed as

$$W = \begin{bmatrix} 0.0015 & -0.0017 & 0 \\ 0 & 0.0441 & 0 \\ -0.0320 & -0.0045 & 0.0032 \end{bmatrix}. \tag{7.24}$$

Since W is an M-matrix, the system $\hat{\mathbf{S}}^{0.04}$ is stable.

The obtained result is trivial if we recognize the fact that A and B both have a lower block triangular (LBT) structure. This special form was not intended here, but could have been discovered by LBT decomposition using Algorithm 6.13.

To evaluate the performance of the closed-loop system obtained above, the degree of suboptimality is computed. First, we take locally optimal system as reference, and get from (3.62) the value

$$\mu^* = 0.8115. \tag{7.25}$$

This implies that less than 20% deterioration of locally optimal performance is caused by the epsilon coupling terms. When the centrally optimal system is taken as reference, the degree of suboptimality is

$$\bar{\mu}^* = 0.8139, \tag{7.26}$$

and we conclude that the locally optimal performance is less than 20% off of the best possible case we could achieve without decomposition. The epsilon terms have minor effects on the overall performance of the system, which justifies the epsilon decomposition design of decentralized feedback control system.

7.4. Nested Connective Stability

It has been argued repeatedly that complexity in systems, be it in biology or physics, society or technology, takes the form of a hierarchy. A considerable number of examples in these diverse fields have been provided by Simon (1962) to show that complex systems evolve by putting together stable parts and subassemblies (components and subsystems) on a number of levels in a hierarchy. Bronowsky (1970) described the process of creation of carbon from helium to demonstrate "a physical model which shows how simple units come together to make more complex configurations; how these configurations, if they are stable, serve as units to make higher configurations; and how these configurations again, provided they are stable, serve as units to build still more complex ones, and so on." He calls this phenomena "stratified stability" and argues that it is fundamental in the evolution of living systems.

Inspired by the notion of stratified stability, Ikeda and Šiljak (1985) proposed a concept of hierarchical Liapunov functions. They serve to establish stability of *nested hierarchies* composed of components and subsystems, and are especially suitable for the type of nested hierarchies produced by epsilon decompositions.

First, Liapunov functions are constructed to show stability of each individual component (block of $\bar{A}_0$) on the zeroth level of the hierarchy. Then, the components are interconnected (using the matrix $\epsilon_1 \bar{A}_1$) to form subsystems. Stability of the subsystems is established on a higher hierarchical level by subsystem Liapunov functions composed of the component Liapunov functions. The subsystems can be further interconnected (by $\epsilon_2 \bar{A}_2$) into still larger subassemblies, and stability can be shown using the subsystem Liapunov functions, and so on, until the overall system is integrated into a whole ($\bar{\bar{A}}$), and its stability is determined by a single *hierarchical Liapunov function.*

A remarkable consequence of a hierarchically constructed Liapunov function is the fact that it establishes connective stability on each level of the hierarchy, that is, *nested connective stability.* During the process of interconnecting stable dynamic elements on any level of a nested hierarchy, the system can fall apart in precisely the way it was constructed and stability would be preserved. The evolution of the system can then restart immediately towards more complex forms by interconnecting back the stable subsystems and subassemblies. In this way, we mimic in the context of interconnected dynamic systems what Simon (1962) argues goes on in the natural evolution of complex systems, which exhibits inherent reliability and robustness to environmental perturbations.

The notion of nested connective stability is introduced *via* interconnection matrices that describe the nestedness on two levels involving subsystems and their components. A system is described as

$$\mathbf{S}\colon\ \dot{x} = f(t, x), \tag{7.27}$$

where $x(t) \in \mathbf{R}^n$ is the state of $\mathbf{S}$ at time $t \in \mathbf{R}$, and the function $f\colon\ \mathbf{R} \times \mathbf{R}^n \to \mathbf{R}^n$ is smooth enough so that solutions $x(t;\, t_0,\, x_0)$ of (7.27) exist for all initial conditions $(t_0,\, x_0) \in \mathbf{R} \times \mathbf{R}^n$ and time $t \in \mathbf{R}$. Furthermore, we assume that $f(t,\, 0) \equiv 0$, and that $x = 0$ is the unique equilibrium state of $\mathbf{S}$.

We assume that the system $\mathbf{S}$ can be decomposed as

$$\mathbf{S}\colon\ \dot{x}_i = g_i(t,\, x_i) + h_i(t,\, x), \qquad i \in \boldsymbol{N}, \tag{7.28}$$

which is an interconnection of N subsystems

$$\mathbf{S}_i\colon\ \dot{x}_i = g_i(t,\, x_i), \qquad i \in \boldsymbol{N}, \tag{7.29}$$

where $x_i(t) \in \mathbf{R}^{n_i}$ is the state of $\mathbf{S}_i$ at time $t \in \mathbf{R}$, $g_i\colon \mathbf{R} \times \mathbf{R}^{n_i} \to \mathbf{R}^{n_i}$ is the subsystem function, $h_i\colon\ \mathbf{R} \times \mathbf{R}^{n} \to \mathbf{R}^{n_i}$ is the interconnection of $\mathbf{S}_i$, and $x_i = 0$ is its unique equilibrium. We assume that the subsystems are disjoint, that is,

$$\mathbf{R}^n = \mathbf{R}^{n_1} \times \mathbf{R}^{n_2} \times \cdots \times \mathbf{R}^{n_N}. \tag{7.30}$$

Our crucial assumption is that the system $\mathbf{S}$ is *nested*, that is, each subsystem $\mathbf{S}_i$ is an interconnection

$$\mathbf{S}_i\colon\ \dot{x}_{ij} = p_{ij}(t,\, x_{ij}) + q_{ij}(t,\, x_i), \qquad j \in \boldsymbol{M}_i, \tag{7.31}$$

of M_i components

$$\mathbf{C}_{ij}\colon\ \dot{x}_{ij} = p_{ij}(t,\, x_{ij}), \qquad j \in \boldsymbol{M}_i, \tag{7.32}$$

where $x_{ij}(t) \in \mathbf{R}^{n_{ij}}$ is the state of the $\mathbf{C}_{ij}$ at time $t \in \mathbf{R}$, $p_{ij} : \mathbf{R}\times\mathbf{R}^{n_{ij}} \to \mathbf{R}^{n_{ij}}$ is the component function, $q_{ij}\colon\ \mathbf{R} \times \mathbf{R}^{n_i} \to \mathbf{R}^{n_{ij}}$ is the interconnection, $x_{ij} = 0$ is the unique equilibrium of $\mathbf{C}_{ij}$, and $\boldsymbol{M} = \{1,\, 2,\, \ldots,\, M_i\}$. Again, the $\mathbf{C}_{ij}$ are disjoint and

$$\mathbf{R}^{n_i} = \mathbf{R}^{n_{i1}} \times \mathbf{R}^{n_{i2}} \times \cdots \times \mathbf{R}^{n_{iMi}}. \tag{7.33}$$

To describe *nested structural perturbations*, we represent the system $\mathbf{S}$ of (7.27) as

$$\mathbf{S}\colon\ \dot{x} = f(t,\, x;\, E,\, L), \tag{7.34}$$

or, in the decomposed form (7.28), as

$$\mathbf{S}\colon\ \dot{x}_i = g_i(t,\, x_i;\, L_i) + h_i(t,\, x;\, E), \qquad i \in \boldsymbol{N}, \tag{7.35}$$

where the $N \times N$ interconnection matrix $E = (e_{ij})$ has elements $e_{ij}\colon \mathbf{R} \to [0,\, 1]$ that are piecewise continuous functions of time. The interconnection functions have the form

$$h_i(t,\, x;\, E) \equiv h_i(t,\, e_{i1}x_1,\, e_{i2}x_2,\, \ldots,\, e_{iN}x_N), \tag{7.36}$$

so that elements e_{ij} quantify the strength of interconnections of the subsystem $\mathbf{S}_i$ with the rest of the system $\mathbf{S}$. The nominal interactions are described by the system fundamental interconnection matrix $\bar{E} = (\bar{e}_{ij})$ (Definition 2.1).

When $E(t) \equiv 0$, the subsystems

$$\mathbf{S}_i\colon\ \dot{x}_i = g_i(t,\, x_i;\, L_i), \qquad i \in \boldsymbol{N}, \tag{7.37}$$

are decoupled from each other. Each subsystem $\mathbf{S}_i$ is connected as an interconnection

$$\mathbf{S}_i\colon\ \dot{x}_{ij} = p_{ij}(t,\, x_{ij}) + q_{ij}(t,\, x_i;\, L_i), \qquad j \in \boldsymbol{M}_i, \tag{7.38}$$

of components

$$\mathbf{C}_{ij}\colon\ \dot{x}_{ij} = p_{ij}(t,\, x_{ij}), \qquad j \in \boldsymbol{M}_i. \tag{7.39}$$

The interconnection functions among the components have the form

$$q_{ij}(t,\, x;\, L_i) \equiv q_{ij}(t,\, \ell^i_{j1}x_{i1},\, \ell^i_{j2}x_{i2},\, \ldots,\, \ell^i_{jM_i}x_{iM_i}), \tag{7.40}$$

where $\ell^i_{jk}\colon\ \mathbf{R} \to [0,\ 1]$ are elements of the $M_i \times M_i$ matrix $L_i = (\ell^i_{jk})$ that are piecewise continuous functions of time. The matrix $L = \text{diag}\{L_1,\, L_2,\, \ldots,\, L_N\}$ defines the interconnection structure of components in $\mathbf{S}$, with L_i describing the interconnections among the components $\mathbf{C}_{ij}$ constituting the subsystem $\mathbf{S}_i$. A critical assumption is the block diagonal structure of L, which rules out interconnections between components of two distinct subsystems. These interconnections, if they exist, are a part of the interconnections among the subsystems.

The nominal structure describing the interactions among the components in each $\mathbf{S}_i$ is specified by the subsystem fundamental interconnection matrix $\bar{L}_i = (\bar{\ell}^i_{jk})$. The $M_i \times M_i$ binary occurrence matrix is defined as

$$\bar{\ell}^i_{jk} = \begin{cases} 1, & x_{ik} \text{ occurs in } q_{ij}(t,\, x_i;\, L_i), \\ 0, & x_{ik} \text{ does not occur in } q_{ij}(t,\, x_i;\, L_i). \end{cases} \tag{7.41}$$

Again, $L_i(t) \leq \bar{L}_i$ element-by-element, that is, $L_i(t) \in \bar{L}_i$ (see Section 2.1), and the ℓ^i_{jk} quantify the individual interconnection of components $\mathbf{C}_{ij}$ in $\mathbf{S}_i$ relative to a maximal (nominal) interconnection strength. A nominal interconnection structure of components in $\mathbf{S}$ is, therefore, defined by the first-level fundamental interconnection matrix $\bar{L} = \text{diag}\{\bar{L}_1,\, \bar{L}_2,\, \ldots,\, \bar{L}_N\}$, while the second-level fundamental interconnection matrix $\bar{E}$ describes a nominal structure of interconnections among the subsystems.

We give a definition of nested connective stability:

7.11. DEFINITION. A system $\mathbf{S}$ is nested connectively stable if:

(*i*) For $E(t) \equiv 0$, the equilibrium $x_i = 0$ of $\mathbf{S}_i$ is asymptotically stable in the large for all $L_i(t) \in \bar{L}_i$, $i \in \boldsymbol{N}$; and

(*ii*) For $L(t) \equiv \bar{L}$, the equilibrium $x = 0$ of $\mathbf{S}$ is asymptotically stable in the large for all $E(t) \in \bar{E}$.

When compared with the standard notion of connective stability (Definition 2.3), Definition 7.11 introduces two levels of connective stability, each level having two distinct types of structural perturbations: one on the component level, and the other on the subsystem level. It is crucial to note that the two levels are not independent. When a subsystem $\mathbf{S}_i$ is decoupled from the rest of the system $\mathbf{S}$, its components $\mathbf{C}_{ij}$ can be disconnected and again connected in various ways during operation of $\mathbf{S}_i$—this is part (*i*) of Definition 7.11. When, however, the interconnection structure among the subsystems $\mathbf{S}_i$ is changing, the interactions among all $\mathbf{C}_{ij}$ have to stay fixed at their nominal strengths—this is part (*ii*) of Definition 7.11.

In order to establish connective stability of a nested system using multilevel Liapunov functions, we start with the assumption that each component $\mathbf{C}_{ij}$ is stable and we have a Liapunov function $v_{ij}(t, x_{ij})$ to prove it. Using these functions, we construct a Liapunov function $\nu_i(t, x_i)$ for each subsystem $\mathbf{S}_i$ as

$$\nu_i(t, x_i) = \sum_{j=1}^{M_i} d_{ij} v_{ij}(t, x_{ij}), \tag{7.42}$$

where the d_{ij} are all positive numbers. We also assume that each function v_{ij}: $\mathbf{R} \times \mathbf{R}^{n_{ij}} \to \mathbf{R}_+$ is continuously differentiable on the domain $\mathbf{R} \times \mathbf{R}^{n_{ij}}$, as well as positive definite, decrescent, and radially unbounded, and satisfies the inequality

$$\dot{v}_{ij}(t, x_{ij})_{(7.39)} \leq -2\pi_{ij}\psi_{ij}^2(x_{ij}), \qquad \forall (t, x_{ij}) \in \mathbf{R} \times \mathbf{R}^{n_{ij}}, \tag{7.43}$$

where π_{ij} is a positive number, and ψ_{ij}: $\mathbf{R}^{n_{ij}} \to \mathbf{R}_+$ is a positive definite function.

We assume that

$$\|\text{grad } v_{ij}(t, x_{ij})\| \leq 2\kappa_{ij}\psi_{ij}(x_{ij}), \qquad \forall (t, x_{ij}) \in \mathbf{R} \times \mathbf{R}^{n_{ij}}, \tag{7.44}$$

where the $\kappa_{ij} > 0$ are known numbers, and constrain the interconnections as

$$\|q_{ij}(t, x_i; L_i)\| \leq \sum_{k=1}^{M_i} \ell^i_{jk}(t)\xi^i_{jk}\psi_{ik}(x_{ik}), \qquad \forall (t, x_i) \in \mathbf{R} \times \mathbf{R}^{n_i}, \qquad (7.45)$$

where the $\xi^i_{jk} \geq 0$ are known numbers. An $M_i \times M_i$ test matrix $W_i = (w^i_{jk})$ is defined as

$$w^i_{jk} = \begin{cases} \pi_{ij} - \kappa_{ij}\ell^i_{jj}\xi^i_{jj}, & j = k, \\ -\kappa_{ij}\ell^i_{jk}\xi^i_{jk}, & j \neq k. \end{cases} \qquad (7.46)$$

As in Section 2.2, we conclude connective stability of $\mathbf{S}_i$ by testing the M-matrix property of $\bar{W}_i$. Assuming this property of all $\mathbf{S}_i$, we construct a second level Liapunov function

$$V(t, x) = \sum_{i=1}^{N} d_i\nu_i(t, x_i), \qquad (7.47)$$

where the d_i are all positive numbers. We have already the functions $\nu_i(t, x_i)$, so that we can compute

$$\dot{\nu}_i(t, x_i)_{(7.35)} \leq -2\pi_i\psi_i^2(x_i), \qquad \forall (t, x_i) \in \mathbf{R} \times \mathbf{R}^{n_i}, \qquad (7.48)$$

where π_i is a positive number and $\psi_i\colon\ \mathbf{R}^{n_i} \to \mathbf{R}_+$ is a positive definite function. Again, we assume

$$\|\text{grad}\ \nu_i(t, x_i)\| \leq 2\kappa_i\psi_i(x_i), \qquad \forall (t, x_i) \in \mathbf{R} \times \mathbf{R}^{n_i}, \qquad (7.49)$$

with $\kappa_i > 0$, and

$$\|h_i(t, x)\| \leq \sum_{j=1}^{N} e_{ij}(t)\xi_{ij}\psi_j(x_j), \qquad \forall (t, x) \in \mathbf{R} \times \mathbf{R}^n, \qquad (7.50)$$

where $\xi_{ij} \geq 0$, and compute the $N \times N$ test matrix $W(t) = [w_{ij}(t)]$ having the elements

$$w_{ij}(t) = \begin{cases} \pi_i - \kappa_i e_{ii}(t)\xi_{ii}, & i = j, \\ -\kappa_i e_{ij}(t)\xi_{ij}, & i \neq j. \end{cases} \qquad (7.51)$$

By $\bar{W}_i$, $i \in \boldsymbol{N}$, and $\bar{W}$, we denote the test matrices corresponding to the nominal interconnections defined by $\bar{L}_i$, $i \in \boldsymbol{N}$, and $\bar{E}$, and prove the following:

7.12. THEOREM. If $\bar{W}_i$, $i \in \boldsymbol{N}$, and $\bar{W}$ are M-matrices, then the system $\mathbf{S}$ is nested connectively stable.

Proof. We start at the component level, use the description (7.38) of $\mathbf{S}_i$, and from (7.43)–(7.45) compute

$$\nu_i(t, x_i)_{(7.38)} \leq -z_i^T(x_i)(\bar{W}_i^T D_i + D_i \bar{W}_i) z_i(x_i),$$

$$\forall (t, x_i) \in \mathbf{R} \times \mathbf{R}^{n_i}, \qquad \forall L_i(t) \in \bar{L}_i, \tag{7.52}$$

where $D_i = \text{diag}\{d_{i1}, d_{i2}, \ldots, d_{iM_i}\}$ and $z_i: \mathbf{R}^{n_i} \to \mathbf{R}_+^{M_i}$ is defined as $z_i(x_i) = [\psi_{i1}(x_{i1}), \psi_{i2}(x_{i2}), \ldots, \psi_{iM_i}(x_{iMi})]^T$. When $\bar{W}_i$ is an M-matrix, the matrix $\bar{W}_i^T D_i + D_i \bar{W}_i$ is positive definite for some D (Theorem 2.4), and $\mathbf{S}_i$ is connectively stable with respect to its components $\mathbf{C}_{ij}$, $j \in \boldsymbol{M}_i$. This is part (i) of Definition 7.11.

Part (ii) of Definition 7.11 is shown by noting that, when $L(t) \equiv \bar{L}$, all the interconnections among the components are fixed, and we have the standard case of connective stability again. The inequality is

$$\dot{V}(t, x)_{(7.34)} \leq -z^T(x)(\bar{W}^T D + D\bar{W}) z(x),$$

$$\forall (t, x) \in \mathbf{R} \times \mathbf{R}^n, \qquad \forall E(t) \in \bar{E}, \tag{7.53}$$

where $D = \text{diag}\{d_1, d_2, \ldots, d_N\}$ and $z: \mathbf{R}^n \to \mathbf{R}_+^N$ is $z(x) = [\psi_1(x_1), \psi_2(x_2), \ldots, \psi_N(x_N)]^T$. Since $\bar{W}$ is an M-matrix, $\bar{W}^T D + D\bar{W}$ is positive definite and $\mathbf{S}$ is connectively stable with respect to $\bar{E}$. Q.E.D.

Theorem 7.12 is a straightforward consequence of Theorem 2.9, and hierarchical Liapunov functions would not be so interesting if it were not for the fact that the proposed multilevel construction may show stability where a straightforward vector Liapunov function fails. We show this fact using a simple example:

7.13. EXAMPLE. Let us consider a system

$$\mathbf{S}: \; \dot{x} = \left[\begin{array}{cc|c} -3 & -2 & 2 \\ 3 & -4 & 1 \\ \hline 3 & 3 & -4 \end{array}\right] x, \tag{7.54}$$

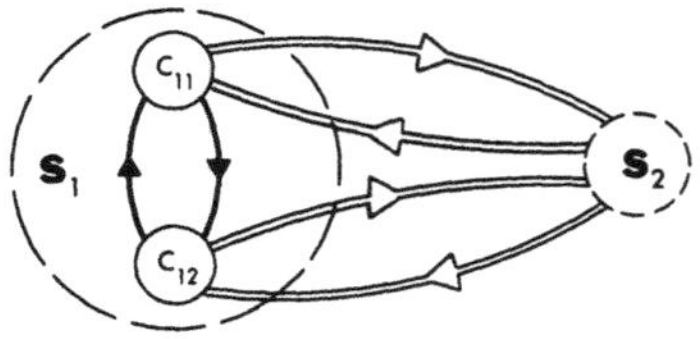

Fig. 7.5. System structure.

which is decomposed along the solid lines into two subsystems

$$\begin{aligned} \mathbf{S}_1\colon\ \dot{x}_1 &= \begin{bmatrix} -3 & -2 \\ 3 & -4 \end{bmatrix} x_1, \\ \mathbf{S}_2\colon\ \dot{x}_2 &= -4x_2. \end{aligned} \tag{7.55}$$

The subsystem $\mathbf{S}_1$ is further decomposed into two components along the dotted lines as

$$\begin{aligned} \mathbf{C}_{11}\colon\ \dot{x}_{11} &= -3x_{11}, \\ \mathbf{C}_{12}\colon\ \dot{x}_{12} &= -4x_{12}. \end{aligned} \tag{7.56}$$

The structure of $\mathbf{S}$ is shown in Figure 7.5, where single lines denote first level interactions among components $\mathbf{C}_{11}$ and $\mathbf{C}_{12}$, while double lines denote second level interconnections between subsystems $\mathbf{S}_1$ and $\mathbf{S}_2$.

On the component level we choose the functions

$$\begin{aligned} v_{11}(x_{11}) &= x_{11}^2, & \psi_{11}(x_{11}) &= |x_{11}|, \\ v_{12}(x_{12}) &= x_{12}^2, & \psi_{12}(x_{12}) &= |x_{12}|, \end{aligned} \tag{7.57}$$

and compute the numbers

$$\begin{aligned} &\pi_{11} = 3, \quad \kappa_{11} = 1, \quad \pi_{12} = 4, \quad \kappa_{12} = 1, \\ &\xi'_{11} = 0, \quad \xi'_{12} = 2, \quad \xi'_{21} = 3, \quad \xi'_{22} = 0, \end{aligned} \tag{7.58}$$

to get the matrix

$$\bar{W}_1 = \begin{bmatrix} 3 & -2 \\ -3 & 4 \end{bmatrix},$$

which is an M-matrix. Obviously, the functions

$$\nu_1(x_1) = 3x_{11}^2 + 2x_{12}^2, \qquad \nu_2(x_2) = x_2^2, \tag{7.59}$$

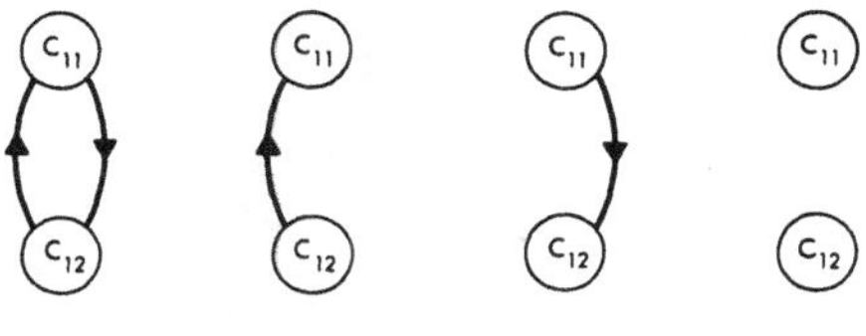

Fig. 7.6. Structural perturbations: Level I.

are Liapunov functions for $\mathbf{S}_1$ and $\mathbf{S}_2$, and connective stability under structural perturbations on the level of components (Figure 7.6) is established.

By choosing

$$\psi_1(x_1) = (x_{11}^2 + x_{12}^2)^{1/2}, \qquad \psi_2(x_2) = |x_2|, \tag{7.60}$$

and using the Euclidean vector norm, we calculate the numbers

$$\begin{aligned} &\pi_1 = 8, \quad &&\kappa_1 = 3, \quad &&\pi_2 = 4, \quad &&\kappa_2 = 1, \\ &\xi_{11} = 0, \quad &&\xi_{12} = \sqrt{5}, \quad &&\xi_{21} = 3\sqrt{2}\,, \quad &&\xi_{22} = 0, \end{aligned} \tag{7.61}$$

and obtain the matrix

$$\bar{W} = \begin{bmatrix} 8 & -3\sqrt{5} \\ -3\sqrt{2} & 4 \end{bmatrix}, \tag{7.62}$$

which is an M-matrix, and

$$V(x) = 3x_{11}^2 + 2x_{12}^2 + x_2^2, \tag{7.63}$$

is a Liapunov function for $\mathbf{S}$. This means that $\mathbf{S}$ is connectively stable under nested structural perturbations affecting subsystems $\mathbf{S}_1$ and $\mathbf{S}_2$ (Figure 7.7) and, thus, nested connectively stable.

We now show that stability of $\mathbf{S}$ cannot be established by a single level construction based on the component Liapunov functions. In other words, the functions $v_{11}(x_{11})$ and $v_{12}(x_{12})$ have to be combined to get the subsystem function $\nu_1(x_1)$ before an overall Liapunov function $V(t, x)$ is attempted. If we consider $\mathbf{C}_{11}$, $\mathbf{C}_{12}$, and $\mathbf{S}_2$ as three interconnected subsystems, then the (standard) function is

$$\nu(x) = d_1 v_{11}(x_{11}) + d_2 v_{12}(x_{12}) + d_3 \nu_2(x_2). \tag{7.64}$$

By following the procedure proposed in Theorem 2.9, we obtain

$$\dot{\nu}(x)_{(7.54)} \leq -z^T(x)(\bar{W}^T D + D\bar{W})z(x), \tag{7.65}$$

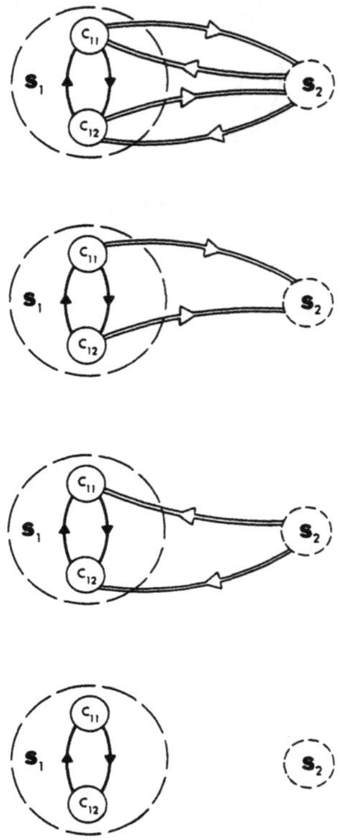

Fig. 7.7. Structural perturbations: Level II.

where $z(x) = (|x_{11}|, |x_{12}|, |x_2|)^T$, and the test matrix is

$$\bar{W} = \begin{bmatrix} 3 & -2 & -2 \\ -3 & 4 & -1 \\ -3 & -3 & 4 \end{bmatrix}. \tag{7.66}$$

$\bar{W}$ is not an M-matrix and we cannot conclude that $\nu(x)$ is a Liapunov function for **S**. The reason we failed is the fact that the single level construction could not exploit the beneficial effect of interconnections among the components $\mathbf{C}_{11}$ and $\mathbf{C}_{12}$ at the subsystem level. The price we have to pay, however, is the hierarchical constraints on allowable structural perturbations. We cannot allow the interconnections among the components to

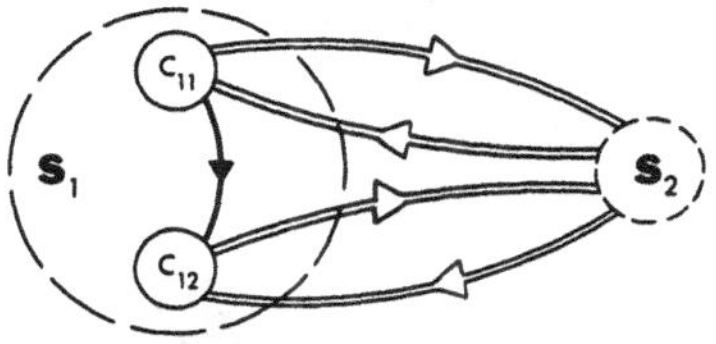

Fig. 7.8. Unstable system.

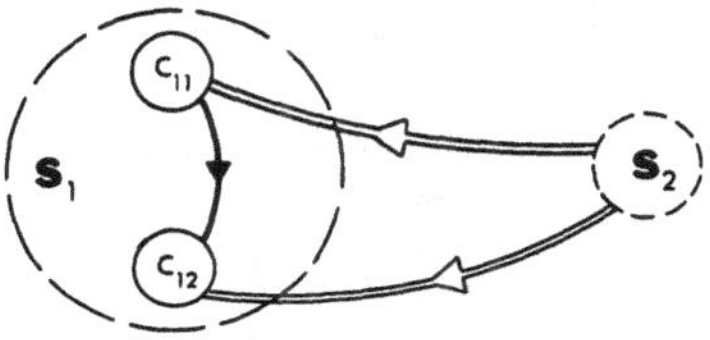

Fig. 7.9. Stable system.

be broken when subsystem interactions are present. The system

$$\mathbf{S}^*\colon\ \dot{x} = \begin{bmatrix} -3 & 0 & 2 \\ 3 & -4 & 1 \\ 3 & 3 & -4 \end{bmatrix} x, \tag{7.67}$$

which has the structure of Figure 7.8, is *unstable*.

We also note that the nested connective stability as stated in Definition 7.11 does not capture the totality of stable configurations. An acyclic structure shown in Figure 7.9 is stable, yet it is a product of simultaneous structural perturbations on both levels of the system **S**.

In concluding this section, we note that the hierarchical Liapunov functions are ideally suited for proving stability of systems having nested epsilon decompositions. The functions can take advantage of nestedness in the same way they are used to establish stability of the system in Example 7.13.

7.5. Block Diagonal Dominance

Epsilon decompositions of a given matrix represent an ideal conditioning procedure for application of block diagonal dominance. A brief exposition of the main results concerning this property is given here. For more details concerning block diagonal dominance, including overlapping blocks, see the paper by Ohta and Šiljak (1985).

Let us consider a complex matrix $A \in \mathbf{C}^{n\times n}$ which is decomposed as

$$A = A_D + A_C, \tag{7.68}$$

where

$$A_D = \operatorname{diag}\{A_1, A_2, \ldots, A_N\}, \tag{7.69}$$

$$A_C = \begin{bmatrix} A_{11} & A_{12} & \cdots & A_{1N} \\ A_{21} & A_{22} & \cdots & A_{2N} \\ \cdots & \cdots & \cdots & \cdots \\ A_{N1} & A_{N2} & \cdots & A_{NN} \end{bmatrix}, \tag{7.70}$$

where $A_i \in \mathbf{C}^{n_i\times n_i}$, $A_{ij} \in \mathbf{C}^{n_i\times n_j}$, $i, j \in \boldsymbol{N}$, $n = \sum_{i=1}^{N} n_i$, are submatrices of A_D and A_C, respectively.

From (Ohta and Šiljak, 1985), we have:

7.14. DEFINITION. A matrix A is said to be quasi-block diagonal dominant (QBDD) if

(*i*) There exist nonsingular matrices $L = \operatorname{diag}\{L_1, L_2, \ldots, L_N\}$ and $R = \operatorname{diag}\{R_1, R_2, \ldots, R_N\}$ such that

$$A_i = L_i R_i, \qquad i \in \boldsymbol{N}. \tag{7.71}$$

(*ii*) There exists a set of matrix norms $\Omega = \{\|\cdot\|_{ij}\colon \mathbf{C}^{n_i\times n_j} \to \mathbf{R}_+;\ i, j \in \boldsymbol{N}\}$ such that

$$\|F_{ik}F_{kj}\| \leq \|F_{ik}\|_{ik}\ \|F_{kj}\|_{kj}, \qquad \forall i, j \in \boldsymbol{N}, \tag{7.72}$$

where $F = (F_{ij}) \in \mathbf{C}^{n\times n}$ and $F_{ij} \in \mathbf{C}^{n_i\times n_j}$ for all $i, j \in \boldsymbol{N}$.

(*iii*) The matrix $W = (w_{ij}) \in \mathbf{R}^{N\times N}$ with elements

$$w_{ij} = \begin{Bmatrix} 1 - \|L_i^{-1}A_{ii}R_i\|_{ii}, & i = j, \\ -\|L_i^{-1}A_{ij}R_j\|_{ij}, & i \neq j, \end{Bmatrix} \tag{7.73}$$

is an M-matrix.

For simplicity, in the following developments, we drop the subscript of $\|\cdot\|_{ij}$, and denote the elements of Ω as $\|\cdot\|$. Given a set of matrix norms Ω, let $G\colon\ \mathbf{C}^{n\times n} \to \mathbf{R}^{N\times N}$ be a matrix function $G(F) = [G_{ij}(F_{ij})]$, where $G_{ij}(F_{ij}) = \|F_{ij}\|$. Note that the matrix W defined by (7.73) can be represented by $W = I - G(\bar{A}_C)$, where $\bar{A}_C = L^{-1}A_C R^{-1}$. For convenience, we use the notation $\bar{A}_{ij} = L_i^{-1}A_{ij}R_j^{-1}$ and $\bar{A}_C = (\bar{A}_{ij})$.

Denoting by $\rho(\bullet)$ the spectral radius and using the properties of M-matrices (Theorem 2.4), it is easy to get:

7.15. THEOREM. The following conditions are equivalent:

(*i*) there exists a monotone matrix norm $\|\bullet\|_m$ such that $\|G(\bar{A}_C\|_m < 1$;

(*ii*) $\rho[G(\bar{A}_C)] < 1$;

(*iii*) $I - G(\bar{A}_C)$ is an M-matrix;

(*iv*) there exist positive numbers d_j, $j \in \boldsymbol{N}$, such that

$$d_j^{-1}\sum_{i=1}^{N} d_i\|\bar{A}_{ij}\| < 1, \qquad \forall j \in \boldsymbol{N}; \tag{7.74}$$

(*v*) there exist positive numbers d_i, $j \in \boldsymbol{N}$, such that

$$d_i^{-1}\sum_{j=1}^{N} d_j\|\bar{A}_{ij}\| < 1, \qquad \forall i \in \boldsymbol{N}. \tag{7.75}$$

By $\|\bullet\|_m$: $\mathbf{R}^{N\times N} \to \mathbf{R}_+$ we denote a monotone matrix norm: For any two matrices $H = (h_{ij})$, $H' = (h'_{ij})$, $h_{ij} \geq |h'_{ij}|$ implies $\|H\|_m \geq \|H'\|_m$. We also note that the function $\|\bullet\|_*$: $\mathbf{C}^{n\times n} \to \mathbf{R}_+$ defined by $\|F\|_* = \|G(F)\|_m$ is a matrix norm (Stewart, 1973), and $A = L(I + \bar{A}_C)R$. Then, from Theorem 7.15, the following result is automatic (Ohta and Šiljak, 1985):

7.16. THEOREM. If A is QBDD, then A is nonsingular.

Now, several remarks are in order:

7.17. REMARK. If A is partitioned as in (7.70) using an epsilon decomposition, we get (7.5) with $K = 1$, $\epsilon_1 = \epsilon$, as $A = A_D + \epsilon A_C$. Then, (7.74) becomes

$$\epsilon d_j^{-1}\sum_{i=1}^{N} d_i\|\bar{A}_{ij}\| < 1, \qquad \forall j \in \boldsymbol{N}, \tag{7.76}$$

which means that epsilon decompositions are conducive to the QBDD property of A in pretty much the same way they are to stability tests of vector

Liapunov functions: The smaller ϵ is, the better is the chance to establish the property of A in question. A similar conclusion holds for stability of A *via* QBDD, which we consider below.

7.18. REMARK. In Definition 7.14, we allow for an arbitrary choice of matrix norms for each individual block $\bar{A}_{ij}$ of $\bar{A}_C = L^{-1}A_C R^{-1}$ separately, as long as (7.72) holds. Moreover, it should be stressed that elements of Ω need not be usual operator norms.

7.19. REMARK. Conditions (7.74) and (7.75) with $d_i = 1$ for all $i \in \boldsymbol{N}$ define the notion of block diagonal dominance introduced by Robert (1969) and Pearce (1974). Okuguchi (1978) generalized this notion by introducing the numbers d_i. Since basic to the notion of diagonal dominance is a normalization of A_C by A_D, Robert (1969) and Pearce (1974) adopted $\bar{A}_C = A_D^{-1}A_C$ or $\bar{A}_C = A_C A_D^{-1}$. In Definition 7.14, we used a more general normalization scheme (Ohta and Šiljak, 1985) where the normalized matrix is $\bar{A}_C = L^{-1}A_C R^{-1}$, and $LR = A_D$. Since $L^{-1}A_C R^{-1} = R(LR)^{-1}A_C R^{-1} = R(A_D^{-1}A_C)R^{-1}$, Definition 7.14 of QBDD is equivalent to that of Okuguchi (1978), save for the additional freedom provided by our definition of the set Ω. Moreover, our normalization scheme turns out to be useful when a choice of the matrix norm is fixed.

7.20. REMARK. Fiedler (1961) defined the notion of block diagonal dominance as follows: A matrix A is said to be block diagonally dominant if there exist positive numbers d_i, $i \in \boldsymbol{N}$, such that

$$d_i^{-1} \sum_{j \neq i}^{N} d_j \|A_{ii}^{-1}\| \; \|A_{ij}\| \leq 1, \qquad \forall i \in \boldsymbol{N}, \tag{7.77}$$

where $\|\bullet\|$ is a usual operator norm; if strict inequality holds in (7.77), then A is strictly block diagonally dominant. In (Feingold and Varga, 1962), $d_i = 1$ for all $i \in \boldsymbol{N}$. Since $\|A_{ii}^{-1}\| \; \|A_{ij}\| \geq \|A_{ii}^{-1}A_{ij}\|$, this definition of Fiedler is more conservative than that of Okuguchi, as long as we are interested in nonsingularity of A. However, Fiedler's definition (7.77) is useful when we consider stability of A (Ohta and Šiljak, 1985).

To consider *stability* of a matrix A, we denote by $\mathrm{Sp}(A)$ its spectrum, and recall that A is stable means that $\mathrm{Sp}(A) \subset \bar{\mathbf{C}}_+$, where $\bar{\mathbf{C}}_+$ denotes the complement of $\mathbf{C}_+$; that is, $A - \lambda I$ is nonsingular for all $\lambda \in \mathbf{C}_+$. From Theorem 7.16 the following result is automatic:

7.21. THEOREM. If $A - \lambda I$ is QBDD for all $\mathbf{C}_+$, then $\mathrm{Sp}(A) \subset \bar{\mathbf{C}}_+$.

For convenience, we say that a matrix A is QBDDS if $A - \lambda I$ is QBDD for all $\lambda = \mathbf{C}_+$ and, hence, A is stable. We should note that this definition of QBDDS is not overly restrictive. That is, $\mathrm{Sp}(A_D) \subset \bar{\mathbf{C}}_+$ and the QBDD property of A alone is not sufficient for $\mathrm{Sp}(A) \subset \bar{\mathbf{C}}_+$. For example, the matrix

$$A = \left[\begin{array}{cc|c} -2 & -6 & -1000 \\ 4 & -1 & -1000 \\ \hline -0.1 & -0.1 & -58 \end{array}\right] = \left[\begin{array}{c|c} A_1 & A_{12} \\ \hline A_{21} & A_{22} \end{array}\right], \qquad A_{11} = 0, \quad A_{22} = 0, \tag{7.78}$$

proves the statement. Choosing $L = \mathrm{diag}\{A_1, A_2\}$ and $R = \mathrm{diag}\{I, 1\}$, and using Ω as a set of spectral norms, we can show that A is QBDD. Moreover, $\mathrm{Sp}(A_1)$, $\mathrm{Sp}(A_2) \subset \bar{\mathbf{C}}_+$. However, $\mathrm{Sp}(A) \not\subset \bar{\mathbf{C}}_+$.

When $\mathrm{Sp}(A_D) \subset \bar{\mathbf{C}}_+$, we can factor $A_D - \lambda I$ as

$$A_D - \lambda I = L(\lambda) R(\lambda), \qquad \forall \lambda \in \mathbf{C}_+, \tag{7.79}$$

where $L(\lambda) = \mathrm{diag}\{L_1(\lambda), L_2(\lambda), \ldots, L_N(\lambda)\}$ and $R(\lambda) = \mathrm{diag}\{R_1(\lambda), R_2(\lambda), \ldots, R_N(\lambda)\}$ are nonsingular matrices. The test matrix $W(\lambda) = [w_{ij}(\lambda)] \in \mathbf{R}^{N \times N}$ has the elements

$$w_{ij}(\lambda) = \begin{cases} 1 - \|L_i^{-1}(\lambda) A_{ii} R_i^{-1}(\lambda)\|, & i = j, \\ -\|L_i^{-1}(\lambda) A_{ij} R_j^{-1}(\lambda)\|, & i \neq j, \end{cases} \tag{7.80}$$

and if $W(\lambda)$ is an M-matrix for all $\lambda \in \mathrm{C}_+$, then A is QBDDS.

Since $W(\lambda)$ is a function of the parameter λ, it may be difficult to test if it is an M-matrix. A sufficient condition for $W(\lambda)$ to be an M-matrix for all $\lambda \in \mathbf{C}_+$ is that $W(\lambda)$ is a quasi-dominant diagonal matrix for all $\lambda \in \mathbf{C}_+$, which is condition *(iii)* of Theorem 2.4. Two types of sufficient conditions have been derived in (Ohta and Šiljak, 1985). The first is:

7.22. THEOREM. Assume that A_D is Hermitian and negative definite. If A is QBDD with $L = -R = (-A_D)^{1/2}$ and Ω is a set of spectral norms, then A is QBDDS with $L(\lambda) = (-A_D)^{1/2} + \lambda(-A_D)^{1/2}$ and $R(\lambda) = -(-A_D)^{-1/2}$.

For the second result, we define a comparison matrix $\hat{A} = (\hat{a}_{ij}) \in \mathbf{R}^{n \times n}$ as

$$\hat{a}_{ij} = \begin{cases} \operatorname{Re} a_{ii}, & i = j, \\ |a_{ij}|, & i \neq j. \end{cases} \tag{7.81}$$

7.23. THEOREM. Assume that $L(\lambda)$ and $R(\lambda)$ are given by

$$L(\lambda) = A_D - \lambda I, \qquad R(\lambda) = I \tag{7.82}$$

and:

(*i*) For each $i \in \boldsymbol{N}$, $-\hat{A}$ is an M-matrix;

(*ii*) The matrix $\hat{W} = (\hat{w}_{ij}) \in \mathbf{R}^{N \times N}$ defined as

$$\hat{w}_{ij} = \begin{cases} 1 - \|\hat{A}_i^{-1}\|_m \, \|I_i\|_m \, \|A_{ii}\|, & i = j, \\ -\|\hat{A}_i^{-1}\|_m \, \|I_j\|_m \, \|A_{ij}\|, & i \neq j, \end{cases} \tag{7.83}$$

is an M-matrix, where I_i is the $n_i \times n_i$ identity matrix. Then, A is QBDDS.

To establish a relationship between the concepts of QBDDS and vector Liapunov functions (Sections 2.2 and 7.4), let us consider a linear time-invariant system.

$$\mathbf{S}\colon\ \dot{x}_i = A_i x_i + \epsilon \sum_{j=1}^{N} A_{ij} x_j, \qquad i \in \boldsymbol{N}, \tag{7.84}$$

which is an interconnection of N subsystems

$$\mathbf{S}_i\colon\ \dot{x}_i = A_i x_i, \qquad i \in \boldsymbol{N}, \tag{7.85}$$

obtained as a product of an epsilon decomposition. We assume that all A_i are stable, that is, $\operatorname{Sp}(A_i) \subset \bar{\mathbf{C}}_{+}$. Then, if in the Liapunov matrix equation

$$A_i^T H_i + H_i A_i = -\bar{A}_i, \qquad i \in \boldsymbol{N}, \tag{7.86}$$

we choose $\bar{A}_i$ as a symmetric negative definite matrix, the solution H_i of

(7.86) is a symmetric positive definite matrix, and the function v_i: $\mathbf{R}^{n_i} \rightarrow \mathbf{R}_+$,

$$v_i(x_i) = x_i^T H_i x_i, \qquad i \in \boldsymbol{N}, \tag{7.87}$$

is a Liapunov function for the decoupled subsystems (7.85).

This process defines an $N \times N$ block matrix

$$\bar{A} = \bar{A}_D + \epsilon \bar{A}_C, \tag{7.88}$$

where

$$\bar{A}_D = \text{diag}\{\bar{A}_1, \bar{A}_2, \ldots, \bar{A}_N\}, \quad \bar{A}_C = (2H_i A_{ij}). \tag{7.89}$$

We use the functions $v_i(x_i)$ of (7.87) as components of a vector Liapunov function $v = (v_1, v_2, \ldots, v_N)^T$, and state the following:

7.24. THEOREM. If $\bar{A}$ is QBDD with respect to Ω consisting of spectral norms and the factorization $L = -(\bar{A}_D)^{1/2}$, $R = -L$, then the equilibrium $x = 0$ of $\mathbf{S}$ is asymptotically stable and $v(x)$ is a vector Liapunov function for $\mathbf{S}$.

Again, the smaller ϵ is, the more likely Theorem 7.24 is to succeed. The vector Liapunov approach, however, offers a possibility of replacing the constant interconnection matrices A_{ij} by matrix functions $A_{ij}(t, x)$, thus allowing for robustness consideration of stability to nonlinear time-varying perturbations of $\mathbf{S}$ (for details, see Ohta and Šiljak, 1985).

7.6. Notes and References

The idea of weakly coupled systems and the kind of nested hierarchies generated by a monotone increase of the coupling threshold was most distinctly articulated by Simon (1962) in his study of complexity in natural and man-made systems. The study is a generalization of an earlier investigation of weak coupling and aggregation of dynamic systems in economic theory (Simon and Ando, 1961). A similarity between this type of result and the decomposition–aggregation idea underlying the method of vector Liapunov functions was identified in (Šiljak, 1978). Use of graphs in this context led to a formulation of the epsilon decomposition scheme (Sezer and Šiljak, 1986), which is presented in Sections 7.1–7.3.

A natural notion of stability for nested hierarchical systems is stratified stability formulated by Bronowski (1970) to provide an explanation of evolution of complex systems in biology. By mimicking this concept in the otherwise entirely different context of dynamic systems, Ikeda and Šiljak (1985: Section 7.4) proposed the notion of stratified (nested) connective stability, and provided hierarchical Liapunov functions to prove it. The framework fits naturally the hierarchies generated by nested epsilon decompositions.

THREE COMMENTS ON EPSILON DECOMPOSITIONS

A system with the matrix

$$M = \begin{bmatrix} * & \epsilon & 0 \\ \epsilon & * & 0 \\ * & * & * \end{bmatrix}, \tag{7.90}$$

where $*$ denotes a relatively large element of M, has a natural decomposition into three components, which an epsilon decomposition would miss due to the presence of the elements m_{31} and m_{32} (the matrix M corresponds to a connected graph). The reason is Algorithm 7.4 does not distinguish between a "hard" zero and a nonzero value smaller than ϵ. To fix this, the original graph should be decomposed into its strong components first, which rearranges M into an *LBT form*, and then epsilon decomposition should be applied to each component separately.

Suppose now that a system matrix is

$$M = \begin{bmatrix} * & \epsilon & \epsilon^3 \\ \epsilon & * & \epsilon^3 \\ \epsilon^{-1} & \epsilon^{-1} & * \end{bmatrix}. \tag{7.91}$$

Then, no epsilon decomposition is possible, as both $\mathbf{D}$ and $\mathbf{D}^\epsilon$ contain a single component. However, a simple *scaling* of M, which consists of multiplying the last row by ϵ^2 and the last column by ϵ^{-2}, modifies the matrix M so that all off-diagonal elements are equal to ϵ, thus resulting in an obvious epsilon decomposition of M into three components. This example illustrates the dependency of the decomposition on scaling of the matrix M. How to resolve the scaling problem is an open question.

Finally, let us consider a system matrix

$$M = \begin{bmatrix} * & * & \epsilon \\ \epsilon & * & \epsilon \\ \epsilon & * & * \end{bmatrix}, \tag{7.92}$$

which has no epsilon decomposition. If we expand the matrix M to get a new matrix

$$\tilde{M} = \left[\begin{array}{cc|cc} * & * & 0 & \epsilon \\ \epsilon & * & 0 & \epsilon \\ \hline \epsilon & 0 & * & \epsilon \\ \epsilon & 0 & * & * \end{array}\right], \tag{7.93}$$

which "includes" M, then from (7.93) we conclude that M has an *overlapping epsilon decomposition* (Arabacioglu *et al.* 1986).

BLOCK DIAGONAL DOMINANCE

Diagonally dominant matrices play an important role in studies of convergence properties of dynamic processes arising in diverse fields of numerical solutions of partial differential equations (Hageman and Young, 1981), mathematical economics (Nikaido, 1968), and a wide variety of mathematical models of complex dynamic systems (Šiljak, 1978). The important fact in these applications is the classical Hadamard result that a diagonally dominant matrix is nonsingular. From this fact, convergence (stability) of a dynamic process follows directly *via* the standard argument involving Gerschgorin circles (*e.g.*, Varga, 1962).

A significant dimension to the range of applications of diagonal dominance has been added by the notion of *block diagonal dominance* introduced independently by Fiedler (1961), Feingold and Varga (1962), and Pearce (1974). Instead of comparing the size of diagonal elements with the sum of magnitudes of the remaining elements in the row or column, the matrix is partitioned into a number of blocks, and the blocks on the main diagonal dominate in magnitude the respective sums of the off-diagonal blocks. Block diagonal dominance implies nonsingularity. With a normalization and an additional negative definiteness restriction on the diagonal blocks, Pearce (1974) has shown that block diagonal dominance implies negativity of the real parts of all eigenvalues of a given matrix, that is, stability of the corresponding dynamic process. Further improvements of the concept have been offered by Okuguchi (1978), who has introduced scaling of the blocks and use of alternative matrix norms in the definition of block diagonal dominance, as well as M-matrix properties for the diagonal blocks as proposed by Feingold and Varga (1962). A considerable increase in flexibility of the block concept has been achieved by introduction of overlapping blocks and the notion of overlapping block diagonal dominance (Ohta and Šiljak, 1985).

How QBDD can be used in a design of decentralized control in the frequency domain, is explained in (Ohta, *et al.* 1986). An application of H^∞ theory in this context is proposed by Wu and Mansour (1988, 1989).

DISCRETE SYSTEMS AND ITERATIVE PROCESSES

Epsilon decompositions apply to discrete systems as well. To see this, we should first show the tradeoff between degree of stability and the size of perturbations as it was done for continuous systems in Section 2.2.

Let us consider a linear constant discrete-time system

$$\mathbf{S}\colon\ x(t+1) = Ax(t), \tag{7.94}$$

where $x(t) \in \mathbf{R}^n$ is the state of $\mathbf{S}$ at time $t \in \mathbf{T}_+ = \{0, 1, 2, \ldots\}$, and A is a constant $n \times n$ matrix. We denote the solution at time t of (7.94) starting at $x_0 = x(0)$ by $x(t; x_0)$. We recall (see Kalman and Bertram, 1960) that $\mathbf{S}$ is asymptotically stable if and only if for any symmetric positive definite matrix G there exists a unique symmetric positive definite matrix H that satisfies the Liapunov equation

$$A^T H A - H = -G. \tag{7.95}$$

Then (Šiljak, 1978), the norm-like function

$$v(x) = (x^T H x)^{1/2} \tag{7.96}$$

is a Liapunov function for $\mathbf{S}$.

If $\mathbf{S}$ is perturbed by an additive function $h\colon\ \mathbf{T}_+ \times \mathbf{R}^n \to \mathbf{R}^n$, we have a perturbed system

$$\tilde{\mathbf{S}}\colon\ x(t+1) = Ax(t) + h[t, x(t)], \tag{7.97}$$

and if the function $h[t, x(t)]$ is bounded as

$$\|h[t, x(t)]\| \le \xi \|x\|, \qquad \forall t \in \mathbf{T}, \quad \forall x \in \mathbf{R}^n, \tag{7.98}$$

where $\xi \ge 0$, we would like to find out under what conditions on ξ stability of the nominal system $\mathbf{S}$ implies stability of the perturbed system $\tilde{\mathbf{S}}$. It has been shown in (Sezer and Šiljak, 1988) that stability is preserved whenever $\xi < \xi^{\oplus}$, where the maximal robustness bound is

$$\xi^{\oplus} = \frac{1}{\lambda_M(H^{\oplus}) + \lambda_M^{1/2}(H^{\oplus})\lambda_M^{1/2}(H^{\oplus} - I)}, \tag{7.99}$$

which is obtained when G in (7.95) is chosen as $G^{\oplus} = I$, and (7.95) is solved to get $H^{\oplus}$. It is interesting that, as in the continuous-time case (Section 2.2), the best choice of G for the robustness bound estimate is the identity matrix.

In the context of epsilon decompositions, we are interested in an interconnected system

$$\mathbf{S}\colon\ x_i(t+1) = A_i x_i(t) + \epsilon h_i[t,\, x(t)], \qquad i \in \boldsymbol{N}, \tag{7.100}$$

where, instead of the linear coupling term $\epsilon \sum_{j=1}^{N} A_{ij} x_j$, we use a nonlinear function $h\colon\ \mathbf{T}_+ \times \mathbf{R}^n \to \mathbf{R}^{n_i}$ to capture the relevance of our results to gradient algorithms (Ladde and Šiljak, 1990). We assume that the interconnections satisfy the inequalities

$$\|h_i(t,\, x)\| \le \sum_{j=1}^{N} \xi_j \|x_j\|, \qquad \forall t \in \mathbf{T}_+, \qquad \forall x \in \mathbf{R}^n, \qquad \forall i \in \boldsymbol{N}, \tag{7.101}$$

for some $\xi_{ij} \ge 0$. In a linear version of $\mathbf{S}$, the numbers ξ_{ij} are chosen as $\lambda_M^{1/2}(A_{ij}^T A_{ij})$.

Taking the function

$$v_i(x_i) = (x_i^T H_i x_i)^{1/2}, \tag{7.102}$$

as a Liapunov function for each (stable) subsystem

$$\mathbf{S}_i\colon\ x_i(t+1) = A_i x_i(t), \tag{7.103}$$

and using the robustness bound $\xi_i^{\oplus}$ computed in (7.99), we arrive at the $N \times N$ test matrix $W^{\oplus} = (w_{ij}^{\oplus})$ defined as

$$w_{ij}^{\oplus} = \begin{cases} \xi_i^{\oplus} - \epsilon \xi_{ii}, & i = j, \\ -\epsilon \xi_{ij}, & i \ne j. \end{cases} \tag{7.104}$$

Then, as shown by Sezer and Šiljak (1988), the system $\mathbf{S}$ is asymptotically stable in the large if $W^{\oplus}$ is an M-matrix.

The above stability result is particularly useful in convergence analysis of iterative processes arising in solving linear equations

$$Ax = b \tag{7.105}$$

by epsilon decompositions (Sezer and Šiljak, 1990), in particular, when Jacoby schemes are employed (Kaszkurewicz *et al.* 1990). We note that, in solving the linear problem (7.105), there is a distinct possibility of using nested multiprocessors in exactly the same pattern as the corresponding epsilon decomposition is formed. Then, the diagonal blocks can be solved in parallel with exchange of information between the blocks with frequency proportional to the size of epsilon.

Control applications of the obtained results can be based upon the work of Kokotović *et al.* (1969), Gajtsgori and Pervozvanski (1979), Peponides and Kokotović (1983), Sezer and Šiljak (1986), Molchanov (1987), and Lu *et al.* (1988).

APPLICATIONS

The most natural application of epsilon decomposition is in *electric power systems.* Under usual assumptions, a system of n machines in post-fault state can be described by the equations

$$m_i\ddot{\delta}_i + d_i\dot{\delta}_i = -\sum_{\substack{j\neq i\\ j=1}}^{n} c_{ij}[\sin(\delta_i - \delta_j) - \sin(\delta_i^e - \delta_j^e)], \qquad i = 1, 2, \ldots, n, \tag{7.106}$$

where δ_i is the rotor angle of the ith generator in the system, δ_i^e is the post-fault equilibrium value of δ_i, and m_i, d_i, and c_{ij} are the system parameters. In this context, it is desirable to identify coherent groups of generators which tend to swing together after an initial perturbation from the equilibrium.

A simple approach to coherency identification is to compare the electrical distances between the generators as measured by the numbers c_{ij}: the larger the numbers c_{ij}, the smaller the distance between the corresponding generators. Generators that are close electrically to each other are grouped together to form subsystems (Lee and Schweppe, 1973; Chow, 1982). Obviously, this is a perfect example to which the epsilon decomposition can be applied directly (see Brucoli *et al.* 1986). For alternative decompositions and clustering, see (Zaborsky *et al.* 1982; Kumar *et al.* 1986; Kusiak and

Chow, 1987; Khorasani and Pai, 1988). For potential use of epsilon decompositions in stability analysis of multimachine systems *via* vector Liapunov functions, see (Araki *et al.* 1982).

Finally, we note that nested epsilon decompositions can be applied to a wide variety of clustering problems in a number of disciplines where near-decomposability concepts are useful (see Courtois, 1977). In particular, the models in economics, queueing networks, and computer systems seem to be the most promising candidates. The near-decomposibility in a broad range of applications in the natural and man-made systems has been most convincingly promoted by Simon (1962).

Bibliography

Arabacioglu, I. M., M. E. Sezer, and O. H. Oral. (1986). Overlapping decomposition of large scale systems into weakly coupled subsystems. *Computational and Combinatorial Methods in System Theory*, C. I. Byrnes and A. Lindquist (eds.), North-Holland, Amsterdam, 135–147.

Araki, M., M. M. Metwally, and D. D. Šiljak (1982). Generalized decompositions for transient stability analysis of multimachine power systems. *Large Scale Systems*, 3, 111–122.

Bronowski, J. (1970). New concepts in the evolution of complexity. *Syntheses*, 21, 228–246.

Brucoli, M., F. Torelli, and M. Trovato (1986). A structural approach for the decomposition of interconnected power systems in stability studies. *Control Theory and Advanced Technology*, 2, 273–292.

Chow, J. H. (1982). *Time-Scale Modeling of Dynamic Networks with Applications to Power Systems.* Springer, New York.

Courtois, P. J. (1977). *Decomposability.* Academic Press, New York.

Feingold, D. G., and R. S. Varga (1962). Block diagonally dominant matrices and generalizations of the Gerschgorin circle theorem. *Pacific Journal of Mathematics*, 12, 1241–1250.

Fiedler, M. (1961). Some estimates of spectra of matrices. *Proceedings of the Symposium on the Numerical Treatment of Ordinary Differential, Integral, and Integro-Differential Equations*, Birkhaüser, Basel, Switzerland, 33–36.

Gajtsgori, V. G., and A. A. Pervozvanski (1979). Perturbation method in the optimal control problems. *Systems Science*, 5, 91–102.

Hageman, L. A., and D. M. Young (1981). *Applied Iterative Methods.* Academic Press, New York.

Ikeda, M., and D. D. Šiljak (1985). Hierarchical Liapunov functions. *Journal of Mathematical Analysis and Applications*, 112, 110–128.

Kalman, R. E., and Bertram (1960). Control system analysis and design *via* the "Second Method" of Lyapunov, Part II: Discrete-time systems. *Transactions of ASME*, Series D, 82, 394–499.

Kaszkurewicz, E., A. Bhaya, and D. D. Šiljak (1990). On the convergence of parallel asynchronous block-iterative computations. *Linear Algebra and Its Applications*, **131**, 139–160.

Khorasani, K., and M. A. Pai (1988). Two time scale decomposition and stability of power systems. *IEE Proceedings*, 135, 205–212.

Kokotović, P. V., W. R. Perkins, and J. B. Cruz (1969). ϵ-coupling method for near-optimum design of large-scale linear systems. *Proceedings of IEE*, 116, 889–892.

Kumar, K. R., A. Kusiak, and A. Vanelli (1986). Grouping parts and components in flexible manufacturing systems. *European Journal of Operations Research*, 24, 387–397.

Kusiak, A., and W. S. Chow (1987). An efficient cluster identification algorithm. *IEEE Transactions*, SMC-17, 696–699.

Ladde, G. S., and D. D. Šiljak (1989). Convergence and stability of distributed stochastic pseudogradient algorithms. *IEEE Transactions*, AC–35, 665–672.

Lee, S. T. Y., and F. C. Schweppe (1973). Distance measure and coherency recognition for transient stability equivalents. *IEEE Transactions*, PAS-92, 1550–1557.

Lu, Q., J. Lu, J. Gao, and G. K. F. Lee (1988). Discrete-time decentralized optimal controllers for multimachine power systems. *International Journal of Control*, 48, 919–928.

Mehlhorn, K. (1984). *Graph Algorithms and NP-Completeness.* Springer, Berlin, F. R. Germany.

Molchanov, A. P. (1987). Liapunov functions for nonlinear discrete-time systems. *Avtomatica i Telemekhanika*, 48, 26–35.

Nikaido, H. (1968). *Convex Structures and Economic Theory.* Academic Press, New York.

Ohta, Y., and D. D. Šiljak (1985). Overlapping block diagonal dominance and existence of Liapunov functions. *Journal of Mathematical Analysis and Applications*, 112, 396–410.

Ohta, Y., D. D. Šiljak, and T. Matsumoto (1986). Decentralized control using quasi-block diagonal dominance of transfer function matrices. *IEEE Transactions*, AC-31, 420–430.

Okuguchi, K. (1978). Matrices with dominant diagonal blocks and economic theory. *Journal of Mathematical Economics*, 5, 43–67.

Park, K. C., and C. A. Felippa (1983). Partitioned analysis of coupled systems. *Computational Methods for Transient Analysis*, T. Belytschko and T. J. R. Hughes (eds.), Elsevier, New York, 157–219.

Pearce, I. F. (1974). Matrices with dominated diagonal blocks. *Journal of Economic Theory*, 9, 159–170.

Peponides, G. M., and P. V. Kokotović (1983). Weak connections, time scales, and aggregation of nonlinear systems. *IEEE Transactions*, CAS-30, 416–422.

Petrovic, B., and Z. Gajić (1988). Recursive solution of linear–quadratic Nash games for weakly interconnected systems. *Journal of Optimization Theory and Applications*, 56, 463–477.

Robert, F. (1969). Blocs-H-matrices et convergence des methodes iteratives classiques par blocs. *Linear Algebra and Applications*, 2, 223–265.

Sezer, M. E., and D. D. Šiljak (1986). Nested ϵ-decomposition and clustering of complex systems. *Automatica*, 22, 321–331.

Sezer, M. E., and D. D. Šiljak (1988). Robust stability of discrete systems. *International Journal of Control*, 48, 2055–2063.

Sezer, M. E., and D. D. Šiljak (1990). Epsilon decomposition of linear systems: Weakly coupled and overlapping blocks. *SIAM Journal on Matrix Analysis and Applications* (to appear).

Šiljak, D. D. (1978). *Large-Scale Dynamic Systems: Stability and Structure.* North-Holland, New York.

Simon, H. A. (1962). The architecture of complexity. *Proceedings of the American Philosophical Society*, 106, 467–482.

Simon, H. A., and A. Ando (1961). Aggregation of variables in dynamic systems. *Econometrica*, 29, 111–138.

Stewart, G. W. (1973). *Introduction to Matrix Computations.* Academic Press, New York.

Varga, R. S. (1962). *Matrix Iterative Analysis.* Prentice-Hall, Englewood Cliffs, New Jersey.

Wu, Q. H., and M. Mansour (1988). An application of H^∞ theory in decentralized control. *Proceedings of the 27th Conference on Decision and Control*, Austin, Texas, 1335–1340.

Wu, Q. H., and M. Mansour (1989). Decentralized robust control using H^∞-optimization technique. *Information and Decision Technologies*, 15, 59–76.

Zaborsky, J., K. W. Whang, G. M. Huang, L. J. Chiang, and S. Y. Lin (1982). A clustered dynamic model for a class of linear autonomous systems using simple enumerative sorting. *IEEE Transactions*, CAS-29, 747–758.

Chapter 8 Overlapping Decompositions

In a wide variety of natural and man-made systems, subsystems share common parts. For either conceptual or computational reasons, it is advantageous to recognize this fact, and build decentralized control and estimation schemes using overlapping information sets. In certain control problems involving traffic regulation, power systems, and large space structures, overlapping decentralized control is the only effective way to go.

It has been known for some time that, by expanding the state space, overlapping subsystems appear as disjoint. Then, standard methods can be used to design decentralized control laws in the expanded space. At the end, the expanded laws are contracted for implementation in the original system. To carry out this expansion–contraction process correctly, certain conditions have to be satisfied that ensure that solutions of the original system are *included* in solutions of the expanded system. The circle of ideas and conditions underlying the process have been organized only recently into a general mathematical framework: *The Inclusion Principle.*

In our presentation of the Principle, we will use a balanced mixture of matrix algebra and geometric concepts to provide a flexible theory that is numerically attractive and, at the same time, one that appeals to intuition. A crucial contribution of the Principle is that it leads to a complete and, it is hoped, definite clarification of the overlapping problem. We will demonstrate this fact using both linear and nonlinear constant systems, but the theory is broad enough to accomodate time-varying, hereditary, and stochastic effects.

Of special interest is the optimization problem with overlapping information sets because of its importance in applications. Expansions and contractions of relevant performance indices are defined for separable quadratic

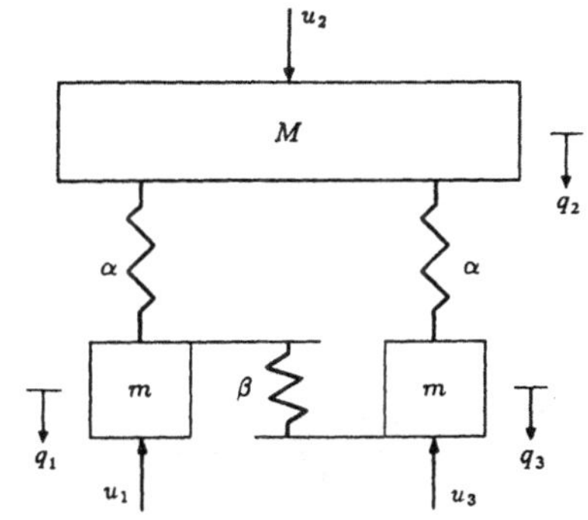

Fig. 8.1. Mass–spring system.

criteria and linear constant systems. This allows for standard decentralized methods (see Chapter 2) to be used in the expanded space to establish suboptimal, but stable, performance of decentralized local controllers obeying the overlapping information structure constraints. These results are of central importance for building reliable control using multiple controller schemes, which is the theme of Chapter 9. Other applications of overlapping optimization methods are described in Notes and References (Section 8.5).

8.1. Preliminaries

To motivate the reader for overlapping decompositions, let us consider the mechanical system in Figure 8.1, which consists of two identical mass and spring elements attached to a rigid structure. The equations of motion are

$$\begin{aligned} m\ddot{q}_1 &= -\alpha(q_1 - q_2) - \beta(q_1 - q_3) - u_1, \\ M\ddot{q}_2 &= -\alpha(q_2 - q_1) - \alpha(q_2 - q_3) + u_2, \\ m\ddot{q}_3 &= -\alpha(q_3 - q_2) - \beta(q_3 - q_1) - u_3. \end{aligned} \tag{8.1}$$

By choosing the state vector $x = (q_1, \dot{q}_1, q_2, \dot{q}_2, q_3, \dot{q}_3)^T$, we rewrite the

Equations (8.1) in the matrix form as the system

$$
\mathbf{S}\colon\ \dot{x} = \begin{bmatrix} 0 & 1 & 0 & 0 & 0 & 0 \\ -\frac{\alpha+\beta}{m} & 0 & \frac{\alpha}{m} & 0 & \frac{\beta}{m} & 0 \\ 0 & 0 & 0 & 1 & 0 & 0 \\ \frac{\alpha}{M} & 0 & -2\frac{\alpha}{M} & 0 & \frac{\alpha}{M} & 0 \\ 0 & 0 & 0 & 0 & 0 & 1 \\ \frac{\beta}{m} & 0 & \frac{\alpha}{m} & 0 & -\frac{\alpha+\beta}{m} & 0 \end{bmatrix} x
+ \begin{bmatrix} 0 & 0 & 0 \\ -\frac{1}{m} & 0 & 0 \\ 0 & 0 & 0 \\ 0 & \frac{1}{M} & 0 \\ 0 & 0 & 0 \\ 0 & 0 & -\frac{1}{m} \end{bmatrix} u, \qquad (8.2)
$$

where $u = (u_1, u_2, u_3)^T$ is the control vector of $\mathbf{S}$.

The underlying assumption, which calls for an overlapping decomposition of $\mathbf{S}$, is that the mass M of the rigid structure is much larger than the mass m of the attached elements. Furthermore, the stiffness β is small so that the coupling between the elements is weak. Under these assumptions, an obvious decomposition of $\mathbf{S}$ into three subsystems, where each mass corresponds to a subsystem, is not a natural decomposition. The reason is that the strong coupling of size α/m of the structure to the elements are exposed as interconnections between the subsystems. This makes a decentralized control strategy, which is based on three independent inputs u_1, u_2, and u_3, unjustified. A better way to partition the system $\mathbf{S}$ is to pair the structure with each element separately and form two subsystems that share the structure as a common part—this is an *overlapping decomposition.* It is indicated by dashed lines in (8.2).

Although the two overlapping subsystems are well defined, the interconnections between them are not. To identify the interconnections, we *expand* the system $\mathbf{S}$ by repeating the two state equations describing the structure.

This leads to the expanded system

$$
\tilde{\mathbf{S}}\colon\ \dot{\tilde{x}} = \left[\begin{array}{cccc|cccc}
0 & 1 & 0 & 0 & 0 & 0 & 0 & 0 \\
-\frac{\alpha+\beta}{M} & 0 & \frac{\alpha}{m} & 0 & 0 & 0 & \frac{\beta}{m} & 0 \\
0 & 0 & 0 & 1 & 0 & 0 & 0 & 0 \\
\frac{\alpha}{M} & 0 & -2\frac{\alpha}{M} & 0 & 0 & 0 & 0 & 0 \\
\hline
0 & 0 & 0 & 0 & 0 & 1 & 0 & 0 \\
\frac{\alpha}{M} & 0 & 0 & 0 & -2\frac{\alpha}{M} & 0 & \frac{\alpha}{M} & 0 \\
0 & 0 & 0 & 0 & 0 & 0 & 0 & 1 \\
0 & \frac{\beta}{m} & 0 & 0 & \frac{\alpha}{M} & 0 & -\frac{\alpha+\beta}{M} & 0
\end{array}\right] \tilde{x}
+ \left[\begin{array}{cc|cc}
0 & 0 & 0 & 0 \\
-\frac{1}{m} & 0 & 0 & 0 \\
0 & 0 & 0 & 0 \\
0 & \frac{1}{M} & 0 & 0 \\
\hline
0 & 0 & 0 & 0 \\
0 & 0 & \frac{1}{M} & 0 \\
0 & 0 & 0 & 0 \\
0 & 0 & 0 & -\frac{1}{m}
\end{array}\right] u. \tag{8.3}
$$

Now, two subsystems can be formed, resulting into a standard *disjoint decomposition* indicated by dashed lines in (8.3). The important fact is that in the expanded system the interconnections between the two subsystems are well defined as the off-diagonal blocks. Equally important is the obvious weakness of the coupling, which contains only the small numbers α/M and β/m; the large connections corresponding to α/m appear inside the subsystems, and serve to enhance their autonomy. Before we can exploit this weak coupling structure of the expanded system $\tilde{\mathbf{S}}$, we should establish the relationship between $\tilde{\mathbf{S}}$ and the original system $\mathbf{S}$ having a smaller dimension. This we do next.

In order to make transparent a conceptual significance of the unorthodox notion of overlapping subsystems, we start our exposition using linear constant systems without inputs and outputs. As in the preceding example, we consider the basic situation where only two subsystems are involved.

Let us consider a system

$$\mathbf{S}\colon\ \dot{x} = Ax, \tag{8.4}$$

where $x(t) \in \mathbf{R}^n$ is the state of $\mathbf{S}$. We assume that x is composed of three components, $x = (x_1^T, x_2^T, x_3^T)^T$, with each component chosen on the basis of some physical or abstract reasoning. The dimensions of the components are n_1, n_2, and n_3 so that $n = n_1 + n_2 + n_3$. This partition of the state x induces a partition of the $n \times n$ matrix A as

$$A = \left[\begin{array}{cc:c} A_{11} & A_{12} & A_{13} \\ \hdashline A_{21} & A_{22} & A_{23} \\ \hdashline A_{31} & A_{32} & A_{33} \end{array}\right], \tag{8.5}$$

where the submatrices have appropriate dimensions.

Now, we arrange the three components of the state x into two overlapping components $\tilde{x}_1 = (x_1^T, x_2^T)^T$ and $\tilde{x}_2 = (x_2^T, x_3^T)^T$, which we use to form a new vector $\tilde{x} = (\tilde{x}_1^T, \tilde{x}_2^T)^T$, and induce an overlapping decomposition of A indicated by dashed lines in (8.5). The vector $\tilde{x}$ is related to x by a linear transformation

$$\tilde{x} = Vx, \tag{8.6}$$

where V is the $\tilde{n} \times n$ matrix

$$V = \begin{bmatrix} I_1 & 0 & 0 \\ 0 & I_2 & 0 \\ 0 & I_2 & 0 \\ 0 & 0 & I_3 \end{bmatrix}, \tag{8.7}$$

and $\tilde{n} = n_1 + 2n_2 + n_3$. In (8.7), I_1, I_2, and I_3 are identity matrices with dimensions corresponding to the vector components x_1, x_2, and x_3 of x. Transformation (8.6) defines the expanded system

$$\tilde{\mathbf{S}}\colon\ \dot{\tilde{x}} = \tilde{A}\tilde{x}, \tag{8.8}$$

with the state $\tilde{x}(t) \in \mathbf{R}^{\tilde{n}}$ and the $\tilde{n} \times \tilde{n}$ matrix

$$\tilde{A} = VAU + M. \tag{8.9}$$

The matrix $U = (V^TV)^{-1}V^T$ is chosen as the pseudoinverse of V:

$$U = \begin{bmatrix} I_1 & 0 & 0 & 0 \\ 0 & \frac{1}{2}I_2 & \frac{1}{2}I_2 & 0 \\ 0 & 0 & 0 & I_3 \end{bmatrix}; \tag{8.10}$$

and $UV = I_n$, which is the identity matrix of order n. The matrix M is termed a *complementary matrix*, and the choice

$$M = \begin{bmatrix} 0 & \frac{1}{2}A_{12} & -\frac{1}{2}A_{12} & 0 \\ 0 & \frac{1}{2}A_{22} & -\frac{1}{2}A_{22} & 0 \\ 0 & -\frac{1}{2}A_{22} & \frac{1}{2}A_{22} & 0 \\ 0 & -\frac{1}{2}A_{32} & \frac{1}{2}A_{32} & 0 \end{bmatrix} \tag{8.11}$$

produces a matrix $\tilde{A}$ in the form

$$\tilde{A} = \left[\begin{array}{cc|cc} A_{11} & A_{12} & 0 & A_{13} \\ A_{21} & A_{22} & 0 & A_{23} \\ \hline A_{21} & 0 & A_{22} & A_{23} \\ A_{31} & 0 & A_{32} & A_{33} \end{array}\right]. \tag{8.12}$$

By comparing the original matrix A with the expanded matrix $\tilde{A}$, we note that under the transformation (8.6), the description of the overlapping diagonal blocks remained invariant due to the choice (8.11) of the complementary matrix M. This invariant property of the expansion is crucial because the identity of overlapping subsystems is preserved in $\tilde{A}$, and they now appear as disjoint.

To take advantage of the form (8.12) of the expanded matrix $\tilde{A}$, we need to establish the relation between the motions $x(t; x_0)$ and $\tilde{x}(t; \tilde{x}_0)$ of the two systems $\mathbf{S}$ and $\tilde{\mathbf{S}}$, which start at $t = 0$ with x_0 and $\tilde{x}_0$. It is easy to show that $\tilde{x}_0 = Vx_0$ implies $x(t; x_0) = U\tilde{x}(t; \tilde{x}_0)$ for all $t \geq 0$. This means that U is a projection of the motion space of $\tilde{\mathbf{S}}$ onto the motion space of $\mathbf{S}$. We express this fact by saying that $\tilde{\mathbf{S}}$ *includes* $\mathbf{S}$, or that $\mathbf{S}$ *is included by* $\tilde{\mathbf{S}}$. A circle of ideas underlying the concept of inclusion became known as the *Inclusion Principle* (*e.g.*, Ikeda *et al.* 1984a). Before we engage in a formalization of this concept, let us take stock of what we have done so far. It is intuitively obvious that the stability of $\tilde{\mathbf{S}}$ implies the stability of $\mathbf{S}$. What is not so obvious is that proving stability of $\tilde{\mathbf{S}}$ may be easier than proving stability of the original system $\mathbf{S}$. This fact makes overlapping decompositions of dynamic systems attractive indeed.

Let us use a simple example to show that overlapping decompositions can be used in the context of vector Liapunov functions to establish stability of a system when disjoint decompositions fail.

8.1. EXAMPLE. Let us consider a stable system

$$\mathbf{S}\colon\ \dot{x} = \begin{bmatrix} -1 & -26 & 6 \\ 3 & -20 & 3 \\ 6 & -26 & -1 \end{bmatrix} x. \tag{8.13}$$

To show stability by using vector Liapunov functions (Section 2.3), we use first a disjoint decomposition such that

$$\mathbf{S}\colon\ \begin{bmatrix} \dot{x}_1 \\ \dot{x}_2 \end{bmatrix} = \left[\begin{array}{c|cc} -1 & -26 & 6 \\ \hline 3 & -20 & 3 \\ 6 & -26 & -1 \end{array}\right] \begin{bmatrix} x_1 \\ x_2 \end{bmatrix}, \tag{8.14}$$

where x_1 and x_2 are the states of the two subsystems

$$\begin{aligned} \mathbf{S}_1\colon\ \dot{x}_1 &= -x_1, \\ \mathbf{S}_2\colon\ \dot{x}_2 &= \begin{bmatrix} -20 & 3 \\ -26 & -1 \end{bmatrix} x_2. \end{aligned} \tag{8.15}$$

For subsystem Liapunov functions, we choose

$$v_1(x_1) = x_1^2, \qquad v_2(x_2) = x_2^T H_2 x_2, \tag{8.16}$$

where H_2 is a positive definite matrix, which is the solution of the matrix equation

$$\begin{bmatrix} -20 & -26 \\ 3 & -1 \end{bmatrix} H_2 + H_2 \begin{bmatrix} -20 & 3 \\ -26 & -1 \end{bmatrix} = -G_2, \tag{8.17}$$

for a symmetric positive definite matrix G_2. The derivatives of the functions (8.16) are computed with respect to the system **S** written as two interconnected subsystems:

$$\mathbf{S}\colon\ \dot{x}_1 = -x_1 + [\,-26 \quad 6\,]\, x_2, \tag{8.18a}$$

$$\dot{x}_2 = \begin{bmatrix} -20 & 3 \\ -26 & -1 \end{bmatrix} x_2 + \begin{bmatrix} 3 \\ 6 \end{bmatrix} x_1. \tag{8.18b}$$

We get

$$\dot{v}_1(x_1)_{(8.18a)} = -2x_1^2 + 2x_1 \,[-26 \quad 6]\, x_2,$$

$$\dot{v}_2(x_2)_{(8.18b)} = -x_2^T G_2 x_2 + 2x_2^T H_2 \begin{bmatrix} 3 \\ 6 \end{bmatrix} x_1. \tag{8.19}$$

By choosing the function

$$\nu(x) = d_1 v_1(x_1) + d_2 v_2(x_2) \tag{8.20}$$

as a Liapunov function for **S** and using (8.19), we compute

$$\dot{\nu}(x)_{(8.13)} \leq -\frac{1}{2} z^T (W^T D + DW)\, z,$$

where $z = (z_1,\, z_2)^T$, $z_1 = |x_1|$, $z_2 = (x_2^T Q_2 x_2)^{1/2}$, Q_2 is any positive definite matrix, $d = (d_1,\, d_2)^T$ is a positive vector, and

$$W = \begin{bmatrix} 2 & -2\|[-26 \quad 6]\, Q_2^{-1/2}\| \\ -2\left\| Q_2^{-1/2} H_2 \begin{bmatrix} 3 \\ 6 \end{bmatrix} \right\| & \lambda_m\left(Q_2^{-1/2} G_2 Q_2^{-1/2}\right) \end{bmatrix}. \tag{8.21}$$

From Section 2.3, we recall that $\nu(x)$, $-\dot{\nu}(x)_{(8.13)} > 0$, and **S** is stable if W is an M-matrix (Theorem 2.4), which is equivalent to saying that

$$\lambda_m\left(Q_2^{-1/2} G_2 Q_2^{-1/2}\right) - 2\|\,[-26 \quad 6]\, Q_2^{-1/2}\| \left\| Q_2^{-1/2} H_2 \begin{bmatrix} 3 \\ 6 \end{bmatrix} \right\| > 0. \tag{8.22}$$

We show, however, that (8.22) does not hold no matter what is the choice of G_2 and Q_2. For this, we need the relation

$$\lambda_m(Q_2^{-1/2} G_2 Q_2^{-1/2}) = \lambda_M^{-1}(Q_2^{1/2} G_2^{-1} Q_2^{1/2}) = \|Q_2^{1/2} G_2^{-1/2}\|^{-1} \|G_2^{-1/2} Q_2^{1/2}\|^{-1}. \tag{8.23}$$

By denoting the right side of (8.23) by μ and using (8.23), we majorize the left side of (8.22) as

$$\begin{aligned}
&\mu - 2\,\|[26 \quad 6]Q_2^{-1/2}\| \left\| Q_2^{-1/2} H_2 \begin{bmatrix} 3 \\ 6 \end{bmatrix} \right\| \\
&\leq \mu \left(1 - 2\|[26 \quad 6]G_2^{-1/2}\| \left\| G_2^{-1/2} H_2 G_2^{-1/2} G_2^{1/2} \begin{bmatrix} 3 \\ 6 \end{bmatrix} \right\| \right) \\
&\leq \mu \left\{ 1 - 2\lambda_m \left(G_2^{-1/2} H_2 G_2^{-1/2} \right) \|[-26 \quad 6]G_2^{-1/2}\| \left\| G_2^{1/2} \begin{bmatrix} 3 \\ 6 \end{bmatrix} \right\| \right\} \\
&\leq \mu \left\{ 1 - 84\lambda_m \left(G_2^{-1/2} H_2 G_2^{-1/2} \right) \right\}.
\end{aligned} \tag{8.24}$$

Next, we show that

$$\lambda_m \left(G_2^{-1/2} H_2 G_2^{-1/2} \right) > 1/42, \tag{8.25}$$

which implies that the last expression in (8.24) is negative and (8.22) does not hold. For this purpose, we need first to show that if a positive definite matrix H is the solution of the Liapunov matrix equation

$$A^T H + HA = -G, \tag{8.26}$$

for a stable matrix A and a positive definite matrix G, then

$$\lambda_m(G^{-1/2} H G^{-1/2}) > -\frac{1}{2} (\operatorname{tr} A)^{-1}. \tag{8.27}$$

By multiplying (8.26) by $H^{-1/2}$ from the left and from the right, we get

$$H^{-1/2} A^T H^{1/2} + H^{1/2} A H^{-1/2} = -H^{-1/2} G H^{-1/2}. \tag{8.28}$$

Using (8.28), we calculate

$$\begin{aligned}\lambda_m^{-1}(G^{-1/2}HG^{-1/2}) &= \lambda_M(H^{-1/2}GH^{-1/2})\\ &< \sum_i \lambda_i(H^{-1/2}GH^{-1/2})\\ &= \operatorname{tr}(H^{-1/2}GH^{-1/2})\\ &= -2\operatorname{tr}(H^{1/2}AH^{-1/2})\\ &= -2\operatorname{tr} A,\end{aligned} \tag{8.29}$$

which proves (8.27). Applying (8.27) to our example where A is the matrix of system $\mathbf{S}_2$ in (8.6), we establish (8.25). This, in turn, implies that for the disjoint decomposition in (8.14) and the standard choice of Liapunov functions (8.16), the concept of vector Liapunov functions fails to prove stability of the system $\mathbf{S}$. It is easy to show that the three remaining disjoint decompositions of the system $\mathbf{S}$ lead to the same conclusion.

Let us now show that using the overlapping decomposition, we can demonstrate stability of $\mathbf{S}$. We decompose $\mathbf{S}$ along the dashed lines as

$$\mathbf{S}\colon\ \dot{x} = \begin{bmatrix} -1 & -26 & 6 \\ 3 & -20 & 3 \\ 6 & -26 & -1 \end{bmatrix} x, \tag{8.30}$$

and get the expansion

$$\mathbf{S}\colon\ \begin{bmatrix} \dot{\tilde{x}}_1 \\ \dot{\tilde{x}}_2 \end{bmatrix} = \left[\begin{array}{cc|cc} -1 & -26 & 0 & 6 \\ 3 & -20 & 0 & 3 \\ \hline 3 & 0 & -20 & 3 \\ 6 & 0 & -26 & -1 \end{array}\right] \begin{bmatrix} \tilde{x}_1 \\ \tilde{x}_2 \end{bmatrix}, \tag{8.31}$$

which corresponds to the expansion matrix of (8.12). We now choose the subsystem Liapunov functions as

$$\tilde{v}(\tilde{x}_1) = \tilde{x}_1^T \tilde{H}_1 \tilde{x}_1, \qquad \tilde{v}_2(\tilde{x}_2) = \tilde{x}_2^T \tilde{H}_2 \tilde{x}_2, \tag{8.32}$$

where

$$\tilde{H}_1 = \begin{bmatrix} 178 & -377 \\ -377 & 845 \end{bmatrix}, \qquad \tilde{H}_2 = \begin{bmatrix} 845 & -377 \\ -377 & 178 \end{bmatrix}. \tag{8.33}$$

The choice of functions in (8.32) was made using the recipe of Section 2.3, which starts by selecting the matrices $\tilde{H}_1$ and $\tilde{H}_2$ as the identity matrices in the transformed space where the subsystem matrices are diagonal. Using the function

$$\tilde{\nu}(\tilde{x}) = d_1\tilde{v}_1(\tilde{x}_1) + d_2\tilde{v}_2(\tilde{x}_2), \tag{8.34}$$

we compute

$$\dot{\tilde{\nu}}(\tilde{x})_{(8.31)} \leq -\frac{1}{2}\tilde{z}(\tilde{W}^T D + D\tilde{W})\,\tilde{z}, \tag{8.35}$$

where $\tilde{z} = (\tilde{z}_1,\, \tilde{z}_2)^T$, $\tilde{z}_1 = (\tilde{x}_1^T \tilde{H}_1 \tilde{x}_1)^{1/2}$, $\tilde{z}_2 = (\tilde{x}_2^T \tilde{H}_2 \tilde{x}_2)^{1/2}$, and

$$\tilde{W} = \begin{bmatrix} 14 & -6\sqrt{5} \\ -6\sqrt{5} & 14 \end{bmatrix}. \tag{8.36}$$

Because $\tilde{W}$ is an M-matrix, stability of the expansion $\tilde{\mathbf{S}}$ follows. Since $\tilde{\mathbf{S}}$ includes $\mathbf{S}$, stability of $\tilde{\mathbf{S}}$ implies stability of $\mathbf{S}$.

When we compare the disjoint decomposition of $\mathbf{S}$ in (8.14) with the same kind of decomposition of the expansion $\tilde{\mathbf{S}}$ in (8.31), we see why the latter is successful and the former is not. In (8.14) the large connections -26 are exposed in the interconnections among the subsystems, while in the expanded system $\tilde{\mathbf{S}}$ of (8.31) these connections are located inside the subsystems and, for this reason, the subsystems are weakly coupled. This fact makes the method of vector Liapunov functions successful in proving stability of the expanded system $\tilde{\mathbf{S}}$ and, therefore, stability of the original system $\mathbf{S}$.

8.2. The Inclusion Principle

Now we want to formulate rigorously the Inclusion Principle, which provides a proper mathematical framework for the method of overlapping decompositions. We narrow our attention to linear time-invariant systems, but the formulation of the Principle is given in terms of the motions (solutions) of dynamical systems rather than in algebraic terms (matrices). This way, the Principle can be applied to more general types of systems with nonlinearities, time-varying elements, hereditary effects, *etc.* A balanced mixture of matrix algebra and geometric concepts is used to provide a numerically oriented expansion–contraction framework for dynamic systems which, at the same time, offers interpretations and insights of a geometric approach.

We consider two dynamic systems **S** and $\tilde{\mathbf{S}}$ described by the equations

$$\mathbf{S}\colon\ \dot{x} = Ax + Bu, \qquad y = Cx, \tag{8.37}$$

and

$$\tilde{\mathbf{S}}\colon\ \dot{\tilde{x}} = \tilde{A}\tilde{x} + \tilde{B}u, \qquad y = \tilde{C}\tilde{x}. \tag{8.38}$$

Here, $x \in \mathcal{X}$ is an n-vector, $u \in \mathcal{U}$ is an m-vector, $y \in \mathcal{Y}$ is an ℓ-vector; $\mathcal{X}, \mathcal{U}$, and $\mathcal{Y}$ are the state, input, and output spaces of **S**, respectively. Similarly, $\tilde{x} \in \tilde{\mathcal{X}}$ is an $\tilde{n}$-vector, and $\tilde{\mathcal{X}}$, $\mathcal{U}$, and $\mathcal{Y}$ are the state, input, and output spaces of $\tilde{\mathbf{S}}$, respectively. We assume that the input $u(t)$ belongs to the class of piecewise continuous functions. We say that **S** is represented relative to a given basis of the triplet $(A,\ B,\ C)$ of constant matrices having appropriate dimensions. For $\tilde{\mathbf{S}}$, we use the triplet $(\tilde{A},\ \tilde{B},\ \tilde{C})$. Our crucial assumption is $n \leq \tilde{n}$; that is, the dimension of **S** is smaller (or at most equal) to that of $\tilde{\mathbf{S}}$. By $x(t;\ x_0,\ u)$, we denote the unique solution of **S** for the initial time $t = 0$, the initial state $x(0) = x_0$, and a *fixed* control input $u(t)$ defined on the time interval $[0,\ t]$. For brevity, we use the notation $x(t)$ to indicate the solution for arbitrary $x_0 \in \mathcal{X}$ and admissible $u(t)$, and the notation x to indicate the vector $x(t)$ at $t \geq 0$. A similar convention is applied to solutions $\tilde{x}(t;\ \tilde{x}_0,\ u)$ of $\tilde{\mathbf{S}}$.

8.2. DEFINITION. We say that a system $\tilde{\mathbf{S}}$ includes a system **S** (equivalently, a system **S** is included by a system $\tilde{\mathbf{S}}$) if there exists an ordered pair of matrices $(U,\ V)$ such that $UV = I$, and for any initial state $x_0 \in \mathcal{X}$ of **S** and any fixed input $u(t)$, we have

$$x(t;\ x_0,\ u) = U\tilde{x}(t;\ Vx_0,\ u), \qquad y[x(t)] = y[\tilde{x}(t)], \quad \forall t \geq 0. \tag{8.39}$$

Condition (8.39) implies that system $\tilde{\mathbf{S}}$ contains all the necessary information about the behavior of system **S**. More importantly, we can extract any property of **S** from $\tilde{\mathbf{S}}$ and, alternatively, evaluate properties of $\tilde{\mathbf{S}}$ using **S** as a reduced order model of $\tilde{\mathbf{S}}$. This is the underlying feature of the Inclusion Principle.

With an abuse of standard notation, by $\tilde{\mathbf{S}} \supset \mathbf{S}$, we mean that $\tilde{\mathbf{S}}$ includes $\mathbf{S}$; that is, $\mathbf{S}$ is included by $\tilde{\mathbf{S}}$. When $\tilde{\mathbf{S}} \supset \mathbf{S}$, we say that $\mathbf{S}$ is a *contraction* of $\tilde{\mathbf{S}}$, or $\tilde{\mathbf{S}}$ is an *expansion* of $\mathbf{S}$.

The inclusion property is topological (independent of the choice of basis). This is easily seen by considering an arbitrary change of basis by $\bar{x} = Tx$, $\bar{\tilde{x}} = \tilde{T}^{-1}\tilde{x}$. Then, the choice $\bar{V} = \tilde{T}^{-1}VT^{-1}$, $\bar{U} = TU\tilde{T}$ satisfies $\bar{x}(t;\, \bar{x}_0,\, u) = \bar{U}\bar{\tilde{x}}(t;\, \bar{V}\bar{x}_0,\, u)$ whenever (8.39) holds.

Before we proceed to a general treatment of the inclusion concept, we consider two important special cases. We first use a matrix V that is monic (full column rank) and assume that, for any $x_0 \in \mathcal{X}$ and any fixed $u = u(t)$,

$$\begin{aligned} \tilde{x}(t;\, Vx_0,\, u) &= Vx(t;\, x_0,\, u), \\ y[x(t)] &= y[\tilde{x}(t)], \qquad \forall t \geq 0, \end{aligned} \tag{8.40}$$

holds. Then, the subspace $\operatorname{Im} V \subset \tilde{\mathcal{X}}$ is invariant in the sense that the solutions $\tilde{x}(t)$ of $\tilde{\mathbf{S}}$ starting at $\tilde{x}_0 \in \operatorname{Im} V$ stay in $\operatorname{Im} V$ for any input $u(t)$. Such solutions $\tilde{x}(t)$ are represented by the solutions $x(t)$ of $\mathbf{S}$ as in (8.40). Since V is monic, there exists a matrix U which is epic (full row rank) such that $UV = I$ and we have (8.39), implying that $\tilde{\mathbf{S}}$ includes $\mathbf{S}$. Whenever the relation (8.40) is satisfied for some monic V, we say that the system $\mathbf{S}$ is a *restriction* of $\tilde{\mathbf{S}}$ to $\operatorname{Im} V$. The use of this term is natural since in the geometric context (Wonham, 1979) it is used to define a map $F|\mathcal{S}$: $\mathcal{S} \rightarrow \mathcal{S}$, where $\mathcal{S}$ is an F-invariant subspace that has the action of F on $\mathcal{S}$, but is not defined off $\mathcal{S}$. The restriction $\mathbf{S}$ is the "dynamic" version of the static notion of a restriction $F|\mathcal{S}$.

If we have some epic U such that

$$\begin{aligned} x(t;\, U\tilde{x}_0,\, u) &= U\tilde{x}(t;\, \tilde{x}_0,\, u), \\ y[x(t)] &= y[\tilde{x}(t)], \qquad \forall t \geq 0, \end{aligned} \tag{8.41}$$

which hold for any $\tilde{x}_0 \in \tilde{\mathcal{X}}$ and $u(t)$, then the system $\mathbf{S}$ is an *aggregation* of $\tilde{\mathbf{S}}$ with respect to the subspace $\operatorname{Im} U^T$. This notion was introduced by Aoki (1971) as a natural dynamic generalization of the static version of aggregation well-known in mathematical economics (*e.g.*, Chipman, 1976). Since U is epic, there exists a monic V such that $UV = I$, and the aggregation condition (8.41) implies condition (8.39) for inclusion.

In order to derive explicit relations between $\mathbf{S}$ and $\tilde{\mathbf{S}}$, we introduce the transform expressions

$$\tilde{A} = VAU + M, \qquad \tilde{B} = VB + N, \qquad \tilde{C} = CU + L, \tag{8.42}$$

where M, N, and L are *complementary matrices* with appropriate dimensions. The following result can be used to establish the aggregation and restriction as duals of each other:

8.3. THEOREM. **S** is a restriction of $\tilde{\mathbf{S}}$ to Im V if and only if

$$MV = 0, \qquad N = 0, \qquad LV = 0. \tag{8.43}$$

S is an aggregation of $\tilde{\mathbf{S}}$ with respect to Im U^T if and only if

$$UM = 0, \qquad UN = 0, \qquad L = 0. \tag{8.44}$$

Proof. We establish only condition (8.43) for restriction. The aggregation condition (8.44) can be proved by a dual argument.

We note that $\tilde{x}(t;\, Vx_0,\, u) = Vx(t;\, x_0,\, u)$ in (8.40) can be rewritten as

$$Ve^{At}x_0 + V\int_0^t e^{A(t-\tau)}\, Bu(\tau)\, d\tau = e^{(VAU+M)t}\, Vx_0$$

$$+ \int_0^t e^{(VAU+M)(t-\tau)}\, (VB+N)u(\tau)\, d\tau \qquad \forall x_0 \in \mathcal{X}, \qquad u(t) \in \mathcal{U}. \tag{8.45}$$

This is true if and only if $MV = 0$ and $N = 0$, as may be seen by taking the power series expansion of the exponentials in (8.45). The requirement $y[\tilde{x}(t)] = y[x(t)]$, that is, $Cx = (CU + L)\tilde{x}$, is equivalent to $L\tilde{x}(t) = LVx = 0$. Therefore, $LV = 0$. Q.E.D.

We note that (8.43) can be restated algebraically as

$$\tilde{A}V = VA, \qquad \tilde{B} = VB, \qquad \tilde{C}V = C, \tag{8.46}$$

or, in geometric terms as

$$\text{Ker}\, M \supset \text{Im}\, V, \qquad \text{Im}\, N = 0, \qquad \text{Ker}\, L \supset \text{Im}\, V. \tag{8.47}$$

Furthermore, if we define $T = [V \;\; W]$, where W is chosen such that $\operatorname{Im} W = \operatorname{Ker} U$, the change of basis $\bar{x} = T^{-1}\tilde{x}$ yields an equivalent representation

$$\bar{\mathbf{S}}\colon\; \dot{\bar{x}} = \begin{bmatrix} A & * \\ 0 & * \end{bmatrix} \bar{x} + \begin{bmatrix} B \\ 0 \end{bmatrix} u, \qquad y = [\,C \quad *\,]\,\bar{x}, \tag{8.48}$$

whenever $(A,\, B,\, C)$ is a restriction of $(\tilde{A},\, \tilde{B},\, \tilde{C})$.

Similarly, the aggregation condition (8.44) is equivalent to

$$U\tilde{A} = AU, \qquad U\tilde{B} = B, \qquad \tilde{C} = CU, \tag{8.49}$$

or

$$\operatorname{Ker} U \supset \operatorname{Im} M, \qquad \operatorname{Ker} U \supset \operatorname{Im} N, \qquad \operatorname{Im} L = 0. \tag{8.50}$$

In other words, $(A,\, B,\, C)$ is an aggregation of $(\tilde{A},\, \tilde{B},\, \tilde{C})$ if and only if $\tilde{\mathbf{S}}$ can be represented in an appropriate state space by

$$\bar{\mathbf{S}}\colon\; \dot{\bar{x}} = \begin{bmatrix} * & * \\ 0 & A \end{bmatrix} \bar{x} + \begin{bmatrix} * \\ B \end{bmatrix} u, \qquad y = [\,0 \quad C\,]\,\bar{x}. \tag{8.51}$$

We note that aggregation and restriction are mutually exclusive conditions in that aggregation specifies the behavior of a projection of the state trajectory of $\tilde{\mathbf{S}}$ for arbitrary initial conditions $\tilde{x}_0$, while restriction specifies the entire state trajectory of $\tilde{\mathbf{S}}$ for a limited set of initial conditions $\tilde{x}_0 \in \operatorname{Im} V \subset \tilde{\mathcal{X}}$. In either case, the system $\tilde{\mathbf{S}}$ *includes* all of the information concerning the behavior of $\mathbf{S}$. Inclusion, however, need not imply either of these conditions as demonstrated by the following:

8.4. THEOREM. $\tilde{\mathbf{S}} \supset \mathbf{S}$ if and only if

$$UM^iV = 0, \qquad UM^{i-1}N = 0, \qquad LM^{i-1}V = 0, \qquad LM^{i-1}N = 0, \tag{8.52}$$

for all $i \in \tilde{\mathbf{n}}$, $\tilde{\mathbf{n}} = \{1,\, 2,\, \ldots,\, \tilde{n}\}$, where $\tilde{n} = \dim \tilde{\mathcal{X}}$.

Proof. Using the explicit solution form as in the proof of Theorem 8.3 and power series expansions of the matrix exponentials, one establishes (8.52). For details, see (Ikeda and Šiljak, 1980a). Q.E.D.

Relations (8.52) can be restated as

$$U\tilde{A}^iV = A^i, \qquad U\tilde{A}^i\tilde{B} = A^iB, \qquad \tilde{C}\tilde{A}^iV = CA^i, \qquad \tilde{C}\tilde{A}^i\tilde{B} = CA^iB, \tag{8.53}$$

for $i = 0, 1, 2, \ldots$; or geometrically as

$$\begin{aligned} \text{Ker } U &\supset M\{M|\text{Im } V\}, \qquad \text{Ker } U \supset \{M|\text{Im } N\}, \\ \text{Ker } L &\supset \{M|\text{Im } V\}, \qquad \text{Ker } L \supset \{M|\text{Im } N\}. \end{aligned} \tag{8.54}$$

We also note that the first relation above is equivalent to $U\tilde{A}^iV = A^i$ for all $i \in \mathbf{n} = \{1, 2, \ldots, n\}$.

Now, we show that a contraction $\mathbf{S}$ of a given system $\tilde{\mathbf{S}}$ can be obtained *via* aggregation and restriction of $\tilde{\mathbf{S}}$, that is, contraction is an "intersection" of aggregation and restriction of $\tilde{\mathbf{S}}$. For this purpose, let us introduce a system

$$\begin{aligned} \bar{\mathbf{S}}\colon \ \dot{\bar{x}} &= \bar{A}\bar{x} + \bar{B}u, \\ y &= \bar{C}\bar{x}, \end{aligned} \tag{8.55}$$

and prove the following:

8.5. THEOREM. $\tilde{\mathbf{S}} \supset \mathbf{S}$ if and only if there exists $\bar{\mathbf{S}}$ such that

$$\tilde{\mathbf{S}} \supset \bar{\mathbf{S}} \supset \mathbf{S}, \tag{8.56}$$

where $\bar{\mathbf{S}}$ is a restriction (aggregation) of $\tilde{\mathbf{S}}$ and $\mathbf{S}$ is an aggregation (restriction) of $\bar{\mathbf{S}}$.

Proof. We assign a pair (U_1, V_1) to $\tilde{\mathbf{S}} \supset \bar{\mathbf{S}}$ and a pair (U_2, V_2) to $\bar{\mathbf{S}} \supset \mathbf{S}$. Now, the *sufficiency* is trivial, for restriction and aggregation are special cases of inclusion, which is a transitive relation since $\tilde{\mathbf{S}} \supset \bar{\mathbf{S}}$ and $\bar{\mathbf{S}} \supset \mathbf{S}$ implies $\tilde{\mathbf{S}} \supset \mathbf{S}$ (choose $U = U_2U_1$, $V = V_1V_2$).

The *necessity* proceeds constructively by assuming $\tilde{\mathbf{S}} \supset \mathbf{S}$ and choosing $V_1 = [W \ \ V]$, where $W = (I - VU)\hat{W}$, and $\hat{W}$ is any basis matrix for $\mathcal{W} = \left(\{\tilde{A}|\text{Im } V\} + \{\tilde{A}|\text{Im } N\}\right) \cap (\text{Im } V)^{\perp}$. From

$$[W \ \ V] = [\hat{W} \ \ V]\begin{bmatrix} I & 0 \\ -U\hat{W} & I \end{bmatrix} \tag{8.57}$$

and $UV = I$, we see that V_1 is monic and Im $V_1 = \{\tilde{A}|\text{Im } V\} + \{\tilde{A}|\text{Im } N\}$, Ker $U \supset$ Im W. This fact guarantees the existence of W^L such that the choice

$$U_1 = \begin{bmatrix} W^L \\ U \end{bmatrix} \tag{8.58}$$

implies $U_1 V_1 = I$.

Now, let us show that $\bar{A} = U_1 \tilde{A} V_1$, $\bar{B} = U_1 \tilde{B}$, and $\bar{C} = \tilde{C} V_1$, so that $\bar{\mathbf{S}} = (\bar{A}, \bar{B}, \bar{C})$ is a restriction of $\tilde{\mathbf{S}} = (\tilde{A}, \tilde{B}, \tilde{C})$ to Im V_1. We first note that Im V_1 is $\tilde{A}$ invariant and

$$\tilde{A} = V_1 \bar{A} U_1 + M_1, \tag{8.59}$$

where $M_1 = \tilde{A} - V_1 \bar{A} U_1$ and $M_1 V_1 = \tilde{A} V_1 - V_1 \bar{A} = (I - V_1 U_1) \tilde{A} V_1 = 0$. We also note that Im $\tilde{B} \subset$ Im V + Im $N \subset$ Im V_1, which implies

$$\tilde{B} = V_1 \bar{B} + N_1, \tag{8.60}$$

where $N_1 = \tilde{B} - V_1 \bar{B} = (I - V_1 U_1) \tilde{B} = 0$. We then conclude that

$$\tilde{C} = \bar{C} U_1 + L_1, \tag{8.61}$$

where $L_1 = \tilde{C} - \bar{C} U_1$ and $L_1 V_1 = \tilde{C}(I - V_1 U_1) V_1 = 0$, proving $\bar{\mathbf{S}}$ is a restriction of $\tilde{\mathbf{S}}$ to Im V_1 by Theorem 8.3.

To show that $\mathbf{S} = (A, B, C)$ is an aggregation of $\bar{\mathbf{S}} = (\bar{A}, \bar{B}, \bar{C})$, we choose $V_2 = [0 \;\; I]^T$ and $U_2 = [0 \;\; I]$. Since $U\tilde{A}^i V = (U\tilde{A}V)^i = A^i$ and Ker $U \supset \{\tilde{A}|\text{Im } N\}$, we can show that Ker $[U\tilde{A}(I - VU)] \supset \{\tilde{A}|\text{Im } V\} + \{\tilde{A}|\text{Im } N\}$, which together with $\{\tilde{A}|\text{Im } N\} + \{\tilde{A}|\text{Im } N\} \supset$ Im $\hat{W}$ implies $U\tilde{A}W = U\tilde{A}(I - VU)\hat{W} = 0$. Then, we observe that

$$\bar{A} = V_2 A U_2 + M_2, \tag{8.62}$$

where $M_2 = \bar{A} - V_2 A U_2$ and $U_2 M_2 = U\tilde{A}[W \;\; V] - A[0 \;\; I] = 0$. We have

$$\bar{B} = V_2 B + N_2, \tag{8.63}$$

where $N_2 = \bar{B} - V_2 B$, and the inclusion relation $U\tilde{B} = B$ implies $U_2 N_2 = 0$. Finally, from the inclusion relation Ker $L \supset \{\tilde{A}|\text{Im } V\} + \{\tilde{A}|\text{Im } N\} \supset$ Im V_1 and Ker $U \supset$ Im W, we see that

$$\bar{C} = C U_2 + L_2, \tag{8.64}$$

and $L_2 = \bar{C} - C U_2 = (CU + L)[W \;\; V] - C[0 \;\; I] = 0$ which, in turn, implies that $\mathbf{S}$ is an aggregation of $\bar{\mathbf{S}}$ to Im U_2^T. Q.E.D.

The construction procedure of finding first a restriction (aggregation) and then an aggregation (restriction) to get a suitable contraction implies that whenever $\tilde{\mathbf{S}} \supset \mathbf{S}$, the system can be represented in the *canonical form*

$$\tilde{\mathbf{S}}\colon\ \dot{\tilde{x}} = \begin{bmatrix} * & * & * \\ 0 & A & * \\ 0 & 0 & * \end{bmatrix} \tilde{x} + \begin{bmatrix} * \\ B \\ 0 \end{bmatrix} u, \qquad (8.65)$$

$$y = [0 \quad C \quad *]\, \tilde{x},$$

using an appropriate coordinate frame. Comparing the canonical form with the two previous representations (8.48) and (8.51), we conclude that contractions arise naturally as "intersections" of restrictions and aggregations. This fact is expressed in algebraic terms by the following corollary to Theorem 8.5:

8.6. COROLLARY. $\tilde{\mathbf{S}} \supset \mathbf{S}$ if and only if $M = M_1 + M_2$ such that

$$\begin{aligned} &M_1V = 0, \quad && M_1N = 0, \quad && UM_2 = 0, \quad && LM_2 = 0, \\ &M_1M_2 = 0, && UN = 0, && LV = 0, && LN = 0. \end{aligned} \qquad (8.66)$$

Proof. Assume $\tilde{\mathbf{S}} \supset \mathbf{S}$, so that by Theorem 8.5 we have $U = U_1U_2$ and $V = V_2V_1$, that is, $\bar{\mathbf{S}}$ is an aggregation of $\tilde{\mathbf{S}}$ and $\mathbf{S}$ is a restriction of $\bar{\mathbf{S}}$. It follows that

$$\begin{aligned} &\bar{A} = V_1AU_1 + \bar{M}, \quad && \bar{M}V_1 = 0, \quad && \bar{B} = V_1B, \\ &\bar{C} = CU_1 + \bar{L}, && \bar{L}V_1 = 0, && \end{aligned} \qquad (8.67)$$

and

$$\begin{aligned} &\tilde{A} = V_2\bar{A}U_2 + M_2, \quad && U_2M_2 = 0, \quad && \tilde{B} = V_2\bar{B} + N, \\ &U_2N = 0, && \tilde{C} = \bar{C}U_2. && \end{aligned} \qquad (8.68)$$

Defining $M_1 = V_2\bar{M}U_2$, and $L = \bar{L}U_2$, and substituting (8.67) into (8.68), we conclude that (8.66) holds.

The sufficiency follows by the substitution of $M = M_1 + M_2$ into (8.52) and using (8.66). Q.E.D.

Stability in the context of inclusion is obvious from the canonical form (8.65): eigenvalues of the matrix A must at the same time be the eigenvalues of the matrix $\tilde{A}$. This fact means that stability of $\tilde{\mathbf{S}}$ implies stability of $\mathbf{S}$ if $\tilde{\mathbf{S}} \supset \mathbf{S}$. The rest of the eigenvalues can be obtained from the following:

8.7. THEOREM. If $\tilde{\mathbf{S}} \supset \mathbf{S}$, then

$$\pi(\tilde{A}) = \pi(A)\pi(\hat{U}M\hat{V}), \tag{8.69}$$

where $\pi(\bullet)$ denotes the characteristic polynomial, $\hat{V}$ is any basis matrix for Ker U, and $\hat{U}$ is the unique left inverse of $\hat{V}$ such that Ker $\hat{U} =$ Im V.

Proof. Define a nonsingular matrix $T = [V \ \hat{V}]$ and form

$$T^{-1}\tilde{A}T = \begin{bmatrix} A & UM\hat{V} \\ \hat{U}MV & \hat{U}M\hat{V} \end{bmatrix}, \tag{8.70}$$

where $T^{-1} = \begin{bmatrix} U \\ \hat{U} \end{bmatrix}$. Then,

$$\begin{aligned} \det(\lambda I - \tilde{A}) &= \det(\lambda I - A)\det[\lambda I - \hat{U}M\hat{V} - \hat{U}MV(\lambda I - A)^{-1}UM\hat{V}] \\ &= \det(\lambda I - A)\det(\lambda I - \hat{U}M\hat{V}) \\ &\quad \times \det[I - UM\hat{V}(\lambda I - \hat{U}M\hat{V})^{-1}\hat{U}MV(\lambda I - A)^{-1}]. \end{aligned} \tag{8.71}$$

But,

$$UM\hat{V}(\lambda I - \hat{U}M\hat{V})^{-1}(\hat{U}MV) = \sum_{i=0}^{\infty} \lambda^{-i-1} UM(\hat{V}\hat{U}M\hat{V}\hat{U})^i MV = 0 \tag{8.72}$$

is implied by $\hat{V}\hat{U} = (I - VU)$ and $UM^iV = 0, \ i \in \tilde{\mathbf{n}}$. Q.E.D.

Finally, we turn our attention to the problem of *feedback control* in the framework of inclusion. Referring back to the expanded system $\tilde{\mathbf{S}}$ in (8.3), we see that it is in an input-decentralized form. We can use the standard decentralized control strategies of Chapter 2 to stabilize $\tilde{\mathbf{S}}$, then contract the obtained control law, and implement it in the original system $\mathbf{S}$. Alternatively, a model reduction scheme can be used to simplify the design of feedback (Skelton, 1988). Then, the control law resulting from a reduced order model **S** must be expanded for implementation in the original system $\tilde{\mathbf{S}}$ (Sezer and Šiljak, 1982).

We relate the two systems **S** and $\tilde{\mathbf{S}}$ under constant feedback

$$u = -Kx + v, \qquad u = -\tilde{K}\tilde{x} + v, \tag{8.73}$$

where $v \in \mathcal{U}$ is a reference input. We assume that the $m \times \tilde{n}$ gain matrix $\tilde{K}$ has the form

$$\tilde{K} = KU + F, \tag{8.74}$$

and define the closed-loop system matrices as

$$A_F = A - BK, \qquad \tilde{A}_F = \tilde{A} - \tilde{B}\tilde{K}. \tag{8.75}$$

The constraints on the feedback complementary matrix F guaranteeing that $\tilde{\mathbf{S}}_F = (\tilde{A}_F, \tilde{B}, \tilde{C})$ includes $\mathbf{S}_F = (A_F, B, C)$ under any choice of feedback gain matrix K are given by the following:

8.8. THEOREM. $\tilde{\mathbf{S}}_F \supset \mathbf{S}_F$ for any feedback gain matrices and for monic B if and only if (8.66) holds and

$$FV = 0, \qquad FM_2 = 0, \qquad FN = 0. \tag{8.76}$$

Proof. By Theorem 8.4, $\tilde{\mathbf{S}}_F \supset \mathbf{S}_F$ is equivalent to

$$UM_F^iV = 0, \qquad UM_F^{i-1}N = 0, \qquad LM_F^{i-1}V = 0, \qquad LM_F^{i-1}N = 0, \tag{8.77}$$

for all $i \in \tilde{\mathbf{n}}$, where $\tilde{\mathbf{n}} = \dim \tilde{\mathcal{X}}$ and

$$M_F = M - VBF - NKU - NF. \tag{8.78}$$

Substituting this expression of M_F into (8.77) and noting that B is monic, we get the requirements

$$FM^{i-1}V = 0, \qquad FM^{i-1}N = 0, \tag{8.79}$$

for all $i \in \tilde{\mathbf{n}}$. Now, note that by Corollary 8.6, (8.52) must be equivalent to (8.66). Since the requirements (8.79) for F and (8.52) for L are identical, we establish (8.76) by Corollary 8.6. Q.E.D.

Because the inclusion concept involves singular transformations, the systems $\tilde{\mathbf{S}}$ and $\mathbf{S}$ are not algebraically equivalent. On the other hand, they are "pseudo-equivalent." That is, from Definition 8.2, we see that for all fixed inputs $u(t)$ and any initial state x_0 of $\mathbf{S}$, there is (at least) one equivalent initial state $\tilde{x}_0$ of $\tilde{\mathbf{S}}$ resulting in identical outputs $y[\tilde{x}(t)]$ and $y[x(t)]$. An opposite statement is generally not true.

When we compare the external behavior of the two systems initially at rest, then a complete *zero-state equivalence* can be established whereby two systems $\mathbf{S}$ and $\tilde{\mathbf{S}}$ are considered equivalent if their input–output behavior is identical whenever they start at their zero states. We denote the transfer functions of $\mathbf{S}$ and $\tilde{\mathbf{S}}$ by

$$G(s) = C(sI - A)^{-1}B, \qquad \tilde{G}(s) = \tilde{C}(sI - \tilde{A})^{-1}\tilde{B}, \tag{8.80}$$

and prove:

8.9. THEOREM. If $\tilde{\mathbf{S}} \supset \mathbf{S}$, then

$$\tilde{G}(s) = G(s). \tag{8.81}$$

Proof. The relations

$$CA^iB = \tilde{C}\tilde{A}^i\tilde{B}, \qquad i = 0, 1, 2, \ldots, \tag{8.82}$$

in the inclusion condition (8.53) are equivalent to (8.81). Q.E.D.

Condition (8.81), however, does not imply $\tilde{\mathbf{S}} \supset \mathbf{S}$. For this we need the following:

8.10. THEOREM. Two systems

$$\tilde{\mathbf{S}}\colon\ \dot{\tilde{x}} = \tilde{A}\tilde{x} + \tilde{B}u, \qquad y = \tilde{C}\tilde{x}, \tag{8.38}$$

with dim $\tilde{\mathcal{X}} = \tilde{n}$, and

$$\bar{\mathbf{S}}\colon\ \dot{\bar{x}} = \bar{A}\bar{x} + \bar{B}u, \qquad y = \bar{C}\bar{x}, \tag{8.83}$$

with dim $\bar{\mathcal{X}} = \bar{n}$, are zero-state equivalent if and only if they have a common contraction $\mathbf{S}$.

Proof. We have already established the *sufficiency* by definition of inclusion, for if $\tilde{\mathbf{S}}$ and $\bar{\mathbf{S}}$ have a common contraction $\mathbf{S}$, then it is obvious that they are realizations of the same transfer function.

To show *necessity*, we assume that $\tilde{\mathbf{S}}$ and $\bar{\mathbf{S}}$ have the same transfer function $G(s)$, and $\mathbf{S}$ is a minimal realization of $G(s)$. Then, from the observability and controllability matrices of $\mathbf{S}$ and $\tilde{\mathbf{S}}$, we define the matrices

$$T = \begin{bmatrix} C \\ CA \\ CA^2 \\ \vdots \\ CA^{\tilde{n}-1} \end{bmatrix}, \qquad \tilde{T} = \begin{bmatrix} \tilde{C} \\ \tilde{C}\tilde{A} \\ \tilde{C}\tilde{A}^2 \\ \vdots \\ \tilde{C}\tilde{A}^{\tilde{n}-1} \end{bmatrix}, \tag{8.84}$$

$$W = [\, B \quad AB \quad A^2B \quad \cdots \quad A^{\tilde{n}-1}B \,],$$

$$\tilde{W} = [\, \tilde{B} \quad \tilde{A}\tilde{B} \quad \tilde{A}^2\tilde{B} \quad \cdots \quad \tilde{A}^{\tilde{n}-1}\tilde{B} \,].$$

From relation (8.82), which is necessary and sufficient for zero-state equivalence, we get

$$TA^iW = \tilde{T}\tilde{A}^i\tilde{W}, \qquad i = 0,\, 1,\, 2,\, \ldots\,. \tag{8.85}$$

By the assumption of minimality of $\mathbf{S}$, we know that T is epic and W is monic, so that the left inverse T^L of T and the right inverse W^R of W both exist. From (8.85) for $i = 0$, we see that if $V = \tilde{W}W^R$, then $U = T^L\tilde{T}$ is a left inverse of V. Then, from (8.85), we further obtain

$$U\tilde{A}^iV = A^i, \qquad U\tilde{A}^i\tilde{B} = A^iB, \qquad \tilde{C}\tilde{A}^iV = CA^i, \tag{8.86}$$

for all $i = 0,\, 1,\, 2,\, \ldots$. Noting that these conditions together with (8.82) are equivalent to the inclusion condition (8.53), we conclude that $\tilde{\mathbf{S}} \supset \mathbf{S}$. Similarly, we can show that $\bar{\mathbf{S}} \supset \mathbf{S}$ if $\bar{\mathbf{S}}$ is a realization of $G(s)$, that is, $\tilde{\mathbf{S}}$ and $\bar{\mathbf{S}}$ are zero-state equivalent. Q.E.D.

It is now appropriate to give a general interpretation of overlapping decompositions introduced in Section 8.2. When we partition the state vector $x \in \mathbf{R}^n$ of the system

$$\begin{aligned} \mathbf{S}\colon\ \dot{x} &= Ax + Bu, \\ y &= Cx, \end{aligned} \tag{8.37}$$

as $x = (x_1^T, x_2^T, \ldots, x_N^T)^T$, we induce a disjoint decomposition of the matrices A, B, C as

$$A = \begin{bmatrix} A_{11} & A_{12} & \ldots & A_{1N} \\ A_{21} & A_{22} & \ldots & A_{2N} \\ \cdots & \cdots & \cdots & \cdots \\ A_{N1} & A_{N2} & \ldots & A_{NN} \end{bmatrix}, \qquad B = \begin{bmatrix} B_1 \\ B_2 \\ \vdots \\ B_N \end{bmatrix},$$

$$C = [C_1 \quad C_2 \quad \ldots \quad C_N], \tag{8.87}$$

and $\mathbf{S}$ can be represented as an interconnection of N subsystems:

$$\mathbf{S}\colon \; \dot{x}_i = A_{ii}x_i + B_i u + \sum_{\substack{j=1 \\ j \neq i}} A_{ij}x_j,$$

$$y = \sum_{i=1}^{N} C_i x_i, \qquad i \in \boldsymbol{N}. \tag{8.88}$$

State trajectories of $\mathbf{S}$ are regarded as projections of the solution $x(t; x_0, u)$ on the subspaces $\mathcal{X}_i$, $i \in \boldsymbol{N}$, such that

$$\mathcal{X} = \mathcal{X}_1 \oplus \mathcal{X}_2 \oplus \cdots \oplus \mathcal{X}_N, \tag{8.89}$$

that is, $\mathcal{X}$ is the *direct sum* of the subspaces $\mathcal{X}_i$. These subspaces are uniquely specified by the partitioning of x. Conversely, a specification of subspaces corresponds uniquely to a partitioning of the state vector x.

Now we introduce a partitioning of $\mathcal{X}$ as

$$\mathcal{X} = \mathcal{X}_1 + \mathcal{X}_2 + \cdots + \mathcal{X}_N, \tag{8.90}$$

where $\mathcal{X}$ is an *ordinary sum* of the subspaces $\mathcal{X}_i$, which are not necessarily disjoint. Hence, this partitioning cannot be associated with the standard disjoint partitioning of the state x. In order to achieve a disjoint decomposition, we have to "expand" the space $\mathcal{X}$. With each $\mathcal{X}_i$ in (8.90), we associate a mapping V_i: $\mathcal{X} \to \mathcal{X}_i$, $i \in \boldsymbol{N}$, and construct an expansion map V: $\mathcal{X} \to \tilde{\mathcal{X}}$, where $\dim \tilde{\mathcal{X}} = \sum_{i=1}^{N} \dim \mathcal{X}_i$, as $V = \left[V_1^T \; V_2^T \; \cdots \; V_N^T\right]^T$, which is necessarily monic. The underlying decomposition of the space $\tilde{\mathcal{X}}$ of $\tilde{\mathbf{S}}$ is

$$\tilde{\mathcal{X}} = \tilde{\mathcal{X}}_1 \oplus \tilde{\mathcal{X}}_2 \oplus \cdots \oplus \tilde{\mathcal{X}}_N, \tag{8.91}$$

which is a direct sum of subspaces $\tilde{\mathcal{X}}_i$. It is essential that $\tilde{\mathcal{X}}_i \simeq \mathcal{X}_i$, which means that the identity of the subsystems is preserved in the expansion $\tilde{\mathbf{S}}$; compare (8.2) with (8.3), or (8.5) with (8.12). In this way, we obtain a direct sum of subspaces from an ordinary sum preserving isomorphism of subspaces, yet allowing for standard disjoint methods to be used for the study of overlapping decompositions. The expansion $\tilde{\mathbf{S}}$ has the same disjoint form as $\mathbf{S}$ in (8.88):

$$\tilde{\mathbf{S}}\colon\ \dot{\tilde{x}}_i = \tilde{A}_{ii}\tilde{x}_i + \tilde{B}_i u + \sum_{\substack{j=1 \\ j \neq i}}^{N} \tilde{A}_{ij}\tilde{x}_j,$$
$$y = \sum_{i=1}^{N} \tilde{C}_i \tilde{x}_i, \qquad i \in \boldsymbol{N}. \tag{8.92}$$

Now, a control design can be performed in $\tilde{\mathcal{X}}$ by using standard methods for disjoint decompositions. Then, the resulting control law can be contracted to $\mathcal{X}$ for implementation in the original system $\mathbf{S}$.

Finally, we should note that if we want to get, in (8.92), the subsystems with independent inputs u_i and outputs y_i instead of u and y, we have to expand the input and output spaces as well (Ikeda and Šiljak, 1986; Iftar, 1988; Iftar and Özgüner, 1990). This additional effort is not necessary if the system is input and output decentralized to start with (see Example 8.22).

8.3. Optimization

To address optimization problems in the context of overlapping decompositions, we need to define expansions and contractions of performance indices. We will do this for linear systems with a separable quadratic criterion related to overlapping subsystems. Decentralized controllers are designed to be locally optimal in the expanded space when the interconnections are ignored. Then, the obtained control laws are contracted for implementation in the original system. Suboptimality of the decentralized controllers is determined with respect to the decoupled optimal system as reference (see Section 3.1). Most importantly, we can show that, under relatively mild conditions, suboptimality implies and is implied by stability of the overall closed-loop system.

We assume that in linear system

$$\mathbf{S}\colon\ \dot{x} = Ax + Bu, \tag{8.37'}$$

we can measure the whole state, and consider the performance index

$$J(x_0, u) = \int_0^\infty (x^T Q x + u^T R u)\, dt, \tag{8.93}$$

where Q is a nonnegative definite matrix and R is a positive definite matrix. With the pair $(\mathbf{S}, J)$, we associate another pair $(\tilde{\mathbf{S}}, \tilde{J})$, where

$$\tilde{\mathbf{S}}\colon\ \dot{\tilde{x}} = \tilde{A}\tilde{x} + \tilde{B}u, \tag{8.38$'$}$$

and

$$\tilde{J}(\tilde{x}_0, u) = \int_0^\infty (\tilde{x}^T \tilde{Q} \tilde{x} + u^T \tilde{R} u)\, dt, \tag{8.94}$$

with $\tilde{Q}$ being a nonnegative definite matrix.

8.11. DEFINITION. We say that a pair $(\tilde{\mathbf{S}}, \tilde{J})$ includes a pair $(\mathbf{S}, J)$ if there exists a pair of matrices (U, V) such that $UV = I$ and for any initial state x_0 of $\mathbf{S}$ and any fixed input $u(t)$, we have

$$x(t;\, x_0,\, u) = U\tilde{x}(t;\, Vx_0,\, u), \qquad \forall t \geq 0, \tag{8.95a}$$

$$J(x_0,\, u) = \tilde{J}(Vx_0,\, u). \tag{8.95b}$$

This definition was introduced in (Ikeda *et al.* 1981a) to establish an optimal control problem $(\tilde{\mathbf{S}}, \tilde{J})$ in the expanded space that provides a solution to the original problem $(\mathbf{S}, J)$. If we define the relations

$$\begin{aligned} \tilde{A} &= VAU + M, & \tilde{B} &= VB + N, \\ \tilde{Q} &= U^T Q U + M_Q, & \tilde{R} &= R + N_R, \end{aligned} \tag{8.96}$$

where we assume that $UV = I$, then inclusion conditions for $(\tilde{\mathbf{S}}, \tilde{J})$ and $(\mathbf{S}, J)$ are formulated as follows:

8.12. THEOREM. $(\tilde{\mathbf{S}}, \tilde{J}) \supset (\mathbf{S}, J)$ if either

(*i*) $MV = 0, \quad N = 0, \quad V^T M_Q V = 0, \quad N_R = 0,$ (8.97)

or

(*ii*)

$$\begin{aligned} &UM^iV = 0, \qquad M_Q M^{i-1} N = 0, \qquad M_Q M^{i-1} V = 0, \\ &UM^{i-1}N = 0, \qquad N_R = 0, \qquad \forall i \in \tilde{\mathbf{n}}. \end{aligned} \tag{8.98}$$

Proof. We first prove part (i). From the proof of Theorem 8.3, we have the first two relations in (8.97), which establish (8.40) and then (8.95a). Equation (8.95b) follows from

$$\tilde{x}^T(t;\, Vx_0,\, u)\tilde{Q}\tilde{x}(t;\, Vx_0,\, u) = x^T(t;\, x_0,\, u)Qx(t;\, x_0,\, u), \qquad u^T\tilde{R}u = u^TRu, \tag{8.99}$$

which is implied by (8.40) and the rest of part (i).

To prove part (ii), we note first that (8.95a) is implied by the first two Equations of (8.98)—see Theorem 8.4. Equation (8.95b) follows from (8.99), which is implied by (8.95a), $N_R = 0$, and

$$M_Q\tilde{x}(t;\, Vx_0,\, u) = 0, \qquad \forall t \geq 0, \tag{8.100}$$

for any fixed $u(t)$. This equation follows from $M_QM^{i-1}N = 0$ and $M_QM^{i-1}N = 0$ for all $i \in \tilde{\mathbf{n}}$. Q.E.D.

Parts (i) and (ii) of Theorem 8.12 are independent of each other. They are two different sets of conditions for $(\tilde{\mathbf{S}},\, \tilde{J})$ to be an expansion of $(\mathbf{S},\, J)$, that is, for $(\mathbf{S},\, J)$ to be a contraction of $(\tilde{\mathbf{S}},\, \tilde{J})$.

Our basic intention is to choose a linear feedback law

$$u = -Kx, \tag{8.101}$$

and to optimize $(\mathbf{S},\, J)$ by choosing appropriately a control law

$$u = -\tilde{K}\tilde{x} \tag{8.102}$$

that optimizes $(\tilde{\mathbf{S}},\, \tilde{J})$ in the expanded space. To do this, we need conditions under which the control $-\tilde{K}\tilde{x}$ is *contractible* to $-Kx$, that is,

$$Kx(t;\, x_0,\, u) = \tilde{K}\tilde{x}(t;\, Vx_0,\, u), \qquad \forall t \geq 0, \tag{8.103}$$

for any fixed input $u(t)$. From the proof of Theorem 8.8, we have the following:

8.13. COROLLARY. A control law $-\tilde{K}\tilde{x}$ is contractible to the control law $-Kx$ if and only if

$$FM^{i-1}V = 0, \qquad FM^{i-1}N = 0, \qquad \forall i \in \tilde{\mathbf{n}}. \tag{8.104}$$

It is important to note that testing of contractibility conditions (8.104) can be avoided if $\mathbf{S}$ is a restriction of $\tilde{\mathbf{S}}$ (Theorem 8.3).

8.14. COROLLARY. If

$$MV = 0, \qquad N = 0, \tag{8.105}$$

then any control law $-\tilde{K}\tilde{x}$ is contractible to the control law $-Kx$, and

$$K = \tilde{K}V. \tag{8.106}$$

8.15. EXAMPLE. To make transparent the announced control strategy involving overlapping decompositions, let us reconsider the system $\mathbf{S}$ of (8.4) and its expansion $\tilde{\mathbf{S}}$ of (8.8) when control is present and a quadratic optimization is desired. We start from

$$\mathbf{S}: \begin{bmatrix} \dot{x}_1 \\ \dot{x}_2 \\ \dot{x}_3 \end{bmatrix} = \begin{bmatrix} A_{11} & A_{12} & A_{13} \\ A_{21} & A_{22} & A_{23} \\ A_{31} & A_{32} & A_{33} \end{bmatrix} \begin{bmatrix} x_1 \\ x_2 \\ x_3 \end{bmatrix} + \begin{bmatrix} B_{11} & B_{12} \\ B_{21} & B_{22} \\ B_{31} & B_{32} \end{bmatrix} \begin{bmatrix} u_1 \\ u_2 \end{bmatrix}. \tag{8.107}$$

By choosing V, U, and M as in (8.7), (8.10), and (8.11), and $N = 0$, we get

$$\tilde{\mathbf{S}}: \begin{bmatrix} \dot{\tilde{x}}_1 \\ \dot{\tilde{x}}_2 \end{bmatrix} = \left[\begin{array}{cc|cc} A_{11} & A_{12} & 0 & A_{13} \\ A_{21} & A_{22} & 0 & A_{23} \\ \hline A_{21} & 0 & A_{22} & A_{23} \\ A_{31} & 0 & A_{32} & A_{33} \end{array}\right] \begin{bmatrix} \tilde{x}_1 \\ \tilde{x}_2 \end{bmatrix} + \left[\begin{array}{c|c} B_{11} & B_{12} \\ B_{21} & B_{22} \\ \hline B_{21} & B_{22} \\ B_{31} & B_{32} \end{array}\right] \begin{bmatrix} u_1 \\ u_2 \end{bmatrix}. \tag{8.108}$$

Now, $\tilde{\mathbf{S}}$ can be represented as two interconnected *disjoint* subsystems

$$\begin{aligned} \tilde{\mathbf{S}}: \ \dot{\tilde{x}}_1 &= \tilde{A}_1\tilde{x}_1 + \tilde{B}_1 u_1 + \tilde{A}_{12}\tilde{x}_2 + \tilde{B}_{12}u_2, \\ \dot{\tilde{x}}_2 &= \tilde{A}_2\tilde{x}_2 + \tilde{B}_2 u_2 + \tilde{A}_{21}\tilde{x}_1 + \tilde{B}_{21}u_1, \end{aligned} \tag{8.109}$$

where

$$\begin{aligned} \tilde{A}_1 &= \begin{bmatrix} A_{11} & A_{12} \\ A_{21} & A_{22} \end{bmatrix}, \qquad \tilde{A}_2 = \begin{bmatrix} A_{22} & A_{23} \\ A_{32} & A_{33} \end{bmatrix}, \\ \tilde{B}_1 &= \begin{bmatrix} B_{11} \\ B_{21} \end{bmatrix}, \qquad \tilde{B}_2 = \begin{bmatrix} B_{22} \\ B_{32} \end{bmatrix} \end{aligned} \tag{8.110}$$

are matrices corresponding to the two (decoupled) subsystems

$$\begin{aligned}\tilde{\mathbf{S}}_1\colon\ \dot{\tilde{x}}_1 &= \tilde{A}_1\tilde{x}_1 + \tilde{B}_1 u_1,\\ \tilde{\mathbf{S}}_2\colon\ \dot{\tilde{x}}_2 &= \tilde{A}_2\tilde{x}_2 + \tilde{B}_2 u_2,\end{aligned} \tag{8.111}$$

and

$$\begin{aligned}\tilde{A}_{12} &= \begin{bmatrix} 0 & A_{13} \\ 0 & A_{23} \end{bmatrix}, \qquad \tilde{A}_{21} = \begin{bmatrix} A_{21} & 0 \\ A_{31} & 0 \end{bmatrix},\\ \tilde{B}_{12} &= \begin{bmatrix} B_{12} \\ B_{22} \end{bmatrix}, \qquad \tilde{B}_{21} = \begin{bmatrix} B_{21} \\ B_{31} \end{bmatrix}\end{aligned} \tag{8.112}$$

are the interconnection matrices. For control purposes, with subsystems $\tilde{\mathbf{S}}_1$ and $\tilde{\mathbf{S}}_2$ we associate performance indices

$$\begin{aligned}\tilde{J}_1(\tilde{x}_{10},\, u) &= \int_0^\infty (\tilde{x}_1^T\tilde{Q}_1\tilde{x}_1 + u_1^T\tilde{R}_1u_1)\, dt,\\ \tilde{J}_2(\tilde{x}_{20},\, u) &= \int_0^\infty (\tilde{x}_2^T\tilde{Q}_2\tilde{x}_2 + u_2^T\tilde{R}_2u_2)\, dt,\end{aligned} \tag{8.113}$$

where $\tilde{x}_{10}$ and $\tilde{x}_{20}$ are the initial states of $\tilde{\mathbf{S}}_1$ and $\tilde{\mathbf{S}}_2$, and $\tilde{Q}_1$, $\tilde{R}_1$, $\tilde{Q}_2$, $\tilde{R}_2$ are appropriate matrices. This choice of local indices is consistent with our assumption that subsystems are physical units whose autonomy is desirable (see Section 3.1). Then, the joint performance index for the overall system $\tilde{\mathbf{S}}$ is simply

$$\tilde{J}_D(\tilde{x}_0,\, u) = \int_0^\infty (\tilde{x}^T\tilde{Q}_D\tilde{x} + u^T\tilde{R}_Du)\, dt, \tag{8.114}$$

where $\tilde{Q}_D = \text{diag}\{\tilde{Q}_1,\, \tilde{Q}_2\}$ and $\tilde{R}_D = \text{diag}\{\tilde{R}_1,\, \tilde{R}_2\}$. By part ($i$) of Theorem 8.12, $\tilde{J}$ is an expansion of

$$J_D(x_0,\, u) = \int_0^\infty (x^TQ_Dx + u^TR_Du)\, dt, \tag{8.115}$$

when $Q_D = V^T\tilde{Q}_DV$ and $R_D = \tilde{R}_D$, and J_D is a performance index for the original system $\mathbf{S}$.

A decentralized control law

$$u_1 = -\tilde{K}_1\tilde{x}_1, \qquad u_2 = -\tilde{K}_2\tilde{x}_2, \tag{8.116}$$

can now be chosen to optimize the individual overlapping subsystems $\tilde{\mathbf{S}}_1$ and $\tilde{\mathbf{S}}_2$ with respect to the indices $\tilde{J}_1$ and $\tilde{J}_2$ because they appear in $\tilde{\mathbf{S}}$ as disjoint. The resulting gain matrix is block diagonal:

$$\tilde{K}_D = \left[\begin{array}{c|c} \tilde{K}_1 & 0 \\ \hline 0 & \tilde{K}_2 \end{array}\right], \tag{8.117}$$

where

$$\tilde{K}_1 = \tilde{R}_1^{-1}\tilde{B}_1^T\tilde{P}_1, \qquad \tilde{K}_2 = \tilde{R}_2^{-1}\tilde{B}_2^T\tilde{P}_2, \tag{8.118}$$

and $\tilde{P}_1$, $\tilde{P}_2$ are solutions of the corresponding Riccati equations. The decentralized control law

$$u_D^{\odot} = -\tilde{K}_D\tilde{x} \tag{8.119}$$

is suboptimal with respect to $(\tilde{\mathbf{S}}, \tilde{J}_D)$.

Finally, to get an overlapping decentralized control law

$$u_D^{\odot} = -K_D x \tag{8.120}$$

of the same quality for implementation in the original system $\mathbf{S}$, we rewrite $\tilde{K}_D$ as

$$\tilde{K}_D = \left[\begin{array}{cc|cc} \tilde{K}_{11} & \tilde{K}_{12} & 0 & 0 \\ \hline 0 & 0 & \tilde{K}_{23} & \tilde{K}_{24} \end{array}\right] \tag{8.121}$$

to conform with representation (8.108) of $\tilde{\mathbf{S}}$. By using Corollary 8.14, we contract $\tilde{K}_D$ of (8.121) to get the gain matrix

$$K_D = \left[\begin{array}{ccc} \tilde{K}_{11} & \tilde{K}_{12} & 0 \\ \hline 0 & \tilde{K}_{23} & \tilde{K}_{24} \end{array}\right] \tag{8.122}$$

for the overlapping decentralized control law (8.120). The basic reason for expanding a pair $(\mathbf{S}, J_D)$ into $(\tilde{\mathbf{S}}, \tilde{J}_D)$ is that the concept of suboptimality of Section 3.3 can be directly applied to the pair $(\tilde{\mathbf{S}}, \tilde{J}_D)$: the overlapping subsystems of $\mathbf{S}$ appear as disjoint in $\tilde{\mathbf{S}}$. Since the suboptimality index $\tilde{\mu}$ is defined for $(\tilde{\mathbf{S}}, \tilde{J}_D)$, it is important to note that the contracted control law (8.119) with K_D of (8.122) is suboptimal for the pair $(\mathbf{S}, J_D)$ with the same degree of suboptimality $\tilde{\mu}$, where J_D is a contraction of $\tilde{J}_D$. That is, the inequality

$$\tilde{J}_D^{\oplus}(\tilde{x}_0) \le \tilde{\mu}^{-1}\tilde{J}_D^{\odot}(\tilde{x}_0) \tag{8.123}$$

implies (but is not implied by) the inequality

$$J_D^{\oplus}(x_0) \le \tilde{\mu}^{-1}J_D^{\odot}(x_0), \tag{8.124}$$

where $J_D^{\oplus}(x_0)$ denotes the value of $J_D(x_0, u)$ for **S** under the control law (8.120), which is equal to $\tilde{J}_D^{\oplus}(Vx_0)$, and

$$J_D^{\odot}(x_0) = \tilde{J}_D^{\odot}(Vx_0). \tag{8.125}$$

To illustrate further some of the issues discussed in this example, let us consider a numerical version of **S** in (8.107) given as

$$\mathbf{S}\colon \quad \begin{bmatrix} \dot{x}_1 \\ \dot{x}_2 \\ \dot{x}_3 \end{bmatrix} = \left[\begin{array}{c|c|c} 1 & -24 & 1 \\ \hline 1 & -20 & 1 \\ \hline 1 & -24 & 1 \end{array}\right] \begin{bmatrix} x_1 \\ x_2 \\ x_3 \end{bmatrix} + \left[\begin{array}{c|c} 2 & 0 \\ \hline 0 & 0 \\ \hline 0 & 2 \end{array}\right] \begin{bmatrix} u_1 \\ u_2 \end{bmatrix}, \tag{8.126}$$

where the subsystems are indicated by the dashed lines. The expansion of **S** is

$$\tilde{\mathbf{S}}\colon \quad \begin{bmatrix} \dot{\tilde{x}}_1 \\ \dot{\tilde{x}}_2 \end{bmatrix} = \left[\begin{array}{cc|cc} 1 & -24 & 0 & 1 \\ 1 & -20 & 0 & 1 \\ \hline 1 & 0 & -20 & 1 \\ 1 & 0 & -24 & 1 \end{array}\right] \begin{bmatrix} \tilde{x}_1 \\ \tilde{x}_2 \end{bmatrix} + \left[\begin{array}{c|c} 2 & 0 \\ 0 & 0 \\ \hline 0 & 0 \\ 0 & 2 \end{array}\right] \begin{bmatrix} u_1 \\ u_2 \end{bmatrix}. \tag{8.127}$$

With $\tilde{\mathbf{S}}$, we associate a "decentralized" performance index

$$\tilde{J}_D(\tilde{x}_0, u) = \int_0^{\infty} (\tilde{x}_1^T \tilde{x}_1 + \tilde{x}_2^T \tilde{x}_2 + u_1^2 + u_2^2)\, dt. \tag{8.128}$$

The optimization problem $(\tilde{\mathbf{S}}, \tilde{J}_D)$ was solved in Example 3.20 as a disjoint decomposition problem, and a suboptimality index

$$\tilde{\mu}^* = 0.95 \tag{8.129}$$

was achieved for $(\tilde{\mathbf{S}}, \tilde{J}_D)$ and, thus, for $(\mathbf{S}, J_D)$. The decentralized control law, calculated by local optimization procedure, was found to be also stabilizing.

8.16. REMARK. In the above example, we used Corollary 8.14 to construct a contractible control law. When the expansion $\tilde{\mathbf{S}}$ is generated using part (*ii*) of Theorem 8.12, not every control law for $\tilde{\mathbf{S}}$ is contractible for **S**. A class of contractible laws is defined when

$$UM = 0, \qquad UN = 0. \tag{8.130}$$

Then, if $\tilde{K}_D$ satisfies the condition

$$\tilde{K}_D = \tilde{K}_D V U, \tag{8.131}$$

the control law $-\tilde{K}_D \tilde{x}$ is contractible to $-K_D x$ with K_D defined in (8.106), that is, $K_D = \tilde{K}_D V$.

8.17. EXAMPLE. In the so-called *particular case* (Ikeda *et al.* 1981a), when the overlapping part in **S** is not asymptotically stable, $\tilde{\mathbf{S}}$ is not asymptotically stable even if **S** is. For instance, if

$$\mathbf{S}\colon\quad \begin{bmatrix} \dot{x}_1 \\ \dot{x}_2 \\ \dot{x}_3 \end{bmatrix} = \left[\begin{array}{cc:c} a_{11} & a_{12} & 0 \\ \hdashline a_{21} & 0 & a_{23} \\ \hdashline 0 & a_{32} & a_{33} \end{array}\right] \begin{bmatrix} x_1 \\ x_2 \\ x_3 \end{bmatrix} + \left[\begin{array}{c:c} b_{11} & 0 \\ \hdashline 0 & 0 \\ \hdashline 0 & b_{32} \end{array}\right] \begin{bmatrix} u_1 \\ u_2 \end{bmatrix}, \tag{8.132}$$

then, choosing V, U, and M as in Section 8.2, we obtain

$$\tilde{\mathbf{S}}\colon\quad \begin{bmatrix} \dot{\tilde{x}}_1 \\ \dot{\tilde{x}}_2 \end{bmatrix} = \left[\begin{array}{cc:cc} a_{11} & a_{12} & 0 & 0 \\ a_{21} & 0 & 0 & a_{23} \\ \hdashline a_{21} & 0 & 0 & a_{23} \\ 0 & 0 & a_{32} & a_{23} \end{array}\right] \begin{bmatrix} \tilde{x}_1 \\ \tilde{x}_2 \end{bmatrix} + \left[\begin{array}{c:c} b_{11} & 0 \\ 0 & 0 \\ \hdashline 0 & 0 \\ 0 & b_{32} \end{array}\right] \begin{bmatrix} u_1 \\ u_2 \end{bmatrix}. \tag{8.133}$$

Due to a special structure of the matrices, the expansion $\tilde{\mathbf{S}}$ can never be stabilized by state feedback; the corresponding closed-loop system matrix would always have two identical rows resulting in a fixed zero eigenvalue. For this reason, the corresponding matrix $\tilde{H}$ would not exist causing suboptimality to fail.

To recover suboptimality (and asymptotic stability) in particular cases, we have to contract the control and test suboptimality of the closed-loop system in the original space. Of course, a contraction requires an additional effort that is not required in the standard case.

Let us assume that a particular case results in a decomposition

$$\tilde{\mathbf{S}}\colon\ \dot{\tilde{x}}_i = \tilde{A}_i \tilde{x}_i + \tilde{B}_i u_i + \sum_{j=1}^{N} \tilde{A}_{ij} \tilde{x}_j + \sum_{j=1}^{N} \tilde{B}_{ij} u_j, \qquad i \in \boldsymbol{N}, \tag{8.134}$$

where $\tilde{x}(t) \in \mathbf{R}^{\tilde{n}}$ and $u(t) \in \mathbf{R}^m$ are the state and input of $\tilde{\mathbf{S}}$ at $t \in \mathbf{R}$. $\tilde{A}_i$, $\tilde{A}_{ij}$, $\tilde{B}_i$, $\tilde{B}_{ij}$ are constant matrices of appropriate dimensions. A compact representation of $\tilde{\mathbf{S}}$ is

$$\tilde{\mathbf{S}}\colon\ \dot{\tilde{x}} = \tilde{A}_D\tilde{x} + \tilde{B}_D u + \tilde{A}_C\tilde{x} + \tilde{B}_C u, \tag{8.135}$$

which is the standard form adapted for $\tilde{\mathbf{S}}$ in (3.57). Following the development of Section 3.2, we associate with $\tilde{\mathbf{S}}$ a performance index

$$\tilde{J}_D(\tilde{x}_0, u) = \int_0^\infty (\tilde{x}^T\tilde{Q}_D\tilde{x} + u^T\tilde{R}_D u)\, dt. \tag{8.136}$$

A decentralized control

$$u_D^\odot = -\tilde{K}_D\tilde{x}, \tag{8.137}$$

is provided by optimizing $\tilde{J}_D$ with respect to

$$\tilde{\mathbf{S}}_D\colon\ \dot{\tilde{x}} = \tilde{A}_D\tilde{x} + \tilde{B}_D u. \tag{8.138}$$

We get

$$\tilde{K}_D = \tilde{R}_D^{-1}\tilde{B}_D^T\tilde{P}_D, \tag{8.139}$$

where

$$\tilde{A}_D^T\tilde{P}_D + \tilde{P}_D\tilde{A}_D - \tilde{P}_D\tilde{B}_D\tilde{R}_D^{-1}\tilde{B}_D^T\tilde{P}_D + \tilde{Q}_D = 0. \tag{8.140}$$

By contracting the control law $u_D^\odot$ of (8.137) and implementing it in the original system

$$\mathbf{S}\colon\ \dot{x} = Ax + Bu, \tag{8.37$'$}$$

we get the closed-loop system

$$\hat{\mathbf{S}}\colon\ \dot{x} = \hat{A}x, \tag{8.141}$$

where

$$\hat{A} = A - BK_D, \qquad K_D = \tilde{K}_D V, \tag{8.142}$$

and the contracted performance index is computed as

$$J_D^\oplus(x_0) = x_0^T H x_0, \tag{8.143}$$

where

$$H = \int_0^\infty \exp(\hat{A}^T t)\, G_D \exp(\hat{A}t)\, dt, \qquad G_D = Q_D + K_D^T R_D K_D,$$
$$Q_D = V^T\tilde{Q}_D V, \qquad R_D = \tilde{R}_D. \tag{8.144}$$

Now, we use the notation

$$J_D^{\odot}(x_0) = \tilde{J}_D^{\odot}(Vx_0), \tag{8.145}$$

and state the following (Ikeda *et al.* 1981a):

8.18. DEFINITION. We say that a control law

$$u_D^{\odot} = -\tilde{R}_D^{-1}\tilde{B}_D^T\tilde{P}_D V x \tag{8.146}$$

is suboptimal for the pair ($\mathbf{S}$, J_D) if there exists a positive number μ such that

$$J_D^{\oplus}(x_0) \leq \mu^{-1} J_D^{\odot}(x_0), \qquad \forall x_0 \in \mathbf{R}^n. \tag{8.147}$$

Imitating Theorem 3.18 of Section 3.2, we arrive at the following result:

8.19. THEOREM. A control law $u_D^{\odot}$ is suboptimal with the degree of suboptimality

$$\mu^* = \lambda_M^{-1}[H(V^T\tilde{P}_D V)^{-1}] \tag{8.148}$$

for the pair ($\mathbf{S}$, J_D) if and only if the matrix H is finite.

Proof. We note that the matrix $V^T\tilde{P}_D V$ is positive definite because V is monic (full column rank). Then, using the argument of the proof of Theorem 3.3, we establish (8.148). Q.E.D.

8.20. REMARK. The number μ^* computed by Theorem 8.19 is the largest (best) for which (8.147) is satisfied. If $\hat{A}$ is stable, the matrix H can be computed from the Liapunov equation

$$\hat{A}^T H + H\hat{A} = -G_D, \tag{8.149}$$

and μ^* is obtained using (8.148). This means that stability implies suboptimality. The converse result can be established as in Theorem 3.4 if we assume that ($\hat{A}$, $G_D^{1/2}$) is observable.

So far, the reference system for suboptimality was a collection of optimal decoupled subsystems. This choice is justified when complete autonomy of each subsystem is regarded as an ideal situation. Alternatively, we may be interested in comparing the performance of a locally optimal strategy with the centrally optimal system, which is the best one can achieve by LQ control.

Let us consider again the system

$$\mathbf{S}\colon\ \dot{x} = Ax + Bu, \tag{8.37'}$$

with the performance index

$$J_D(x_0,\, u) = \int_0^\infty (x^T Q_D x + u^T R_D u)\, dt. \tag{8.93'}$$

We assume that the pair $(A,\, B)$ is controllable and the pair $(A,\, Q_D^{1/2})$ is observable. The optimal control is

$$u^{\odot} = -R_D^{-1} B^T P x, \tag{8.150}$$

resulting in the value of the performance index given as

$$J^{\odot}(x_0) = x_0^T P x_0, \tag{8.151}$$

where P is the symmetric positive definite solution of the Riccati equation

$$A^T P + PA - PBR_D^{-1} B^T P + Q_D = 0. \tag{8.152}$$

If instead of the centrally optimal control $u^{\odot}$, we use the locally optimal control from the expanded space,

$$u_D^{\odot} = -\tilde{R}_D^{-1} \tilde{B}_D^T \tilde{P}_D V x, \tag{8.146}$$

suboptimality can be evaluated by computing $J_D^{\oplus}(x_0)$ defined by (8.143) and comparing the obtained value with that of $J^{\odot}(x_0)$ defined by (8.151). A positive number $\bar{\mu}$, such that the inequality

$$J_D^{\oplus}(x_0) \le \bar{\mu}^{-1} J^{\odot}(x_0) \tag{8.153}$$

holds for all x_0, is the suboptimality index. Obviously, $\bar{\mu}$ exists if H of (8.144) is finite. In this case, the largest number $\bar{\mu}^*$ for which inequality (8.153) is satisfied can be calculated as

$$\bar{\mu}^* = \lambda_M^{-1}(HP^{-1}), \tag{8.154}$$

which is a similar expression to that of μ^* in (8.148).

8.21. REMARK. The similarity of μ^* and $\bar{\mu}^*$ implies that once suboptimality of a control law is established with respect to one reference system, it is suboptimal with respect to another provided the corresponding observability conditions are satisfied. The degree of suboptimality is of course different in general, involving different computations as well. For example, to get $\bar{\mu}^*$, we have to compute the centrally optimal control, which may be unattractive due to a dimensionality problem. It is usually simpler to work with μ^*, because only optimization of subsystems is required (see Section 3.2).

To illustrate the application of overlapping decentralized control, we consider the problem of Automatic Generation Control (AGC).

8.22. EXAMPLE. Because of the size and complexity of present-day power systems, decentralized control schemes for AGC are more attractive than alternative centralized designs. The main reason is an excessive information requirement imposed by a globally optimal AGC. This has been a prime motivation for building AGC schemes of the decentralized type, where control is confined to individual power areas employing the standard tie-line bias control concept and noninteraction principle, which preserve the steady-state area autonomy and economy of operation (Ćalović *et al.* 1978; Šiljak, 1978). In this kind of control scheme, the areas are represented as overlapping subsystems with the tie-lines being the overlapping parts.

Let us consider a two-area model described in (Ćalović *et al.* 1978) by the equation

$$\mathbf{S}\colon\quad \begin{bmatrix} \Delta\dot{x}_1 \\ \Delta\dot{v}_1 \\ \Delta\dot{p}_e \\ \Delta\dot{v}_2 \\ \Delta\dot{x}_2 \end{bmatrix} = \begin{bmatrix} A_1 & 0 & a_{t1} & 0 & 0 \\ d_1^T & 0 & 1 & 0 & 0 \\ m_{12}^T & 0 & 0 & 0 & -m_{21}^T \\ 0 & 0 & -1 & 0 & d_2^T \\ 0 & 0 & -a_{t2} & 0 & A_2 \end{bmatrix} \begin{bmatrix} \Delta x_1 \\ \Delta v_1 \\ \Delta p_e \\ \Delta v_2 \\ \Delta x_2 \end{bmatrix} + \begin{bmatrix} b_1 & 0 \\ 0 & 0 \\ 0 & 0 \\ 0 & 0 \\ 0 & b_2 \end{bmatrix} \begin{bmatrix} u_1 \\ u_2 \end{bmatrix} + \begin{bmatrix} f_1 & 0 \\ 0 & 0 \\ 0 & 0 \\ 0 & 0 \\ 0 & f_2 \end{bmatrix} \begin{bmatrix} z_1 \\ z_2 \end{bmatrix}, \tag{8.155}$$

where the dashed lines identify the two areas; Δx_1 and Δx_2 are the deviations of states from their nominal values; Δv_1 and Δv_2 are the variables introduced to achieve the integral control; and Δp_e is the variation of the total power exchange between the areas. In (8.155), u_1 and u_2 are scalar inputs, and z_1 and z_2 are unmeasurable disturbances. The matrices and vectors indicated in (8.155) are numerically specified as follows:

$$A_1 = \begin{bmatrix} -0.2 & 0 & 0 & 0 & 0 & 0 & 0 & -4 \\ 4.75 & -5 & 0 & 0 & 0 & 0 & 0 & 0 \\ 0 & 0.1667 & -0.1667 & 0 & 0 & 0 & 0 & 0 \\ 0 & 0 & 2 & -2 & 0 & 0 & 0 & 0 \\ 0 & -0.08 & -0.0747 & -0.112 & -3.994 & 10 & -0.928 & -9.1011 \\ 0 & 0 & 0 & 0 & 0.2 & -0.5 & 0 & 0 \\ 0 & 0 & 0 & 0 & 1.3194 & 0 & -1.3889 & -0.2778 \\ 0 & 0.01 & 0.0093 & 0.014 & -0.0632 & 0 & 0.1160 & -01124 \end{bmatrix},$$

$$A_2 = \begin{bmatrix} -0.2 & 0 & 0 & 0 & 0 & 0 & 0 & -4 \\ 4.75 & -5 & 0 & 0 & 0 & 0 & 0 & 0 \\ 0 & 0.1667 & -0.1667 & 0 & 0 & 0 & 0 & 0 \\ 0 & 0 & 2 & -2 & 0 & 0 & 0 & 0 \\ 0 & -0.1 & -0.0933 & -0.14 & -4.096 & 10 & -0.7442 & -9.1079 \\ 0 & 0 & 0 & 0 & 0.2 & -0.5 & 0 & 0 \\ 0 & 0 & 0 & 0 & 1.3194 & 0 & -1.3889 & -0.2778 \\ 0 & 0.0125 & 0.0117 & 0.0175 & -0.0506 & 0 & 0.0928 & -0.1115 \end{bmatrix},$$

$$\begin{aligned} a_{t1} &= a_{t2} = [0,\ 0,\ 0,\ 0,\ 0.6667,\ 0,\ 0,\ -0.0833]^T, \\ d_1^T &= d_2^T = [0,\ 0,\ 0,\ 0,\ 0,\ 0,\ 0,\ 10], \\ m_{12}^T &= m_{21}^T = [0,\ 0,\ 0,\ 0,\ 0,\ 0,\ 0,\ 22.2144], \\ b_1 &= [1.6,\ 0,\ 0,\ 0,\ 6,\ 0,\ 0,\ 0]^T, \\ b_2 &= [2,\ 0,\ 0,\ 0,\ 5,\ 0,\ 0,\ 0]^T, \\ f_1 &= a_{t1}, \qquad f_2 = a_{t2}. \end{aligned} \tag{8.156}$$

To achieve autonomy of each power area, we use the performance indices

$$
\begin{aligned}
&J_1[x_1(0),\ \Delta v_1(0),\ \Delta p_e(0),\ u_1] \\
&\qquad = \int_0^\infty [\Delta x_1^T Q_1 \Delta x_1 + 10(\Delta v_1)^2 + 10(\Delta p_e)^2 + u_1^2]\, dt, \\
&J_2[\Delta x_2(0),\ \Delta v_2(0),\ \Delta p_e(0),\ u_2] \\
&\qquad = \int_0^\infty [\Delta x_2^T Q_2 \Delta x_2 + 10(\Delta v_2)^2 + 10(\Delta p_e)^2 + u_2^2]\, dt.
\end{aligned}
\tag{8.157}
$$

where

$$
Q_1 = Q_2 = \operatorname{diag}\{1,\ 1,\ 1,\ 1,\ 1,\ 1,\ 1,\ 10\}. \tag{8.158}
$$

This is equivalent to associating the performance indices J_1 and J_2 with the (decoupled) overlapping subsystems delineated by the dashed lines in (8.155).

By following the optimization procedure presented in the preceding section, we expand $\mathbf{S}$, optimize the decoupled subsystems, calculate the local feedback gains, and contract the block diagonal gain matrix to get the overlapping decentralized control law

$$
u_D^{\odot} = -\begin{bmatrix} k_{11}^T & k_{12} & k_{13} & 0 & 0 \\ 0 & 0 & k_{23} & k_{22} & k_{21}^T \end{bmatrix} \begin{bmatrix} \Delta x_1 \\ \Delta v_1 \\ \Delta p_e \\ \Delta v_2 \\ \Delta x_2 \end{bmatrix}, \tag{8.159}
$$

where

$$
\begin{aligned}
k_{11}^T &= [1.353,\ 0.31,\ 2.47,\ 0.597,\ 0.75,\ 0.021,\ 6.56,\ 89.92], \\
k_{21}^T &= [1.631,\ 0.406,\ 3.019,\ 0.792,\ 0.551,\ -0.63,\ 5.501,\ 94.84], \\
k_{12} &= 3.162, \qquad k_{13} = -2.668, \qquad k_{22} = 3.162, \qquad k_{23} = 2.929.
\end{aligned}
\tag{8.160}
$$

The degree of suboptimality μ^* is computed, using (8.148), as

$$
\mu^* = 0.376. \tag{8.161}
$$

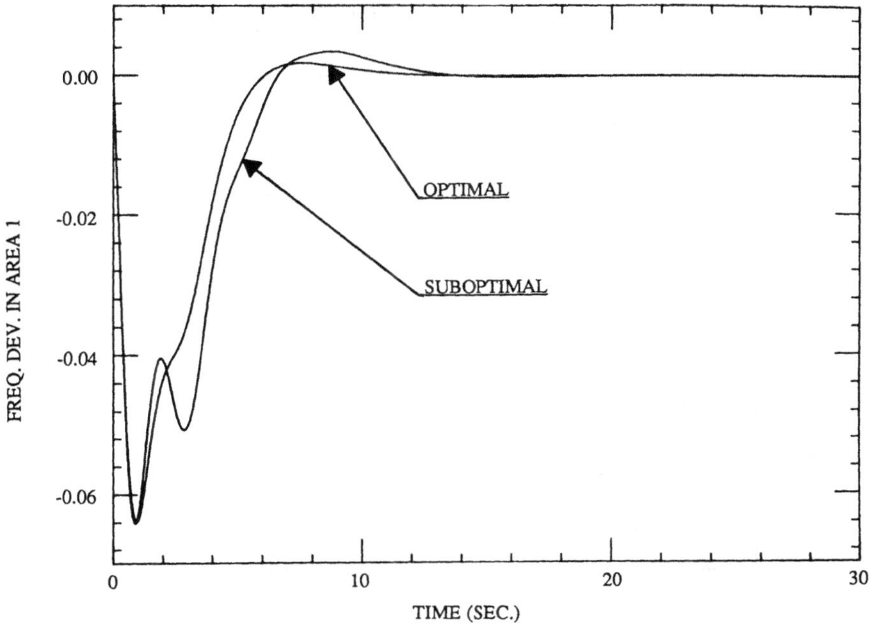

Fig. 8.2. Frequency deviation in Area 1.

When the centrally optimal system is chosen as reference, the degree of suboptimality is computed as

$$\bar{\mu}^* = 0.820. \tag{8.162}$$

The control law $u_D^{\odot}$ of (8.159) is also stabilizing. This follows from the fact that Q_1 and Q_2 of (8.158) and the rest of the terms in J_1 and J_2 of (8.157), which correspond to the states of $\mathbf{S}$, are all positive definite.

Finally, we present the results of simulation in Figures 8.2–4, where responses to the unit-step disturbance

$$z_1(t) = \begin{cases} 0, & t < 0, \\ 1, & t \geq 0, \end{cases} \tag{8.163}$$

of the frequency deviations Δf_1 and Δf_2 (the eighth element of Δx_1 and Δx_2, respectively) and the power exchange Δp_e are shown. Both the centralized optimal control with global information and the decentralized suboptimal control with overlapping information sets are represented for comparison. The centralized control produces smoother and somewhat faster

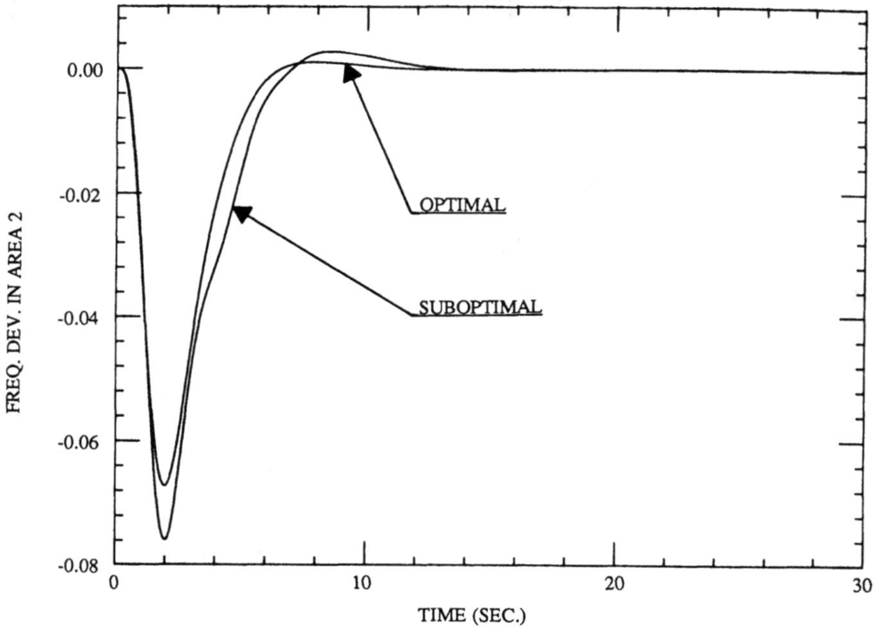

Fig. 8.3. Frequency deviation in Area 2.

responses, but the overlapping decentralized control should be judged satisfactory to the extent of distinctly weaker information structure requirements.

8.4. Nonlinear Systems

The inclusion concept can be generalized to nonlinear systems with similar benefits, but with more constraints than in the linear case. Special care has to be taken of the equilibrium states, and inclusion conditions are only sufficient. For simplicity, we consider systems with no inputs or outputs.

Let us consider two dynamic systems

$$\mathbf{S}\colon\ \dot{x} = f(t, x), \tag{8.164}$$

and

$$\tilde{\mathbf{S}}\colon\ \dot{\tilde{x}} = \tilde{f}(t, \tilde{x}), \tag{8.165}$$

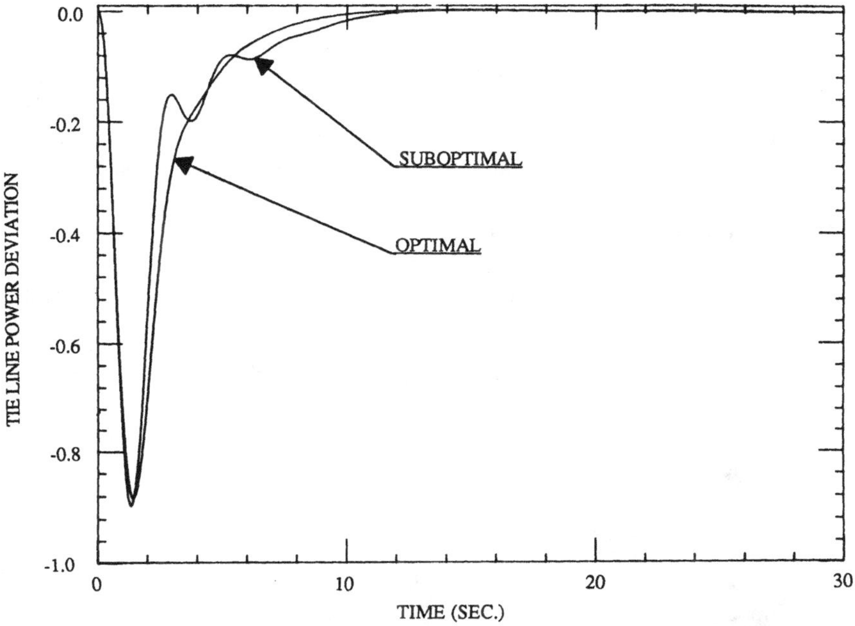

Fig. 8.4. Tie-line power deviation.

where $x(t) \in \mathbf{R}^n$ and $\tilde{x}(t) \in \mathbf{R}^{\tilde{n}}$ are the states of $\mathbf{S}$ and $\tilde{\mathbf{S}}$ at $t \in \mathbf{R}$, and $n \le \tilde{n}$. The functions f: $\mathbf{R} \times \mathbf{R}^n \to \mathbf{R}^n$ and $\tilde{f}$: $\mathbf{R} \times \mathbf{R}^{\tilde{n}} \to \mathbf{R}^{\tilde{n}}$ are assumed to be sufficiently smooth, so that solutions $x(t;\, t_0,\, x_0)$ and $\tilde{x}(t;\, t_0,\, \tilde{x}_0)$ of $\mathbf{S}$ and $\tilde{\mathbf{S}}$ exist and are unique for all initial conditions $(t_0,\, x_0) \in \mathbf{R} \times \mathbf{R}^n$ and $(t_0,\, \tilde{x}_0) \in \mathbf{R} \times \mathbf{R}^{\tilde{n}}$, and all $t \in \mathbf{T}_0 = [t_0,\, +\infty)$.

We use linear transformations

$$\tilde{x} = Vx, \qquad x = U\tilde{x}, \tag{8.166}$$

where V is an $\tilde{n} \times n$ constant matrix with full column rank and U is an $n \times \tilde{n}$ constant matrix with full row rank. The inclusion concept is stated as follows:

8.23. DEFINITION. $\tilde{\mathbf{S}} \supset \mathbf{S}$ if there exists an ordered pair of matrices $(U,\, V)$ such that $UV = I$, and for any $(t_0,\, x_0) \in \mathbf{R} \times \mathbf{R}^n$, $\tilde{x}_0 = Vx_0$ implies

$$x(t;\, t_0,\, x_0) = U\tilde{x}(t;\, t_0,\, Vx_0), \qquad \forall t \in \mathbf{T}_0. \tag{8.167}$$

To derive conditions for inclusion, we represent the function $\tilde{f}(t, \tilde{x})$ as

$$\tilde{f}(t, \tilde{x}) = Vf(t, U\tilde{x}) + \tilde{m}(t, \tilde{x}), \tag{8.168}$$

where $\tilde{m}\colon\ \mathbf{R} \times \mathbf{R}^{\tilde{n}} \to \mathbf{R}^{\tilde{n}}$ is called a *complementary function.* For $\tilde{\mathbf{S}}$ to include $\mathbf{S}$, $\tilde{m}$ is required to satisfy certain restrictions as in the following:

8.24. THEOREM. $\tilde{\mathbf{S}} \supset \mathbf{S}$ if either

$$\tilde{m}(t, Vx) = 0, \qquad \forall (t, x) \in \mathbf{R} \times \mathbf{R}^{n}, \tag{8.169}$$

or

$$U\tilde{m}(t, \tilde{x}) = 0, \qquad \forall (t, \tilde{x}) \in \mathbf{R} \times \mathbf{R}^{\tilde{n}}. \tag{8.170}$$

Proof. We first prove $\tilde{\mathbf{S}} \supset \mathbf{S}$ if (8.169) holds. Combining (8.168) and (8.169), we get

$$\dot{\tilde{x}} - V\dot{x} = V[f(t, U\tilde{x}) - f(t, x)] + \tilde{m}(t, \tilde{x}). \tag{8.171}$$

It is easy to see that $\tilde{x} - Vx = 0$ is the trivial solution of (8.171). This means that $\tilde{x}_0 = Vx_0$ implies

$$\tilde{x}(t;\, t_0, \tilde{x}_0) = Vx(t;\, t_0, x_0), \qquad \forall t \in \mathbf{T}_0, \tag{8.172}$$

which, in turn, implies that (8.167) holds.

To prove $\tilde{\mathbf{S}} \supset \mathbf{S}$ if (8.170) holds, we use (8.168) and (8.169) to get

$$U\dot{\tilde{x}} - \dot{x} = f(t, U\tilde{x}) - f(t, x). \tag{8.173}$$

Obviously, the trivial solution of this equation is $U\tilde{x} - x = 0$ and, therefore, $\tilde{x}_0 = Vx_0$ implies (8.167). Q.E.D.

8.25. REMARK. The two conditions of Theorem 8.24 are two independent conditions. They correspond to restriction and aggregation in linear systems (Theorem 8.3).

We turn our attention to stability properties of the two systems $\mathbf{S}$ and $\tilde{\mathbf{S}}$. Obviously, if $\tilde{\mathbf{S}} \supset \mathbf{S}$ then stability of $\tilde{\mathbf{S}}$ implies stability of $\mathbf{S}$. Unlike linear systems, this statement can be used only if the equilibrium of $\mathbf{S}$ is preserved in $\tilde{\mathbf{S}}$ under the linear transformation (8.166). More precisely, if

$$f(t, x_e) = 0, \qquad \forall t \in \mathbf{R}, \tag{8.174}$$

and x_e is an equilibrium of $\mathbf{S}$, then we require that

$$\tilde{f}(t, Vx_e) = 0, \qquad \forall t \in \mathbf{R}, \tag{8.175}$$

so that $\tilde{x}_e = Vx_e$ is the corresponding equilibrium of $\tilde{\mathbf{S}}$. It is easy to see that this takes place if and only if

$$\tilde{m}(t, Vx_e) = 0, \qquad \forall t \in \mathbf{R}. \tag{8.176}$$

This condition is implied by the inclusion condition (8.169) of Theorem 8.24, but is not implied by condition (8.170).

Stability is resolved by the following:

8.26. THEOREM. Suppose $\tilde{\mathbf{S}} \supset \mathbf{S}$ and $\tilde{x}_e = Vx_e$. Then, stability of the equilibrium $\tilde{x}_e$ of $\tilde{\mathbf{S}}$ implies stability of the equilibrium x_e of $\mathbf{S}$.

Proof. We prove the theorem with respect to global asymptotic stability in the sense of Liapunov, which is stated here for an easy reference in the proof of the theorem.

An equilibrium x_e of $\mathbf{S}$ is globally asymptotically stable if the following two conditions hold:

(*i*) It is stable, that is, for any initial time $t_0 \in \mathbf{R}$ and positive number ϵ, there exists a positive number $\delta(t_0, \epsilon)$ such that $\|x_0 - x_e\| \leq \delta$ implies

$$\|x(t; t_0, x_0) - x_e\| \leq \epsilon, \qquad \forall t \in \mathbf{T}_0; \tag{8.177}$$

and

(*ii*) It is attractive in the large, that is,

$$\lim_{t \to +\infty} x(t; t_0, x_0) = x_e, \qquad \forall (t_0, x_0) \in \mathbf{R} \times \mathbf{R}^n. \tag{8.178}$$

When $\tilde{x}_e = Vx_e$ is stable in the sense of (*i*) above, then, for any t_0 and $\tilde{\epsilon} > 0$, there exists a $\tilde{\delta}(t_0, \tilde{\epsilon}) > 0$ such that $\|\tilde{x}_0 - Vx_e\| \leq \tilde{\delta}$ implies

$$\|\tilde{x}(t; t_0, \tilde{x}_0) - Vx_e\| \leq \tilde{\epsilon}, \qquad \forall t \in \mathbf{T}_0. \tag{8.179}$$

To prove stability (*i*) of the equilibrium x_e of $\mathbf{S}$, we note that for any $\epsilon > 0$ we can choose $\delta = \|V\|^{-1}\tilde{\delta}$, where $\tilde{\delta} = \tilde{\delta}(t_0, \tilde{\epsilon})$ and $\tilde{\epsilon} = \|U\|^{-1}\epsilon$. Then, $\|x_0 - x_e\| \leq \delta$ implies $\|Vx_0 - Vx_e\| \leq \tilde{\delta}$. Therefore, $\|\tilde{x}(t; t_0, Vx_0) - Vx_e\| \leq \tilde{\epsilon}$, and $\|x(t; t_0, x_0) - x_e\| \leq \|U\| \, \|\tilde{x}(t; t_0, Vx_0) - Vx_e\| \leq \epsilon$ for all $t \in \mathbf{T}_0$, which implies stability of x_e.

The global attractivity (*ii*) of x_e follows directly from the global attractivity of $\tilde{x}_e = Vx_e$ and the relation $x(t;\, t_0,\, x_0) - x_e = U[\tilde{x}(t;\, t_0,\, Vx_0) - Vx_e]$. Therefore, if the equilibrium $\tilde{x}_e = Vx_e$ of the expansion $\tilde{\mathbf{S}}$ is globally asymptotically stable, then the equilibrium x_e of $\mathbf{S}$ is likewise globally asymptotically stable. Q.E.D.

The above proof can easily be modified to accommodate other types of stability in the sense of Liapunov. It should also be noted that the converse of Theorem 8.26 is not true in general. It has been shown (Ikeda and Šiljak, 1980a) that the converse can be established for linear systems provided additional conditions hold (see Theorem 8.7).

In Example 8.1, we applied the Inclusion Principle to show that in linear systems overlapping decompositions can be used to show stability where disjoint decompositions fail. Now, we show that the nonlinear version of the Principle can be utilized for the same purpose. The object is a *Lotka–Volterra model*, which is a popular mathematical representation of multispecies communities in mathematical biology (*e.g.*, May, 1974; Svirezhev and Logofet, 1983).

Let us consider the Lotka–Volterra equations in the vector form

$$\mathbf{S}\colon\ \dot{x} = X(a + Ax), \tag{8.180}$$

where $x = (x_1,\, x_2,\, \ldots,\, x_n)^T \in \mathbf{R}^n$ is the state vector, $X = \mathrm{diag}\{x_1,\, x_2,\, \ldots,\, x_n\}$, $a \in \mathbf{R}^n$ is a constant vector, and $A = (a_{ij})$ is an $n \times n$ constant matrix. The equilibria of the system (8.180) are given as solutions of the algebraic equation

$$X(a + Ax) = 0. \tag{8.181}$$

If every principal minor of A is different from zero, then (8.181) has 2^n solutions, which are not necessarily distinct. When it has a positive solution x_e, the solution is unique and is given by $x_e = -A^{-1}a$. We assume the existence of the positive equilibrium x_e, and consider the global asymptotic stability of this equilibrium with respect to the invariant region $\mathbf{R}^n_+$ and $t \in \mathbf{T}_0 = [0,\, +\infty)$.

In order to use the overlapping decompositions, we need an expansion of $\mathbf{S}$ defined as

$$\tilde{\mathbf{S}}\colon\ \dot{\tilde{x}} = \tilde{X}(\tilde{a} + \tilde{A}\tilde{x}), \tag{8.182}$$

which contains the equilibrium x_e. This takes place if the transformation matrix V of (8.166) is such that every row contains exactly one unit element. Moreover, $\tilde{a}$ and $\tilde{A}$ are given as

$$\tilde{a} = Va, \qquad \tilde{A} = VAU + M, \tag{8.183}$$

where M is an $\tilde{n} \times \tilde{n}$ constant matrix satisfying $MV = 0$. To see this, we compute from (8.168)

$$\begin{aligned} m(\tilde{x}) &= \tilde{X}(\tilde{a} + \tilde{A}\tilde{x}) - VX(\tilde{x})(a + AU\tilde{x}) \\ &= [\tilde{X}V - VX(\tilde{x})](a + AU\tilde{x}) + \tilde{X}M\tilde{x}, \end{aligned} \tag{8.184}$$

and use the condition (8.169) of Theorem 8.24. In (8.184), $X(\tilde{x})$ is an $n \times n$ diagonal matrix, which is obtained by substituting the relation $x = U\tilde{x}$ into the corresponding elements of X.

We note here that once the overlapping parts of the system $\mathbf{S}$ are chosen as a result of either modeling or computational considerations, the exact form of the transformation matrix V is automatic (see Section 8.2). The matrix V has the characteristic required after Equation (8.182), which ensures that the positive orthant $\mathbf{R}_+^n$ of the state space of $\mathbf{S}$ is preserved in the positive orthant $\mathbf{R}_+^{\tilde{n}}$ of the state space of $\tilde{\mathbf{S}}$. This fact is essential, since we want to establish stability properties of $\mathbf{S}$ with respect to $\mathbf{R}_+^n$ from the same properties of $\tilde{\mathbf{S}}$ with respect to $\mathbf{R}_+^{\tilde{n}}$.

Let us assume that we have expanded a system $\mathbf{S}$ with overlapping subsystems into a system

$$\tilde{\mathbf{S}}\colon\ \dot{\tilde{x}}_i = \tilde{X}_i\left(\tilde{a}_i + \tilde{A}_i\tilde{x}_i + \sum_{j=1}^{N} \tilde{A}_{ij}\tilde{x}_j\right), \qquad i \in \boldsymbol{N}, \tag{8.185}$$

where the subsystems of $\mathbf{S}$ are now subsystems

$$\tilde{\mathbf{S}}_i\colon\ \dot{\tilde{x}}_i = \tilde{X}_i(\tilde{a}_i + A_i\tilde{x}_i), \qquad i \in \boldsymbol{N}, \tag{8.186}$$

which appear as disjoint subsystems of $\tilde{\mathbf{S}}$. The state $\tilde{x}_i(t) \in \mathbf{R}_i^{\tilde{n}}$ of $\tilde{\mathbf{S}}_i$ at $t \in \mathbf{R}$ has the form $\tilde{x}_i = \{\tilde{x}_{i1}, \tilde{x}_{i2}, \ldots, \tilde{x}_{i\tilde{n}_i})^T$, $\tilde{X}_i = \text{diag}\{\tilde{x}_{i1}, \tilde{x}_{i2}, \ldots, \tilde{x}_{i\tilde{n}_i}\}$, $\tilde{a}_i \in \mathbf{R}^{\tilde{n}_i}$, and $\tilde{A}_i$, $\tilde{A}_{ij}$ are matrices of the proper dimensions. We denote the positive equilibrium $\tilde{x}_e$ of $\tilde{\mathbf{S}}$ as $\tilde{x}_e = [(\tilde{x}_1^e)^T, (\tilde{x}_2^e)^T, \ldots, (\tilde{x}_N^e)^T]^T$, where $\tilde{x}_i^e = (\tilde{x}_{i1}^e, \tilde{x}_{i2}^e, \ldots, \tilde{x}_{i\tilde{n}_i}^e)^T$, which corresponds to the equilibrium x_e of $\mathbf{S}$ *via* $\tilde{x}_e = Vx_e$. From $\tilde{a} + \tilde{A}\tilde{x}_e = 0$, which implies

$$\tilde{a}_i + \tilde{A}_i\tilde{x}_i^e + \sum_{j=1}^{N} A_{ij}\tilde{x}_j^e = 0, \qquad \forall i \in \boldsymbol{N}, \tag{8.187}$$

we can rewrite $\tilde{\mathbf{S}}$ of (8.185) as

$$\tilde{\mathbf{S}}\colon\ \dot{\tilde{x}}_i = \tilde{X}_i \left[\tilde{A}_i(\tilde{x}_i - \tilde{x}_i^e) + \sum_{j=1}^{N} \tilde{A}_{ij}(\tilde{x}_j - \tilde{x}_j^e) \right]. \tag{8.188}$$

To establish stability of the equilibrium $\tilde{x}_e$ using the concept of vector Liapunov functions (Section 2.3), we associate with each subsystem $\tilde{\mathbf{S}}_i$ a Volterra-type function (*e.g.*, Goh, 1980)

$$\tilde{v}_i(\tilde{x}_i) = \sum_{k=1}^{\tilde{n}_i} \tilde{d}_{ik}[\tilde{x}_{ik} - \tilde{x}_{ik}^e - \tilde{x}_{ik}^e \ln(\tilde{x}_{ik}/\tilde{x}_{ik}^e)], \tag{8.189}$$

where the $\tilde{d}_{ik}$ are positive numbers. The function $\tilde{v}_i$: $\mathbf{R}_+^{\tilde{n}_i} \to \mathbf{R}_+$ is continuously differentiable and positive definite on $\mathbf{R}_+^{\tilde{n}_i}$, and $\tilde{v}_i(\tilde{x}_i^e) = 0$. The time derivative of $\tilde{v}_i(\tilde{x}_i)$ with respect to the system $\tilde{\mathbf{S}}$ of (8.188) is

$$\dot{\tilde{v}}_i(\tilde{x}_i)_{(8.188)} = -(\tilde{x}_i - \tilde{x}_i^e)^T \tilde{G}_i(\tilde{x}_i - \tilde{x}_i^e) + \sum_{j=1}^{N} (\tilde{x}_i - \tilde{x}_i^e)^T \tilde{D}_i \tilde{A}_{ij}(\tilde{x}_j - \tilde{x}_j^e), \tag{8.190}$$

where the $\tilde{n}_i \times \tilde{n}_i$ matrices $\tilde{G}_i$ and $\tilde{D}_i$ are defined by

$$\tilde{G}_i = -\frac{1}{2}(\tilde{A}_i^T \tilde{D}_i + \tilde{D}_i \tilde{A}_i), \qquad \tilde{D}_i = \operatorname{diag}\{\tilde{d}_{i1}, \tilde{d}_{i2}, \ldots, d_{i\tilde{n}_i}\}. \tag{8.191}$$

We assume that a positive diagonal matrix $\tilde{D}_i$ is available such that $\tilde{G}_i$ is positive definite. This assumption means that all the equilibria $\tilde{x}_i^e$ of decoupled subsystems $\tilde{\mathbf{S}}_i$ in (8.186) are stable. This fact can be verified on the level of subsystems and the question arises: Under what conditions does stability of the subsystems $\tilde{\mathbf{S}}_i$ implies stability of the overall expanded system $\tilde{\mathbf{S}}$? An answer to this question is provided by the vector Liapunov function $\tilde{v}$: $\mathbf{R}^{\tilde{n}} \to \mathbf{R}_+^N$ having the form $\tilde{v} = (\tilde{v}_1, \tilde{v}_2, \ldots, \tilde{v}_N)^T$.

Let us use the function $\tilde{v}(\tilde{x})$ to form a scalar function $\tilde{\nu}$: $\mathbf{R}^{\tilde{n}} \to \mathbf{R}_+$ defined as

$$\tilde{\nu}(\tilde{x}) = \sum_{i=1}^{N} \tilde{d}_i \tilde{v}_i(\tilde{x}_i). \tag{8.192}$$

This is our candidate for a Liapunov function for the system $\tilde{\mathbf{S}}$, where the $\tilde{d}_i$ are positive (yet unspecified) numbers. Utilizing $\dot{\tilde{v}}_i(\tilde{x}_i)_{(8.188)}$ of (8.190), we compute $\dot{\tilde{\nu}}(\tilde{x})_{(8.188)}$ and majorize the result to get

$$\dot{\tilde{\nu}}(\tilde{x})_{(8.188)} \leq -\tilde{u}^T(\tilde{x})\tilde{G}\tilde{u}(\tilde{x}), \tag{8.193}$$

where $\tilde{G}$ is an $N \times N$ matrix:

$$\tilde{G} = \frac{1}{2}(\tilde{W}^T \tilde{D} + \tilde{D}\tilde{W}). \tag{8.194}$$

The $N \times N$ matrix $\tilde{W} = (\tilde{w}_{ij})$ is defined by

$$\tilde{w}_{ij} = \begin{cases} \lambda_m(\tilde{G}_i) - \|\tilde{D}_i \tilde{A}_{ii}\|, & i = j, \\ -\|\tilde{D}_i \tilde{A}_{ij}\|, & i \neq j, \end{cases} \tag{8.195}$$

where $\tilde{D} = \text{diag}\{\tilde{d}_1, \tilde{d}_2, \ldots, \tilde{d}_N\}$. The function $\tilde{u}$: $\mathbf{R}^{\tilde{n}} \to \mathbf{R}_+^N$ is defined as

$$\tilde{u}(\tilde{x}) = (\|\tilde{x}_1 - \tilde{x}_1^e\|, \|\tilde{x}_2 - \tilde{x}_2^e\|, \ldots, \|\tilde{x}_N - \tilde{x}_N^e\|)^T. \tag{8.196}$$

Inequality (8.193) implies the following:

8.27. THEOREM. The positive equilibrium $\tilde{x}_e$ of the system $\tilde{\mathbf{S}}$ is globally asymptotically stable with respect to the region $\mathbf{R}_+^{\tilde{n}}$ if the matrix $\tilde{W}$ is an M-matrix.

Proof. Due to a special structure of the Lotka–Volterra Equation (8.180), which gives rise to multiple equilibria and the fact that the solutions $\tilde{x}(t; \tilde{x}_0)$ of $\tilde{\mathbf{S}}$ that start from $\tilde{x}_0$ at $t_0 = 0$ in the positive orthant $\mathbf{R}_+^{\tilde{n}}$ stay there for all time $t \in \mathbf{T}_0$ or diverge to infinity in finite time, we adopt the following modification (Ikeda and Šiljak, 1980c) of the standard definition (see the proof of Theorem 8.26):

A positive equilibrium $\tilde{x}_e$ of the system $\tilde{\mathbf{S}}$ is said to be globally asymptotically stable with respect to the region $\mathbf{R}_+^{\tilde{n}}$ if:

(*i*) $\tilde{x}_e$ is stable with respect to $\mathbf{R}_+^{\tilde{n}}$, that is, for every $\tilde{\epsilon} > 0$ there exists $\tilde{\delta}(\tilde{\epsilon}) >$ such that $\|\tilde{x}_0 - \tilde{x}_e\| \leq \tilde{\delta}$ and $\tilde{x}_0 \in \mathbf{R}_+^{\tilde{n}}$ imply

$$\|\tilde{x}(t; \tilde{x}_0) - \tilde{x}_e\| \leq \tilde{\epsilon} \qquad \forall t \in \mathbf{T}_0; \tag{8.197}$$

and

(*ii*) $\tilde{x}_e$ is attractive with respect to $\mathbf{R}_+^{\tilde{n}}$, that is, $\tilde{x}_0 \in \mathbf{R}_+^{\tilde{n}}$ implies

$$\lim_{t \to +\infty} \tilde{x}(t; \tilde{x}_0) = \tilde{x}_e \qquad \forall \tilde{x}_0 \in \mathbf{R}_+^{\tilde{n}}. \tag{8.198}$$

Now, we start the proof of the theorem by defining the region

$$\Omega(\alpha) = \{\tilde{x} \in \mathbf{R}_+^{\tilde{n}}\colon \ \tilde{\nu}(\tilde{x}) \leq \alpha\}, \tag{8.199}$$

where $\tilde{\nu}(\tilde{x})$ is the scalar function defined in (8.192), and α is a positive number. We note that $\alpha_1 < \alpha_2$ implies $\Omega(\alpha_1) < \Omega(\alpha_2)$.

To establish (i) of the above definition, we show that for any given $\tilde{\epsilon} > 0$, the choice

$$\tilde{\delta} = \min_{\tilde{\nu}(\tilde{x})=\alpha(\tilde{\epsilon})} \|\tilde{x} - \tilde{x}_e\| \tag{8.200}$$

suffices, where

$$\alpha(\tilde{\epsilon}) = \max_{\|\tilde{x}-\tilde{x}_e\|=\tilde{\epsilon}} \tilde{\nu}(\tilde{x}). \tag{8.201}$$

We define the region

$$\Gamma(\rho) = \{\tilde{x} \in \mathbf{R}_+^{\tilde{n}}\colon\ \|\tilde{x} - \tilde{x}_e\| \leq \rho\}, \tag{8.202}$$

where ρ is a positive number, and note that

$$\Gamma(\tilde{\delta}) \subseteq \Omega[\alpha(\tilde{\epsilon})] \subseteq \Gamma(\tilde{\epsilon}). \tag{8.203}$$

Since $\tilde{W}$ is an M-matrix, from Theorem 2.4 we know that there exists a positive diagonal matrix $\tilde{D}$ such that the matrix $\tilde{G}$ is positive definite. Then, from (8.193), it follows that $\dot{\tilde{\nu}}(\tilde{x})_{(8.188)} < 0$ for all $\tilde{x} \in \mathbf{R}_+^{\tilde{n}} - \{\tilde{x}_e\}$ and, therefore, any solution $\tilde{x}(t;\, \tilde{x}_0)$ that starts in $\Gamma(\tilde{\delta})$ does not leave $\Omega[\alpha(\tilde{\epsilon})]$ and, hence, remains in $\Gamma(\tilde{\epsilon})$ for all $t \in \mathbf{T}_0$. This means that $\tilde{x}_e$ is stable with respect to $\mathbf{R}_+^{\tilde{n}}$.

For part (ii) of the above definition, we use (8.193) to get the inequality

$$\dot{\tilde{\nu}}(\tilde{x})_{(8.188)} \leq -\lambda_m(\tilde{G})\, \|\tilde{x} - \tilde{x}_e\|^2. \tag{8.204}$$

For any positive $\tilde{\epsilon}$, we can choose $\tilde{\delta}$ as in (8.200). Let us assume that the solution $\tilde{x}(t;\, \tilde{x}_0)$ that starts in $\mathbf{R}_+^{\tilde{n}}$ does not enter $\Gamma(\tilde{\delta})$. Then, from (8.204), we get

$$\tilde{\nu}[\tilde{x}(t;\, \tilde{x}_0)] = \tilde{\nu}(\tilde{x}_0) + \int_0^t \dot{\tilde{\nu}}[\tilde{x}(\tau;\, \tilde{x}_0)]\, d\tau \leq \tilde{\nu}(\tilde{x}_0) - \lambda_m(\tilde{G})\tilde{\delta}^2 t. \tag{8.205}$$

From the fact that $\tilde{x}(t;\, \tilde{x}_0) \subset \mathbf{R}_+^{\tilde{n}}$ for all $t \in \mathbf{T}_0$, we conclude from (8.205) that $\tilde{\nu}(x)$ cannot remain nonnegative when $t \to +\infty$, which is a contradiction of the definition (8.192) of $\tilde{\nu}(\tilde{x})$. Therefore, every solution that starts in $\mathbf{R}_+^{\tilde{n}}$ outside $\Gamma(\tilde{\delta})$ must enter $\Gamma(\tilde{\delta})$ in finite time and stay there for all future times. Since $\tilde{\epsilon}$ can be chosen arbitrarily small, we have $\tilde{x}(t;\, \tilde{x}_0) \to \tilde{x}_e$ as $t \to +\infty$. Q.E.D.

At this point it is of interest to present an example that demonstrates the additional flexibility offered by the overlapping decompositions *via* the Inclusion Principle.

8.28. EXAMPLE. Let us consider a third order Lotka–Volterra system

$$\mathbf{S}\colon\ \begin{bmatrix}\dot{x}_1\\ \dot{x}_2\\ \dot{x}_3\end{bmatrix} = \begin{bmatrix}x_1 & 0 & 0\\ 0 & x_2 & 0\\ 0 & 0 & x_3\end{bmatrix}\left(\begin{bmatrix}1\\ 6\\ 1\end{bmatrix} + \begin{bmatrix}-4 & 5 & 0\\ -8 & -12 & -8\\ 0 & 5 & -4\end{bmatrix}\begin{bmatrix}x_1\\ x_2\\ x_3\end{bmatrix}\right), \tag{8.206}$$

which has a positive equilibrium $x_e = (21/64,\ 1/16,\ 21/64)^T$. We start with a disjoint decomposition of $\mathbf{S}$ into two subsystems

$$\mathbf{S}_1\colon\ \begin{bmatrix}\dot{x}_1\\ \dot{x}_2\end{bmatrix} = \begin{bmatrix}x_1 & 0\\ 0 & x_2\end{bmatrix}\left(\begin{bmatrix}1\\ 6\end{bmatrix} + \begin{bmatrix}-4 & 5\\ -8 & -12\end{bmatrix}\begin{bmatrix}x_1\\ x_2\end{bmatrix} + \begin{bmatrix}0\\ -8\end{bmatrix} x_3\right),$$

$$\mathbf{S}_2\colon\ \dot{x}_3 = \left(1 - 4x_3 + \begin{bmatrix}0 & 5\end{bmatrix}\begin{bmatrix}x_1\\ x_2\end{bmatrix}\right). \tag{8.207}$$

With $\mathbf{S}_1$ and $\mathbf{S}_2$, we associate scalar functions (8.189) as

$$\begin{aligned} v_1(x_1,\ x_2) &= d_{11}[x_1 - (21/64) - (21/64)\ \ln(64\,x_1/21)] \\ &\quad + [x_2 - (1/16) - (1/16)\ \ln(16\,x_2)], \\ v_2(x_3) &= x_3 - (21/64) - (21/64)\ \ln(64\,x_3/21), \end{aligned} \tag{8.208}$$

where d_{11} is a positive number. Without loss of generality, we set $d_{12} = d_{21} = 1$ and choose $D_1 = \text{diag}\{d_{11},\ 1\}$, $D_2 = 1$. Matrix W is

$$W = \begin{bmatrix}\lambda_m(G_1) & -8\\ -5 & 4\end{bmatrix}, \tag{8.209}$$

where

$$G_1 = -\frac{1}{2}\left(\begin{bmatrix}-4 & -8\\ 5 & -12\end{bmatrix}\begin{bmatrix}d_{11} & 0\\ 0 & 1\end{bmatrix} + \begin{bmatrix}d_{11} & 0\\ 0 & 1\end{bmatrix}\begin{bmatrix}-4 & 5\\ -8 & -12\end{bmatrix}\right). \tag{8.210}$$

It is easy to show that $\lambda_m(G_1) < 10$ for any $d_{11} > 0$, the matrix W is not an M-matrix, and we fail to establish stability.

Now, we expand $\mathbf{S}$ as

$$
\tilde{\mathbf{S}}\colon\ \begin{bmatrix} \dot{\tilde{x}}_{11} \\ \dot{\tilde{x}}_{12} \\ \dot{\tilde{x}}_{21} \\ \dot{\tilde{x}}_{22} \end{bmatrix} = \left[\begin{array}{cc:cc} \tilde{x}_{11} & 0 & & \\ 0 & \tilde{x}_{12} & \multicolumn{2}{c}{\bigcirc} \\ \hdashline & & \tilde{x}_{21} & 0 \\ \multicolumn{2}{c:}{\bigcirc} & 0 & \tilde{x}_{22} \end{array}\right]
$$

$$
\times \left(\begin{bmatrix} 1 \\ 6 \\ 6 \\ 1 \end{bmatrix} + \left[\begin{array}{cc:cc} -4 & 5 & 0 & 0 \\ -8 & -12 & 0 & -8 \\ \hdashline -8 & 0 & -12 & -8 \\ 0 & 0 & 5 & -4 \end{array}\right] \begin{bmatrix} \tilde{x}_{11} \\ \tilde{x}_{12} \\ \tilde{x}_{21} \\ \tilde{x}_{22} \end{bmatrix} \right), \tag{8.211}
$$

using

$$
V = \begin{bmatrix} 1 & 0 & 0 \\ 0 & 1 & 0 \\ 0 & 1 & 0 \\ 0 & 0 & 1 \end{bmatrix}, \qquad U = \begin{bmatrix} 1 & 0 & 0 & 0 \\ 0 & 1/2 & 1/2 & 0 \\ 0 & 0 & 0 & 1 \end{bmatrix},
$$

$$
M = \begin{bmatrix} 0 & 5/2 & -5/2 & 0 \\ 0 & -6 & 6 & 0 \\ 0 & 6 & -6 & 0 \\ 0 & -5/2 & 5/2 & 0 \end{bmatrix}. \tag{8.212}
$$

$\tilde{\mathbf{S}}$ has a positive equilibrium $\tilde{x}_e = (21/64,\ 1/16,\ 1/16,\ 21/64)^T$ corresponding to the positive equilibrium x_e of the original system $\mathbf{S}$.

Next, we decompose $\tilde{\mathbf{S}}$ into two disjoint subsystems $\tilde{\mathbf{S}}_1$ and $\tilde{\mathbf{S}}_2$, which are indicated by the dashed lines in (8.211). This decomposition corresponds to the overlapping decomposition of the state vector $x = (x_1,\ x_2,\ x_3)^T$ of the original system $\mathbf{S}$ into two state vectors $\tilde{x}_1 = (x_1,\ x_2)^T = (\tilde{x}_{11},\ \tilde{x}_{12})^T$ and $\tilde{x}_2 = (x_2,\ x_3)^T = (\tilde{x}_{21},\ \tilde{x}_{22})^T$ of the subsystems $\tilde{\mathbf{S}}_1$ and $\tilde{\mathbf{S}}_2$, with x_2 as the overlapping part. With this choice of decomposition, the form of V in (8.212) is automatic.

Finally, we choose the scalar functions

$$\tilde{v}_1(x_{11}, x_{12}) = 3[\tilde{x}_{11} - (21/64) - (21/64)\ln(64\tilde{x}_{11}/21)]$$
$$+ [\tilde{x}_{12} - (1/16) - (1/16)\ln(16\tilde{x}_{12})],$$
$$\tilde{v}_2(\tilde{x}_{21}, \tilde{x}_{22}) = [\tilde{x}_{21} - (1/16) - (1/16)\ln(16\tilde{x}_{21})]$$
$$+ 3[\tilde{x}_{22} - (21/64) - (21/64)\ln(64\tilde{x}_{22}/21)], \quad (8.213)$$

and we get the matrix

$$\tilde{W} = \begin{bmatrix} 17/2 & -8 \\ -8 & 17/2 \end{bmatrix}, \quad (8.214)$$

which is an M-matrix, and the equilibrium $\tilde{x}_e$ of $\tilde{\mathbf{S}}$ is globally asymptotically stable with respect to the region $\mathbf{R}^4_+$. Since each initial state $x_0 \in \mathbf{R}^3_+$ of the original system $\mathbf{S}$ is preserved in the region $\mathbf{R}^4_+$ of the state space of $\tilde{\mathbf{S}}$ under the transformation $\tilde{x} = Vx$, we can conclude by Theorem 8.26 that the positive equilibrium x_e of $\mathbf{S}$ is also globally asymptotically stable with respect to the region $\mathbf{R}^3_+$.

8.29. REMARK. Examples used in this chapter suggest that an overlapping decomposition can succeed where disjoint decompositions fail, when critical interconnections can be incorporated inside the overlapping subsystems to enhance stability of the subsystems. In this way, critical interconnections are integral parts of the subsystems and do not appear as interconnections at all; thus, vector Liapunov functions can be used effectively to show stability of the overall system.

8.5. Notes and References

Overlapping decompositions can be traced back to the work of H. A. Schwarz (1890), where he showed that the solutions to Laplace's equation with Dirichlet boundary condition, which are obtained on the two overlapping subregions, are equivalent to solutions to the same problem on the union of the subregions. Most interestingly, he showed that the solution procedure, which alternates from one subregion to the other, converges to the solution of the original problem. In the context of this chapter, we defined an overlapping decomposition of a dynamic (iterative) process and established (implicitly) stability of an expanded system. Numerical procedures based upon the Schwarz's alternating procedure have been applied successfully to more general equations and boundary conditions (Miller, 1965; Rodrigue and Saylor, 1985).

A suitable mathematical framework for stability analysis of systems with overlapping subsystems has been initiated in the context of vector Liapunov functions (Šiljak, 1978). The two formulations of the concept, proposed independently by Matrosov (1962) and Bellman (1962), are different in the way the components of a function are defined. While in Bellman's formulation the components are associated with *disjoint* subspaces of the underlying space, Matrosov allows for *overlapping*. The significance of the difference has remained obscure until, in applications of the concept to decentralized control, it has been shown to be essential in dealing with overlapping information sets (Šiljak, 1979). Expansions and contractions of dynamic systems have been formulated for this purpose, and the Inclusion Principle has emerged as a powerful formalization of these ideas. Our presentation of the Principle in this chapter follows closely several papers (Ikeda and Šiljak, 1980a, b, c; Ikeda *et al.* 1981a, b, 1984a, b), where additional results, examples, and applications can be found. The Principle appears as a natural tool for the design of overlapping decentralized control in as diverse fields as economics (Aoki, 1976) and electric power systems (Šiljak, 1978; Ćalović *et al.* 1978; Ćalović, 1986), traffic control (Levine and Athans, 1966; Athans *et al.* 1967; Isaksen and Payne, 1973; Papageorgiou, 1984) and large space structures (Yousuff and Patel, 1987; Yousuff, 1988; Young, 1987).

STOCHASTIC INCLUSION PRINCIPLE

In order to use the decentralized LQG scheme of Chapter 3 for overlapping decompositions, a stochastic version of the Inclusion Principle was formulated in (Hodžić *et al.* 1983). By expanding the original LQG problem into a larger space, the overlapping information sets become disjoint and the expanded LQG problem can be solved by standard decentralized methods. Under the inclusion conditions, which are given in terms of means and covariances of the solution process, the solution of the expanded LQG problem can be contracted for implementation into the original system.

The most important part of this procedure is the fact that the contracted control law satisfies the overlapping information structure constraints.

When decentralized estimators and controllers are to be designed for overlapping subsystems, both aggregations and restrictions need to be used; aggregations for estimators and restrictions for controllers. The entire design process is formulated for discrete systems in (Hodžić and Šiljak, 1986; Salters and Jamshidi, 1986).

REDUCED-ORDER DESIGN

The Inclusion Principle can be used in reduced-order design if we reverse the direction of the design process to be; contract, formulate the control law for the small system, and expand the control law for implementation in the original system. While there is an infinite number of ways we can expand a given system, the number of contractions is finite. This fact makes reduced-order designs different from those of overlapping decentralized control, as explained in deterministic terms by Sezer and Šiljak (1982). The stochastic version of the Inclusion Principle offers a broader scope of model reduction by allowing for "imperfect reductions" in a probabilistic sense.

INPUT–OUTPUT INCLUSION

A considerable increase of flexibility is provided by allowing for expansions and contractions of inputs and outputs. For example, by expanding the input space in a model representing a string of moving vehicles, one can obtain the input matrix in a block-diagonal form, that is, input-decentralized form. This is not possible to achieve without expanding the input space as well. Dynamic output feedback becomes decentralized, too (for details, see Ikeda and Šiljak, 1986; Iftar, 1988; Iftar and Özgüner, 1990).

OVERLAPPING DIAGONAL DOMINANCE

Well-known diagonal dominance conditions (Section 7.5) are reformulated to consider overlapping diagonal blocks (Ohta and Šiljak, 1985). This can be useful in diverse fields such as economics (Pearce, 1974; Okuguchi, 1978) and model ecosystems (May, 1974; Šiljak, 1978; Svirezhev and Logofet, 1983), iterative numerical schemes (Robert, 1969; Hageman and Young, 1981) and computers (Courtois, 1977). We should note that besides overlapping, the diagonal dominance, as defined by Ohta and Šiljak

(1985), provides additional flexibility by normalization, scaling, and alternative norm utilization. A link between this new version of diagonal dominance and vector Liapunov functions is also established.

INCLUSION OF TRANSFER FUNCTIONS

Having formulated a flexible diagonal dominance condition, a natural place to use it is in the control design framework of Rosenbrock (1974). We were able to produce a generalization of the Nyquist array method to blockwise overlapping decompositions (Ohta *et al.* 1986). In this context, it is suitable to synthesize overlapping decentralized controllers *via* the inclusion principle involving transfer functions (see also, Iftar and Özgüner, 1987a).

HEREDITARY SYSTEMS

A more realistic model of AGC in Example 8.22 would be one that includes a time delay in power exchange over the tie-line. This is especially so if the two power areas are geographically far apart. For such a model, we developed an Hereditary Inclusion Principle (Ohta and Šiljak, 1984). The Principle is formulated for nonlinear functional equations and applied to the stability of linear and nonlinear interconnected systems with time-delays and overlapping subsystems. The obtained theorems offer added flexibility even if the subsystems are disjoint, or the system is considered as a whole (Ladde, 1976; Anderson, 1979; Sinha, 1980; Lewis and Anderson, 1980; Shigui, 1988). Decentralized feedback for systems with time delays and overlapping information sets could be developed following the results of Ikeda and Šiljak (1980d).

POWER SYSTEMS

Natural overlapping decompositions appear most prominently in stability analysis of models in electric power engineering. In Example 8.22, the tie-line is a logical overlapping part between two power areas. This fact has been investigated in some detail in (Ćalović *et al.* 1978; Šiljak, 1978; Ikeda, Šiljak, and White, 1981a; Ćalović, 1986). In transient stability analysis, the reference machine is a natural overlapping part; the fact was first noted by Pai and Narayana (1975), and later used in a wide variety of models by many people (see, Šiljak, 1978; Araki *et al.* 1982; Ribbens–Pavella and Evans, 1985).

MECHANICAL SYSTEMS

The Inclusion Principle has been applied to standard systems in the Bogolyubov sense, and general results have been obtained relating the stability of the expanded and the original system (Martynyuk, 1984, 1986). The Principle has also been generalized for matrix second order systems, and used to simplify numerical procedures involving finite elements methods (Yousuff and Patel, 1987), as well as controlled component synthesis (Young, 1987). Especially interesting are the results of Yousuff (1988) and Yousuff and Ikeda (1988) pertaining to approximate versions of aggregation and restrictions (see also, Sezer and Šiljak, 1982; Lidner and Babendreier, 1988; Iftar, 1988; Lidner, 1988).

STRING OF MOVING VEHICLES

Control of a string of moving vehicles involves overlapping subsystems (Levine and Athans, 1966). When the system is expanded into a large space, the substring appears as decoupled subsystems. The modes of the expanded system, which are not shared by the corresponding contraction, are precisely the uncontrollable modes of the subsystems. This means that, under the information structure constraint imposed by the overlapping subsystems, we may assign arbitrarily the eigenvalues of the contracted closed-loop system. For details, see (Ikeda *et al.* 1981b; Ikeda and Šiljak, 1986).

Bibliography

Anderson, B. D. O. (1979). Time delays in large scale systems. *Proceedings of the 18th IEEE Conference on Decision and Control*, Fort Lauderdale, Florida, 655–660.

Aoki, M. (1971). Aggregation. *Optimization Methods for Large-Scale Systems with Applications.* D. A. Wismer (ed.). McGraw-Hill, New York, 191–232.

Aoki, M. (1976). On decentralized stabilization and dynamic assignment problems. *Journal of International Economics*, 6, 143–171.

Araki, M., M. M. Metwally, and D. D. Šiljak (1982). Generalized decompositions for transient stability analysis of multimachine power systems. *Large Scale Systems*, 3, 111–122.

Athans, M., M. Levine, and A. H. Levis (1967). A system for the optimal and suboptimal position and velocity control for a string of high-speed vehicles. *Proceedings of the 5th AICA Congress*, Lausanne, Switzerland, 1–15.

Bellman, R. (1962). Vector Lyapunov functions. *SIAM Journal of Control*, 1, 32–34.

Benkherouf, A., and A. Y. Allidina (1987). Sensor fault detection using overlapping decomposition. *Large Scale Systems*, 12, 3–21.

Brucoli, M., M. LaScala, F. Torelli, and M. Trovato (1987). Overlapping decomposition for small disturbance stability analysis of interconnected power networks. *Large Scale Systems*, 13, 115–129.

Ćalović, M. S. (1986). Recent developments in decentralized control of generation and power flows. *Proceedings of the 25th Conference on Decision and Control*, Athens, Greece, 1192–1197.

Ćalović, M., M. Djorović, and D. D. Šiljak (1978). Decentralized approach to automatic generation control. *Proceedings of the International Conference on Large High-Voltage Electric Systems* (CIGRE), Paris, 32:06, 1–12.

Chipman, J. S. (1976). Estimation and aggregation in econometrics: An application of the theory of generalized inverses. *Generalized Inverses and Applications*, M. Z. Nashed (ed.), Academic Press, New York, 549–769.

Courtois, P. J. (1977). *Decomposability*. Academic Press, New York.

Erol, Y., and K. A. Loparo (1982). Decentralized deadbeat control of cascaded production—inventory systems. *IEEE Transactions*, SMC-12, 924–927.

Goh, B. S. (1980). *Management and Analysis of Biological Populations*. Elsevier, Amsterdam, The Netherlands.

Hageman, L. A., and D. M. Young (1981). *Applied Iterative Methods*. Academic Press, New York.

Hassan, M. F., M. I. Younis, and M. A. Sultan (1989). A decentralized controller for cold rolling mills. *Information and Decision Technologies*, 15, 1–31.

Hodžić, M., and D. D. Šiljak (1986). Decentralized estimation and control with overlapping information sets. *IEEE Transactions*, AC-31, 83–86.

Hodžić, M., R. Krtolica, and D. D. Šiljak (1983). A stochastic inclusion principle. *Proceedings of the 22th IEEE Control and Decision Conference*, San Antonio, Texas, 17–22.

Iftar, A. (1988). Robust controller design for large scale systems. *Ph.D. Thesis*. The Ohio State University, Columbus, Ohio.

Iftar, A., and F. Khorrami (1989). A comparison of multiply time-scale analysis and overlapping decomposition. *IEEE Transactions*, SMC-19, 1296–1300.

Iftar, A., and Ü. Özgüner (1984). Closed-loop balanced realizations in the analysis of suboptimality and stability of decentralized systems. *Proceedings of the 23rd IEEE Conference on Decision and Control*, Las Vegas, Nevada, 143–148.

Iftar, A., and Ü. Özgüner (1987a). Decentralized LQG/LTR controller design for interconnected systems. *Proceedings of the American Control Conference*, Minneapolis, Minnesota, 1682–1687.

Iftar, A., and Ü. Özgüner (1987b). Local LQG/LTR controller design for decentralized systems. *IEEE Transactions*, AC-32, 926–930.

Iftar, A., and Ü. Özgüner (1990). Contractible controller design and optimal control with state and input inclusion. *Automatica*, 26, 593–597.

Ikeda, M., and D. D. Šiljak (1980a). Overlapping decompositions, expansions, and contractions of dynamic systems. *Large Scale Systems*, 1, 29–38.

Ikeda, M., and D. D. Šiljak (1980b). Generalized decompositions and stability of nonlinear systems. *Proceedings of the 18th Allerton Conference*, Monticello, Illinois, 726–734.

Ikeda, M., and D. D. Šiljak (1980c). Lotka–Volterra equations: Decomposition, stability, and structure. Part I: Equilibrium analysis. *Journal of Mathematical Biology*, 9, 65–83.

Ikeda, M., and D. D. Šiljak (1980d). Decentralized stabilization of large scale systems with time delay. *Large Scale Systems*, 1, 273–279.

Ikeda, M., and D. D. Šiljak (1981). Generalized decompositions of dynamic systems and vector Liapunov functions. *IEEE Transactions*, AC-26, 1118–1125.

Ikeda, M., and D. D. Šiljak (1982). Lotka–Volterra equations: Decomposition, stability, and structure. Part II: Nonequilibrium Analysis. *Nonlinear Analysis, Theory, Methods, and Applications*, 6, 487–501.

Ikeda, M., and D. D. Šiljak (1986). Overlapping decentralized control with input, state, and output inclusion. *Control-Theory and Advanced Technology*, 2, 155–172.

Ikeda, M., and D. D. Šiljak (1987). Stability of reduced-order models *via* vector Liapunov functions. *Proceedings of the American Control Conference*, Minneapolis, Minnesota, 482–489.

Ikeda, M., D. D. Šiljak, and D. E. White (1981a). Decentralized control with overlapping information sets. *Journal of Optimization Theory and Applications*, 34, 279–310.

Ikeda, M., D. D. Šiljak, and D. E. White (1981b). On decentralized control with overlapping information sets. *Proceedings of the 8th IFAC Congress*, Kyoto, Japan, 9, 32–38.

Ikeda, M., D. D. Šiljak, and D. E. White (1984a). An inclusion principle for dynamic systems. *IEEE Transactions*, AC-29, 244–249.

Ikeda, M., D. D. Šiljak, and D. E. White (1984b). Overlapping decentralized control of linear time-varying systems. *Advances in Large Scale Systems*, vol. 1, J. B. Cruz, Jr. (ed.), JAI Press, Greenwich, Connecticut, 93–116.

Isaksen, L., and H. J. Payne (1973). Suboptimal control of linear systems by augmentation with application to freeway traffic regulation. *IEEE Transactions*, AC-18, 210–219.

Ladde, G. S. (1976). Systems of functional differential inequalities and functional differential systems. *Pacific Journal of Mathematics*, 66, 161–171.

Leela, S. (1984). Large-scale systems, cone-valued Lyapunov functions and quasi-solutions. *Trends in Theory and Practice of Nonlinear Differential Equations*, V. Lakshmikantham (ed.), Marcel Dekker, New York.

Levine, W. S., and M. Athans (1966). On the optimal error regulation of a string of moving vehicles. *IEEE Transactions*, AC-11, 355–361.

Lewis, R. M., and B. D. O. Anderson (1980). Necessary and sufficient conditions for delay-independent stability of linear autonomous systems. *IEEE Transactions*, AC-25, 735–739.

Lidner, D. K. (1988). Near aggregation, the dual GHR and pole-zero cancellation. *International Journal of Control*, 48, 705–727.

Lidner, D. K., and J. Babendreier (1988). A trajectory analysis of near aggregation. *IEEE Transactions*, AC-33, 474–477.

Malinowski, K., and M. G. Singh (1985). Controllability and observability of expanded systems with overlapping decompositions. *Automatica*, 21, 203–208.

Martynyuk, A. A. (1984). The inclusion principle for standard systems. *Dokladi Academii Nauk SSSR*, 276, 34–37.

Martynyuk, A. A. (1986). Expansion of the state space of dynamic systems and stability problems. *Prikladnaya Mekhanika*, 22, 10–25.

Matrosov, V. M. (1962). On the theory of stability of motion. *Prikladnaya Matematika i Mekhanika*, 26, 992–1002.

May, R. M. (1974). *Stability and Complexity in Model Ecosystems*, 2nd ed., Princeton University Press, Princeton, New Jersey.

Miller, K. (1965). Numerical analogs of the Schwarz alternating procedure. *Numerische Mathematik*, 7, 91–103.

Ohta, Y., and D. D. Šiljak (1984). An inclusion principle for hereditary systems. *Journal of Mathematical Analysis and Applications*, 98, 581–598.

Ohta, Y., and D. D. Šiljak (1985). Overlapping block diagonal dominance and existence of Liapunov functions. *Journal of Mathematical Analysis and Applications*, 112, 396–410.

Ohta, Y., D. D. Šiljak, and T. Matsumoto (1986). Decentralized control using quasi-block diagonal dominance of transfer function matrices. *IEEE Transactions*, AC-31, 420–430.

Okuguchi, K. (1978). Matrices with dominant diagonal blocks and economic theory. *Journal of Mathematics Economics*, 5, 43–52.

Pai, M. A., and C. I. Narayana (1975). Stability of large scale power systems. *Proceedings of the 6th IFAC Congress*, Boston, Massachusetts, 31.6:1–10.

Papageorgiou, M. (1984). Multilayer control system design applied to freeway traffic. *IEEE Transactions*, AC-29, 482–490.

Pearce, I. F. (1974). Matrices with dominant diagonal blocks. *Journal of Economic Theory*, 9, 159–170.

Ribbens–Pavella, M., and F. J. Evans (1985). Direct methods for studying dynamics of large-scale electric power systems: A survey. *Automatica*, 21, 1–21.

Robert, F. (1969) Blocs-H-matrices et convergence des methodes iteratives classiques par blocs. *Linear Algebra and Its Applications*, 2, 223–265.

Rodrigue, G., and P. Saylor (1985). Inner–outer iterative methods and numerical Schwarz algorithm, Part II. *Report UCRL-92077-II*, Lawrence Livermore National Laboratory, Livermore, California.

Rosenbrock, H. H. (1974). *Computer-Aided Control System Design*. Academic Press, New York.

Salters, R. E., and M. Jamshidi (1986). On the aggregation of large-scale stochastic systems with multiplicative noise. *Large Scale Systems*, 11, 31–42.

Schwarz, H. A. (1890). *Gesammelte Mathematische Abhandlungen*. Vol. 2, Springer, Berlin, 133–143.

Sezer, M. E., and D. D. Šiljak (1982). Validation of reduced-order models for control systems design. *Journal of Guidance, Control, and Dynamics*, 5, 430–437.

Shigui, R. (1988). Connective stability for large scale systems described by functional differential equations. *IEEE Transactions*, AC-33, 198–200.

Šiljak, D. D. (1978). *Large-Scale Dynamic Systems: Stability and Structure*. North-Holland, New York.

Šiljak, D. D. (1979). Overlapping decentralized control. *Large Scale Systems Engineering Applications*. M. Singh and A. Titli (eds.), North-Holland, Amsterdam, The Netherlands, 145–166.

Šiljak, D. D. (1980). Reliable control using multiple control systems. *International Journal of Control*, 31, 303–329.

Singh, M., M. F. Hassan, Y. L. Chen, and Q. R. Pan (1983). New approach to failure detection in large scale systems. *IEE Proceedings*, Part. D., 130, 243–249.

Sinha, A. S. C. (1980). Lyapunov functions for a class of large-scale systems. *IEEE Transactions*, AC-25, 558–560.

Skelton, R. E. (1988). *Dynamic Systems Control*. Wiley, New York.

Svirezhev, Yu. M., and D. O. Logofet (1983). *Stability of Biological Communities*. Mir, Moscow, USSR.

Wonham, W. M. (1979). *Linear Multivariable Control: A Geometric Approach*. Springer, New York.

Young, K. D. (1987). Distributed finite element modeling and controls. *Report No. UCID-20950-87*, Lawrence Livermore National Laboratory, Livermore, California, 412–455.

Yousuff, A. (1988). Application of inclusion principle to mechanical systems. *Proceedings of the American Control Conference*, Atlanta, Georgia, 1516–1520.

Yousuff, A., and M. Ikeda (1988). Overlapping decomposition and expansion of mechanical systems. *Proceedings of the* 1988 *AIAA Conference on Guidance, Navigation, and Control*, Paper No. 88-4169-CP, 958–963.

Yousuff, A., and N. Patel (1987). Reduction of finite element models employing inclusion principle. *Proceedings of the* 28*th IEEE Conference on Decision and Control*, Los Angeles, California, 486–488.

Chapter 9 Reliable Control

Due to increasing complexity of the present-day technology, reliability of control has become an essential requirement in the design of large systems. It has been recognized that the degree of reliability required in a high-performance design cannot be achieved merely by diligent application of standard control engineering practice, that is, by using high-quality components in otherwise standard optimization techniques. For this reason, there has been a considerable effort to invent new control schemes with some form of built-in reliability enhancement. The major objective has been to synthesize a control structure so that the system performs satisfactorily under faulty conditions.

In the early developments of reliable control, it was demonstrated that decentralized control schemes are superior to their centralized counterparts. By now, there are a considerable number of results discussed throughout this book that show explicitly a high reliability of decentralized control strategies for plants under structural perturbations. The strategies are robust to a wide range of nonlinearities in both the subsystems and their couplings as well. Reliability of decentralized schemes, however, is not satisfactory with regard to perturbations of feedback connections and controller failures. If a local controller breaks down, it is quite likely that the whole system may do the same. Replacement of a faulty controller by a standby, or a disconnection of the corresponding subsystem for the purpose of preventing a system break-down, may be impossible or equally undesirable due to design constraints. From the point of view of reliability theory (*e.g.*, Barlow and Proschan, 1975), decentralized control schemes are *series connections* of controllers and a control configuration is as reliable as the least reliable controller.

"In building complex computing machines, unreliability of components should be overcome not by making the components more reliable, but by

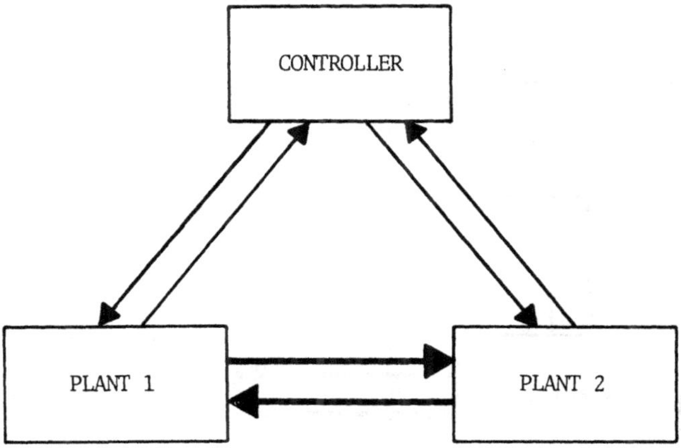

Fig. 9.1. Centralized control.

organizing them in such a way that the reliability of the whole computer is greater than the reliability of its parts," is the von Neumann's dictum for building reliable computers. The dictum was formulated in the context of automata theory (von Neumann, 1956) and was subsequently utilized as a major principle in synthesizing reliable circuits (Moore and Shannon, 1956) and systems (Barlow and Proschan, 1975). It is in the spirit of von Neumann's dictum, but otherwise using entirely different techniques, that *multiple control schemes* have been proposed (Šiljak, 1978b) for the design of reliable control systems. An essential redundancy has been introduced by multiplicity of controllers in a *parallel connection*, allowing for a highly reliable design using less reliable controllers.

9.1. Multiple Control Systems

Let us start by contrasting two basic control structures: centralized structure in Figure 9.1 and decentralized structure in Figure 9.2. In the centralized configuration, one controller monitors two separate but interconnected plants, which may represent two parts of a single large plant. When a decentralized structure is used for the same plant, both "subplants" have a separate controller. A subplant and its controller constitute a subsystem. Although systems appear in much more complex form, the two generic structures can best illustrate the reliability aspect of control.

We distinguish two basic failure modes: plant failures and controller failures. Plant failures can further be classified as failures of individual interconnections among the subplants, or total disconnection of a subplant

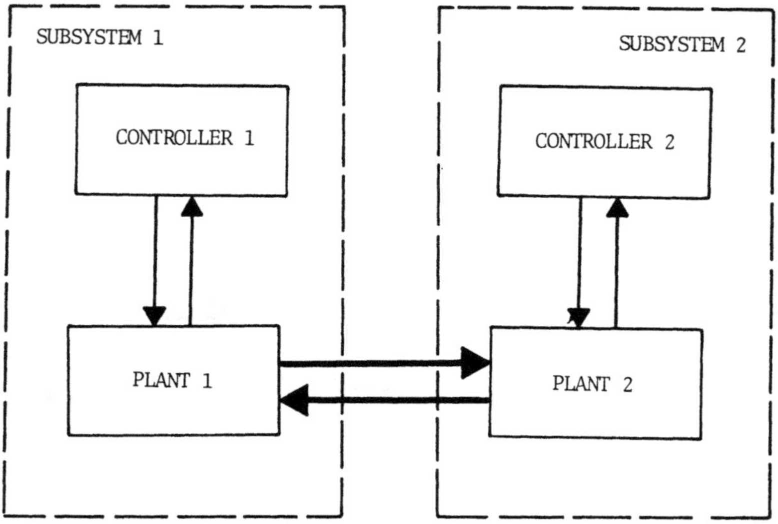

Fig. 9.2. Decentralized control.

from the rest of the system. Structural perturbations due to plant failures are illustrated on Figures 9.3 and 9.4 for centralized and decentralized control structures, respectively. We have demonstrated so far a good understanding of plant failures. For example, it was demonstrated by a simple control system (Šiljak, 1978a) that such a failure can cause instability and, thus, a breakdown of a centralized system. On the other hand, decentralized control schemes described in this book display fault-tolerance to plant and interconnection failures shown in Figure 9.4.

Control failures can also be either partial or total failures, that is, either some control connections between a controller and a plant fail, or an entire controller is disconnected from a subplant. Such disconnections are shown in Figure 9.5 for both centralized and decentralized control structures. None of the two control schemes have shown to be reliable to controller failures. For this reason, *multiple control schemes* were initiated (Šiljak, 1978b), and proved to be promising in design of reliable control.

A way to introduce redundancy in an essential way into control design is to assign more than one controller to a plant. A generic case of a multiple control system is shown in Figure 9.6, where two controllers control a single plant. When a controller failure occurs, as illustrated in Figure 9.7, the other controller is able to carry on the plant without any interruption of the functioning of the overall system. This is what makes the multiple control concept so attractive: Each controller is a *hot spare* for the other.

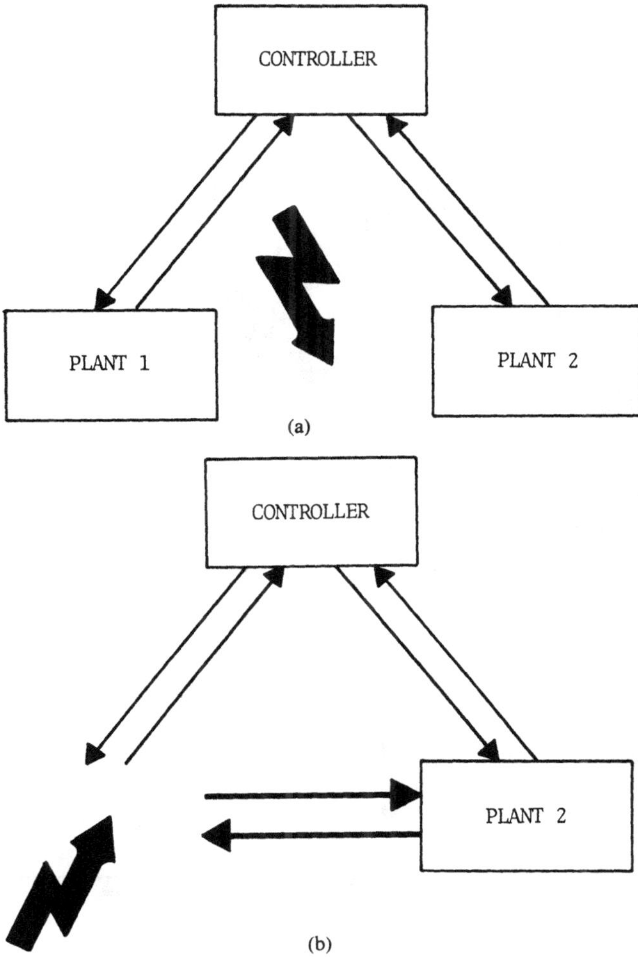

Fig. 9.3. Centralized control: (a) interconnection failure; (b) plant failure.

To show in quantitative terms how the multiple control schemes can improve reliability, we follow the basic concepts of reliability theory (Barlow and Proschan, 1975) and assign to each controller a binary number

$$c_i = \begin{cases} 1, & \text{if the } i\text{th controller is functioning,} \\ 0, & \text{if the } i\text{th controller is failed,,} \end{cases} \tag{9.1}$$

for $i \in \mathbf{M} = \{1, 2, \dots, M\}$, where M is the number of controllers in the system. Assumption (9.1) means that we consider *dichotomic control*

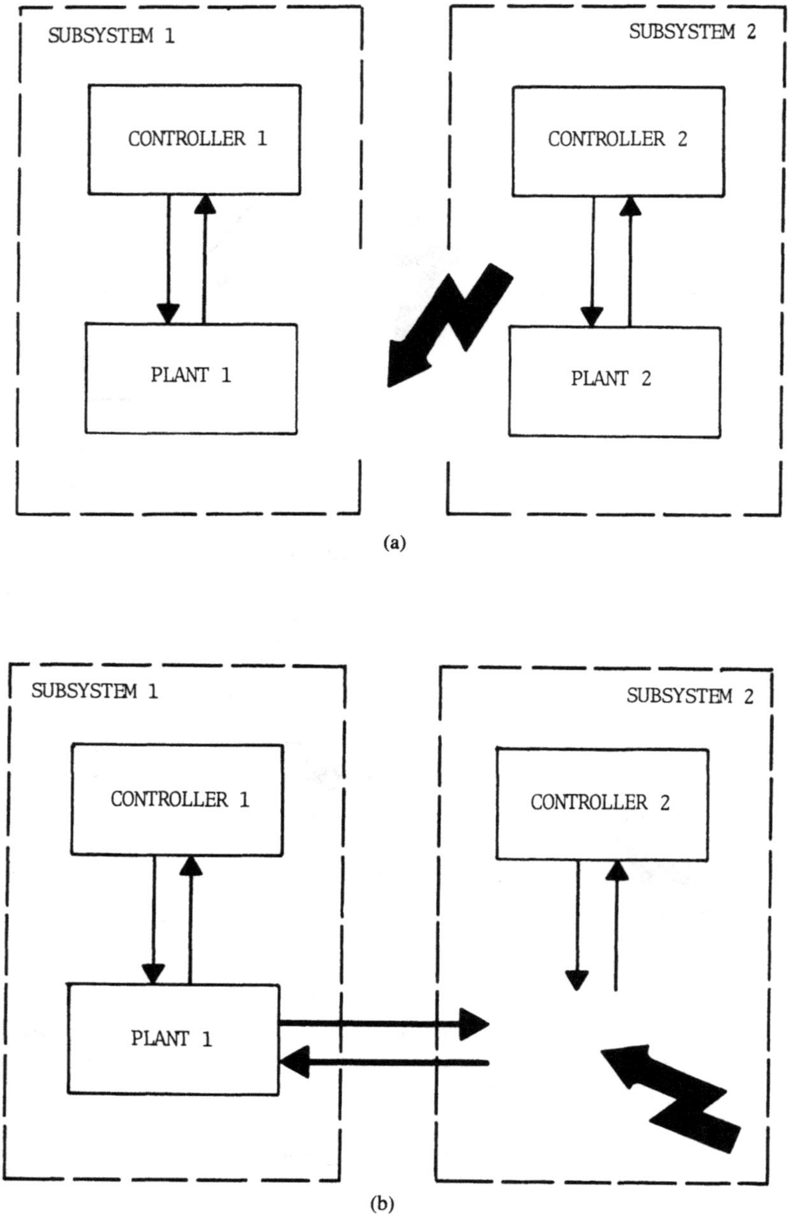

Fig. 9.4. Decentralized control: (a) interconnection failure; (b) plant failure.

structures only, which may appear to be an overly restrictive requirement. It is a well-known fact that controllers are capable of a continuous range of performance quality; from perfect performance to total failure. At present, the assumption (9.1) is a necessary idealization to get us started. It should be noted here that some of our results apply to control structures with partial performance.

With the overall control system, we associate a binary indicator

$$\phi = \begin{cases} 1, & \text{if the control system is functioning,} \\ 0, & \text{if the control system is failed.} \end{cases} \tag{9.2}$$

It is essential to define what we mean by "functioning." If there are no parts of an interconnected plant that are not "covered" by a controller, then we say that the control system is functioning. For example, in the generic case of Figure 9.6, where two on-line controllers $\mathbf{C}_1$ and $\mathbf{C}_2$ simultaneously control a single plant $\mathbf{P}$, there are four possible control configurations as shown in Figure 9.8. The loss of a controller is described by decoupling the corresponding light circle from the black circle representing the plant. We consider the first three structures $\mathbf{C}_1\mathbf{C}_2$, $\not{\mathbf{C}}_1\mathbf{C}_2$, $\mathbf{C}_1\not{\mathbf{C}}_2$ as functioning, while the state $\not{\mathbf{C}}_1\not{\mathbf{C}}_2$ is considered as failure. This is a reasonable assumption because we expect the plant to have unstable elements and, if it is left alone, the system becomes unstable and nonfunctioning. On the other hand, we will show that one can design controllers $\mathbf{C}_1$ and $\mathbf{C}_2$ so that the functioning structures of Figure 9.8 are all stable.

States of the controllers are determined by the vector $c = (c_1, c_2, \dots, c_N)^T$, and the state of the control system is given by the function

$$\phi = \phi(c), \tag{9.3}$$

which is referred to simply as *control structure* ϕ. For example, the system in Figure 9.2 has a *series control* structure

$$\phi(c) = c_1 c_2 = \min\{c_1, c_2\}. \tag{9.4}$$

In case of N controllers in series, the control structure is described as

$$\phi(c) = \prod_{i=1}^{N} c_i = \min\{c_1, c_2, \dots, c_N\}. \tag{9.5}$$

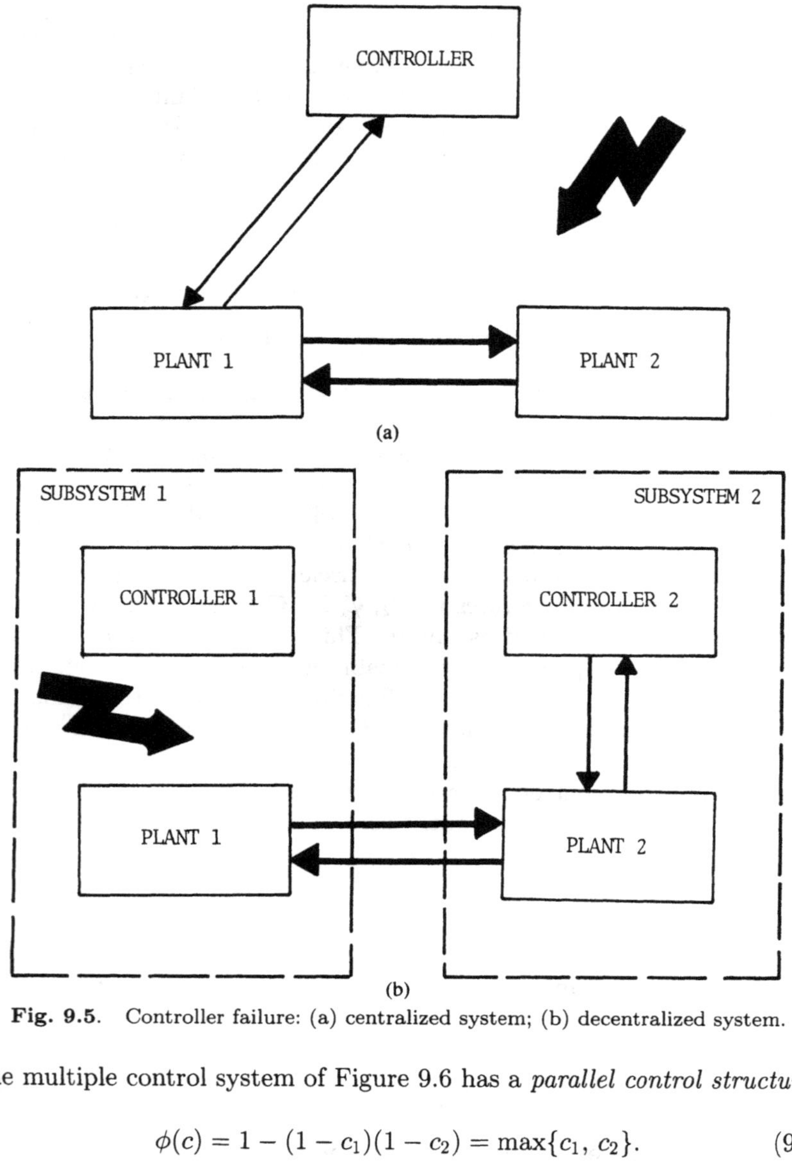

Fig. 9.5. Controller failure: (a) centralized system; (b) decentralized system.

The multiple control system of Figure 9.6 has a *parallel control structure*

$$\phi(c) = 1 - (1 - c_1)(1 - c_2) = \max\{c_1, c_2\}. \tag{9.6}$$

For N controllers in parallel,

$$\phi(c) = 1 - \prod_{i=1}^{N}(1 - c_i) = \max\{c_1, c_2, \dots, c_N\}. \tag{9.7}$$

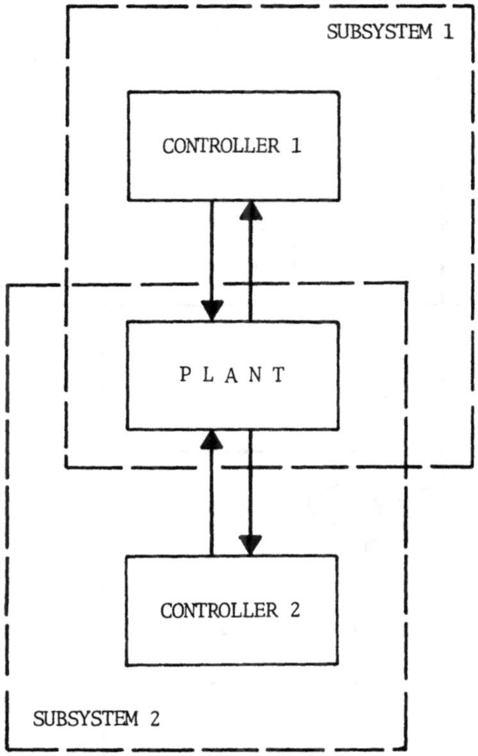

Fig. 9.6. Multiple control system.

A control system can be designed as a combination of the two generic configurations: the decentralized control of Figure 9.2, and the multiple control of Figure 9.6; that is, as a combination of controllers connected in series and parallel arrangements. In this context, the *coherent structures* emerge as natural objects of reliability analysis, since in these structures improvement of reliability on the component level implies a reliability improvement on the overall system level. In the deterministic analysis of control structures, the coherent systems are described by a monotonically increasing function $\phi(c)$. Other properties of coherent structures can be found in (Barlow and Proschan, 1975).

We assume now that all controllers of a given system perform or fail at random and independently of each other, that is, the state c_i of the ith controller is a random variable with the probability distribution

$$P\{c_i = 1\} = p_i = \mathcal{E}c_i, \qquad i \in \boldsymbol{N}, \tag{9.8}$$

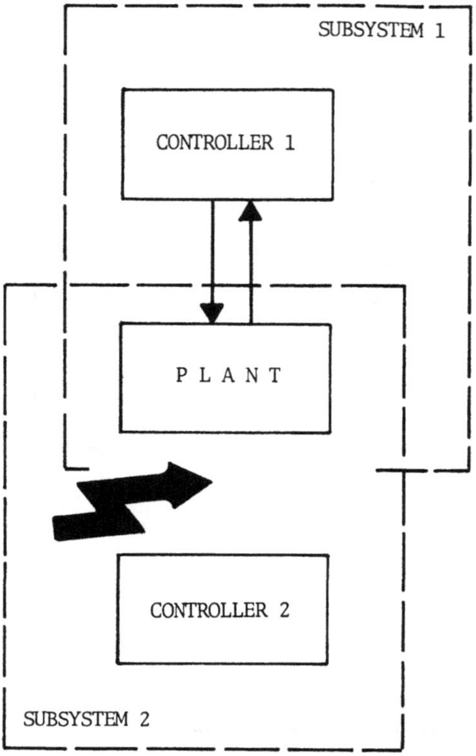

Fig. 9.7. Controller failure.

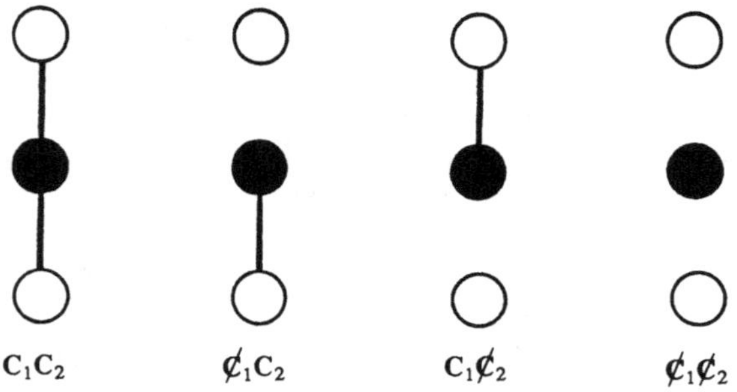

Fig. 9.8. Control structures.

where $\mathcal{E}c_i$ denotes the expected value of the random variable c_i. We refer to p_i as the reliability of the ith controller. The reliability of the overall control system is

$$P\{\phi(c) = 1\} = \chi = \mathcal{E}\phi(c). \tag{9.9}$$

Since the controllers are independent, the system reliability can be represented as a function of controller reliabilities by the *reliability function*

$$\chi = \chi(p), \tag{9.10}$$

where $p = (p_1, p_2, \dots, p_N)^T$.

In Figure 9.9 we show three control structures with the following reliability functions:

$$\begin{aligned} &(a) \quad \text{series structure:} && \chi_a(p) = p^3; \\ &(b) \quad \text{parallel structure:} && \chi_b(p) = 1 - (1-p)^3; \\ &(c) \quad \text{two-out-of-three structure:} && \chi_c(p) = 3p^2(1-p) + p^3. \end{aligned} \tag{9.11}$$

It is assumed that all controllers have the same reliability p. The reliability functions are plotted in Figure 9.10. The S shaped curve (c) for the two-out-of-three structures is most interesting, because it confirms the fact that *we can achieve a desirable reliability of a control system using a sufficient number of low quality controllers.* This is an exact copy of the fact observed by Moore and Shannon (1956) in the context of relay circuits. From curve (c) in Figure 9.10, we see that if the controller reliability p is greater than $1/2$, the reliability function $\chi(p)$ is above the diagonal line $\chi(p) = p$, and the overall structure is more reliable than a single controller.

From Figure 9.10, we see that the purely decentralized structure (a) is least reliable with regard to control failures, but it is the most reliable structure under perturbations in the plant interconnections, that is, under plant failures. Therefore, a combination of decentralized and multiple controllers may be used to account for both interconnection and controller failures. This brings about the problem of optimum redundancy allocation, which should be attempted in the proposed framework using the reliability theory (*e.g.*, Barlow and Proschan, 1975).

There are several reasons why the multiple control configurations are superior to their alternatives. The most important of all is the recent trend of the microprocessors technology, which makes the redundant decentralized controllers an appealing design concept. When a single centralized controller is used instead, duplication is performed on the lower (compo-

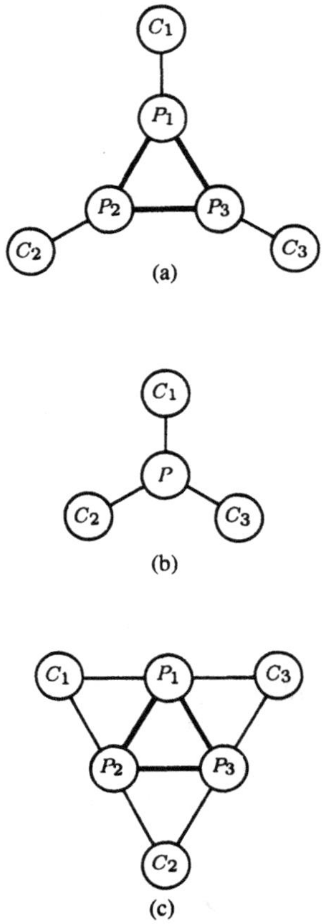

Fig. 9.9. Three-controller structure: (a) series structure; (b) parallel structure; (c) 2-out-3 structure.

nent) level, and the failed component cannot be replaced during operation. When parallel redundancy is achieved using multiple control schemes, the failed controllers can be disconnected, tested, and repaired or replaced without any interfering with the functioning of the overall system. Most of the present microprocessor designs have built-in functions disconnect-test-reconnect in multi-unit configurations.

Another alternative to multiple controllers is the possibility of implementing a reconfiguration of the control structure each time a structural change occurs and is detected by a failure-detection scheme. This solution is usually unattractive because the failure detections and reconfiguration schemes are quite complex themselves, which can decrease to an unknown degree the reliability of the overall system.

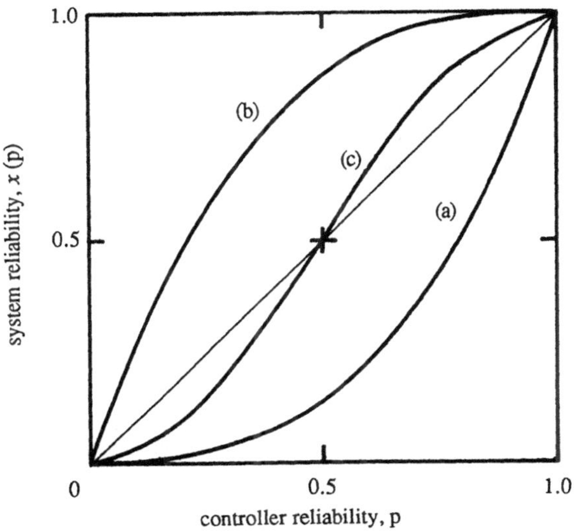

Fig. 9.10. System reliability.

Still another possibility to improve reliability of the overall system is to use identical controllers in a standby redundant configuration. Generally, mechanisms to switch on- and off-line are similarly complex and can fail to perform. Furthermore, in many cases, it cannot be assumed that the off-line controllers are immune to failure until they are activated to on-line status. These and other similar facts are well-known in fault-tolerant computer technology and avionics (Bernard, 1980).

9.2. Reliability of Control Structures

A widely accepted definition of reliability is:

"Reliability is the probability of a device performing its purpose adequately for the period of time intended under the operating conditions encountered." (IEEE Dictionary of Electrical and Electronic Terms).

In order to replicate this definition in the control context, the status of the multiple control system should be described by a Markov process as proposed in (Šiljak, 1981): Each controller failure or recovery results in a transition to a new state that depends only on the current state of the control structure. The states of the control configuration, which correspond

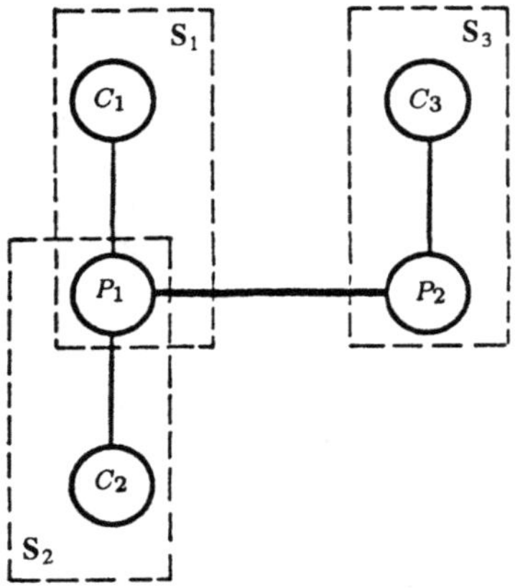

Fig. 9.11. Multiple control system.

to a stable system, are considered as acceptable. The mean time spent by the structure in those states is a measure of system reliability. A Markov process is associated with the dynamics of the overall system *via* connective stability: stability of a multiple control system is established for all of its acceptable states. In this way, the above reliability definition is approximated in a natural way by the new notion of control reliability.

First, our definition of control reliability is given in *probabilistic terms.* Second, the *adequate performance* is stability of the overall system. Third, the "period of time" is specified by the mean time before a failed (unstable) state of the control structure is reached. Fourth, "the operating conditions encountered" are assumed as controller and plant failures. These facts open up a real possibility of using numerous results of reliability theory (Bernard, 1980) in building fault-tolerant control schemes for complex dynamic systems.

Structural configurations arising from controller failures are considered as states of a Markov process **M** (*e.g.*, Feller, 1968) formulated as

$$\mathbf{M}\colon\ \dot{\phi} = \Pi\phi, \tag{9.12}$$

where $\phi(t) \in \mathbf{R}^N$ is the state probability vector at time $t \in \mathbf{R}$, and $\Pi = (\pi_{ij})$ is the constant transition rate $N \times N$ matrix. For example, in the case of the multiple control system of Figure 9.11 with permissible control structures

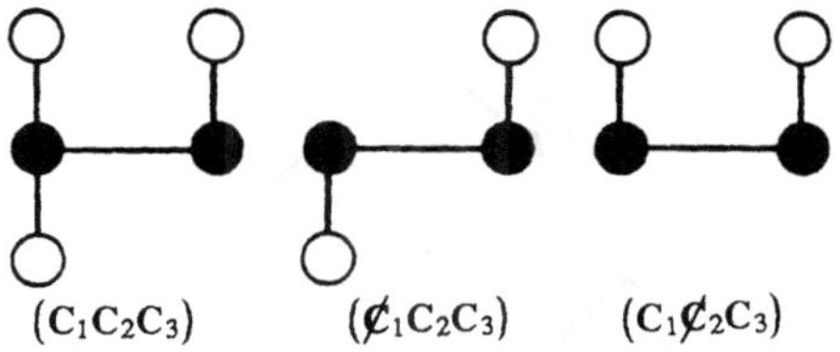

Fig. 9.12. Permissible control structures.

in Figure 9.12, the matrix Π is

$$\Pi = \begin{array}{c} \\ \end{array}\begin{matrix} 1 & 2 & 3 & 4 \\ \left[\begin{matrix} -3\lambda & 0 & 0 & 0 \\ \lambda & -2\lambda & 0 & 0 \\ \lambda & 0 & -2\lambda & 0 \\ \lambda & 2\lambda & 2\lambda & 0 \end{matrix}\right] \end{matrix} \begin{matrix} 1 \\ 2 \\ 3 \\ 4 \end{matrix}, \tag{9.13}$$

where by ϕ_4 we denote the super-state which corresponds to five failed states. We have assumed identical constant failure rate λ for all three controllers which fail independently of each other. We also have assumed no maintenance of the failed controllers so that the failed state ϕ_4 is the *trapping state*. The corresponding transition rate diagram is given in Figure 9.13. If we assume that at time $t = 0$ all controllers are in a working order, the initial conditions are $\phi(0) = (1, 0, 0, 0)^T$. To determine the reliability function $\chi(t)$ of the control structure in Figure 9.11, we solve Equation (9.12) for the specified ϕ_0, and combine the probabilities of being an acceptable state to get

$$R(t) = 2e^{-2\lambda t} - e^{-3\lambda t}, \tag{9.14}$$

which is the probability of the control structure surviving at time t. One of the standard measures of reliability (*e.g.*, Barlow and Proschan, 1975) is the Mean Time To Failure (MTTF),

$$\text{MTTF} = \int_0^\infty R(t)\, dt = \frac{2}{3\lambda}, \tag{9.15}$$

which is the mean time for control system to reach the failure state ϕ_4 of $\mathbf{M}$ starting from the structural state ϕ_1 when all controllers are functioning.

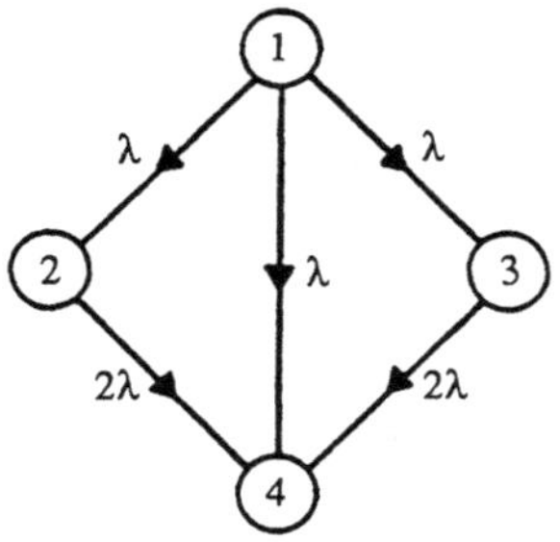

Fig. 9.13. Transition rate diagram.

We note that no repair of controllers has been considered. To include maintenance, all we need is to introduce repair rates of controllers above the main diagonal in Π of (9.13). If the rates are placed in the last column of Π, then the failed state ϕ_4 is no longer a trapping state. Intuitively, this seems an acceptable situation provided that the system does not stay "too long" in the failed state, because the plant runs open-loop, and instability, if present, may destroy the system. This case is considered in Section 9.4.

9.3. Design of Reliable Control

In order to design reliable control using multiple controllers, we should augment the structural reliability considerations with a formulation of control laws that makes the system stable when it has any of the permissible structures. For this, we need the overlapping decompositions and the Inclusion Principle from the preceding chapter.

Let us consider the generic case of the multiple control system in Figure 9.6, and assume that it is described as

$$\begin{aligned} &\text{CONTROLLER 1:} && \dot{x}_1 = A_{11}x_1 + A_{12}x_2 + B_1u_1, \\ &\text{PLANT:} && \dot{x}_2 = A_{21}x_1 + A_{22}x_2 + A_{23}x_3, \qquad (9.16) \\ &\text{CONTROLLER 2:} && \dot{x}_3 = A_{32}x_2 + A_{33}x_3 + B_2u_2, \end{aligned}$$

where $x_2(t) \in \mathbf{R}^{n_2}$, $x_1(t) \in \mathbf{R}^{n_1}$, and $x_3(t) \in \mathbf{R}^{n_3}$ are the states of the plant P and the controllers $\mathbf{C}_1$ and $\mathbf{C}_2$; $u_1(t) \in \mathbf{R}^{m_1}$ and $u_2(t) \in \mathbf{R}^{m_2}$ are the inputs to the controllers; and the matrices in (9.16) are constant and have proper dimensions. We can put (9.16) into a matrix form

$$\mathbf{S}\colon\ \dot{x} = Ax + Bu, \tag{9.17}$$

where $x = (x_1^T,\, x_2^T,\, x_3^T)^T$, and the matrices A and B have the block form

$$A = \left[\begin{array}{c:c:c} A_{11} & A_{12} & 0 \\ \hdashline A_{21} & A_{22} & A_{23} \\ \hdashline 0 & A_{32} & A_{33} \end{array}\right], \qquad B = \left[\begin{array}{c:c} B_1 & 0 \\ \hdashline 0 & 0 \\ \hdashline 0 & B_2 \end{array}\right]. \tag{9.18}$$

The plant is an overlapping part of $\mathbf{S}$. By expanding $\mathbf{S}$ as in the Example 8.15, we get an expansion (Section 8.1),

$$\begin{aligned} \tilde{\mathbf{S}}\colon\ \dot{\tilde{x}}_1 &= \begin{bmatrix} A_{11} & A_{12} \\ A_{21} & A_{22} \end{bmatrix} \tilde{x}_1 + \begin{bmatrix} B_1 \\ 0 \end{bmatrix} u_1 + e_{12} \begin{bmatrix} 0 & 0 \\ 0 & A_{23} \end{bmatrix} \tilde{x}_2, \\ \dot{\tilde{x}}_2 &= \begin{bmatrix} A_{22} & A_{23} \\ A_{32} & A_{33} \end{bmatrix} \tilde{x}_2 + \begin{bmatrix} 0 \\ B_2 \end{bmatrix} u_2 + e_{21} \begin{bmatrix} A_{21} & 0 \\ 0 & 0 \end{bmatrix} \tilde{x}_1, \end{aligned} \tag{9.19}$$

which is an interconnection of two subsystems

$$\begin{aligned} \tilde{\mathbf{S}}_1\colon\ \dot{\tilde{x}}_1 &= \tilde{A}_1 \tilde{x}_1 + \tilde{B}_1 u_1, \\ \tilde{\mathbf{S}}_2\colon\ \dot{\tilde{x}}_2 &= \tilde{A}_2 \tilde{x}_2 + \tilde{B}_2 u_2, \end{aligned} \tag{9.20}$$

with states $\tilde{x}_1 = (x_1^T,\, x_2^T)^T$ and $\tilde{x}_2 = (x_2^T,\, x_3^T)^T$. These two subsystems correspond to the two middle configurations in Figure 9.8 when one of the controllers is failed. The first configuration corresponds to both controllers functioning, and is described by (9.19) as

$$\begin{aligned} \tilde{\mathbf{S}}\colon\ \dot{\tilde{x}}_1 &= \tilde{A}_1 \tilde{x}_1 + \tilde{B}_1 u_1 + e_{12} \tilde{A}_{12} \tilde{x}_2, \\ \dot{\tilde{x}}_2 &= \tilde{A}_2 \tilde{x}_2 + \tilde{B}_2 u_2 + e_{21} \tilde{A}_{21} \tilde{x}_2. \end{aligned} \tag{9.21}$$

In (9.19) and (9.21), we introduced the elements of the 2×2 interconnection matrix

$$E = \begin{bmatrix} 0 & e_{12} \\ e_{21} & 0 \end{bmatrix}, \tag{9.22}$$

which can effectively describe the three permissible structures in Figure 9.8:

$$E_1 = \begin{bmatrix} 0 & 1 \\ 1 & 0 \end{bmatrix}, \qquad E_2 = \begin{bmatrix} 0 & 0 \\ 1 & 0 \end{bmatrix}, \qquad E_3 = \begin{bmatrix} 0 & 1 \\ 0 & 0 \end{bmatrix}. \tag{9.23}$$

Now, reliable control design is reduced to connective stabilization of $\tilde{\mathbf{S}}$ in (9.21) under structural perturbations (9.23), which we know how to do.

We choose the decentralized feedback control law

$$u_1 = -\tilde{K}_1 \tilde{x}_1, \qquad u_2 = -\tilde{K}_2 \tilde{x}_2, \tag{9.24}$$

and get the closed-loop system

$$\tilde{\mathbf{S}}_F\colon\ \dot{\tilde{x}} = \tilde{A}_F(E)\tilde{x}, \tag{9.25}$$

where $\tilde{x} = (\tilde{x}_1^T, \tilde{x}_2^T)^T$ and

$$\tilde{A}_F = \begin{bmatrix} \tilde{A}_1 - \tilde{B}_1\tilde{K}_1 & e_{12}\tilde{A}_{12} \\ e_{21}\tilde{A}_{21} & \tilde{A}_2 - \tilde{B}_2\tilde{K}_2 \end{bmatrix}. \tag{9.26}$$

If $\tilde{\mathbf{S}}_F$ is connectively stable, it means that all three permissible control structures of Figure 9.8 are *simultaneously stable* (Šiljak, 1980).

To be more specific, let us consider a "scalar" version of the multiple control system (9.16).

9.1. EXAMPLE.

$$\begin{aligned} &\text{CONTROLLER 1:} && \dot{x}_1 = u_1, \\ &\text{PLANT:} && \dot{x}_2 = a_{21}x_1 + a_{22}x_2 + a_{23}x_3, \\ &\text{CONTROLLER 2:} && \dot{x}_3 = u_2. \end{aligned} \tag{9.27}$$

The structure of the system is shown in Figure 9.14, where dashed lines indicate the two overlapping subsystems. By defining the state vector $x = (x_1, x_2, x_3)^T$ and the input vector $u = (u_1, u_2)^T$, we rewrite (9.27) as the system

$$\mathbf{S}\colon\ \dot{x} = \left[\begin{array}{cc:c} 0 & 0 & 0 \\ \hdashline a_{21} & a_{22} & a_{23} \\ \hdashline 0 & 0 & 0 \end{array}\right] x + \left[\begin{array}{c:c} 1 & 0 \\ \hdashline 0 & 0 \\ \hdashline 0 & 1 \end{array}\right] u. \tag{9.28}$$

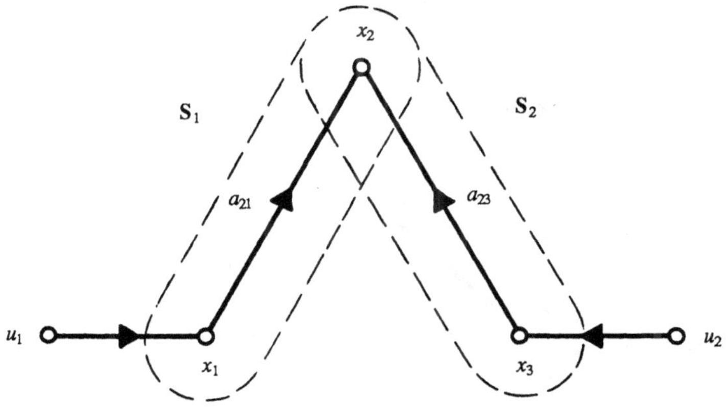

Fig. 9.14. System structure.

By using the transformation

$$\tilde{x} = \begin{bmatrix} 1 & 0 & 0 \\ 0 & 1 & 0 \\ 0 & 1 & 0 \\ 0 & 0 & 1 \end{bmatrix} x, \tag{9.29}$$

we expand the system $\mathbf{S}$ to get

$$\tilde{\mathbf{S}}\colon\ \dot{\tilde{x}} = \left[\begin{array}{cc|cc} 0 & 0 & 0 & 0 \\ a_{21} & a_{22} & 0 & a_{23} \\ \hline a_{21} & 0 & a_{22} & a_{23} \\ 0 & 0 & 0 & 0 \end{array}\right] \tilde{x} + \left[\begin{array}{c|c} 1 & 0 \\ 0 & 0 \\ \hline 0 & 0 \\ 0 & 1 \end{array}\right] u, \tag{9.30}$$

which is composed of two disjoint subsystems (Figure 9.15):

$$\begin{aligned} \tilde{\mathbf{S}}_1\colon\ \dot{\tilde{x}}_1 &= \begin{bmatrix} 0 & 0 \\ a_{21} & a_{22} \end{bmatrix} \tilde{x}_1 + \begin{bmatrix} 1 \\ 0 \end{bmatrix} u_1, \\ \tilde{\mathbf{S}}_2\colon\ \dot{\tilde{x}}_2 &= \begin{bmatrix} a_{22} & a_{23} \\ 0 & 0 \end{bmatrix} \tilde{x}_2 + \begin{bmatrix} 0 \\ 1 \end{bmatrix} u_2. \end{aligned} \tag{9.31}$$

To illustrate simultaneous stabilization, we choose a decentralized control law

$$u_1 = -[\,\tilde{k}_{11} \quad \tilde{k}_{12}\,]\,\tilde{x}_1, \qquad u_2 = -[\,\tilde{k}_{21} \quad \tilde{k}_{22}\,]\,\tilde{x}_2. \tag{9.32}$$

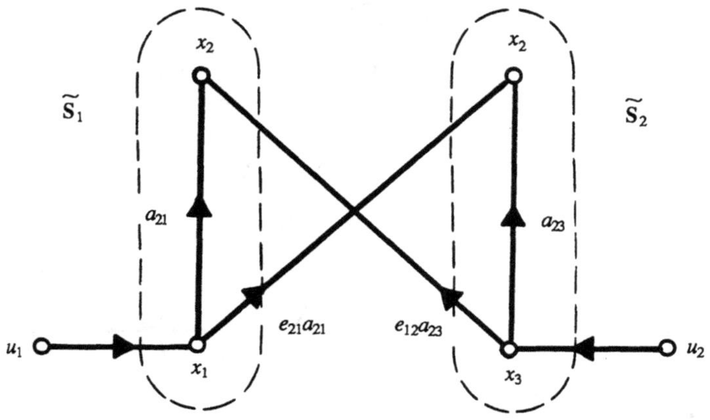

Fig. 9.15. Expansion structure.

The closed-loop system is now described as

$$\tilde{\mathbf{S}}_F\colon \ \dot{\tilde{x}} = \left[\begin{array}{cc|cc} -\tilde{k}_{11} & -\tilde{k}_{12} & 0 & 0 \\ a_{21} & a_{22} & 0 & e_{12}a_{23} \\ \hline e_{21}a_{21} & 0 & a_{22} & a_{23} \\ 0 & 0 & -\tilde{k}_{21} & -\tilde{k}_{22} \end{array}\right] \tilde{x}, \tag{9.33}$$

the structure of which is shown in Figure 9.16. This structure corresponds to $e_{12} = e_{21} = 1$, which is E_1 of (9.23). After any one of the controllers fail, we have E_2 or E_3 in (9.23), which is equivalent to $e_{12} = 0$ or $e_{21} = 0$, that is,

$$\begin{aligned} \tilde{\mathbf{S}}_1^F\colon \ \dot{\tilde{x}}_1 &= \begin{bmatrix} -\tilde{k}_{11} & -\tilde{k}_{12} \\ a_{21} & a_{22} \end{bmatrix} \tilde{x}_1, \\ \tilde{\mathbf{S}}_2^F\colon \ \dot{\tilde{x}}_2 &= \begin{bmatrix} a_{22} & a_{23} \\ -\tilde{k}_{21} & -\tilde{k}_{22} \end{bmatrix} \tilde{x}_2 \end{aligned} \tag{9.34}$$

are decoupled, as shown in Figure 9.17. If we assume for simplicity that $a_{22} = a_{21} = a_{23} = a$, $\tilde{k}_{11} = \tilde{k}_{22} = k_1$, and $\tilde{k}_{12} = \tilde{k}_{21} = k_2$, the characteristic

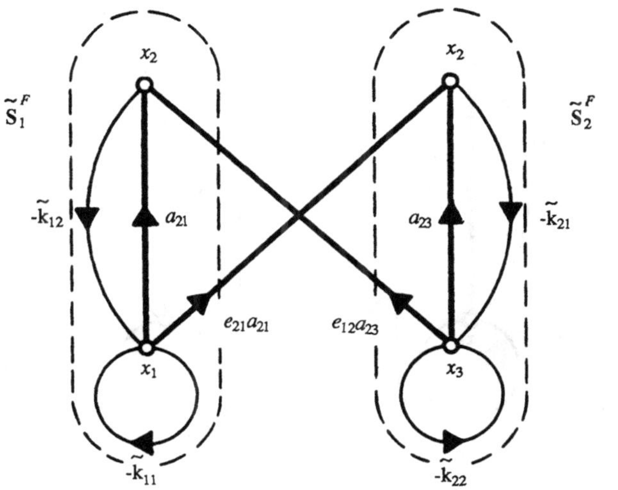

Fig. 9.16. Closed-loop system.

equations become

$$\text{SUBSYSTEMS} \quad \tilde{\mathbf{S}}_1^F, \tilde{\mathbf{S}}_2^F\colon \; s^2 + (k_1 - a)s + (k_2 - k_1)a = 0,$$
$$\text{SYSTEM} \quad \tilde{\mathbf{S}}_F\colon \; (s-a)(s+k_1)[s^2 + (k_1 - a)s + (2k_2 - k_1)a] = 0, \tag{9.35}$$

and the simultaneous stabilization conditions, which are necessary and sufficient, are

$$a < 0, \quad k_1 > 0, \quad k_2 < \frac{1}{2}k_1. \tag{9.36}$$

Under these conditions, the system $\tilde{\mathbf{S}}_F$ is connectively stable.

From the stability conditions (9.36), we conclude that if the plant is unstable ($a > 0$), then $\tilde{\mathbf{S}}_F$ is unstable no matter what we choose for feedback gains k_1 and k_2. The system $\tilde{\mathbf{S}}$ has a fixed mode at $s = a$, and we have the singular case as in Example 8.17. In this case, we have to contract $\tilde{\mathbf{S}}_F$ to $\mathbf{S}_F$ using the transformation

$$x = \begin{bmatrix} 1 & 0 & 0 & 0 \\ 0 & 1/2 & 1/2 & 0 \\ 0 & 0 & 0 & 1 \end{bmatrix} \tilde{x}, \tag{9.37}$$

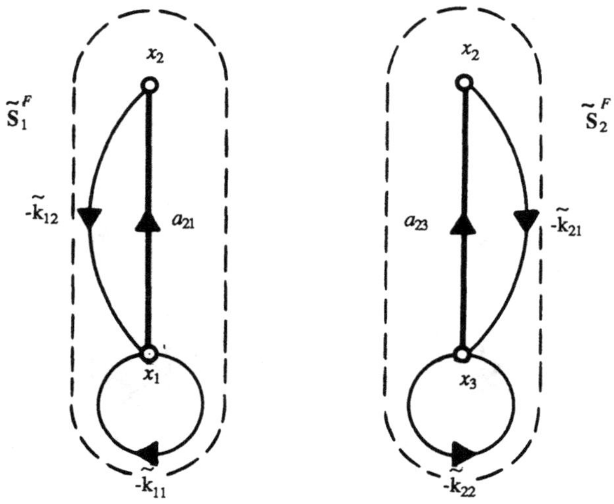

Fig. 9.17. Controller failure.

and obtain

$$\mathbf{S}_F\colon\ \dot{x} = \begin{bmatrix} -k_1 & -k_2 & 0 \\ a & a & a \\ 0 & -k_2 & -k_1 \end{bmatrix} x. \tag{9.38}$$

The characteristic equations are

$$\begin{aligned} &\text{SUBSYSTEMS}\quad \tilde{\mathbf{S}}_1^F,\ \tilde{\mathbf{S}}_2^F\colon\ s^2 + (k_1 - a)s + (k_2 - k_1)a = 0, \\ &\text{SYSTEM}\quad \mathbf{S}_F\colon\ (s + k_1)[s^2 + (k_1 - a)s + (2k_2 - k_1)a] = 0, \end{aligned} \tag{9.39}$$

and $\mathbf{S}_F$, $\tilde{\mathbf{S}}_1^F$, and $\tilde{\mathbf{S}}_2^F$ are simultaneously stable if and only if

$$a < k_1 < 2k_2, \tag{9.40}$$

where it was assumed that $a > 0$.

9.2. REMARK. If, in reliability design, we apply the method of Wang and Davison (1973), we may get a closed-loop system

$$\mathbf{S}_F\colon\ \dot{x} = \left[\begin{array}{cc:c} -k_{11} & -k_{12} & 0 \\ a & a & a \\ \hdashline 0 & 0 & -k_{22} \end{array}\right] x, \tag{9.41}$$

which is stabilized by choosing sequentially the gains k_{11}, k_{12}, and k_{22} for the two blocks in (9.41). It is easy to see that when $a > 0$, the decoupled closed-loop subsystem

$$\tilde{\mathbf{S}}_2^F\colon\ \dot{\tilde{x}}_2 = \begin{bmatrix} a & a \\ 0 & -k_{22} \end{bmatrix} \tilde{x}_2 \tag{9.42}$$

is unstable no matter what we choose for the gain k_{22} and the element a. This means that the method of Wang and Davison (1973) is not suitable when a system is composed of interconnected subsystems and is subject to structural perturbations.

In order to formulate a definition of reliable control, we consider a system

$$\mathbf{S}\colon\ \dot{x} = Ax + Bu, \tag{9.43}$$

with N controllers in a multicontroller configuration. We expand the system $\mathbf{S}$ to get

$$\tilde{\mathbf{S}}\colon\ \dot{\tilde{x}} = \tilde{A}_D\tilde{x} + \tilde{B}_D u + \tilde{A}_C\tilde{x}, \tag{9.44}$$

which is a system composed of N subsystems. In (9.44), $\tilde{x}(t) \in \mathbf{R}^{\tilde{n}}$ and $u(t) \in \mathbf{R}^m$ are the state and input of $\tilde{\mathbf{S}}$ at time $t \in \mathbf{R}$. The matrices are, as usual,

$$\begin{aligned} \tilde{A}_D &= \text{diag}\{\tilde{A}_1, \tilde{A}_2, \ldots, \tilde{A}_N\}, \\ \tilde{B}_D &= \text{diag}\{\tilde{B}_1, \tilde{B}_2, \ldots, \tilde{B}_N\}, \\ \tilde{A}_C &= (e_{ij}A_{ij}), \end{aligned} \tag{9.45}$$

where the blocks have dimensions corresponding to the decomposition $\tilde{x} = (\tilde{x}_1^T, \tilde{x}_2^T, \ldots, \tilde{x}_N^T)^T$.

Finally, we consider a multiple control system as a pair $\boldsymbol{\Sigma} = (\mathbf{S}_F, \mathbf{M})$, where $\mathbf{S}_F$ is a contraction of $\tilde{\mathbf{S}}_F$ and $\mathbf{M}$ is a Markov process defined in (9.12), and state the following:

9.3. DEFINITION. A multiple control system $\boldsymbol{\Sigma}$ is reliable with degree τ if:

(*i*) The process $\mathbf{M}$ has MTTF $> \tau$;
(*ii*) The equilibrium $x = 0$ of $\tilde{\mathbf{S}}_F$ is connectively asymptotically stable in the large.

So far only controller failures have been considered. Plant failures represented by disconnections can be included as well. We show this fact by the following:

9.4. EXAMPLE. Let us reconsider the multiple control system of Figure 9.11, which is described as follows

$$
\begin{aligned}
&\text{CONTROLLER 1:} \quad && \dot{x}_1 = u_1 \quad u_1 = -k_{11}x_1 - k_{12}x_2, \\
&\text{PLANT 1:} \quad && \dot{x}_2 = a_{21}x_1 + a_{22}x_2 + a_{23}x_3 + a_{24}x_4, \\
&\text{CONTROLLER 2:} \quad && \dot{x}_3 = u_2, \quad u_2 = -k_{21}x_2 - k_{22}x_3, \\
&\text{PLANT 2:} \quad && \dot{x}_4 = a_{42}x_2 + u_3, \\
&\text{CONTROLLER 3:} \quad && u_3 = -k_3x_4.
\end{aligned}
\tag{9.46}
$$

The open-loop system $\mathbf{S}$ can be put into a matrix form as

$$
\mathbf{S}\colon\ \dot{x} = \begin{bmatrix} 0 & 0 & 0 & 0 \\ a_{21} & a_{22} & a_{23} & a_{24} \\ 0 & 0 & 0 & 0 \\ 0 & a_{42} & 0 & 0 \end{bmatrix} x + \begin{bmatrix} 1 & 0 & 0 \\ 0 & 0 & 0 \\ 0 & 1 & 0 \\ 0 & 0 & 1 \end{bmatrix} u, \tag{9.47}
$$

where the state is $x = (x_1, x_2, x_3, x_4)^T$ and the input is $u = (u_1, u_2, u_3)^T$.

Choosing the transformation and complementary matrix as

$$V = \begin{bmatrix} 1 & 0 & 0 & 0 \\ 0 & 1 & 0 & 0 \\ 0 & 1 & 0 & 0 \\ 0 & 0 & 1 & 0 \\ 0 & 0 & 0 & 1 \end{bmatrix}, \qquad M = \begin{bmatrix} 0 & 0 & 0 & 0 & 0 \\ 0 & \frac{1}{2}a_{22} & -\frac{1}{2}a_{22} & 0 & 0 \\ 0 & -\frac{1}{2}a_{22} & \frac{1}{2}a_{22} & 0 & 0 \\ 0 & 0 & 0 & 0 & 0 \\ 0 & \frac{1}{2}a_{42} & -\frac{1}{2}a_{42} & 0 & 0 \end{bmatrix}, \tag{9.48}$$

we get the expansion

$$\tilde{\mathbf{S}}\colon\ \dot{\tilde{x}} = \left[\begin{array}{cc:cc:c} 0 & 0 & 0 & 0 & 0 \\ a_{21} & a_{22} & 0 & a_{23} & a_{24} \\ \hdashline a_{21} & 0 & a_{22} & a_{23} & a_{24} \\ 0 & 0 & 0 & 0 & 0 \\ \hdashline 0 & a_{42} & 0 & 0 & 0 \end{array}\right] \tilde{x} + \left[\begin{array}{c:c:c} 1 & 0 & 0 \\ 0 & 0 & 0 \\ \hdashline 0 & 0 & 0 \\ 0 & 1 & 0 \\ \hdashline 0 & 0 & 1 \end{array}\right] u, \tag{9.49}$$

where the state is $\tilde{x} = (x_1,\ x_2,\ x_2,\ x_3,\ x_4)^T$ and input remains the same. The closed-loop system is

$$\tilde{\mathbf{S}}_F\colon\ \dot{\tilde{x}} = \left[\begin{array}{cc:cc:c} -k_{11} & -k_{12} & 0 & 0 & 0 \\ a_{21} & a_{22} & 0 & e_{12}a_{23} & e_{13}a_{24} \\ \hdashline e_{21}a_{21} & 0 & a_{22} & a_{23} & e_{23}a_{24} \\ 0 & 0 & -k_{21} & -k_{22} & 0 \\ \hdashline 0 & e_{31}a_{42} & e_{32}a_{42} & 0 & -k_3 \end{array}\right] \tilde{x}, \tag{9.50}$$

where, for compatibility of (9.50) with (9.47), we impose $e_{31}e_{32} = 0$. If $a_{22} < 0$, then connective stability of $\tilde{\mathbf{S}}_F$ is assured by choosing positive values for the feedback gains k_{11}, k_{12}, k_{21}, k_{22}, and k_3, and by verifying the inequalities

$$k_{11} > k_{12}, \qquad k_{22} > k_{21}, \qquad k_3 > |a_{42}|, \qquad |a_{22}| > |a_{21}| + |a_{23}| + |a_{24}|. \tag{9.51}$$

Since (9.51) implies that the matrix of (9.50) is diagonally dominant [condition (*iii*) of Theorem 2.4 when $d_i = 1$, $i \in \boldsymbol{N}$], the system $\tilde{\mathbf{S}}_F$ is

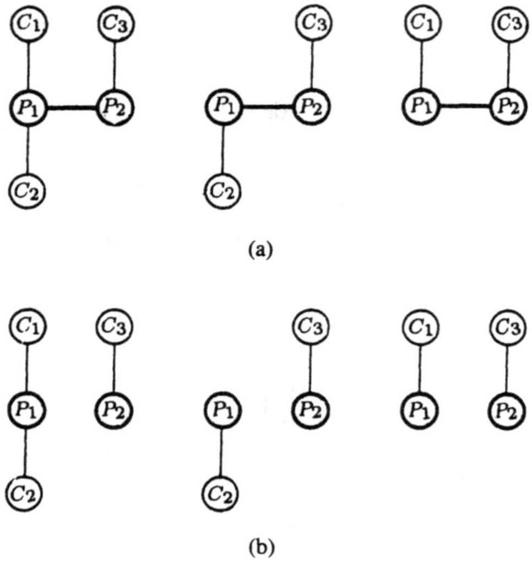

Fig. 9.18. Control configurations: (a) controller failures, (b) interconnection failures.

connectively stable for the fundamental interconnection matrix

$$\bar{E} = \begin{bmatrix} 0 & 1 & 1 \\ 1 & 0 & 0 \\ 1 & 0 & 0 \end{bmatrix}. \tag{9.52}$$

This simply means that all structural configurations shown in Figure 9.18 are all stable, with those in Figure 9.18b being the result of interconnection failures. We also note that the system includes both dynamic ($\mathbf{C}_1$ and $\mathbf{C}_2$) and static ($\mathbf{C}_3$) controllers.

The obvious weakness of Definition 9.3 is that it is suitable only for nonmaintained control systems. This limits the scope of reliable control considerably, and we turn our attention to systems where controllers are repairable after failure.

9.4. Maintained Control Systems

To include maintenance of controllers, we have to alter our definition of control reliability in a crucial way. We have to allow a control system to alternate between operating (stable) and failure (unstable) structures for an indefinite period of time. To capture the effect of these transitions, we have to redefine reliability in the context of stochastic stability, and use the Liapunov-type results of Kats and Krasovskii (1960) to derive conditions for stability under Markovian structural perturbations.

Let us first derive a general result concerning stability of a linear system subject to Markovian perturbations. The most interesting feature of our stability analysis is the fact that we can treat a system in any of the Markov states as a separate subsystem. This opens up the possibility of using the stochastic version (Ladde *et al.* 1974) of vector Liapunov functions and establishing reliability of a control law in the framework of interconnected systems.

A stochastic system $\boldsymbol{\Sigma}$ is described by a pair of equations

$$\begin{aligned} \boldsymbol{\Sigma}\colon \ \dot{x} &= \hat{A}[\eta(t)]\, x \qquad (\hat{\mathbf{S}}), \\ \dot{p} &= \Pi p \qquad\qquad (\mathbf{M}), \end{aligned} \tag{9.53}$$

where $\hat{\mathbf{S}}$ is a linear dynamic system with the state $x \in \mathbf{R}^n$, and $\mathbf{M}$ describes a Markov process $\eta(t)$ with finite number of states E_k, $k \in \boldsymbol{N}$. The system $n \times n$ matrix $\hat{A} = (\hat{a}_{ij})$ is defined by

$$\hat{a}_{ij}(E_k) = e_{ij}(k)\alpha_{ij}, \tag{9.54}$$

where the e_{ij}: $\boldsymbol{N} \to \{0, 1\}$ are elements of a binary $n \times n$ interconnection matrix $E(k)$, and the α_{ij} are given numbers. Each state E_k of the Markov process $\eta(t)$ is associated with an interconnection matrix $E(k)$ which, in turn, represents a certain structure of the dynamic system $\hat{\mathbf{S}}$. In this way, $\mathbf{M}$ describes a *structural dynamics* underlying the dynamic process $\hat{\mathbf{S}}$. In (9.53), $p \in \mathbf{R}^N$ is the probability state vector, and $\Pi = (\pi_{ij})$ is a constant state intensity matrix of appropriate dimension.

Let us first explain the role of $\mathbf{M}$ *via* a matrix

$$\Pi = \begin{array}{c} \begin{array}{cccc} E_1 & E_2 & E_3 & E_4 \end{array} \\ \left[\begin{array}{cccc} -(\lambda_1+\lambda_2) & \mu_1 & \mu_2 & 0 \\ \lambda_1 & -(\lambda_2+\mu_1) & 0 & \mu_2 \\ \lambda_2 & 0 & -(\lambda_1+\mu_2) & \mu_1 \\ 0 & \lambda_2 & \lambda 1 & -(\mu_1+\mu_2) \end{array}\right] \end{array} \begin{array}{c} \\ E_1 \\ E_2 \\ E_3 \\ E_4 \end{array} \tag{9.55}$$

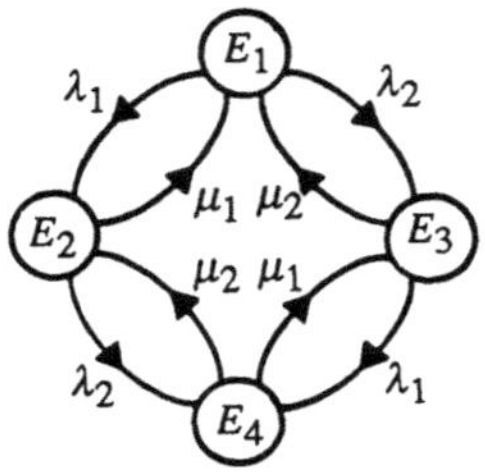

Fig. 9.19. Transition rate diagram.

which corresponds to the transition rate diagram shown in Figure 9.19. Each state E_i of $\mathbf{M}$ corresponds to a distinct linear system $\hat{\mathbf{S}}$ defined in (9.53). The system $\hat{\mathbf{S}}$ is jumping like a frog from one state of $\mathbf{M}$ to another according to the rates λ_i and μ_i. At each state, the dynamics of $\hat{\mathbf{S}}$ evolves as determined by the system defined at that state. Intuitively, the longer the system $\hat{\mathbf{S}}$ resides at "stable" states of $\mathbf{M}$, the more tendency it has to stay stable overall. The question is how long is long enough?

9.5. DEFINITION. A system $\boldsymbol{\Sigma} = (\hat{\mathbf{S}}, \mathbf{M})$ is said to be connectively stable if the equilibrium $x = 0$ of $\hat{\mathbf{S}}$ is globally asymptotically stable in the mean.

To derive conditions for stability of $\hat{\mathbf{S}}$, we introduce a positive definite matrix $H(E_k)$ and a matrix $G(E_k)$ satisfying the equation

$$\bar{A}^T(E_k)H(E_k) + H(E_k)\bar{A}(E_k) = -G(E_k), \qquad k \in \boldsymbol{N}, \tag{9.56}$$

where

$$\bar{A}(E_k) = \hat{A}(E_k) - \frac{1}{2}\sum_{\substack{\ell=1 \\ \ell \neq k}}^{N} \pi_{\ell k} I_n, \tag{9.57}$$

where I_n is the $n \times n$ identity matrix. We also define an $nN \times nN$ matrix

$$G = \begin{bmatrix} G(E_1) & -\pi_{12}H(E_1) & \dots & -\pi_{1N}H(E_1) \\ -\pi_{21}H(E_2) & G(E_2) & \dots & -\pi_{2N}H(E_2) \\ \dots & \dots & \dots & \dots \\ -\pi_{N1}H(E_N) & -\pi_{N2}H(E_N) & \dots & G(E_N) \end{bmatrix}, \tag{9.58}$$

and prove the following (Ladde and Šiljak, 1983):

9.6. THEOREM. A system $\boldsymbol{\Sigma} = (\hat{\mathbf{S}}, \mathbf{M})$ is connectively stable if G is positive definite.

Proof. As a candidate for a Liapunov function, we choose $V\colon \mathbf{R} \times \mathbf{R}^n \to \mathbf{R}_+$ defined as

$$V[x, \eta(t)] = x^T H[\eta(t)]x, \tag{9.59}$$

where $H[\eta(t)]$ is an $n \times n$ symmetric positive definite matrix for all E_k, $k \in \boldsymbol{N}$, of $\mathbf{M}$. We define

$$D^+\mathcal{E}\{V[x, \eta(t)]\} = \limsup_{h \to 0+} \frac{1}{h}(\mathcal{E}\{V[x+hA[\eta(t)]x, \eta(t+h)]|\eta(t)\} - V[x, \eta(t)]). \tag{9.60}$$

Using (9.57) and (9.58), and choosing

$$H = \sum_{k=1}^{N} H(E_k), \tag{9.61}$$

we compute

$$\begin{aligned} D^+\mathcal{E}\{V[x, \eta(t)]\} &= D^+\mathcal{E}\left\{\sum_{k=1}^{N} x^T H(E_k)x\right\} \\ &\leq \sum_{k=1}^{N}\left\{x^T[\hat{A}^T(E_k)H(E_k) + H(E_k)\hat{A}(E_k)]x\right. \\ &\qquad \left. + \sum_{\substack{\ell=1 \\ \ell \neq k}}^{N} \pi_{\ell k}[x^T H(E_\ell)x - x^T H(E_k)x]\right\} \\ &\leq -\sum_{k=1}^{N}\left\{x^T G(E_k)x - \sum_{\substack{\ell=1 \\ \ell \neq k}}^{N} \pi_{\ell k} x^T H(E_\ell)x\right\} \\ &= -X^T G X, \qquad \forall X \in \mathbf{R}^{nN}. \end{aligned} \tag{9.62}$$

Positive definiteness of G in (9.62) and Theorem 4.1 of Ladde and Lakshmikantham (1980) imply global asymptotic stability in the mean of the equilibrium $x = 0$ of $\hat{\mathbf{S}}$ and, thus, connective stability of $\boldsymbol{\Sigma}$. Q.E.D.

Testing positive definiteness of the matrix G in (9.58) may be difficult due to the high dimension nN of G. For this reason, an alternative test can be derived using the concept of vector Liapunov functions (Section 2.1). To show this, we introduce an $N \times N$ matrix $W = (w_{k\ell})$ defined as

$$w_{k\ell} = \begin{cases} -\lambda_m\{G(E_k)H^{-1}(E_k)\}, & k = \ell, \\ \pi_{k\ell}, & k \neq \ell, \end{cases} \tag{9.63}$$

and say that W is row-sum dominant diagonal if

$$\lambda_m\{G(E_\ell)H^{-1}(E_\ell)\} > \sum_{\substack{k=1 \\ k\neq\ell}}^{N} \pi_{\ell k}, \qquad \forall \ell \in \boldsymbol{N}, \tag{9.64}$$

where $\lambda_m(\bullet)$ denotes the minimum eigenvalue.

9.7. THEOREM. A system $\boldsymbol{\Sigma} = (\hat{\mathbf{S}}, \mathbf{M})$ is connectively stable if the matrix W is row-sum dominant diagonal.

Proof. From the second inequality in (9.62), we get

$$\begin{aligned} D^+\mathcal{E}\{V[x, \eta(t)]\} &\leq -\sum_{k=1}^{N} \left[\lambda_m\{G(E_k)H^{-1}(E_k)\}v(x, k) - \sum_{\substack{\ell=1 \\ \ell\neq k}}^{N} \pi_{\ell k}v(x, \ell) \right] \\ &= -\sum_{\ell=1}^{N} \left[\lambda_m\{G(E_\ell)H^{-1}(E_\ell)\} - \sum_{\substack{k=1 \\ k\neq 1}}^{N} \pi_{\ell k} \right] v(x, \ell), \\ & \qquad \forall x \in \mathbf{R}^n, \end{aligned} \tag{9.65}$$

where $v(x, k) = x^T H(E_k)x$ is considered as the kth component $v_k = v(x, k)$ of a vector Liapunov function $v = (v_1, v_2, \ldots, v_N)^T$. Now, if the matrix W of (9.63) is row-sum dominant diagonal, then, from Theorem 4.1 of Ladde and Lakshmikantham (1980), we conclude global asymptotic stability in the mean of the equilibrium $x = 0$ of $\hat{\mathbf{S}}$. Q.E.D.

This general result is what we need in the analysis of multiple control schemes with controller repairs. We illustrate this fact by the following:

9.8. EXAMPLE. Let us consider again the generic case of Figure 9.6. The system $\hat{\mathbf{S}}$ is described as in (9.27):

$$\begin{aligned} &\text{CONTROLLER 1:} && \dot{x}_1 = u_1, \\ &\text{PLANT:} && \dot{x}_2 = 1.5x_1 + 1.5x_3, \\ &\text{CONTROLLER 2:} && \dot{x}_3 = u_2, \end{aligned} \tag{9.66}$$

with $a_{21} = a_{23} = 1.5$, and $a_{22} = 0$. That is,

$$\mathbf{S}\colon\ \dot{x} = Ax + Bu, \tag{9.43}$$

where

$$A = \begin{bmatrix} 0 & 0 & 0 \\ 1.5 & 0 & 1.5 \\ 0 & 0 & 0 \end{bmatrix}, \qquad B = \begin{bmatrix} 1 & 0 \\ 0 & 0 \\ 0 & 1 \end{bmatrix}. \tag{9.67}$$

We choose the overlapping decentralized control as before:

$$u_1 = -k_1x_1 - k_2x_2, \qquad u_2 = -k_2x_2 - k_1x_1, \tag{9.68}$$

and get the closed-loop system as

$$\hat{\mathbf{S}}\colon\ \dot{x} = \begin{bmatrix} -k_1 & -k_2 & 0 \\ 1.5 & 0 & 1.5 \\ 0 & -k_2 & -k_1 \end{bmatrix} x. \tag{9.69}$$

Now, because $a_{22} = 0$, we have the singular case. This means that the structure $\not{C}_1\not{C}_2$ in Figure 9.8 is unstable and represents the failure state. We assume, however, that the controllers can be repaired and the failure state is not a trapping state. The structural dynamics is described by the transition rate diagram of Figure 9.19, that is, by the state intensity matrix in (9.55), where λ_1 and λ_2 are the failure rates of $\mathbf{C}_1$ and $\mathbf{C}_2$, and μ_1 and μ_2 denote the repair rates of the same controllers. The four system matrices corresponding to the four states E_i, $i = 1, 2, 3, 4$, of the Markov process are

$$\bar{A}_1 = \begin{bmatrix} -k_1 - \lambda & -k_2 & 0 \\ 1.5 & -\lambda & 1.5 \\ 0 & -k_2 & -k_1 - \lambda \end{bmatrix},$$

$$\bar{A}_2 = \begin{bmatrix} -k_1 - \frac{1}{2}(\lambda + \mu) & 0 & 0 \\ 0 & -\frac{1}{2}(\lambda + \mu) & 1.5 \\ 0 & -k_2 & -k_1 - \frac{1}{2}(\lambda + \mu) \end{bmatrix},$$

$$\bar{A}_3 = \begin{bmatrix} -k_1 - \frac{1}{2}(\lambda + \mu) & -k_2 & 0 \\ 1.5 & -\frac{1}{2}(\lambda + \mu) & 0 \\ 0 & 0 & -k_1 - \frac{1}{2}(\lambda + \mu) \end{bmatrix}, \tag{9.70}$$

$$\bar{A}_4 = \begin{bmatrix} -k_1 - \mu & & 0 \\ 0 & -\mu & 0 \\ 0 & & -k_1 - \mu \end{bmatrix},$$

where $\lambda_1 = \lambda_2 = \lambda$ and $\mu_1 = \mu_2 = \mu$.

The most remarkable fact is that $\bar{A}_4$ in (9.70) is stable due to maintenance of controllers ($\mu \neq 0$). Furthermore, in order to satisfy stability conditions (9.64), the repair rate μ should be greater than the failure rate λ, that is, $\mu > \lambda$. If $\lambda = 0.4$ and $\mu = 0.55$, then with $k_1 = 2.85$ and $k_2 = 0.33$ the condition (9.64) can be satisfied by a proper choice of the matrices $G(E_k)$. This choice is made as in Section 2.2. We use a suitable transformation to get $\lambda_m\{\tilde{G}(E_k)\tilde{H}^{-1}(E_k)\} = 2\sigma_M(\bar{A}_k)$, which is the maximum value that $\lambda_m\{G(E_k)H^{-1}(E_k)\}$ can achieve. Upon returning to the original space, we note that $\lambda_m\{G(E_k)H^{-1}(E_k)\} = \lambda_m\{\tilde{G}(E_k)\tilde{H}^{-1}(E_k)\}$, which is the largest possible value for the diagonal elements in the inequalities (9.64). This choice satisfies the inequalities, and the stochastic system $\boldsymbol{\Sigma} = (\hat{\mathbf{S}}, \mathbf{M})$ has the equilibrium $x = 0$ which is globally asymptotically stable in the mean.

It is of interest to recognize the important fact that Theorems 9.6 and 9.7 are general and can be applied to any linear control system with closed-loop matrix $\hat{A}(E_k)$, where E_k represent sensor or actuator failures. In other words, we do not have to choose decentralized control (9.68) in Example 9.8—any linear feedback could be tried. Multiple controllers, however, allow for a decentralized reliability design of control that is in tune with microprocessor technology and modern fail-safe concepts in computer and system engineering.

9.5. Notes and References

Stability analysis of dynamic systems containing random parameters has been initiated by Kats and Krasovskii (1960) in the context of Liapunov's direct method. Parameter variations were represented by a Markov process, which provided a suitable framework for reliability design of control systems with random structures (Krasovskii and Lidskii, 1961). The design was performed with respect to the expected value of a quadratic performance index. Wonham (1968) and Sworder (1976) used the Riccati-type approach of Kalman to come up with the optimal control law.

Inspired by von Neumann's approach to building a reliable computer, a multiple control system was invented (Šiljak, 1978b) in the context of overlapping decentralized control. Multiplicity of controllers provided the necessary redundancy and, at the same time, eliminated the need for reconfiguration of control each time the system enters another state of the Markov process which was a drawback of the early Riccati-type designs. By considering the dynamic system at each state as a separate subsystem of a large system, the concept of vector Liapunov functions was shown (Ladde and Šiljak, 1983) to be suitable for ensuring stability of the overall control system. The use of several Liapunov functions was already suggested by Kats (1964) in a somewhat simpler problem.

Recently, nonswitching optimal control laws have been proposed (Mariton and Bertrand, 1986) for multiple control systems with and without measurable states or modes. The numerical problem is to find a solution of interconnected Liapunov and Riccati equations, which is not easy, but can be done using homotopy algorithms (Mariton and Bertrand, 1985). Estimation can be added (Mariton, 1987a, b), but the estimator and control must be designed simultaneously, which makes the numerical problem twice as hard. For further discussion on the subject, including fault detection and isolation schemes with reconfiguration control devices, see the report and book by Mariton (1988b, 1990).

Finally, we should note that reliability design using multiple controllers (Šiljak, 1978a) is essentially a dual of the *simultaneous stabilization problem* (Vidysagar and Viswanadham, 1982; Saeks and Murray, 1982). Instead of having a multiplicity of controllers that are supposed to stabilize a given plant, one is interested in determining if a single controller exists, which stabilizes simultaneously each member of a family of plants. The reason is that the plants are various versions of a single plant undergoing structural perturbations. If one thinks of a set of decentralized controllers as constituting a single controller, then connective stabilization of Chapter 2, which was initiated back in the early 1970s, is essentially the same concept. From this circle of ideas, a purely deterministic design of reliable control has evolved (Vidyasagar and Viswanadham, 1985), whereby two controllers can be found that stabilize a single plant individually and acting together. All that is required is a strongly stabilizable plant. Initial results in these directions were obtained by Youla *et al.* (1976), who provided an important characterization of *all* controllers that stabilize a given plant. For a review of this type of results, see the books by Vidyasagar (1985) and Viswanadham *et al.* (1987).

Bibliography

Barlow, R. E., and F. Proschan (1975). *Statistical Theory of Reliability and Life Testing.* Holt, Rinehart, and Winston, New York.

Bernard, R. (1980). The 'no-downtime' computer. *Spectrum*, 17, 33–37.

Cho, Y. J., and Z. Bien (1989). Reliable control *via* an additive redundant controller. *International Journal of Control*, 50, 385–398.

Feller, W. (1968). *An Introduction to Probability Theory and Its Applications.* Vol. I, *3rd* edition, Wiley, New York.

Kats, I. Ya. (1964). On the stability of stochastic systems in the large. *Prikladnaya Matematika i Mekhanika*, 28, 366–372.

Kats, I. Ya., and N. N. Krasovskii (1960). On the stability of systems with random parameters. *Prikladnaya Matematika i Mekhanika*, 24, 809–823.

Krasovskii, N. N., and E. A. Lidskii (1961). Analytical design of controllers in systems with random attributes. *Avtomatika i Telemekhanika*, 22, 1021–1025, 1141–1146, and 1289–1294.

Krtolica, R. (1984). A singular perturbation model of reliability in system control. *Automatica*, 20, 51–57.

Ladde, G. S., and V. Lakshmikantham (1980). *Random Differential Inequalities.* Academic Press, New York.

Ladde, G. S., and D. D. Šiljak (1983). Multiplex control systems: stochastic stability and dynamic reliability. *International Journal of Control*, 38, 515–524.

Ladde, G. S., and D. D. Šiljak (1989). Convergence and stability of distributed stochastic pseudogradient algorithms. *IEEE Transactions*, AC-35, 665–672.

Ladde, G. S., V. Lakshmikantham, and P. T. Liu (1974). Differential inequalities and stability and boundedness of stochastic differential equations. *Journal of Mathematical Analysis and Applications*, 48, 341–352.

Mariton, M. (1987a). Joint estimation and control of jump linear systems with multiplicative noises. *ASME Transactions, Journal on Dynamics, Measurement, and Control*, 109, 24–28.

Mariton, M. (1987b). Jump linear quadratic control with random state discontinuities. *Automatica*, 23, 237–240.

Mariton, M. (1988a). Almost sure and moments stability of jump linear systems. *Systems and Control Letters*, 11, 393–397.

Mariton, M. (1988b). Control of nonlinear systems with Markovian parameter changes. *Laboratoire des Signaux et Systemes*, CNRS-ESE, Gif-sur-Yvette, Report No. 86-1391.

Mariton, M. (1989). On systems with non-Markovian regime changes. *IEEE Transactions*, AC-34, 346–349.

Mariton, M. (1990). *Jump Linear Systems in Automatic Control.* Marcel Dekker, New York.

Mariton, M., and P. Bertrand (1985). A homotopy algorithm for solving coupled Riccati equations. *Optimal Control Applications and Methods*, 6, 351–357.

Mariton, M., and P. Bertrand (1986). Improved multiplex control systems: dynamic reliability and stochastic optimality. *International Journal of Control*, 44, 219–234.

Moore, E. F., and C. E. Shannon (1956). Reliable circuits using less reliable relays. *Journal of the Franklin Institute*, 262, 191–208 and 281–297.

Mori, K., K. Sano, and H. Ihara (1981). Autonomous controllability of decentralized system aiming at fault-tolerance. *Proceedings of the 8th IFAC Congress*, Kyoto, Japan, 12, 129–134.

Pakshin, P. V. (1983). Stability of discrete systems with random structure under steadily acting disturbances. *Avtomatica i Telemekhanika*, 44, 74–84.

Saeks, R., and J. Murray (1982). Fractional representation, algebraic geometry and simultaneous stabilization problem. *IEEE Transactions*, AC-27, 895–903.

Sandler, G. H. (1963). *System Reliability Engineering.* Prentice-Hall, Englewood Cliffs, New Jersey.

Šiljak, D. D. (1978a). *Large-Scale Dynamic Systems: Stability and Structure.* North-Holland, New York.

Šiljak, D. D. (1978b). Dynamic reliability using multiple control systems. *Proceedings of the 2nd Lawrence Symposium on Systems and Decision Sciences*, Berkeley, California, 173–187.

Šiljak, D. D. (1980). Reliable control using multiple control systems. *International Journal of Control*, 31, 303–329.

Šiljak, D. D. (1981). Dynamic reliability of multiplex control systems. *Proceedings of the 8th IFAC Congress*, Kyoto, Japan, 12, 110–115.

Šiljak, D. D. (1987). Reliability of Control. *Systems and Control Encyclopedia*, M. G. Singh (ed.), Pergamon Press, Oxford, UK, 4008–4011.

Sworder, D. D. (1976). Control of systems subject to sudden change in character. *Proceedings of the IEEE*, 64, 1219–1225.

Tsui, C. C. (1989). On the solution to the state failure detection problem. *IEEE Transactions*, AC-34, 1017–1018.

Vidyasagar, M. (1985). *Control System Synthesis: A Factorization Approach.* MIT Press, Cambridge, Massachusetts.

Vidyasagar, M., and N. Viswanadham (1982). Algebraic design techniques for reliable stabilization. *IEEE Transactions*, AC-27, 1085–1095.

Vidysagar, M., and N. Viswanadham (1985). Reliable stabilization using multi-controller configuration. *Automatica*, 21, 599-602.

Viswanadham, N., V. V. S. Sarma, and M. G. Singh (1987). Reliability of computer and control systems. North-Holland, Amsterdam, The Netherlands.

von Neumann, J. (1956). Probabilistic logic and the synthesis of reliable organisms from unreliable components. *Annals of Mathematics*, No. 34, C. E. Shannon and E. F. Moore (eds.), Princeton University Press, Princeton, New Jersey, 43–95.

Wang, S. H., and E. J. Devison (1973). On the stabilization of decentralized control systems. *IEEE Transactions*, AC-18, 473–478.

Wonham, W. M. (1966). Liapunov criteria for weak stochastic stability. *Journal of Differential Equations*, 2, 195–207.

Wonham, W. M. (1968). On a matrix equation of stochastic control. *SIAM Journal of Control*, 6, 681–697.

Wonham, W. M. (1970). Random differential equations in control theory. *Probabilistic Methods in Applied Mathematics*, II, A. T. Bharucha-Reid (ed.), Academic Press, New York, 131–212.

Youla, D. C., H. A. Jabr, and J. J. Bongiorno, Jr. (1976). Modern Wiener–Hopf design of optimal controllers. Part II: The multivariable case. *IEEE Transactions*, AC-21, 319–338.

Appendix | Graph-Theoretic Algorithms

In this appendix, we wish to present a collection of the basic graph-theoretic algorithms that underlie our analysis of structured systems throughout the book. Our main objective is to demonstrate the inherent efficiency of these algorithms and their easy implementation in algol-type languages, such as PASCAL and C. We shall also comment on the computational complexity of some of these algorithms, because it is this feature that makes the graph-theoretic approach to complex systems so attractive in applications. At the end of the appendix, we will describe a software package for graph-theoretic analysis of dynamic systems, which is presently being developed for control design of complex systems. Finally, we should note that, in this section, we assume some familiarity of the reader with data structures (*e.g.*, Horowitz and Sahni, 1990) and algorithms in the graph-theoretic framework (*e.g.*, Swamy and Thulasiraman, 1981).

Depth-First Search

In structural analysis and decompositions of complex systems, the fundamental role is played by input and output reachability: can each state vertex of a system digraph be reached from at least one input vertex and can each state vertex reach at least one output vertex (Section 1.2)? An ideal method for verifying these properties is the depth-first search (Tarjan, 1972; see also Swamy and Thulasiraman, 1981).

To determine reachability properties of a system digraph $\mathbf{D} = (\mathtt{V}, \mathtt{E})$ by Depth-First Search (DFS), all we have to do is construct so called DFS trees rooted at the input vertices and see if they span $\mathbf{D}$. To show how this is done, we need some notation. An edge of $\mathbf{D}$ is an ordered pair (v, w) of vertices, where v is the *tail* and w is the *head* of the edge. We also recall (*e.g.*, Harary, 1969) that a directed (rooted) *tree* $\mathbf{T}$ is a digraph, which is connected (directions of the edges are ignored), has one vertex which is the head of no edges called the *root*, and all other vertices are the head of exactly one edge. Specifically, if (v, w) is an edge of $\mathbf{T}$ we denote it by $v \to w$ and call v the *father* of w and w the *son* of v. In a path $\mathbf{P}_w^v$ of $\mathbf{T}$ from v to w, v is an *ancestor* of w and w is a *descendant* of v. A tree is a *spanning tree* of $\mathbf{D}$ if $\mathbf{T}$ is a subgraph of $\mathbf{D}$ and $\mathbf{T}$ contains all the vertices of $\mathbf{D}$. A disjoint set $\mathbf{F}$ of trees is called a *forest.* A collection of trees containing all vertices of $\mathbf{D}$ is said to be a *spanning forest of* $\mathbf{D}$.

To represent a system digraph, we use the *adjacency list*, which requires a minimal storage space; only the existing edges of the digraph are recorded. This fact is especially significant in large sparse systems. Furthermore, the adjacency list representations are ideal for the use of *linked list* data structures to maximize the speed of the proposed algorithms. As is standard (*e.g.*, Horowitz and Sahni, 1990), the nodes in a list are nodes adjacent to the head node of the list. Each node contains three fields: CH for the type of node, VERTEX for the number of the node, and LINK for the *pointer* to nodes in the list. The head node is used for sequential random access to any vertex in the adjacency list of the digraph.

We do not include the self-loops of $\mathbf{D}$ and we start DFS at any vertex u, which we call a *root of* DFS *tree.* This vertex is now visited. We pick any edge incident on u and traverse it to visit an adjacent vertex v. The edge $u \to v$ is now examined and becomes an edge of a DFS tree rooted in u. At v we choose an edge incident on v and traverse it to another vertex w. If w has not been visited yet, then we continue the search from w. Otherwise, we go back to v, the father of w, and try another unexamined edge incident on v. If none exists, we go back upstream to the father of v and repeat the search. In this way, DFS generates a tree over the digraph $\mathbf{D}$. The search from the root terminates when no unvisited vertex can be reached from any of the visited ones. Then, the next vertex in the list is selected as a root of another DFS tree, and the procedure is repeated. The DFS procedure terminates when no vertex of a given digraph is left unsearched and a spanning DFS forest of the graph is obtained.

Whenever a vertex w is visited for the first time, it is included in the DFS tree and assigned a *depth-first number* DFN $(w) = i$. This number is a distinct integer, which indicates that w is the ith vertex of the DFS tree (or forest). If w has already been visited, then the edge (v, w) is a forward, back, or cross edge. If w is a descendant of v in a DFS tree (there is a directed path of the tree from v to w), then (v, w) is a *forward edge.* If w is an ancestor of v (that is, v is a descendant of w), then (v, w) is a *back edge.* If v and w are not related in the DFS forest and DFN$(v) >$ DFN(w), then $(v\,w)$ is a *cross edge.*

A.1. ALGORITHM (DFS).

```
BEGIN
     PROCEDURE DFS
     {given a digraph D = (V, E) with n vertices by the
     adjacency list; initialize Boolean array VISITED to FALSE}
            BEGIN
                 INTEGER i
                 FOR i: = 1 TO n DO
                      VISITED(I): = FALSE;
                 FOR i: = 1 TO n DO
                      BEGIN
                           IF VISITED(i) = FALSE;
                           THEN SEARCHFROM(i);
                      END;
            END; {end DFS tree}
     PROCEDURE SEARCHFROM(v);
     {carries out DFS from unvisited vertex v}
            BEGIN
                 VISITED(v): = TRUE;
                 FOR w in the adjacency list on v DO
                      SEARCHFROM(w);
            END; {searchfrom}
END; {end DFS spanning forest}
```

An example of DFS over a digraph is shown in Fig. A.1. DFS numbers are shown next to each vertex. The edges of the two trees (a forest) are shown as solid lines, while the other three types of edges are drawn as dashed lines.

Strong Components

All graph-theoretic algorithms for dynamic systems can be greatly simplified by finding first the strong components of the system digraph. This fact was utilized throughout Chapter 1 starting with Section 1.3, where strong components were defined, and was subsequently used in decomposition schemes of Chapters 6 and 7. It was also mentioned there that Tarjan's version of the depth-first search can detect the strong components in $O(\max\{n, e\})$ time for a digraph of n vertices and e edges. In this section, we describe this algorithm following the original paper of Tarjan (1972). A more detailed presentation of the same algorithm can be found in (Swamy and Thulasiraman, 1981).

Let us consider a digraph $\mathbf{D} = (\mathtt{V}, \mathtt{E})$. Let v and w lie in the same strong component $\mathbf{K}$. If $\mathbf{F}$ is the DFS forest of $\mathbf{D}$, then v and w have a

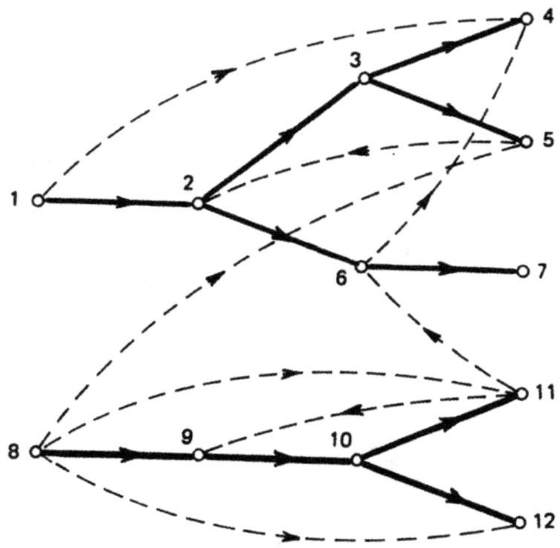

Fig. A.1. DFS tree.

common ancestor in **F**. Furthermore, if u is the highest numbered ancestor of v and w, then u lies in **K**. The vertices of **K** define a subtree in **F**, the root of which is called the root of **K**. In this way, Tarjan reduced the problem of finding the strongly connected components of **D** to the problem of finding the roots of these components. A necessary new notion in this context is LOWLINK (v), which is the lowest numbered vertex in the same component as v and which is reachable from v by a directed path containing at most one back edge or one cross edge. Furthermore, all the edges of such a directed path lie necessarily in the same component. Then, as shown by Tarjan, the vertex v is the root of some strong component **K** of **D** if and only if LOWLINK $(v) = v$.

Most importantly, LOWLINK can be computed by DFS. For a vertex, which has been reached for the first time, we set LOWLINK (v) = DFN (v) and store it on a stack. If a cross edge (v, w) is examined, when both v and w are in the same strong component, then LOWLINK (v) is set to the minimum of its current value and of the DFN (w). When a back edge (v, w) is explored, then LOWLINK (v) is set again to the minimum of its current value and DFN (w). After the search of the son w of v is completed and DFS returns to v, we set LOWLINK (v) equal to the minimum of its current value and LOWLINK (w). Forward edges are ignored altogether.

A.2. ALGORITHM (STRONGCONNECT).

```
BEGIN
   INTEGER i;
   PROCEDURE STRONGCONNECT
        BEGIN
           i: = i + 1
           LOWLINK (v): = DFN (v): = i;
              put v on stack of points
           FOR w in the adjacency list of v DO
                BEGIN
                   IF w is not yet numbered THEN
                      BEGIN
                         {(v, w) is a tree arc}
                         STRONGCONNECT(w);
                         LOWLINK(v):  = min (LOWLINK(v), LOWLINK(w));
                      END;
                   ELSE IF DFN(w) < DFN(v) DO
                            BEGIN
                                  {(v, w) is a back or cross edge}
                                  IF w is on stack THEN
                                     LOWLINK(v):  = min(LOWLINK(v), DFN(w));
                            END;
                END;
           IF(LOWLINK(v) = DFN(v)) THEN
                BEGIN
                   {v is the root of as strong component}
                   start a new strong component;
                   WHILE  w on top of stack satisfies
                            DFN(w) ≥ DFN(v) DO
                        delete w from stack and put w in current component;
                END;
        END;
     i: = 0
     empty stack of points
     FOR w a vertex IF w is not yet numbered THEN STRONGCONNECT(w);
END;
```

In Fig. A.2, we show the LOWLINK and DFN assignments. The strong components are $\{1\}$, $\{2\}$, $\{3, 4, 5\}$, and $\{6, 7, 8, 9, 10\}$.

Reachability

It is obvious that DFS is ideal for use in testing input and output reachability defined in Section 1.2. To determine if a digraph **D** is input reachable, all one has to do is apply DFS to input vertices to see if all state vertices can be reached from at least one input vertex.

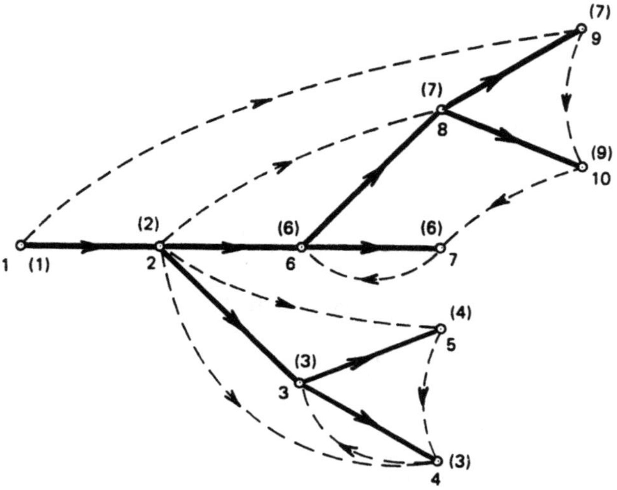

Fig. A.2. DFS tree with LOWLINK numbers.

A.3. ALGORITHM (INREACH).

```
BEGIN
        initialize VISIT_X to FALSE
     FOR all state vertices x DO
          VISIT_X(x): = FALSE;
     PROCEDURE INREACH
          BEGIN
               FOR all input vertices u DO
                    BEGIN
                         DFS(u):
                         FOR all state vertices x DO
                              IF VISIT_X(x) = FALSE
                              THEN GOTO 1;
                         WRITE ('digraph is input reachable');
                         GOTO 2;
1)                  END;
               FOR all state vertices x DO
                    IF VISIT_X(x) = FALSE;
                    THEN WRITE ('state x is not input reachable');
2)        END;
```

Output reachability can be tested by applying Algorithm A.3 to the dual of **D** at the output vertices. This algorithm is called OUTREACH and is easily implemented using an inverse adjacency list for **D**.

When, in addition to input and output reachability, DFS is applied to unvisited state vertices, then the reachability matrix R of (1.21) can be computed with minimal effort.

A.3. ALGORITHM (MATREACH).

```
PROCEDURE MATREACH
    FOR all input vertices u DO
        BEGIN
          {the 'marking' variables VISIT_X & VISIT_Y
           are set FALSE before each search starting from
           an input vertex u}
          FOR all state vertices x DO
              VISIT_X(x) = FALSE;
          FOR all output vertices y DO
              VISIT_Y(y) = FALSE;
              DFS(u);
                  {establish paths between input & states, that is, matrix G}
          FOR all state vertices x DO
              IF VISIT_X(x) = TRUE
              THEN u → x;
                  {establish paths between inputs & outputs, that is, matrix θ}
          FOR all output vertices y DO
              IF VISIT_Y(y) = TRUE
              THEN u → y;
        END;
    FOR all state vertices x DO
        BEGIN
          {the 'marking' variables VISIT_X & VISIT_Y
           are set FALSE before each search starting from
           a state vertex x}
          FOR all state vertices x DO
              VISIT_X(x): = FALSE;
          FOR all ouput vertices y DO
              VISIT_Y(y): = FALSE;
              DFS(x):
                  {establish paths between states, that is, matrix F}
          FOR all state vertices x DO
              IF VISIT_X(x) = TRUE
              THEN x → x;
                  {establish paths between states & outputs, that is, matrix H}
          FOR all output vertices y DO
              IF VISIT_Y(y) = TRUE
              THEN x → y;
        END;
END;
```

The reachability matrix R is essential for the hierarchical (LBT) decompositions (Algorithm 6.12).

Generic Rank

We recall from Section 1.4 that generic rank of a structured matrix is equal to its maximum number of nonzero elements, no two of which lie in the same row or column. These nonzeros can be made to appear on the diagonal of a permuted matrix to form the *maximum transversal.* The problem of selecting the transversal has been studied for a long time in various fields under different names: systems of distinct representatives, maximum assignment, an output set, maximum or perfect matching, *etc.* The long history of the problem was reviewed by Duff (1981a), who proposed an algorithm to determine the maximum transversal of a given matrix, that is, its generic rank. The algorithm, which is based upon DFS, is described in this section. It is used in structural controllability (Section 1.4), in structurally fixed modes (Section 1.6, Lemma 1.55), and in Chapter 5 (Theorem 5.15), when we need a test for the existence of a suitable matching (see Remarks 5.16 and 19).

Following Duff (1981a), we obtain the transversal recursively in n steps, where n is the order of the matrix. At the kth step we have a transversal for a submatrix of order k. At each step, Duff associates an unorthodox digraph with the matrix in such a way that every vertex corresponds to a distinct row of the matrix, and an edge exists from vertex i_0 to vertex i_1 if there is a column, say j_1, so that the nonzero element (i_1, j_1) is currently in the transversal and, in addition, the element (i_0, j_1) is nonzero. When the current transversal includes the nonzeros (i_1, j_1), (i_2, j_2), $\dots$, (i_k, j_k), we consider a sequence of nonzeros (i_0, j_1), (i_1, j_2), $\dots$, (i_{k-1}, j_k). If there is a nonzero at (i_k, j_{k+1}) and if none of the nonzeros in the i_0th row or the j_{k+1}th column are in the current transversal, we can perform a *reassignment* in the transversal by replacing the nonzeros (i_ℓ, j_ℓ), $\ell = 1, 2, \dots, k$, by the nonzeros $(i_\ell, j_{\ell+1})$, $\ell = 1, 2, \dots, k$. The reassignment increases the length of the transversal by one. A reassignment chain is illustrated in Fig. A.3, where the three nonzeros shown in squares are replaced by the four encircled nonzero elements.

The most interesting part of Duff's algorithm is the use of DFS with a look-ahead technique to find an appropriate reassignment chain. If (i_2, j_2), say, is a present assignment and (i_1, j_2) is a nonzero, we apply DFS to the above defined digraph starting at any unassigned row (vertex) i_0, regularly the next row of the matrix. DFS search along a path terminates when we reach a nonzero (i_k, j_{k+1}) and when j_{k+1} is an unassigned column of the

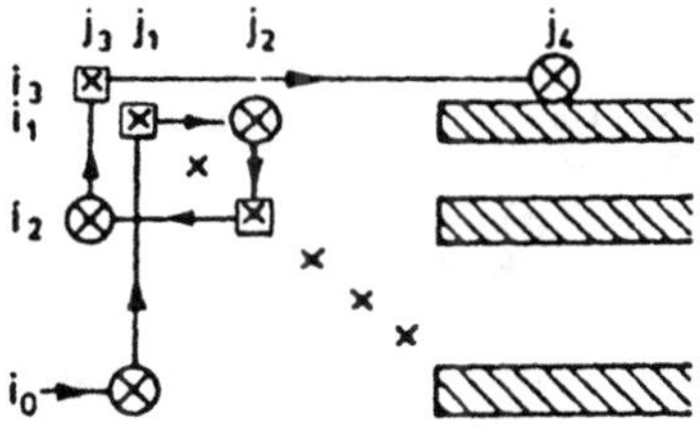

Fig. A.3. A reassignment chain.

matrix; the vertex i_k is a free vertex. By backtracking from i_k to i_0 , we obtain the reassignment chain and complete the kth step of the transversal algorithm. In performing the DFS search it is crucial that at each vertex we check all of the unvisited vertices that can be reached from the current one, if any are free. If a free vertex is found we got a cheap assignment. This short-cut in the Duff's algorithm is called the *look-ahead technique.* The cheap assignment would have taken place in the chain of Figure A.3 if the shaded areas had not been all zeros. If e is the number of nonzeros of the matrix, then the additional effort due to the look-ahead technique is only $O(e)$. This is a small price to pay for a considerable savings in time, provided an auxiliary pointer array is used for the cheap assignment phase of the algorithm to make sure that no more than e nonzeros are visited during each phase.

For a pseudo-code presentation of Duff's algorithm, which is termed GENRANK, we introduce the variable NUMNX. This variable represents the length of the transversal, that is, the generic rank of the matrix. The Boolean variable ASSIGN(i) checks if the row i has been assigned already in the DFS search.

A.4. ALGORITHM (GENRANK).

```
BEGIN
   NUMNX: INTEGER
   NUMNX: = 0
   PROCEDURE GENRANK
      BEGIN
           FOR w the next examined row DO
                BEGIN
   1)              i: = 0;
                   REPEAT i: = i + 1
                      UNTIL ((A(w, i) <> 0) AND
                      (ASSIGN(i): = FALSE))
                      OR (i > n);
```

```
                IF i ≥ n THEN
                   BEGIN
                      make this first nonzero an assignment
                      AND perform reassignment on all rows in path
                      ASSIGN(i): = TRUE;
                      GENRANK(A);
                   END;
2)              IF there are nonzeros in row w that have not been examined
                THEN
                   BEGIN
                      FOR v the next nonzero in row w DO
                          IF v has not been accessed during this pass THEN
                          THEN
                               BEGIN
                                   put row y with assignment column v on path
                                   w: = y
                                   GOTO 1
                               END;
                          ELSE GOTO 2;
                   END;
                ELSE
                   BEGIN
                      IF there exists a previous row x on path
                      THEN
                          BEGIN
                               x: = w;
                               GOTO 2;
                          END;
                      ELSE
                          {the matrix has not full generic rank
                           but we continue to maximize assignment}
                           GENRANK(A);
                   END; {ELSE corresponds to line marked 2}
            END; {FOR w the next unexamined row}
   END; {GENRANK}
 IF NUMNX = n THEN matrix has full generic rank;
END;
```

The worst case bound of this algorithm is $O(ne)$, where n is the order of the matrix and e is the number of its nonzero elements. Duff (1981a) claims that in realistic examples, the complexity is much better and equals $O(n) + O(e)$. He provides a comparison with other existing algorithms and, in a separate paper (Duff, 1981b), gives a code in FORTRAN.

Decompositions

The LBT and EPSILON decompositions where described in Chapters 6 and 7, respectively. The LBT decompositions are suitable for sparse systems, where they can take advantage of zero elements and produce hierarchically ordered subsystems with independent inputs and outputs. Decentralized controllers and estimators can then be built piece-by-piece with assured stability and satisfactory performance. On the other hand, the EPSILON decompositions can be applied to dense matrices as well, but stability of a subsequent decentralized design would depend heavily on the size of the chosen value for the threshold epsilon. As explained in Section 7.6, in a control problem of large dimension, both decompositions should be used, with the LBT decomposition being applied first.

The LBT Algorithm (6.12) can be sped up considerably by using the condensation of the original digraph, which is obtained by STRONGCONNECT (Algorithm A.2) following Remark 6.14. We also note (Remark 6.15) that the algorithm can be used to provide either input or output LBT decompositions, which is a much easier task than finding input–output decompositions (see Remark 6.16). Finally, we recall (Remark 6.18) that structural controllability and observability of the overall system should be tested using input and output LBT decompositions. The reason is that an LBT reachable decomposition is structurally controllable (observable) if and only if the corresponding hierarchical composite matrices of the subsystems have full generic rank. In other words, the generic rank condition for the overall system can be tested recursively on the subsystem levels, which can reduce considerably the dimensionality of controllability and observability tests.

With regard to the EPSILON decomposition, we provide here additional information about the corresponding Algorithm 7.4 and give a pseudocode. We first note that to identify the weakly coupled subsystems by connected (not strongly connected!) components of $\mathbf{D}^\epsilon$, we have to obtain the undirected version of the original digraph $\mathbf{D}$. This we do by forming a new interconnection matrix $\tilde{E} = (\tilde{e}_{ij})$ from the original matrix $E = (e_{ij})$ such that $\tilde{e}_{ij} = 1$ if and only if $e_{ij} = 1$ or $e_{ji} = 1$. Then, we construct a new adjacency list corresponding to $\tilde{E}$ and $\tilde{\mathbf{D}}$. To detect the components of $\mathbf{D}^\epsilon$ we identify the strong components of $\tilde{\mathbf{D}}^\epsilon$ using STRONGCONNECT (Algorithm A.2).

In Section 7.3, when we considered control applications, only state and inputs were involved. If not all states are available for measurement and we have an output equation, then epsilon decomposition can be carried out with three different values of ϵ, namely ϵ_A, ϵ_B, and ϵ_C. As in the case of hierarchical decompositions, the input and output decompositions can be performed independently using the pairs (ϵ_A, ϵ_B) and (ϵ_A, ϵ_C). The output decomposition using (ϵ_A, ϵ_C) is obviously a dual of the input decomposition and the discussion in Section 7.3 applies to observer designs in a straightforward way. All these different options are made available in the program. The user chooses different values for ϵ_A, ϵ_B, and ϵ_C. At the start, ϵ_B and ϵ_C

are set to zero and an ϵ_A decomposition is obtained first. Then, ϵ_B and ϵ_C are chosen to minimize the sets $\boldsymbol{N}_{xu}$ and $\boldsymbol{N}_{xy}$, or a set $\boldsymbol{N}_{uxy} = \boldsymbol{N}_u \cap \boldsymbol{N}_x \cap \boldsymbol{N}_y$ if a joint controller–observer design is attempted.

For every chosen triple $(\epsilon_A, \epsilon_B, \epsilon_C)$, we check the decoupled subsystems $(\epsilon_A = \epsilon_B = \epsilon_C = 0)$ for reachability. If subsystems are not reachable, then they would remain unreachable for any larger values of ϵ_A, ϵ_B, and ϵ_C. One can continue to increase these values if reachability can be restored by combining unreachable and reachable subsystems.

After a satisfactory grouping of states is achieved, the values of ϵ_B and ϵ_C can be chosen to minimize the sets $\boldsymbol{N}_{xu}$ and $\boldsymbol{N}_{xy}$, while retaining reachability. We can also remove some inputs or outputs from state clusters that have their own inputs and outputs, and keep them for states that have no other inputs and outputs. The procedures that attach the inputs and outputs to states are CONDU and CONDY. A pseudo-code for epsilon decomposition with fixed ϵ_A, ϵ_B, and ϵ_C is given by the following:

A.5.   ALGORITHM (EDEC).

```
PROCEDURE EDEC
    BEGIN
            {arrays A, B, C have the same dimensions as A, B, C}.
            FOR all i, j DO
                BEGIN
                        IF |aij| ≤ εA THEN aεij = 0 ELSE aεij = 1
                        IF |bij| ≤ εB THEN bεij = 0 ELSE bεij = 1
                        IF |cij| ≤ εC THEN cεij = 0 ELSE cεij = 1
                END;
            {find connected components of states}
            FOR all i, j DO
                IF (Aε[i, j] = 1 OR Aε[j, i] = 1)
                THEN construct edge (xj, xi);
            FOR i: = 1 to # of states DO
                STRONGCONNECT(i);
            {insert back the original edges into components and test reachability}
            FOR each connected components DO
                BEGIN
                        FOR i and j in component p DO
                            IF A[i, j] ≠ 0 THEN A[i, j] = 1;
                        {associate inputs and outputs with component p}
                        CONDU (U set);
                        CONDY (Y set);
                        {test reachability of component p}
                        INREACH (U set);
                        OUTREACH (Y set);
                END;
        END;
```

Nested epsilon decompositions, which call EDEC, are obtained by the following:

A.6. ALGORITHM (ENEST).

BEGIN
REAL
PROCEDURE ENEST
BEGIN
ϵ_A: $= \min |a_{ij}|$; ϵ_B: $= 0.0$; ϵ_C: $= 0.0$
1) EDEC
Print out ϵ_A, ϵ_B, ϵ_C and subsystems;
ϵ_A: $= \epsilon_A + \delta_A$
IF unreachable
THEN combine unreachable clusters of states with reachable cluster of states;
IF (($\epsilon_A < \max |a_{ij}|$, $i \neq j$) AND (reachable))
THEN GOTO 1;
{based on size of cluster user reads in value of E_A}
READ (ϵ_A);
EDEC
{vary values of ϵ_B and ϵ_C}
ϵ_B: $= \min |b_{ij}|$; ϵ_C: $= \min |c_{ij}|$;
EDEC;
IF reachable THEN
BEGIN
Print out ϵ_A, ϵ_B, ϵ_C and subsystem
ϵ_B: $= \epsilon_B + \delta_B$; ϵ_C: $= \epsilon_C + \delta_C$;
END;
END;
END;

It is important to note that the EPSILON decomposition procedure is very much user-dependent. Finding the appropriate values for the triple (ϵ_A, ϵ_B, ϵ_C) may require considerable interactive use of the ENEST algorithm.

Graphpack

An interactive software package (GRAPHPACK) is presently under development (LaMont, 1990), with the objective of integrating the above algorithms to offer the user a suitable environment for graph-theoretic analysis of dynamic systems. The package is menu-driven to allow a casual or new user to utilize the package immediately, as well as to enable a serious user to monitor various algorithms in a step-by-step fashion. In the interactive mode, the central role is played by the supervisor. It is designed to provide all the necessary linkage between individual modules in the package, when a sequence of operations is executed. The supervisor controls the operation of the preprocessors and postprocessors, as well as the individual algorithms, thus creating a medium for the optimal utilization of power and speed of the graph-theoretic approach to control design of complex systems.

Bibliography

Duff, I. S. (1981a). On algorithms for obtaining a maximum transversal. *ACM Transactions on Mathematical Software*, 7, 315–330.

Duff, I. S. (1981b). ALGORITHM 575: Permutations for a zero-free diagonal [F1]. *ACM Transactions on Mathematical Software*, 7, 387–390.

Harary, F. (1969). *Graph Theory*. Addison-Wesley, Reading, Massachusetts.

Horowitz, E., and S. Sahni (1990). *Fundamentals of Data Structures in Pascal*. Computer Science Press, New York.

LaMont, D. (1990). GRAPHPACK: *A Software Package for Graph-Theoretic Analysis of Control Structures*. Ph.D. Thesis, Santa Clara University, Santa Clara, California.

Swamy, M. N. S., and K. Thulasiraman (1981). *Graphs, Networks, and Algorithms*, Wiley, New York.

Tarjan, R. (1972). Depth-first search and linear graph algorithms. *SIAM Journal on Computing*, 1, 146–160.

Index

Errata

	For	Read This
p.21		
Line 1 from below	$\tilde{B}$	$\tilde{C}$
p.22		
Line 1 from below	$\tilde{B}$	$\tilde{B}_2$
p.25		
Line 3 from above	$[b\ Ab]$	$[b\ Ab\ A^2b\ A^3b]$
Line 6 from above	$[b\ Ab]$	$[b\ Ab\ A^2b\ A^3b]$
p.33		
Line 8 from above	$(\theta_1, \dot{\theta}_1, \theta_2, \dot{\theta}_2)$	$(\theta_1, \dot{\theta}_1, \theta_2, \dot{\theta}_2)^T$
p.35		
Eq. (1.63)	$+e\begin{bmatrix} 0 & 0 \\ -\gamma & 0 \end{bmatrix}$	$+\begin{bmatrix} 0 & 0 \\ -\gamma & 0 \end{bmatrix}$
	$+e\begin{bmatrix} 0 & 0 \\ \gamma & 0 \end{bmatrix}$	$+\begin{bmatrix} 0 & 0 \\ \gamma & 0 \end{bmatrix}$
p.38		
Eq. (1.78)	$[C_1\ C_2\ \ldots\ C_N]$	$[C_1^T\ C_2^T\ \ldots\ C_N^T]^T$
p.41		
Line 10 from below	Davision	Davison
p.43		
Eq. (1.97)	$\begin{bmatrix} A-\lambda I & B_I \\ C_J & \end{bmatrix}$	$\begin{bmatrix} A-\lambda I & B_I \\ C_J & 0 \end{bmatrix}$
p.51		
Eq. (1.78)	$[C_1\ C_2\ \ldots\ C_N]$	$[C_1^T\ C_2^T\ \ldots\ C_N^T]^T$
p.52		
Eq. (1.122)	$[C_1^T\ C_2^T\ \ldots\ C_N^T]$	$[C_1^T\ C_2^T\ \ldots\ C_N^T]^T$
p.82		
Eq. (2.61)	$\Lambda^T\tilde{H} + \tilde{H}\Lambda = \tilde{G}$	$\Lambda^T\tilde{H} + \tilde{H}\Lambda = -\tilde{G}$
p.84		
Eq. (2.74)	$\tilde{G} = \begin{bmatrix} -1 & 0 \\ 0 & -2 \end{bmatrix}$	$\tilde{G} = 2\begin{bmatrix} -1 & 0 \\ 0 & -2 \end{bmatrix}$
p.90		
Eq. (2.106)	F	Γ
Eq. (2.108)	12	–12
p.91		
Eq. (2.111)	12	–12

Errata

p.132		
Fig. 3.1	(b) $\mu < 2$	(b) $\mu > 2$
p.191		
Line 11 from above	(*i*)	(*ii*)
p.196		
Fig. 4.1	g_i	l_i
p.310		
Line 10 from above	stabilizations	stabilization
p.373		
Line 9 from below	M. E. and	M. E., and
p.404		
Line 6 from above	$\hat{A}$	$\hat{A}_i$
p.410		
Line 5 from above	Jacoby	Jacobi
p.417		
Line 3 from above	$-\frac{\alpha+\beta}{M}$	$-\frac{\alpha+\beta}{m}$
Line 5 from above	0 0 0 0	$0 \quad 0 \quad \frac{\alpha}{M} \quad 0$
Line 9 from above	$0 \quad \frac{\beta}{m}$	$\frac{\beta}{m} \quad 0$
Line 9 from above	$\frac{\alpha}{M}$	$\frac{\alpha}{m}$
Line 9 from above	$-\frac{\alpha+\beta}{M}$	$-\frac{\alpha+\beta}{m}$
p.468		
Line 22 from below	multiply	multiple
p.508		
Line 1 from below	(vw)	(v, w)